PROGRESS IN NUCLEAR ENERGY
The international Review Journal covering all aspects of Nuclear Energy

Advanced Nuclear Energy Systems Toward Zero Release of Radioactive Wastes

Executive Editor
Masaki Saito
Tokyo Institute of Technology, Tokyo, Japan

Editor
Tetsuo Sawada
Tokyo Institute of Technology, Tokyo, Japan

PERGAMON

Published originally as a special issue (Volume 40:3–4) of the international journal *Progress in Nuclear Energy* and also available in hard-bound edition (ISBN: 0 08 0441734)

ELSEVIER SCIENCE Ltd
The Boulevard, Langford Lane
Kidlington, Oxford OX5 1GB, UK

Reprinted from *Progress in Nuclear Energy*, Volume 40 Issue 3–4 2002 (ISSN: 0149-1970)

First edition 2002

Library of Congress Cataloging in Publication Data
A catalog record from the Library of Congress has been applied for.

British Library Cataloguing in Publication Data
A catalogue record from the British Library has been applied for.

ISBN 10: 0 08 044173 4
ISBN 13: 9780080441733

Transferred to digital printing 2007

Advanced Nuclear Energy Systems Towards Zero Release of Radioactive Waste

M. Saito and T. Sawada

Contents

Chapter 9 Self-Consistent Nuclear Energy System (SCNES)

Chapter 10 General Comments Based on Nuclear Physics, Nuclear Reactor Physics, Nuclear Data and Accelerator Technology

Pergamon
www.elsevier.com/locate/pnucene

Progress in Nuclear Energy, Vol. 40, No. 3–4, pp. vii–viii, 2002

0149-1970/02/$ - see front matter

PII: S0149-1970(02)00085-9

FOREWORD

Jiro Kondo
Chairman of Special Committee on High-Level Radioactive Waste Disposal
Director of The Fujihara Foundation of Science

A series of Fujihara Seminars has been held in memory of the great industry man Mr. Ginjiro Fujihara (1869 – 1960) who did more than having developed Oji Paper Co.,Ltd , as a President, to be a world-wide leading company in the field of paper manufacture. He assumed the Minister of Commerce and Industry (1940) as well as the Minister of Munitions (1944) to have endeavored to develop overall the Japan's industries. Furthermore he established The Fujihara Technical College of Engineering in 1939, which is currently the Science and Engineering Faculty of Keio University. In 1959, he founded The Fujihara Foundation of Science by spending his own money. This foundation awards the Fujihara Prizes for two kinds of fields every year to greatly stimulate researchers in science and technology in Japan.

In this year, the 41st award ceremony in the year of 2000 was splendidly held on 17 June which was very close to the birthday of Mr. Fujihara. In addition to the prize, the foundation is holding a series of The Fujihara International Seminar. For the seminar, we select a topic which is currently a world-wide issue in the field of natural science and invite international and domestic scientists who are appropriate to discuss and resolve the issue. Here Professor Masaki Saito hosts the Second Fujihara Seminar. The title of the seminar is International Seminar on Advanced Nuclear Energy Systems Toward Zero Release of Radioactive Wastes.

Professor Saito is a member of Research Laboratory of Nuclear Reactors, Tokyo Institute of Technology, who has been for many years involved in this issue. Besides the distinguished researchers from abroad, a lot of domestic researches have participated in this seminar, who are from universities, atomic energy commission, institutes and private sectors.
In Japan, the research reactor JRR-1 of Japan Atomic Research Institute had performed the criticality on 27 August in 1957. Then the electricity generation by nuclear reactors was launched. Now we are operating 51 nuclear power stations of pressurized water reactors (PWRs) and boiling water reactors (BWRs), which have generated in total 3.313 $\times 10^8$ MWh (1998) to be 36.8 % of the total use in Japan. Japan is an islands country and it is not practical to import electricity from the neighboring countries in case of the shortage in power like European countries. We need to secure the variation in energy resources as far as possible, storing fossil resources against emergencies.

Uranium nuclei will fission by absorbing neutrons, producing other nuclei including plutonium. Then it is expected to transmute nuclear waste materials including plutonium into other nuclei.
I myself was in charge of wrapping up the report concerning waste disposal as the chairman of the Special Committee on High-Level Radioactive Waste Disposal under Science and Technology Agency (STA) in May 1998. The Government passed the bill for the Specified Radioactive Waste Final Disposal in March 2000. The act includes the authorization to collect and deposit Yen 0.14/kwh for the expense of constructing geological disposal site, because the geological disposal is foreseen after 2020.
The nuclear waste reprocessing facilities are under construction at Rokkasyo mura in the most northern part of the Honsyu island. It needs a large investment. I expect that in this seminar you will discuss on the newest theory and technology for processing high-level nuclear wastes.

Ladies and gentlemen, the place of this seminar is located on the foot of Mt. Fuji that is known as the most beautiful mountain in Japan. I guess you will be busy due to the crowded schedule, but you may have a look at the beautiful scenery of Mt. Fuji of which top is covered with snow if you walk around the seminar house. I sincerely hope that your stay here will be a supreme impression with the fruitful discussions and fine prospect for the future.

Pergamon
www.elsevier.com/locate/pnucene

Progress in Nuclear Energy, Vol. 40, No. 3–4, pp. ix–x, 2002

Printed in Great Britain
0149-1970/02/$ - see front matter

PII: S0149-1970(02)00074-4

Preface

This volume is a collection of the papers presented at International Seminar on Advanced Nuclear Energy Systems Toward Zero Release of Radioactive Wastes, which was held at Susono city spreading over the foot of Mt. Fuji, Shizuoka Prefecture, Japan, November 6-9, 2000. The seminar was sponsored by The Fujihara Foundation of Science and supported by Research Laboratory for Nuclear Reactors, Tokyo Institute of Technology, Japan Atomic Energy Research Institute, Japan Nuclear Cycle Development Institute and Japan Atomic Industrial Forum. The seminar was organized by Research Laboratory for Nuclear Reactors, Tokyo Institute of Technology.

The objective of this international seminar is to bring together scientists and engineers working in academia, research organizations and industry for the open discussion on the role and contributions of the nuclear energy to the environmental issues in the new millennium. The seminar is intended to provide a small forum for the open and free discussions in pursuit of the solutions for the reduction of nuclear wastes based on the accelerator and fusion technologies in addition to the advanced fission technology to harmonize the nuclear energy systems with the global environment in the new millennium, and for the promotion of the future international collaboration in the following research fields:

- -The role of nuclear energy in the new millennium
- -Waste management
- -Transmutation of Minor Actinides and Fission Products
- -Advanced fission systems for transmutation of wastes
- -Accelerator driven systems for transmutation of wastes
- -Fusion systems for transmutation of wastes
- -Nuclear database for transmutation of wastes
- -Advanced nuclear fuel cycles for transmutation of wastes

The seminar was attended by 19 foreign scientists from 6 countries and one international organization and 60 Japanese scientists and engineers. The participants from overseas presented 17 papers, while Japanese participants presented 31 papers. At the opening session, Prof. Jiro Kondo, Professor Emeritus of University of Tokyo, Chairman of The Science and Technology Foundation of Japan, made opening address and Prof. Masaki Saito, Chairman of the organizing committee of the seminar made welcome address. Prof. Yoichi Fujii-e, Acting Chairman of Japan Atomic Energy Commission gave a keynote speech titled "Nuclear Energy System for A Sustainable Development Perspective –Self-Consistent Nuclear Energy System-"

In the first technical session " The role of Nuclear Energy in the 21st Century", Dr. V. Kagramanian from IAEA reviewed the major challenges to future nuclear power development and discussed needs for international cooperation in innovative R&D activities to ensure full participation of nuclear power in the world energy market. Dr. J. A. Lake, President of American Nuclear Society, made the presentation titled " The Forth Generation of Nuclear Power". Prof. G. Kessler, former Director of Institute of Neutron Physics and Reactor Technology, FZK, Germany gave the detail outlook of the competitiveness of the nuclear power compared to other energy sources and proposed three requirements to future nuclear power development. Prof. W. Lee, Professor Emeritus of KIST, Korea described Korean Policy and Demand in The Field of Nuclear Energy in Future.

The following technical sessions cover Advanced Fuel Cycle and Waste Management, Role of Transmutation, Objectives , Scenario and R&D Issues of Projects, Hazards of Nuclear Wastes and Impact of Transmutation, Advanced Fission Reactor System for Transmutation, Accelerator-Driven System for Transmutation, Advanced Fuel Cycle for Transmutation, Self-Consistent Nuclear Energy System. At the end of the technical session, Prof. Y. Nagai, Osaka University, Japan, Prof. L. I. Ponomarev, Kurchatov Institute, Russia, Prof. A. Gandini, University of Rome, Italy, Prof. Yu. Korovin, Obninsk Institute of Nuclear Power Engineering, Russia and Dr. H. Takahashi, BNL, USA made the general comments on the transmutation of the nuclear wastes from the viewpoint of Nuclear Physics, Nuclear Reactor Physics, Nuclear Data and Accelerator Technology.

Throughout the seminar, the nuclear energy system is expected to be most promising option for the sustainable growth and the harmonization between the global environments and the human society in the new millennium. To realize such goals it was discussed that the reduction of the high level radioactive wastes, the enhancement of nuclear safety and the economical competitiveness are very important R&D issues in the future. It is also recognized that the problems of the radioactive environmental burdens may not be solved so easily in near future. Therefore, we should continue our R&D efforts in these fields. The promotion of the international collaboration should be accelerated more in near future. The transmission of the science and technologies including philosophy in our age to the next generations is also absolutely necessary.

The seminar was not possible without the financial support by The Fujihara Foundation of Science, and technical supports by Research Laboratory for Nuclear Reactors (Tokyo Institute of Technology), Japan Atomic Energy Research Institute, Japan Nuclear Cycle Development Institute and Japan Atomic Industrial Forum. We are indebted to all of the participants for their fruitful discussion, especially to the authors who spent a lot of time for the preparation of the original papers. We also would like to thank all the members of the organizing committee of this seminar including our students.

Executive Editor
Masaki Saito
Chairman,
Organizing Committee of the Seminar
Associate Professor,
Research Laboratory for Nuclear Reactors,
Tokyo Institute of Technology

Editor
Tetsuo Sawada
Organizing Committee of the Seminar
Assistant Professor,
Research Laboratory for Nuclear Reactors,
Tokyo Institute of Technology

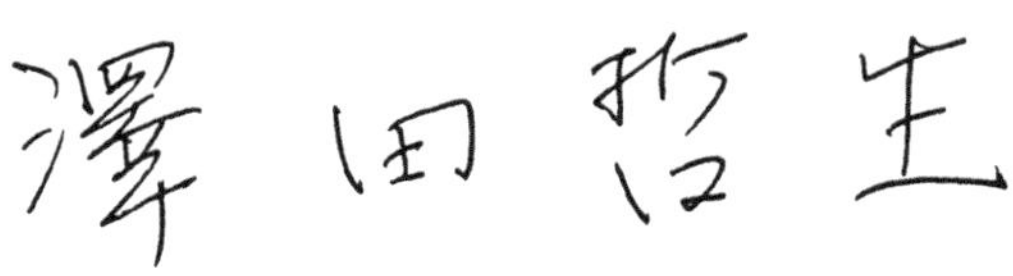

Pergamon

www.elsevier.com/locate/pnucene

Progress in Nuclear Energy, Vol. 40, No. 3–4, pp. 265–283, 2002

Printed in Great Britain
0149-1970/02/$ - see front matter

PII: S0149-1970(02)00021-5

Nuclear Energy System for a Sustainable Development Perspective - Self-Consistent Nuclear Energy System-

Yoichi Fujiie[1], Masao Suzuki [2]

1. Japan Atomic Energy Commission, 3-1-1 Kasumigaseki, Chiyoda-ku, Tokyo, 100-8970 Japan, Phone: +81-3-3581-6688
2. Japan Nuclear Cycle Development Institute, 4002, Narita-cho, O-arai-machi, Ibaraki-ken, 311,Japan,Phone:+81-29-267-6181, Email:suzumasa@oec.jnc.go.jp

1. Introduction

1) Total Picture for Comprehensive Nuclear Science and Engineering

Civilization gives rise to science and engineering, and science and engineering support civilization; through the history of civilization, this inseparable relationship has been maintained. Science and engineering, however, should meet the requirements of civilization.

In developing science and engineering and introducing it to society, it is not easy to view the future by studying technological aspects alone. However, the ongoing dispute over nuclear science and engineering seems to be biased too much toward technological aspects. We live in an age where technology cannot always solve all of our problems. Social or economic benefits are important factors to consider; society will not choose anything that lacks economic benefits. This point should be kept in mind from the beginning, namely, from the development stage.

Although economic or social benefits are given a higher priority than technology, it is still difficult to view the future clearly. Scientific thinking that can develop into abstract or philosophical thinking will be more appropriate for viewing the future. At the turn of the century, scientific thinking should be developed to view the total picture of comprehensive nuclear science and engineering and its long-term perspective. This is the approach that society should support in determining the future direction of nuclear science and engineering as we head into the 21st century.

Nuclear science and engineering deal with the quantum world that has been studied since the end of the nineteenth century. It basically deals with a broad area of science and engineering related to light, charged particles, and neutral particles. Particle accelerators, lasers, nuclear fission reactors and nuclear fusion reactors in future are devices designed to bring quantum-world to society. Nuclear science and engineering have so far focused on the use of nuclear energy as an energy source, and specifically nuclear power generation using light-water reactors. In the non-energy area, various radiation applications have been developed. While looking at civilization as a whole, the scope of those applications should be broadened in terms of comprehensive nuclear science and engineering.

2) Long-term perspective backed by a deep insight into 21st-century civilization

What civilization demanded for science and engineering was, in the ancient age, represented by natural energy such as sunshine, water flow, wind, etc, elements of food, shelter and clothing as material, language for communication as information and tools in living as technology and has increased in both quality and quantity along with the progress of civilization. Those demands have been basically four things: energy, material, information and technology. This will basically remain the same in the future. However, people understand that science and engineering has not been always obedient to civilization nor has done it guarantee the people a decent way of life. Energy-dependent civilizations that came into full bloom during the Industrial Revolution continued developing on the principle that the utilization of resources should be maximized. These civilizations were a prelude to the modern civilization characterized by mass-production, mass-consumption, and mass-disposal. Our civilization is now faced with environmental pollution, global warming and other environmental problems. We are at our turning point.

It is an undeniable fact that science and engineering has been used for military purposes throughout the history of mankind, and nuclear science and engineering is not an exception. It is important to know that there are both positive and negative aspects. To construct a new nuclear civilization, the concept that harmony should take precedence over utilization must be understood and confirmed. Fears and adverse reactions that people display with regard to the enormity and the great level of energy density of nuclear energy must be readily acknowledged. It is important to construct a nuclear science and engineering system that can be accepted and supported by society.

Limited resources and limited capacity on the part of the environment indicate that we must depart from a civilization of mass-consumption and mass-disposal and that we must approach to a recycling based society. Is mankind wise enough to make a shift to a recycling based society in the next century?

Although expectations abound that nuclear science and engineering can play a key role in approaching to this recycling based society, the possibilities of nuclear science and engineering and what follows from its fundamental characteristics must be considered.

It is worth while to note that there seem to be three kinds of spans for the consideration of nuclear science and engineering: a span of decades from political and economic viewpoints, a span of centuries for the research and development of monumental science and engineering to practical use such as fast reactors and fusion reactors, and a span of millennia for the continuation of civilization and the half-life of radioactivity. Therefore, significant factors to discuss are obtaining a clear perspective of the roles these spans play, clarifying realistic goals to reach with a long-term perspective that includes social concerns, and working to reach these goals. We also need to deal with the purposes flexibly and prepare for technological alternatives to solve technological issues.

3) Balanced coexistence of idealism and realism in nuclear policies

Although nuclear science and engineering has developed and gained a significant foothold in society during the century since quantum world was inducted as a subject of scientific research and development, it is appropriate to think that this nuclear science and engineering is still in its development stage, indicating that further research and development are needed. Nuclear science and engineering draw keen interest and expectations in the future whereas some forms of nuclear technology are practical and used in social, economic, and political frameworks. As we know from history, it sometimes takes more than a century for a new science and engineering to be understood and become familiar to the people in general. In introducing nuclear science and engineering to society, therefore, both leadership and accountability are needed.

In formulating nuclear policies that can contribute to the construction of well balanced nuclear science and engineering, the coexistence of the idealism that transcends the times and the realism that seeks solutions to problems should be encouraged. The formulation of realistic policies through clarification of future perspective is important. Nuclear fuel cycle policies must be based on this approach.

Economic benefits available or technological achievements made this moment cannot be ignored when formulating actual policies. Although these two points are important factors, it is not clever, however, to determine actual policies without clarifying future prospects. Here, the balanced coexistence of idealism and realism must rise to meet the future.

Hereafter we will confine arguments on the perspective for nuclear energy development based upon the discussion above.

4) Why we propose a self-consistent nuclear energy system

We should all recognize the fact that a great part of populations in developing countries still live in such a condition that they cannot benefit from the usefulness of electricity enough. According to an estimate, the world energy consumption and the consumption of electricity in the year 2050 would become 2.5 and 3 times as large as that of the year 2000. The world energy consumption in the year 2100 would become 4 times as large, and the electricity consumption in particular would become 5 times as large (J.Holdren , 2000, Ref.1).

We are convinced that nuclear energy continues to be one of essential energy sources in the long future. We should be confident and proud of nuclear technology and actively promote research and development aiming at the construction of nuclear era in the 21st century. However, many of our colleagues and fellow workers have a somewhat cloudy outlook for the future of nuclear energy, and feel the stagnant situation in spite of the glorious and brightening history and achievement in light water reactor development.

It is our understanding that such a cloudy outlook is attributable to the following concerns:

- Significant delays in the development of fast reactors and related nuclear fuel cycle, which was drawn in the 1960's.
- Difficulties in the development of "heavy and huge" technologies, such as nuclear technology, under the present market-driven economy and the free-enterprise system.
- Difficulties in the competitiveness of nuclear energy against advanced fossil energy in power generation costs、 and against co-generation and dispersion system.

In order to break through the circumstances mentioned above, it is necessary to go back to the starting point of nuclear technology development to recognize potential abilities of nuclear energy, reconsider what nuclear energy should be, and clarify the vision that to be aimed at.

Taking into account the circumstances mentioned above, a "Self-Consistent Nuclear Energy System (SCNES)" has been proposed by Dr. Fujiie.(. Fujiie , 1993, Ref.2)

It is possible to achieve, we consider, a breakthrough to reactivate nuclear energy development by establishing a vision that is based on the inherent and essential nature of nuclear energy for the future development. We view the SCNES as a ultimate scientific goal with which the development scenario and concept for R&D are discussed.

This paper presents the results of the arguments of the scientific feasibility of SCNES performed based on the latest information. In addition, long term and near term goals for the element technology development are also presented.

Lastly, the relationship between the SCNES and ongoing R&D activities for the future goals of nuclear energy systems like those in the frame of the Generation IV nuclear energy system (GEN-IV) is discussed.

2. Scientific Feasibility of SCNES

1) What is SCNES?

The SCNES is defined as a system which satisfies simultaneously the following four objectives at least.

Objectives of the SCNES;

1.Energy Production

2.Fuel Generation

3.Environment protection

4.Safety assurance

Figure 1 illustrates the concept of SCNES.

SCNES has the ultimate goal of full utilization of nuclear resources and "zero-release" of radioactive waste by recycling fuel and radioactive material, and "zero-release of radioactive material" by assuring plant safety.

"Zero-release of waste" seems to have a potential to be realized in nuclear energy system comparison with fossil energy system. Never the less, we believe further reduction of radioactive waste is inevitable for nuclear energy development responding to the desire of general public.

2) Scientific Feasibility Evaluation

" Scientific feasibility" here means whether it is feasible for SCNES to be materialized in principle and idealistic condition. The feasibility can be demonstrated by providing the following. First, the ultimate goal can be achieved within the range of

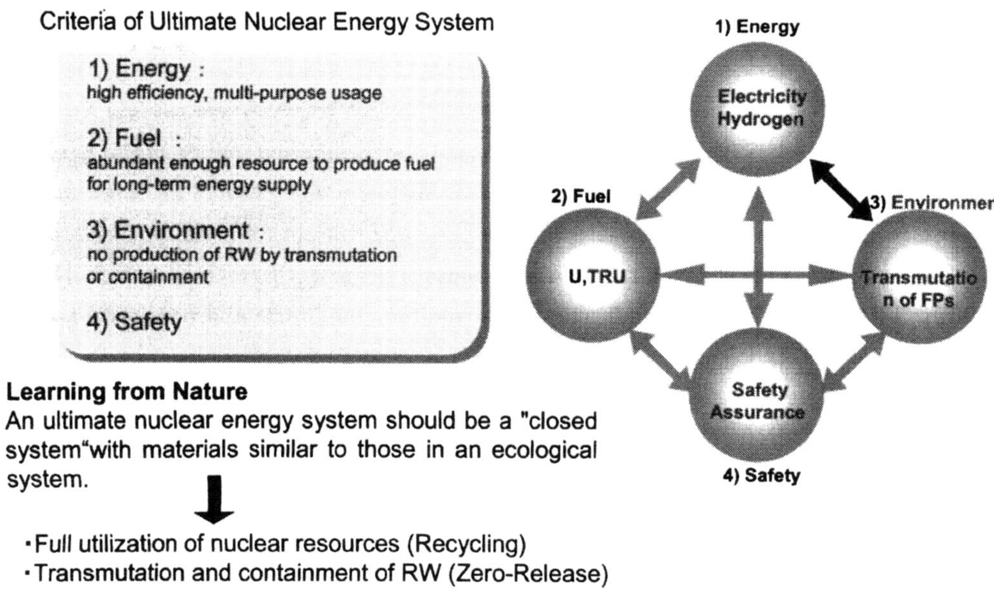

Figure 1 Self-Consistent Nuclear Energy System (SCNES)

Table 1 Neutron Balance in SCNES

Items	MOX Fuel Core	Metal Fuel Core
Required Neutron		
1. For Chain Reaction :		
Fissile (n,fission)	0.80	0.72
Fertile (n,fission)	0.20	0.28
2. For Fuel Generation*1 : Fertile (n,capture)	1.00	0.85
3. For FP*2 Transmutation : FP Target (n,capture)	0.24	0.24
4. Others :		
Fissile (n,capture)	0.20	0.13
Caputure in Structural Material,Sodium etc.	0.25	0.20
Total	2.69	2.42
Generated Neutron by Fission	2.90	2.90
Available Neutron for FP Transmutation*3	0.45	0.72

*1 Fuel Generation Ratio = 1.0
*2 with isotope separation
*3 (Available Neutron for FP Transmutation)
=(Generated Neutron)−[(Required Neutron)−(Neutron for FP Transmutation)]

the neutrons and energy generated by nuclear fission. Assets of a fission reaction are emerging neutron of about three in number and released energy of about 200Mev. Secondly, the required inventory (equilibrium inventory) of fission products for incineration in the reactor core can be within the range to be loadable in the blanket and around the core. That is, as described below, the feasibility can be evaluated by the balances of neutron, energy and mass.

(1) Neutron balance

Fission neutrons ≥ (Neutrons for chain reaction + Fuel breeding + Radioactive FP incineration)

(2) Energy balance

Attainable energy >> Required energy for recycling

(3) Mass balance

Equilibrium inventory of FPs in a reactor ≤Loadable inventory around the core

The feasibility of these points has been investigated by a research team in Research Laboratory for Nuclear Reactors at the Tokyo Institute of Technology, to conclude that feasibility is evident (A. Shimizu, 1994, Ref.3). The Institute of Applied Energy has been studying engineering concepts for SCNES. (M. Mayumi, 1999, Ref.4) The following outlines the results of our feasibility evaluation for the SCNES with fast reactors based on recent information (nuclear cross section and other data).

3)Features of fast reactors

Fast reactors have the following well known features:

- Parasitic capture of fuel is low and fuel breeding is feasible.
- Transuranic elements (TRUs) other than Pu, such as Np, Am, and Cm (so-called minor actinides; MA), existing in the spent fuel can be utilized as fuel.
- By varying the depleted uranium inventory in the blanket around the core, not only fuel generation, but also incineration of transuranic elements (TRUs) can be achieved with a core structure within the same plant.

In fast reactors having these features, the burning (incineration) of transuranic elements can be achieved by actinide recycling that utilizes transuranic elements other than Pu for recycling together with Pu. Consequently, the crucial issue in the SCNES feasibility evaluation is whether the system ensures FP incineration while maintaining fuel generation (fuel self-sustaining) capability and contains radioactive fission products in the system.

From this viewpoint, we have evaluated the feasibility for a typical oxide fuel and metallic fuel fast reactor core.

4) Neutron balance

In an oxide fuel reactor core, 2.9 neutrons are generated per nuclear fission. Among these, 1 neutron is consumed to sustain the chain reaction and another 1 in fuel generation. Considering that 0.45 neutrons are consumed in parasitic capture by fuel and other parasitic absorption, the remaining 0.45 neutrons would be available for the incineration of radioactive FPs produced by the fission. On the other hand, in the case where all the radioactive fission products having a half-life of one year or more are recovered from the spent fuel at 100% recovery rate, processed through isotope separations, placed in the reactor, and then incinerated by neutron irradiation, the number of required neutrons for FP incineration would be 0.24 per fission, which is less than the available neutron number of 0.45. Thus, neutron balance can be achieved. (Table 1)

There are nine FPs, that are I, Tc, Cs, Pd, Zr, Sn, Se, Sr, and Sm, to be incinerated or to be required for long time storage for decay.

In the case of a metallic fuel core, since the neutron absorption by parasitic capture is low, the available neutrons for FP incineration are 0.72. Thus a neutron balance is ensured with sufficient excess neutrons.

However, in the case when isotope separations are not available, the number of neutrons needed for incineration of radioactive FP are greater than 1, so a neutron balance can not be achieved even in a metallic fuel core. Isotope separation technologies for FPs are inevitably necessary to achieve the ultimate goal of SCNES using fast reactors.

5) Energy balance

In nuclear fuel recycling facilities, reprocessing and fuel fabrication processes are considered to be the largest energy consumer. Even when isotope separation of FPs is assumed, the required process energy is 1.9% of that obtained by nuclear fission. Consequently, energy balance can be achieved .

6) Mass balance

The equilibrium inventory that is necessary to make the amount of FPs incineration quantitatively balance with the radioactive FPs generated by fission, reaches an approximate total of 14 tons in a 1 GWe fast reactor. Compared to the case of a conventional 1 GWe fast reactor, in which approximately 40 tons of blanket fuel are installed around the core, the above-mentioned 14 tons of FPs are an amount that can be loaded around the core. That is, mass balance for the radioactive fission products can be achieved .

7) Summary of scientific feasibility

From the above results, SCNES is evaluated to be feasible scientifically. The key word "scientific" is used here since the evaluation is conducted assuming ideal conditions. For example, it is assumed that fission products to be incinerated can be fully recovered, complete isotope separation can be achieved, and only necessary fission products can be selected to be loaded in the reactor. The scientific feasibility does not necessarily mean the engineering feasibility. However, the scientific feasibility is a key to research on the engineering materialization of the SCNES concept and it is one of the factors that encourage us for further research.

3. System Concepts for SCNES

1) Element technology for long term goal

we can draw a system concept for SCNES that is composed of the following element technologies and subsystems achieving the ultimate goal. (Fig.2):

Element technologies and subsystems for SCNES;

Actinide recycling system with fast reactors
Radioactive FP elements recovery
Isotope separation system for FP elements
System that incinerates long half-life FP isotopes in the fast reactor core
Storage system for short life FP elements for decay
It is possible to introduce an accelerator driven systems to help and ease the balances, but discussion made here does not include ADS system for simplicity.

2) Major development items

Establishment of actinide recycling system with fast reactors is the key technical issue and isotope separation technologies for FPs are inevitably necessary to achieve the ultimate goal of SCNES using fast reactors.

In the evaluation for neutron balance, the TRUs and radioactive FPs (30 nuclides) in spent fuel assumed to be recovered at 100% recovery rate. If innovative technologies can be developed that ensure a high recovery rates (6 nines) for the elements recovery and isotope separation processes, the incineration capability of the nuclear reactor core is sufficient, and the radioactivity in high level waste could be reduced to the level of

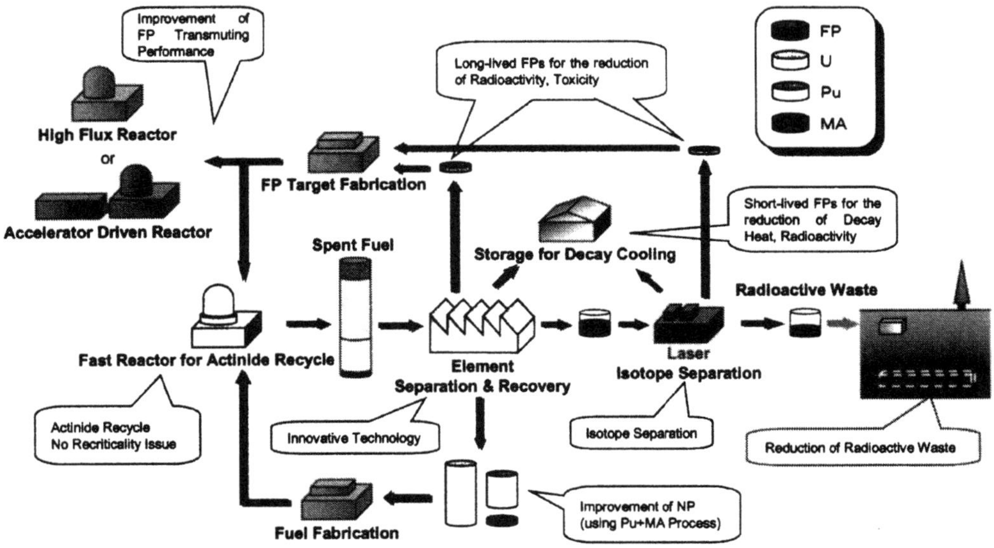

Figure 2 SCNES and Fuel Cycle (Long Term Goal)

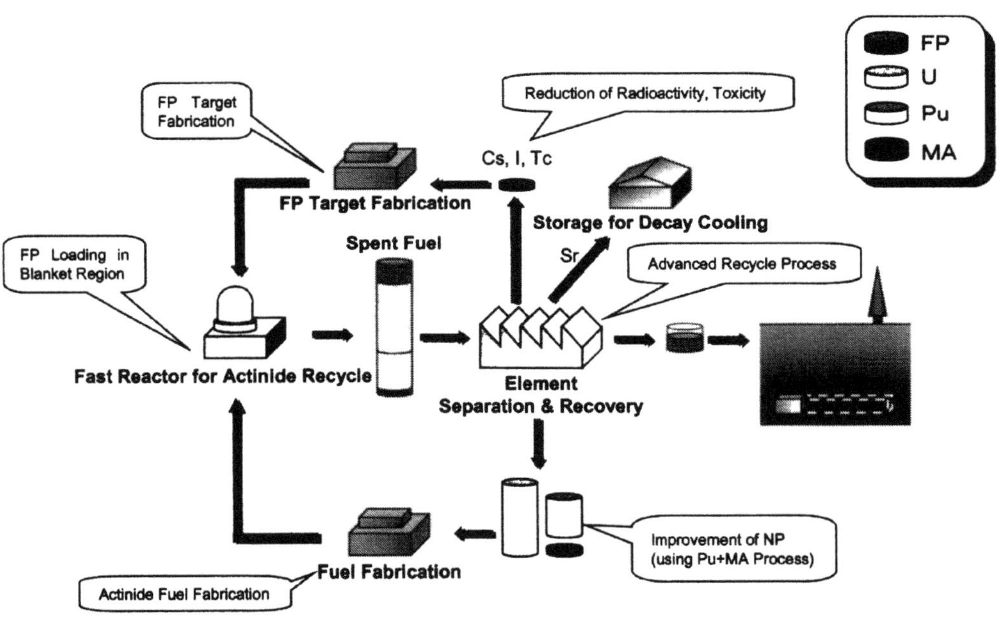

Figure 3 SCNES and Fuel Cycle (Near Term Goal)

natural uranium.

The innovative technologies, which ensure high-efficiency recovery or isotope separation at a rate of 6 nines or more are one of major development items, but they cannot be illustrated in detail from the engineering point of view at present. Although isotope separation technologies, such as Atomic Vapor Laser Isotopic Separation, are now being studied in advanced uranium enrichment technology.

Regarding the elemental separation/recovery technologies, we should not adhere to the development of the technologies, such as PUREX, which pursue recovery of clean Pu alone. Instead, it is necessary to develop a reasonable separation/recovery technology that can be compatible with the capability of impurity acceptance in the reactor.

In order to improve the incineration efficiency of FPs, and to mitigate the purity requirements for recovered FPs, further innovative technologies may be required, such as high-neutron flux FP incineration reactors, and accelerator-driven reactors.

3) Near-Term Goals

Why we need near-term goals

The innovative technologies that will make it possible to implement high-efficiency recovery or isotope separation with 6 nines or more of efficiency cannot be identified at present stage. Accordingly, investigation of near-term development scenarios requires the setting of near term goals.

Viewpoint for near-term goal setting

We discussed the near-term goals by adopting the following requirements:

- Scenarios can be drafted for development and technology selection toward goal achievement. (The feasibility of achievement can be forecast relatively easily.)
- Significant effects can be expected from incineration.
- Nuclear fuel recycling technology that does not place special importance on Pu can be achieved.

In order to select the fission products to be incinerated in the near-term goals, the evaluation from the above-mentioned viewpoints has been conducted for key FPs. As a result, we have selected I, Tc and Cs. The following near-term goals are set up.

Near-term goals

Actinide recycling by fast reactor (recovery rate; 3 to 4 nines)
Recovery of I, Tc, Cs and Sr (recovery rate; 2 to 3 nines)
Incineration of I, Tc, Cs by fast reactor

Storage of Sr for decay

4) Feasibility of achieving near-term goals

In order to confirm the feasibility the FP incineration in the near-term goals we evaluated their neutron and mass balances. .

We have assumed that the target FPs for incineration are in a chemical form for loading that is considered appropriate under current stage and that they would not be treated in the isotope separation process. We evaluated them in terms of the required number of neutrons including the amount for parasitic capture as compounds and the equilibrium inventory. The number of required neutrons is 0.32 and the equilibrium mass is 12.4 tons. Thus, both neutron balance and mass balance are found to be achievable.

Further, an evaluation of specific recovery processes has revealed that in the improved PUREX process, dry oxide process, and dry metallic process, the target recovery rate could be achieved by adding processes or modifying the processes.

5) SCNES to meet the near-term goals

A system concept for SCNES that meets the near term goals is shown in Fig.3. The near term SCNES will reduce high-level waste radioactivity and potential toxicity (1/100 to 1/1000 of one-through level as shown in Fig. 4).

The SCNES realizes an actinide recycling system that does not regard Pu as a special substance, which is a desirable system in the view of nuclear nonproliferation

6) The relation between the reduction of the radioactive materials and the geological disposal

Achievement of reduction in high-level waste radioactivity and amount of heat generation does not rule out the necessity of geological disposal of high-level waste. Instead, it will reduce the burden of geological disposal and thus promote public acceptance of geological disposal, leading to improved public acceptance of nuclear energy itself. Therefore, it should be noted that SCNES and geological disposal can complement each other.

4. Global Goals of Nuclear Energy and prospects for international collaboration

4.1 Discussion on GEN-IV goals

Figure 5 shows the goals of GEN-IV comparing with SCNES.

Reduction of HLW by a near term SCNES

Actinide :99.9% Recover, Recycle of the fuel
I, Tc, Cs :99% Recover, Transmuting in Blanket
Sr :99% Recover, Cooling
Relative to once-through core

Item	50yrs later	1000yrs later
Radioactivity	1 / 40	1 / 90
Radioactive Toxicity	1 / 500	1 / 300
Decay Heat	1 / 140	1 / 300

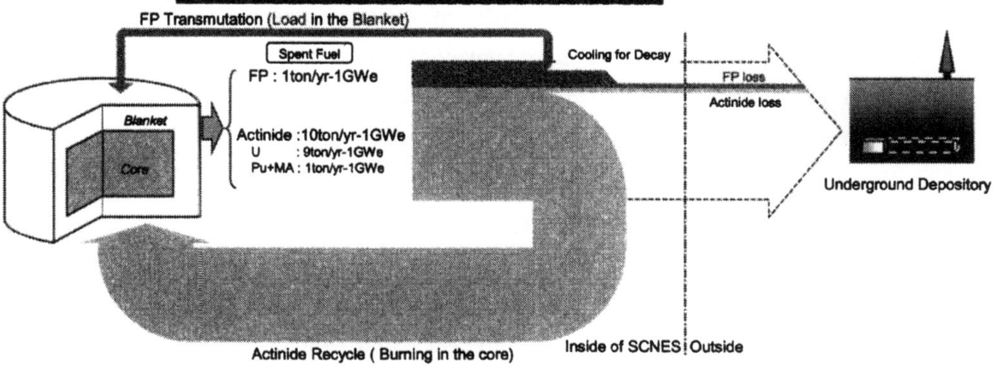

Figure 4 Reduction of Radioactivity and Toxicity

SCNES Goals

Energy
--High efficiency, Multi-purpose usage

Fuel
--Abundant enough resource to produce fuel for long-term energy supply

Environment
--Non production of radioactive waste by trans-mutation or containment

Safety
--Passive safety and re-criticality free core concept

Generation IV Goals (United States)

Economics
--Competitiveness to cost from other sources
--Acceptable risk to capital with respect to other energy development

Sustainability(*1)
-- Advanced sustainable options
High uranium utilization
Flexible use of fissile and fertile resources
Waste minimization etc.

Safety
--Extremely resistant to core damage accident
--Demonstration of safety
--Design afforded radiation protection

Waste
--Complete technical solutions for all waste streams

Proliferation Resistance
--The least attractive route for potential weapons proliferators

*1:under discussion in GEN-IV expert group

Figure 5 SCNES vs. GEN-IV goals

During the discussions about the goals of GEN-IV in the expert group meeting held in Korea in August 2000, "sustainability" was proposed for an additional goal. As a result, GEN-IV and SCNES have the common and consistent concept of goals. The goal of "sustainability" is deeply-rooted in the inherent nature of nuclear energy and is essential as a long-term goal.

There are opinions in the US technical experts that further reduction of radio-waste is unnecessary. Nevertheless, if we consider the span of millennium with the half-life of radioactivity further reduction of radioactivity of nuclear waste lies on the way to the goal. It is recommended for nuclear energy development responding to the desire of human race, because the further reduction of radioactivity is achievable based on the intrinsic nature of nuclear energy.

The discussions on economics and nuclear proliferation resistance aspects, which are not written in SCNES goals explicitly, are shown below.

1) Economic competitiveness

It is important to pursue economic efficiency and competitiveness. It would be necessary to consider how the term "economic efficiency" be defined. That is how to measure the economic efficiency, and how external costs should be included in the comparison with other energy systems. It would be necessary to argue these items.

We should recognize that economic competitiveness is a primary goal but is not all.

2) Nuclear proliferation resistance

As for the SCNES concept we do not adhere to the technologies, such as PUREX, that intends to recover clean Pu alone. Instead, we believe it is necessary to develop a reasonable separation/recovery technology, which is compatible with the capability of impurity acceptance in the reactor. We have selected the actinide recycling process which does not give Pu special substance, taking into account both reduction of the radioactive waste and improvement of the intrinsic nuclear proliferation resistance.

Our fundamental ideas on nuclear proliferation resistance are follows.

a) We consider the prevention of nuclear proliferation as a step towards the ultimate extinction of nuclear weapons. The nuclear states should continue their efforts to reduce nuclear weapons.

b) Development of nuclear energy is essential in terms of getting rid of the motivation and intention of nuclear proliferation because nuclear energy can significantly ease the world energy supply issues, which would cause international conflicts as ever. Especially, the effective utilization of natural uranium and reduction of radioactive waste through recycling, which is one of the advantages of nuclear energy, are the important themes in the future development of nuclear energy.

c) In the meantime, it goes without saying that the developments of reactors and the fuel cycle technologies with higher nuclear proliferation resistance are preferable in order to promote the global utilization of nuclear energy. SCNES which is going not to apply pure separation process as mentioned above can meet the requirements for non-proliferation.

d) However, while the evaluation methods and the criteria regarding the nuclear proliferation resistance have been discussed at the workshop on Technical Opportunities for Proliferation Resistance of Global Civilian Nuclear Power Systems (TOPS, Ref. 6), at present, it will take still some more time to obtain international consensus on these matters. Therefore, we should avoid the statement that the measures to improve the nuclear proliferation resistance that might damage strongly the economic efficiency and other essential advantages inherent to nuclear energy, must be prepared as the absolute conditions of the nuclear development and employment. It is our understanding that the issues on the nuclear proliferation resistance technology is still at the stage of problem definition. So, considerations should be taken so that excessive pursuit of the nuclear proliferation resistance would not jeopardize potential abilities of nuclear energy.

4.2 Activity for "Zero release system"

1) TRU burning and FP transmutation

Table 2 shows research activities for TRU burning and LLFP nuclear transmutation in the world. SPIN (M. Viala, 1994, Ref.7) project in France and DOVITA (A. V. Bychkov, 1997, Ref. 8) project in Russia have been initiated for solving the radioactive waste issues, which are intrinsic to nuclear energy, by reducing the radioactivity contained in nuclear waste. Therefore, both studies are positioned on the half-way to the ultimate goal of SCNES (zero-release goal).

2) Elimination of re-criticality issues in fast reactors

Conservative safety design, which applies the ALARA (As Low As the Reasonably Achievable) principle to normal operation and the " Defense in depth concept " to accidents" has been incorporated and severe accident management measures utilizing the probabilistic safety assessment methodology have been promoted. These measures have been contributed to zero-release in practice.

Table 2 TRU burning and FP transmutation studies

	Organization	Project	Transmutation Device	Target Nuclides	TRU Recovery
USA	LANL	ATW	ADS	U TRU Tc I	
	ANL	IFR	Metal FR	MA	Electro-refining (Metallic Fuel)
France	CEA	CAPRA	FR	Pu	Solvent Extraction with DIAMIDE
		SPIN	FR ADS	MA Tc I Cs Sr	Solvent Extraction with DIAMIDE
EC	CEA EDF ECN FZK IAM ITU	EFTTRA	HFR	Tc I Am	
Switzerland	PSI		LWR ADS	Pu MA	
Russia	RIAR	DOVITA	FR	MA	Electrowinning (Oxide Fuel)
	IPPE		ADS	MA	
Japan	JAERI	OMEGA	ADS	MA Tc I Cs Sr	Solvent Extraction with DIDPA
	JNC		FR		Solvent Extraction with CMPO
	CRIEPI		Metal FR		Electro-refining (Metallic Fuel)

However, in the fast reactor core, a large amount of molten fuel may be centralized during "Hypothetical Core Damage Accidents (HCDAs)", which results in a re-criticality, leading to the generation of mechanical energy.

In order to materialize the zero-release concept, which does not require the emergency evacuation in a HCDA, in a reasonable design concept, it is important to eliminate the possibilities of re-criticality during HCDAs.

In this context, it is necessary to provide design features, which promote discharge of molten fuel outside the core before the centralization of molten fuel occurs in the early stage of CDA, and to verify the effectiveness of the design. As the design feature for the elimination of re-criticality, installation of a duct within the fuel assembly, which facilitates discharge of molten fuel before fuel melt extends to the adjacent fuel assemblies, is conceived. Such design feature can provide a simple countermeasure against re-criticality considering one-dimensional behavior. This concept can be demonstrated by in core tests.

Japan and the Republic of Kazakhstan are now working on a collaborative research program in order to confirm the principle of re-criticality elimination, which is scheduled to be implemented from 1998 to 2004. Research activities are being carried out utilizing Inpile Graphite Reactor (IGR) test facilities and out-of-pile test facilities at Kazak's NNC.(H. Endo, 2000, Ref.9)

3) EPR non-evacuation principle

The European Pressurized Reactor (EPR, (W. Frish, 1997, Ref. 10)) concept also lies in the direction for zero release with applying
non- evacuation principle in case of severe core damage.

4.3 Prospect for international collaboration

As explained above, consistent efforts have been made to achieve the zero-release concept.

In the recent discussions at the TOPS and GEN-IV international workshops, evaluations of the nuclear proliferation resistance for various fuel cycles, including the closed cycle, and the initiation of studies for the fuel cycles with enhanced nuclear proliferation resistance have been insistently recommended. GEN-IV has the common and consistent concept of goals with SCNES. Though GEN-IV would have shorter time span of interest.

As discussed above, future nuclear technologies should aim at recycling and zero-release. Steady progress has been made in the establishment of international consensus on the future goals of nuclear energy development. Therefore, it can be expected that there would be a growing possibility of international collaboration.

5. Conclusion

The scientific feasibility of SCNES, whose ultimate goals are full utilization of nuclear energy resources and zero-release of radioactive material, is presented and its system concepts are illustrated.

SCNES is not just a dream. We view SCNES as a primary future target of development, although a great deal of R&D is required to realize its system concepts.

By comparing with recent GEN-IV goals, the basic directions for the future nuclear energy development are common and consistent. Therefore, a possibility of international collaboration is growing in magnitude.

In implementing long-term R&D like nuclear energy development, it is important to draw the ultimate goals and clarify the vision to aim. As communication with the public is a must issue in a democracy, it is time to present and show potential capabilities and future prospect of nuclear energy to obtain understanding and positive support of the public.

The principles and the visions of nuclear technology development should be further discussed.

Reference:

1.J.Holdren, Energy and Environmental issues in the 21st Century and the Future of Nuclear Power, Presentation at the Tokyo University and Harvard University Symposium on the Future of Nuclear Energy, July(2000)

2. Y. Fujii-e, "An Approach of Self-Consistent Nuclear Energy System - Potential of Fast Reactors," Proc. Design and Safety of Advanced Nuclear Power Plants, Tokyo, October (1992).

3. A. Shimizu and Y. Fujii-e, "Self-Consistent Nuclear Energy Systems," Int'l Symp. Global Environment and Nuclear Energy systems, Susono, Japan, October (1994).

4. M. Mayumi,et.al, "Proposal of an Advanced FBR Plant Concept for the Self-Consistent Nuclear Energy System", Global'99

5. H. Noda, et.al, "Current Status of Fast Reactor Cycle System in Japan", 8th International Conference on Nuclear Engineering, April 4th(2000)

6."International Workshop on Technology Opportunities for Increasing the Proliferation Resistance of Global Nuclear Power Systems", March 29-30, 2000, Washington, DC.

7. M. Viala, M. Salvatores, "An Overview of the Spin Programme" AEN/NEA Actinide and Fission Product Partitioning and Transmutation, CADARACHE, France, 12-14 December 1994
8. A. V. Bychkov, et.al, "Fuel Cycle of Actinide Burner Reactor. Review of Investigations by DOVITA Program" Proceedings of Global'97, p657, Yokohama, Japan, (1997)
9. H. Endo, " Elimination of recriticality Potential for the Self-Consistent- Nuclear Energy System" International Seminar on Advanced Nuclear Energy Systems Toward Zero Release of Radioactive Wastes, Susono Seminar House, Shizuoka, Japan, Nov. 6-9, 2000.
10. W. Frish, and G. Gros, " Key Issues Recently Treated within the French-German Safety Approach," International Topical Meeting on the Safety of Advanced Reactors , ARS97 (1997)P.279.

Pergamon
www.elsevier.com/locate/pnucene

Progress in Nuclear Energy, Vol. 40, No. 3–4, pp. 285–299, 2002

Printed in Great Britain
0149-1970/02/$ - see front matter

PII: S0149-1970(02)00022-7

THE NEED FOR INNOVATIVE NUCLEAR REACTOR AND FUEL CYCLE SYSTEMS: STRATEGY FOR DEVELOPMENT AND FUTURE PROSPECTS

V. MOUROGOV, K. FUKUDA, V. KAGRAMANIAN

Department of Nuclear Energy
International Atomic Energy Agency
E-mail: V.Kagramanian@iaea.org

ABSTRACT

The paper reviews the major nuclear power challenges in safety, waste and non-proliferation areas that are hindering large-scale participation of nuclear power in the world energy market. It reviews also the ongoing R&D in reactor and fuel cycle areas and discusses the need for international cooperation in innovative R&D activities to address the above challenges. The main question is how to address these challenges and at the same time improve economic attractiveness of nuclear power in the global deregulated electricity markets. © 2002 Published by Elsevier Science Ltd.

1. THE NEED FOR INNOVATION

Nuclear power is now at a crossroads, with no consensus concerning its future. A number of factors currently favour nuclear power, including:

- Nuclear energy has grown in only 50 years from a new scientific development to a major part of the energy mix in several of the 32 countries now using nuclear power [1]. The majority of currently operating Nuclear Power Plants (NPPs) perform well, both in terms of economics and safety [2].
- In addition to producing electricity, nuclear power helps meet other important national goals in many countries, including energy independence, clean air, and reduced emissions of greenhouse gases (GHGs).
- Nuclear power is currently the only mature non-carbon electricity generation technology that can significantly contribute to the long-term global sustainable energy mix. Recent scenarios developed by the Intergovernmental Panel on Climate Change (IPCC) foresee a significant potential for the nuclear energy growth – from its current 6%

of primary energy to between 10% and 30% by 2100[3]. This implies an increase in global nuclear power capacity from 349 GW(e) today to 2,000-5,000 GW(e) by 2050[4].

Against this background, one could reasonably expect nuclear power generation to be in the midst of a rising trend extending out into the foreseeable future. Yet, that is not the case. Figure 1 shows average annual new capacity additions, worldwide, for 5-year periods from 1971 to today. As is evident in the Figure, new nuclear additions have been decreasing since the middle of the 1980s and are now an almost vanishingly small part of total capacity additions. Quite simply, over the last 15 years, nuclear power has been losing market share badly in a growing world electricity capacity market.

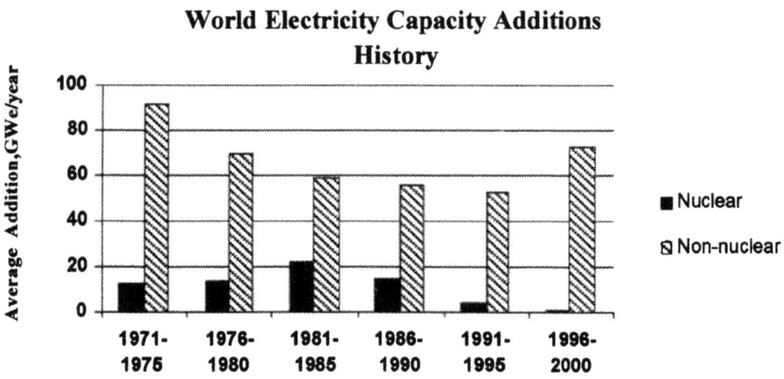

Figure 1

Looking at the developed regions that have principally supported nuclear technology, we find a relatively greater contribution of nuclear power, but not necessarily brighter prospects for the future. In North America (NA), Figure 2 shows that capacity additions declined in the first part of the period presented, driven by energy conservation, economic slowdown, and some overcapacity. Nuclear power maintained a significant share, however, through the 1980s. Now capacity additions have started to grow again, but net additions for nuclear power have dropped first to zero (1991-95) and then below zero (1996-97). Figure 3 shows a similar situation in Western Europe (WE). New additions have begun to rise, but it is clear that nuclear power, despite its substantial role in Western Europe, will not be part of that rise for at least the next decade and quite probably more. Figure 4 presents the history for Eastern Europe (EE). Here the situation is dominated by economic recession. Whatever the cause, however, the picture for nuclear is not promising.

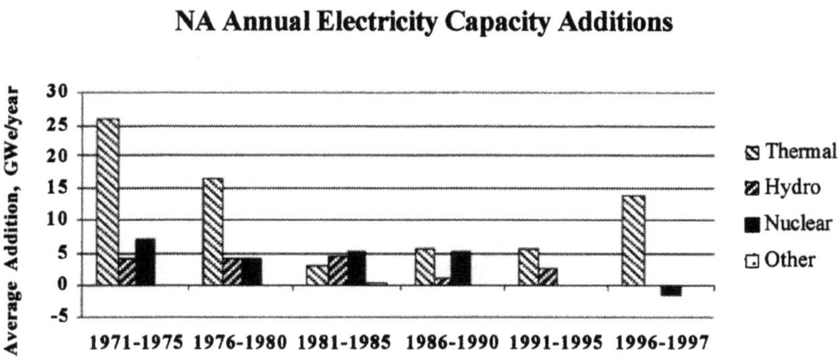

Figure 2

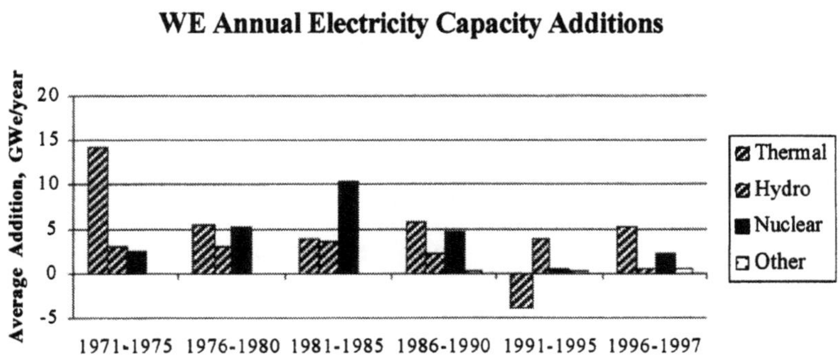

Figure 3

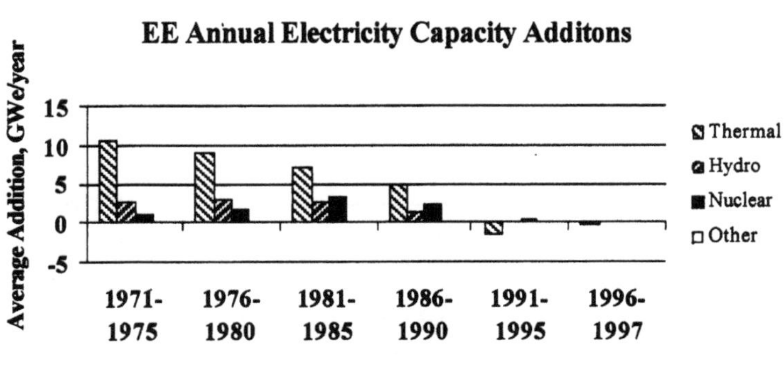

Figure 4

Figure 5 describes the situation in the developing regions, including Latin America, Africa, the Middle East, South and Southeast Asia, and the Pacific. Nuclear's past market has not been in the developing countries. Latin America has some nuclear capacity, but the important point here is that Latin America is a modest part of the rebound since the mid-1980s in new capacity additions, but nuclear power is largely absent. In Africa, there is less of a market, no discernable rebound, and no nuclear since 1985. In the Middle East and South Asia, there is a bit more of a market, new additions seem to be roughly holding steady and there is no nuclear. Again, no nuclear in Southeast Asia and the Pacific, but more of a growth trend in capacity additions in general.

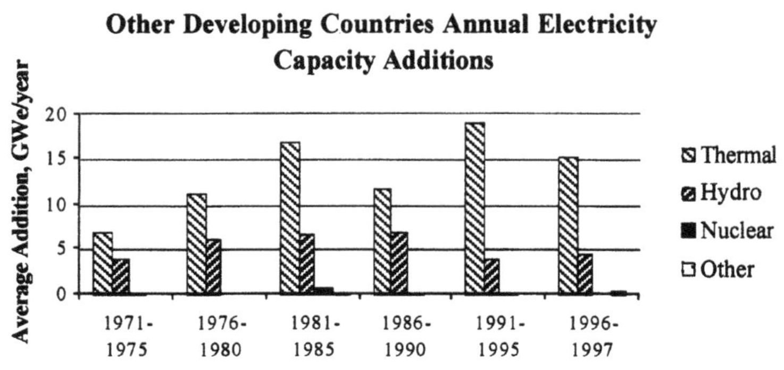

Figure 5

For the Far East (FE), which we usually emphasize as the best current and future market for nuclear, the picture is brightest – see Figure 6. Overall capacity additions are rising fastest and nuclear's share is the most distinctive. But, even here, nuclear capacity additions are basically staying about constant, not growing with the market. Even here, the result is a diminishing market share in a growing market.

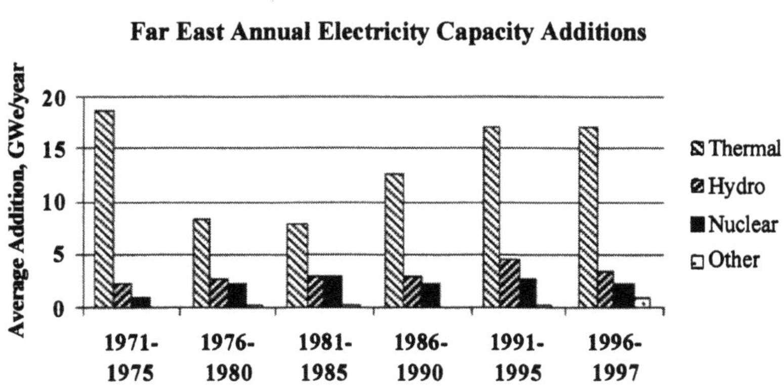

Figure 6

Overall, it is quite a sobering picture. The rate of new electricity generating capacity additions has started to grow in the last 10 years after a prior slowdown. But nuclear additions are hardly keeping pace. Thus, we have a growing market, but a small and shrinking nuclear share.

Looking ahead for the next one to two decades, things do not look much better.

- There is strong current political opposition to nuclear power in some western countries, including those with most expertise in the field. The present governments of Germany, the Netherlands, Belgium and Sweden – countries that have relied heavily on nuclear power – have the intention to phase it out.
- There is also opposition to nuclear power's inclusion among the sustainable energy technologies eligible for Clean Development Mechanism (CDM) projects under the UN Framework Convention on Climate Change (UNFCCC).
- Figure 7 shows that the IAEA's projected capacity additions through 2020. The overall picture for the electricity sector, in both the high and low scenarios, is quite healthy. But not so for nuclear. In the low scenario – and this is globally including Japan, China, India and so forth – net nuclear capacity additions between 2011 and 2020 are *negative*. In the high scenario, they are so small as to be unnoticeable in this picture for the next decade. The OECD/NEA and US DOE statistics show also that there might be no breakthrough in the present situation, unfavourable for nuclear power, in the following 15-20 years. We have been losing market share in a growing market and, according to these projections, we will continue to do so – the only question is how fast.

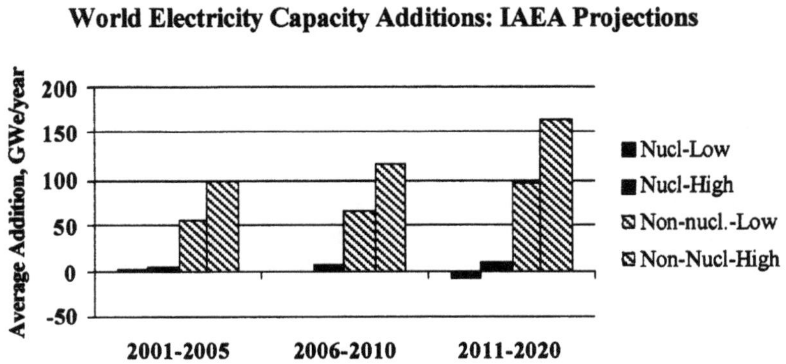

Figure 7

What are the reasons for this "unhealthy" situation with nuclear power?

First, nuclear power development faces serious challenges of an economic and political nature in the two regions with the maximum experience in nuclear power development – North America and Western Europe. The situation was aggravated also because of absence of demand growth for electricity in these countries in the past decade.
Second, in developing countries the main issue are lack of expertise in nuclear technology and its safety culture, lack of adequate infrastructure, and of course economic and financing issues.

And, third, public concerns on nuclear waste, safety and non-proliferation are also hindering the development of nuclear power globally.

If nuclear power is to contribute in significant ways to meeting future energy demands, the above issues, real or perceived, must be addressed.

The nuclear community has done a lot using traditional measures to address the issues mentioned above.

- We have accumulated and distributed statistics illustrating the impressive health and environmental advantages of nuclear power in comparison to alternative electricity generation options. But this does not influence over much the market forces responsible for the choice of new generation capacities. They favour the cheapest options, which are often not nuclear.
- We have projected the impact of possible external developments, such as the introduction of environmental taxes or increased fossil fuel prices, on the attractiveness of nuclear power. But the reality is different. A recent DOE analysis shows that even in the most stringent carbon reduction cases only operation through 2020 may be considered for the predominant part of the existing NPPs. No new NPPs will be needed. The reason is that in most industrialized countries natural gas will be the closest competitor to nuclear. Introduction of symbolic taxes on CO_2 emissions will not change drastically the comparative economics of the nuclear and gas electricity supply options. There is no expectation of the introduction of any taxes on CO_2 emissions in the next 10-20 years in developing countries like China and India where coal is the closest competitor to nuclear. There is also a lack of capital in these countries for large-scale nuclear power deployment.
- We have organized new institutional measures in the areas of safety, waste and non-proliferation, which helped us significantly reduce the risk of severe accident, and proliferation of fissile materials within the existing nuclear power system. These measures are very important and should be further increased. But we should admit that these measures, while they are based on the existing technologies that have their own limits, will not help us much.

- We have improved the performance of existing nuclear facilities. Figure 8 shows improvements in the aggregate availability of nuclear power plants over the last decade. The trend is significant and substantial. It is the equivalent of 28 GW(e) of new capacity – highly cost–effective new capacity. However, this route to success has its limits. Realistically, this trend will have to level out at somewhere around 85% or 90%.

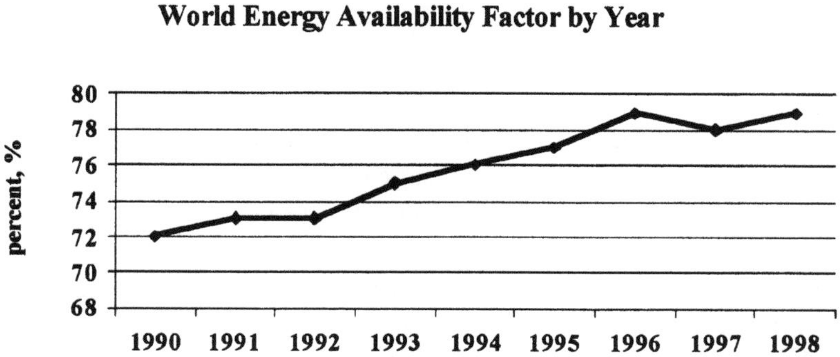

Figure 8

- Lifetime extensions of existing NPPs is another important accomplishment. But, again, this route to success has its limits.
- Figure 9 shows the age distribution of today's operating reactors. In the middle is the "baby boom" from the 1970s and 80s. For these reactors, lifetime extensions promise substantial capacity extensions. But after the "baby boom" reactors, there is not much in the pipeline. The reality is that the impact of lifetime extensions can only diminish.

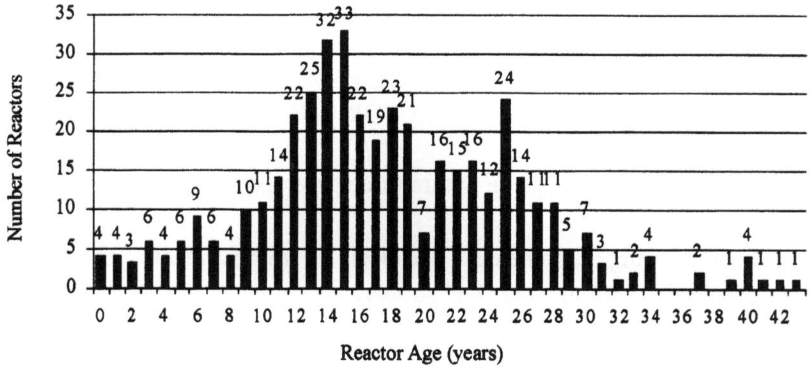

Figure 9

We are working on evolutionary new designs to improve performance even further. But as, was shown in the recent OECD/NEA study, it will be extremely difficult, if not impossible, for new evolutionary advanced nuclear power plants to compete economically with the alternatives such as gas-fired plants in the absence of a change in governmental policies. Today's new designs of advanced LWR reactors may have a market place in a limited number of Asian countries that do not have access to cheap gas from pipelines, like the Republic of Korea and Japan.

In short, we have done a lot, but we understand that this might only stabilize, or slightly increase, the present low share of nuclear power in the world energy mix.

In our understanding, the main reason for this stalemate is that we, in all our doings, continue to rely on nuclear technology developed in the 1950s, which had its roots in military applications which cannot exclude absolutely the possibility of a severe accident and which has reached its limits from an economic point of view. Thus, as we increase nuclear capacity, expand its applications or expand into new countries, we must increase our reliance on human factors and active safety systems, as well as on institutional and organizational measures, to minimize the risk and consequences of severe nuclear accidents, and the risk of proliferation. The level of nuclear power use will also depend to a large extent on many external factors, such as level of fossil fuel market prices, environmental taxes. This reliance has its limits, which in turn limit the use of nuclear power.

Thus, we cannot simply do what we have been doing and expect that external factors (like fossil fuel prices, environmental taxes, public opinion) will become more favourable to us. If we want a different result, we must do something within the nuclear community and we should propose new technological solutions. More of the same won't do. We must look to the future, and we must be innovative. That is our starting point. The main question now is to understand what are innovative technologies: What should be the requirements for them in order to address the main challenges hindering large-scale nuclear power development?

2. WHAT ARE INNOVATIVE SYSTEMS AND WHAT ARE THE REQUIREMENTS FOR THEM?

Technology development is taking place within three general categories:

- Currently Operating Commercial Facilities – Improvements in maintenance, operations, engineering support, fuel supply, etc.

- Evolutionary Designs – Improvements in design and operation for near-term future deployment, involving moderate changes from currently operating commercial facilities.
- Innovative Designs – Advances in design and operation involving major departures from currently operating commercial facilities for long-term future deployment.

The first two categories of developments are critical particularly for the next two decades, in order to keep nuclear power alive while more dramatic innovations are prepared. The subsequent generation of reactors and fuel cycle that must form the basis for large-scale expansion of nuclear power worldwide must depend on innovations in nuclear technology.

The need for innovative R&D has been recognized by the nuclear industry and by those countries that believe in the overall benefits, viability and importance of nuclear power for the long term. Currently, R&D on innovative nuclear fuel cycle and reactor concepts is being performed in Argentina, Canada, China, France, India, Italy, Japan, Republic of Korea, Russia, South Africa, and the US. The US has embarked on a Nuclear Energy Research Initiative in 1999 to develop advanced reactor and fuel cycle concepts and scientific breakthroughs in nuclear technology to overcome obstacles to the expanded use of nuclear energy [5-8].

The IAEA has been assisting Member States for a long time in the development of advanced reactor and fuel cycle technologies through traditional activities within the framework of established International Working Groups on different innovative reactor and fuel cycle systems.

There is a wide diversity of targets used for the development of innovative reactor and fuel cycle concepts. The wide diversity of targets leads to a wide diversity of reactor and fuel cycle concepts. Four main areas of R&D activities on innovative reactor and fuel cycle concepts with the targets used for their justification are listed below.

- **Small modular reactors (HTR, LWR), once-through fuel cycle:** targets: <u>competitiveness, minimum financial risk, period of construction 2-3 years, increased safety level based on passive features, potential for electric and non-electric applications, for use in areas with or without grids.</u>
- **Different Fast Reactor and Closed Fuel Cycle Concepts:** targets: <u>competitiveness, deterministic safety, effective use of uranium resources; minimization of long-lived radioactive wastes through transmutation; increased proliferation resistance.</u>
- **Accelerator driven systems:** targets: <u>minimization of long lived radioactive wastes through transmutation of hazardous minor actinides and fission products.</u>

- **Thorium based reactor and fuel cycle systems:** targets: competitiveness, increase of resource base, increase of proliferation resistance, reduction of minor actinides in waste.

The analysis of the targets used for development of innovative reactor and fuel cycle systems has shown that there are significant differences in the types of requirements, as well as in the definition of a requirement in each area of concern: economics, safety, waste management, non-proliferation and others.

For example, in the economic area, although all concepts aim to be competitive on the future energy market, there are different opinions as to whether they should become competitive taking into account the potential introduction of CO_2 taxes and increases in fossil fuel prices, or not.

A wider diversity exists for the targets in such crucial areas as safety, waste management, non-proliferation, and resource consumption.

For example, in the safety area some believe that today's advanced LWRs are sufficiently safe for large-scale development, because they are neighbour-friendly (no significant release of radioactivity off-site even in the case of severe accident). Others insist that the public will accept large-scale nuclear energy deployment only if we propose a new reactor type with deterministic safety features (no significant release of radioactivity even from fuel), like modular HTRs or innovative FRs.

In the waste management area, some believe that direct underground disposal of small volumes of spent fuel is a sufficiently safe back-end option. To ensure public acceptance, it needs only a practical demonstration. Others insist that only the elimination of nuclear long-lived hazardous nuclides by burning or transmuting them in fast reactors or ADS would gain public support for large-scale nuclear energy development. There is no consensus among the advocates of this view on how to measure the level of success in addressing this challenge. There are different opinions as to precisely which hazardous elements should be eliminated, and to what extent.

In the non-proliferation area, some believe that the present safeguard system with its potential development will be sufficient to safeguard effectively any new reactor and fuel cycle. Some propose to develop special "proliferation resistant" reactor and fuel cycle concepts (new type of fuel, new reprocessing technologies without the extraction of plutonium, new concept of FR, and so on) with increased reliance on technical measures against possible proliferation. There is also no consensus among researchers as to how to measure the level of "proliferation resistance" and to what extent we should increase our reliance on technical measures.

Table 1 presents a view of possible challenges that nuclear power might face in the case of a significant (tenfold) increase in its capacity in the 21st century. In the right column, potential technical solutions are indicated.

From Table 1 we can see that even most stringent challenges for a large-scale nuclear power development may be addressed through technology development. This is a big priority for nuclear power. The main question for nuclear power is how to address the above challenges while at the same time improving economic attractiveness.

A really successful innovative design or fuel cycle is unlikely to be one that is exclusively focused on just the economics, or safety, or proliferation resistance, or waste management. It will have to represent a distinctive improvement on many fronts all at once. This might be a difficult task for national laboratories taking into account significantly reduced level of governmental support for R&D in the nuclear area. An effective international cooperation is becoming crucial.

Moreover, for a new design to become successful on a large scale, and make a substantial contribution to future energy supplies, it must succeed at a global level. A global perspective and understanding is, something that the IAEA is uniquely positioned to offer given its membership, mission and experience.

Table 1: Large-scale Nuclear Power Development: Key Challenges and Potential of Technical Responses

Challenges	**Availability of a Technological Response**
Safety area: Reduce significantly accidents and deterministically eliminate sever core damage	Reactor and fuel cycle facilities based on use of technologies with passive safety features
Waste management: Minimize the volume and toxicity of nuclear waste	Multi recycling of most dangerous long-lived radioactive elements in reactor with fast neutron spectrum or ADS
Non-proliferation: Furtherance of non-proliferation aims, namely that nuclear materials cannot be easily acquired or readily converted for non-peaceful purposes	Recycling of fissile materials, together with radioactive minor actinides: integral nuclear fuel cycle without the stockpile of plutonium. Reactor systems without refuelling within core lifetime
Resource base: Increase significantly the resource base	Multi-recycling of plutonium in fast reactors

3. INTERNATIONAL COOPERATION IN R&D AND THE ROLE OF THE IAEA?

The desirability of coordinating the national activities on innovative reactor and fuel cycle systems has been acknowledged at several meetings held under the auspices of the IAEA. They have recommended that the Agency take steps to facilitate assessment of the potential for, and exchange information on, innovative nuclear reactors and fuel cycles.

The Agency's ongoing activities on innovative reactor and fuel cycle technologies include the activities of several international working groups and coordinated research projects in different areas.

Concerning in particular the waste management issue, which is the main topic of this Forum, the following recent Agency activities may be referred to:

International Conference on safety of Radioactive Waste Management. Cordoba, Spain, 13-17 March 2000

The main conclusions of the Forum were:

- Indefinite surface storage of waste is not a sustainable solution and progress must be made to provide an integrated solution to radioactive waste management that includes disposal.
- International repositories could offer a means for countries with small nuclear power programmes or without nuclear power programmes to dispose of their high activity and long-lived waste without incurring the very high costs of a national repository programme. However, it was clearly recognized that international repositories are controversial and that the disposal of waste is primarily a national responsibility.

Scientific Forum on Radioactive Waste Management: Turning Options into Solutions. Vienna, 19-20 September 2000

The results can be summarized as follows:

- Technological solutions to the safe management of radioactive waste exist, but public acceptance is needed.
- The decision-making process needs to be stepwise with the ability to reverse decisions at a later stage.
- Opinion formers responsible for political decisions have not yet entered the debate.
- International cooperation in the development of technology and safety standards can supplement national programmes.

This IAEA Forum also discussed the use of innovative technologies that did not produce any actinide or transuranic waste, and which transmute long-lived fission products. It was agreed that whilst further research is justified, transmutation on a practical engineering scale is far from certain and the prospect of transmutation should not prevent countries taking positive actions in waste treatment and disposal as soon as possible.

The Coordinated Research Project (CRP) on the Use of Thorium-based Fuel Cycles in Accelerator Driven Systems (ADS) to incinerate Plutonium and to Reduce Long-term Waste Toxicities is now in progress. The purpose of the CRP is to assess the uncertainties of the calculated neutronic parameters of a simple model of thorium or uranium fuelled ADS, in order to get a consensus in the calculational methods and associated nuclear data.

Advisory Group Meting (AGM) on Revue of National accelerator Driven system (ADS) Programs. Taejon, Republic of Korea, 1-4 November 1999.
Its purpose was to review the current R&D programs in the Member States and to assess the progress in the development of hybrid concepts, as well as their potential role relative to both the current status and the future direction of nuclear power worldwide. Further, the AGM participants provided advice and guidance for the IAEA activities in the ADS area.

Coordinated Research Project (CRP) on Safety, environmental and Non-Proliferation Aspects of Partitioning and Transmutation (P&T) of Actinides and Fission Products. The overall objectives of the CRP were to study the possibility of reduction of the long-term hazard arising from the disposal of high-level waste. More specifically, the CRP aimed to identify the critical nuclides to be considered in a P&T strategy, to quantify their radiological importance in a global nuclear fuel cycle analysis and to establish a priority list of radionuclides according to the hazard definition. In the framework of the CRP the radionuclides hazard was studied in order to identify the critical nuclides to be considered in a P&T strategy and to quantify their radiological importance in a global nuclear fuel cycle analysis.

Coordinated Research Program (CRP) on the Potential of Thorium-based Fuel cycles to Constrain Plutonium and to reduce the long-term Waste Toxicity.
This CRP examines the different fuel cycle options in which plutonium can be recycled with thorium to incinerate plutonium. The potential of the thorium-matrix has been examined through computer simulations. The toxicity accumulation and the transmutation potential of thorium-based cycles for current, advanced and innovative nuclear power reactors are investigated. The results of CRP were presented at ICENES 98 and ICENES 2000.

An IAEA Consultancy held in Moscow in July 2000 initiated the preparation of a background report on the use of fusion/fission hybrids for utilization and transmutation of actinides and long-lived fission products, identifying the needs of the R&D groups involved, and thus providing justifications and incentives for the Agency's future initiatives in this area.

The Agency participates also in OECD/NEA's Expert group on "Actinide and fission Product Partitioning and Transmutation (P&T) Study 'Comparative Study of ADS and FR in advanced Nuclear Fuel Cycles'" whose objective is to assess whether ADS deliver distinctive benefits in an advanced fuel cycle that includes P&T as compared to Fast Reactors.

Concerning a global challenge – the development of innovative nuclear technologies for large-scale nuclear energy development, the Agency has recently launched an initiative on innovative reactors and fuel cycles, to be coordinated by a Task Force which will begin its work at the end of November 2000. The initiative will analyse first, the potential demand and user requirements for new nuclear technologies and, second, different proposed technologies including new variations on some that may have been raised earlier, but abandoned as too challenging at the time.

The Secretariat's initial proposal, which will be further elaborated upon during the November meeting, is that the Task Force proceed on two tracks. The first track will involve a global assessment of the user requirements for future reactors and fuel cycles in order to have a better understanding of the demand and potential for the application of innovative reactors and fuel cycles. The second track will involve a compilation and review of technical feature and characteristics of innovative reactor and fuel cycles that could meet projected requirements and demand.

The Task Force might also make recommendations about follow-up actions the Agency and Member States may consider, including ways to facilitate information exchange, co-operative research and development among countries working on similar design concepts, to stimulate the pooling of resources and expertise.

Other current national and international efforts on next-generation nuclear energy systems such as the three-Agency studies on Innovative Reactor Development by OECD/IEA, OECD/NEA and IAEA and the US led Generation IV International Forum have related objectives, with some differences in either scope, time horizon or participation however. Good co-ordination and co-operation with other initiatives have to be pursued in a complementary manner to take full advantage of potential mutual benefits.

What the Agency can contribute, in addition to our global membership, which includes both new technology-user countries, old technology-supplier countries, and everything in between, is also its unique collective expertise in safeguards, safety and technology. I want to close by acknowledging the endorsement of the importance of research on innovative designs both from the US Senate, which recently recommended significant funding on this topic and from Russian President Putin in his initiative at the Millennium Summit in New York. President Putin called upon all countries to join an international project under the auspices of the IAEA to develop new nuclear power technology that would neither require nor produce weapons-grade material – plutonium or highly enriched uranium.

So those are the basic fundamentals of the Agency's planned work on innovative reactor designs and fuel cycles – why this is such an important topic now, and what we believe we have to contribute. Thank you very much for your attention.

References:

1. Nuclear Technology Review 2000, IAEA GOV/INF/2000/5/Part 1.
2. IAEA PRIS Database as of 01 January 2000.
3. IPCC, Working Group III: Mitigation of Climate Change, Special Report on Emission Scenarios. IPCC (2000).
4. V.S. Kagramanian, S.L. Kononov, H.-H. Rogner, Nuclear Energy in the New IPCC Emission Scenarios, IAEA Bulletin Vol. 42/2 (2000).
5. D. Majumdar, J. Kupitz, H.-H. Rogner, T. Shea, F. Niehaus and K. Fukuda. The Need for Innovation. IAEA Bulletin, Vol. 42/2 (2000). Sustainable Energy Development.
6. V.M. Mourogov and J. Kupitz, "Nuclear Energy Issues and the Role of Small and Medium Sized Reactors", 23rd Annual Symposium of Uranium Institute, London, UK, 09-11 September 1998.
7. K. Fukuda, A. Bonne, V.M. Mourogov, Global View on Nuclear Fuel Cycle – Challenges for the 21st Century, Proceedings of the International Conference on Future Nuclear Systems Global'99 "Nuclear Technology".
8. W.D. Magwood, "US Department of Energy Generation IV Nuclear Power Systems", ANS Meeting, 16 November 1999. Bridging the Millennia", 29 August – 3 September 1999, Jackson Hole, Wyoming, USA.

Pergamon
www.elsevier.com/locate/pnucene

Progress in Nuclear Energy, Vol. 40, No. 3–4, pp. 301–307, 2002

Printed in Great Britain
0149-1970/02/$ - see front matter

PII: S0149-1970(02)00023-9

THE FOURTH GENERATION OF NUCLEAR POWER

JAMES A. LAKE, PhD

President
American Nuclear Society
and
Associate Laboratory Director
Nuclear and Energy Systems Engineering
Idaho National Engineering and Environmental Laboratory
P.O. Box 1625-3860
Idaho Falls, ID 83415-3860 USA
Telephone (208)526-7670, Fax (208)526-2930
Email lakeja@inel.gov

ABSTRACT

The outlook for nuclear power in the U.S. is currently very bright. The economics, operations and safety performance of U.S. nuclear power plants is excellent. In addition, both the safety and economic regulation of nuclear power are being changed to produce better economic parameters for future nuclear plant operations and the licenses for plant operations are being extended to 60 years. There is further a growing awareness of the value of clean, emissions-free nuclear power. These parameters combine to form a firm foundation for continued successful U.S. nuclear plant operations, and even the potential for new plant construction.

In order to realize a bright future for nuclear power, we must respond successfully to five challenges:

- Nuclear power must remain economically competitive,
- The public must remain confident in the safety of the plants and the fuel cycle.
- Nuclear wastes and spent fuel must be managed and the ultimate disposition pathways for nuclear wastes must be politically settled.
- The proliferation potential of the commercial nuclear fuel cycle must continue to be minimized, and
- We must assure a sustained manpower supply for the future and preserve the critical nuclear technology infrastructure.

The Generation IV program is conceived to focus the efforts of the international nuclear community on responding to these challenges.

INTRODUCTION, THE ENERGY IMPERATIVE

Plentiful, affordable electrical energy is a critically important commodity to Nations wishing to grow their economy. Energy, and more specifically electricity, is the fuel of economic growth. However, more than one third of the world's population (more than two billion people) live today without access to any electricity. Further, another two billion people in the world exist on less than 100 watts of electricity per capita. By comparison, the large economies of Japan and France use more than 800 watts of electricity per capita, and the United States uses nearly 1500 watts of electricity per capita.

As the Governments of developing Nations strive to improve their economies, and hence the standard of living of their people, electricity use is increasing. Several forecasts of electrical generation growth have concluded that world electricity demand will roughly double in the next 20-25 years, and possibly triple by 2050. This electrical generation growth will occur primarily in the rapidly developing and growing economies in Asia and Latin America. This net growth is in addition to the need for replacement generating capacity in the U.S. and Europe as aging power plants (primarily fossil fueled) are replaced. This very substantial worldwide electricity demand growth places the issue of where this new electricity generation capacity is to come from, squarely in front of the developed countries who have a fundamental desire (if not a moral obligation) to help these developing countries sustain their economic growth and improve their standard of living, while at the same time, protecting the energy (and economic) security of their own countries.

There are currently 435 power reactors generating about 16% of the world's electricity. We know full well that nuclear power shows great promise as an economical, safe, and emissions-free source of electrical energy, but it also carries at least the perception of great problems, from public safety to dealing with radioactive wastes. I will have more to say about this later. For the moment, let me put forth the proposition that nuclear power should (and must) play a role in the future world energy supply, and perhaps should play an increasing role as the only technology capable of large-scale, near-term deployment without greenhouse gas emissions. If there is a moral imperative to assure the world of abundant, affordable, and clean electricity supplies, then there is no less of a moral imperative for us to assure that nuclear power is capable of taking its rightful place in this energy mix.

THE NUCLEAR PARADIGM HAS CHANGED

As we stand on the threshold of the new millennium facing unprecedented energy and economic growth around the world, we need to ask ourselves what state nuclear power is in, what challenges exist that may inhibit growth of nuclear power in the future, and what we need to be doing now to address these challenges.

The United States, as one of the pioneers in the development and application of nuclear power, serves as a very important indicator of the status of nuclear power, and of its future challenges. 103 nuclear power plants generated 20% of U.S. electricity (nearly 730 billion kWhr) in 1999. Although much has been made of the fact that no new nuclear power plant orders have been placed in the U.S. since the early 1970s, the electricity generation from nuclear power has in fact risen 8% per year for the past 20 years. Plants placed on order in the 1970s have been completed (40 since 1980, the last of which was Watts Bar I in 1996), and the plant capacity factors have risen steadily to a high of 88% in 1999. The total electrical output from U.S. nuclear plants has thus risen from something less than 300 billion kWhr in 1980 to 730 billion kWhr today. This increased electrical generation capacity is one of the keys to the excellent economic performance of U.S. nuclear power.

At the same time that nuclear plant economic performance has improved, so to has safety performance. Safety performance indicators published by the World Association of Nuclear Operators (WANO) have shown consistent and steady improvement. These indicators include unplanned automatic shutdowns (where two thirds of U.S. nuclear plants had zero in 1998), industrial safety (U.S. nuclear plants have an industrial accident rate less than one-tenth that of all U.S. industries), and collective radiation exposure to plant workers which are currently 80% lower than 1980 values.

In the U.S., and increasingly around the world, electricity markets are being deregulated in an effort to encourage competition and lower electricity prices for consumers. The early predictions of economic doom for nuclear-generated electricity in a competitive, deregulated U.S. market have been proven wrong. The process leading to deregulation in 24 states has resulted in negotiated agreements related to recovery of the remaining capital costs of nuclear plants. Closure of the capital cost recovery issue has stimulated the financial interest in nuclear power because the remaining nuclear operating costs (operations, maintenance, and fuel) are very competitive with other electricity supplies in the U.S. In 1999, the average non-capital cost of nuclear-generated electricity was about 2 cents/kWhr. This is the low-price market leader in the U.S., approximately the same as coal and substantially lower than natural gas (at about 3.5 cents/kWhr and rising as both natural gas prices and gas turbine capital costs increase).

The improved economic environment for nuclear power in the U.S. has created a desire for acquisition of nuclear assets and a consolidation of ownership of nuclear power plants that is resulting in stronger, more efficient nuclear generating companies. The same consolidation is occurring in the world vendor market and in the nuclear fuel market. This market-driven consolidation, and the strong business interest in U.S. nuclear assets, is a positive indicator of the economic health of the U.S. nuclear industry.

The U.S. Nuclear Regulatory Commission (NRC) is revising the way in which it regulates operations of nuclear power plants. The new regulatory process is performance-based and uses risk-prioritized regulatory criteria. The new process is believed to have the potential to remove undue regulatory (and hence economic) burden without compromising safety. NRC granted the first 20-year license extension to the Baltimore Gas & Electric Calvert Cliffs plant on March 23, 2000, and the Duke Oconee plant followed in May. The efficient processing of these license extension applications in less than 2 years has encouraged another 8 plants to submit license extension applications, and nearly 30 more have announced plans to submit. The industry and NRC ultimately expect that 80% of the U.S. plants will apply for and receive license extensions.

Until very recently, the environmental benefits of clean nuclear energy have gone largely unrecognized and unappreciated. There is now an increasing international dialog about the environmental impacts of various energy sources in light of the growing body of scientific evidence related to health effects of particulate and gaseous emissions from the burning of fossil fuels, and the potential climate effects from rising CO_2 emissions. Environmental quality is becoming an increasingly important part of U.S. energy policy, and continued operation of existing nuclear plants, improvement in the capacity of these plants, and even construction of new nuclear power plants will be an important part of future U.S. plans if we are to balance our economic growth needs with our environmental stewardship responsibilities.

THE CHALLENGES FACING NUCLEAR ENERGY

The set of circumstances affecting the economic, regulatory, operations, safety and environmental performance of nuclear power have changed rather dramatically in the U.S. in the past two or three years. There are signs of similar changes around the world. These changes allow us to have a relatively positive vision for the future of nuclear power, both for the continued operation of existing plants and for new construction. This vision, however, is based on successful solutions being found for five major challenges:

1. Nuclear power must remain economically competitive and must be capable of continuing to improve its economic performance in an increasingly deregulated world electricity market. Whereas the current operating economic parameters for existing nuclear plants is very good, the high capital cost ($1500-$2000/kW) and history of long construction, licensing and commissioning times for new nuclear plants do not stand up to competition from natural gas in the U.S. market.
2. The public must remain confident in the safety of nuclear power plants and their fuel cycle. Although current light water reactor technology is very safe, the heavy reliance on operations and maintenance presents a vulnerability to assuring continued safe operations, especially as the technology is deployed to countries with less sophisticated technical support infrastructures and different safety and work cultures.
3. Nuclear wastes must be managed and the back-end fuel cycle issues resolved. The ongoing political logjam in the efforts to close out the nuclear waste disposition issue in the U.S., whether it involves opening a permanent or interim waste storage facility, can seemingly be resolved when we have the political will, leadership, and consensus to do so.
4. The proliferation potential of the commercial nuclear power fuel cycle must continue to be minimized. As nuclear power becomes more widely deployed worldwide, it is incumbent upon all of the nuclear supplier and operator nations to continually improve the proliferation resistance of the technology.
5. We must assure a sustainable manpower supply for the future and preserve the critical nuclear technology infrastructure around the world. International cooperation is necessary to help assure that a sustainable manpower supply is retained and that the critical technical infrastructure at R&D institutions, National Laboratories, universities, and in industry, are preserved and utilized in an optimum fashion.

GENERATION IV, RESPONDING TO THE CHALLENGES

Nuclear power originated from a first generation of light water cooled plants in the 1950s and 60s. Those plants grew into the larger pressurized and boiling water reactors that are largely deployed around the world today. We are perhaps on the doorstep of the third generation of nuclear power technology that has evolved toward standardized and optimized light water reactor plants with passive safety features. The world community is interested in finding even wider market acceptance for nuclear power in the future, and therefore we need to ask how well the technology can respond to the first four of the five challenges we outlined earlier. At the risk of oversimplifying a very complicated situation, I would offer that the major factor inhibiting expansion of nuclear power tomorrow is going to be cost.

Generation IV nuclear reactor technology will have to be very responsive to the challenges of reduced cost (especially capital cost), improved safety (especially the public perception of safety), minimization of wastes to minimize the long term economic vulnerability to changes in waste disposition policies, and reduced potential for proliferation of nuclear materials. New (possibly revolutionary) reactor technology may be required to meet the capital cost requirements for the 21st century world market, and perhaps new approaches to "manufacturing" and rapidly deploying nuclear plants can play a pivotal role in reducing the capital cost of nuclear plants to future competitive levels. A fundamentally different way to attack the traditional economies of scale is to envision shifting nuclear plant construction from custom field construction, toward more of a manufactured product composed of world components that are assembled or field deployed much as the manufacture of airplanes is different from the design and construction of airports. Such a concept of manufactured nuclear plants probably leads one to look more carefully at smaller (100Mwe) size plants which coincidentally may find better market acceptance where capacity can be added incrementally to a system, more closely paralleling the demand. Several advanced design concepts are already exploring the territory of smaller reactor plants, notably the South African Pebble Bed Modular Reactor and the Argentinean CAREM reactor. Conceptual designs for several small-plant-systems are also being evaluated under the U.S. DOE Nuclear Energy Research Initiative.

In May of this year, DOE sponsored a Workshop attended by nearly 100 U.S. and international experts from the nuclear industry, academia, national laboratories, and international government and non-government organizations. The goal of the Workshop was to develop a first-order set of world design goals that Generation IV nuclear power systems should meet in order to offer a viable and competitive future nuclear energy option for both developing and developed countries. The detailed results of this workshop can be found on http://gen-iv.ne.doe.gov. Briefly, the Workshop concluded:

- The busbar cost of electricity from a Generation IV nuclear system must be competitive with other electricity generation sources in the region or country in which it is deployed (natural gas is the competitive benchmark in the U.S. for example). This competitive cost is in the neighborhood of 3 cents/kWhr in the U.S.
- Generation IV systems must present the smallest possible risk to capital investment. Plant capital costs around $1000/kW and total construction times in the range of 3-4 years are highly desirable.
- Generation IV plants must be capable of demonstrating improved safety margins, not only to regulatory authorities in the country in which they are deployed, but also to the public. As such, a very low likelihood of core damage may be necessary but it is not sufficient. Generation IV designs may have to demonstrate, through integrated reactor testing that is open and transparent, that no severe core damage will result for plausible initiating accidents. This can be accomplished with core fuel and structural materials that do not melt at accident

temperatures, coolant materials that are not reactive, and using passive cooling and heat removal systems that constrain core temperatures in a manageable range under the worst of accident conditions. There should be no credible accident scenario that would require offsite emergency response. Generation IV technology should be designed with today's experience and knowledge of operations and maintenance needs to be highly tolerant of human error.

- The full life cycle from mining to fuel fabrication to reactor operations to waste management, transportation and plant decommissioning and decontamination must be accounted for from the outset in a Generation IV system. In particular, complete solutions should be identified for all waste streams, and Generation IV technology should be designed to minimize the quantities of waste produced (for example, using very high burnup fuels).
- Generation IV advanced reactor systems, and their fuel cycle, should at a minimum preserve the status quo where material from the commercial nuclear fuel cycle is unattractive as a means of proliferation. Further, intrinsic features of the reactor system should improve the proliferation resistant characteristics of the fuel cycle to disadvantage commercial nuclear materials to the point where they are the least attractive path to the acquisition of nuclear weapons. Currently, within DOE, methodologies are being developed to quantify and measure proliferation resistance in order to guide and evaluate Generation IV candidates.

At the present time, the United States is not committed to a particular technical approach to, or reactor concept for, Generation IV. Rather, we are trying to assemble the broad resources of the U.S. and international R&D community at laboratories, universities, and research institutions, along with the world nuclear industry, to build consensus behind the critical performance requirements for 21st century world deployment, and to build a solid technical foundation for a long-term sustainable international design and development program. Among the assembled world nuclear experts are proponents of a wide variety of reactor concepts. Research teams around the world are already examining a wide variety of reactor concepts to compare their performance against the Generation IV requirements. These include high-temperature, gas-cooled reactors in pebble bed or prismatic configurations, liquid metal cooled reactor systems with conventional sodium or lead-alloy coolants, advanced water cooled systems possibly employing supercritical steam, exotic systems such as molten salts that could function as waste burners, and others. Ultra-long-life reactor cores could raise the possibility of small reactors with cartridge cores that would not require refueling and could be field deployed and removed at the end of life to be replaced by a new system.

In each of these cases, Generation IV reactor systems present technical challenges and barriers whose resolution, through focused R&D, can enable the needed system performance. For example, coated particle fuel performance at high temperature and high burnup is a key to the performance of the high temperature, gas-cooled reactors. High temperature materials performance, and particularly corrosion in lead-alloy cooled systems, is an enabling technical issue. DOE intends to build a technology roadmap in 2001 for the leading Generation IV concepts that will allow the U.S. R&D program to focus on the key enabling technical issues to support future selection by the market of candidate Generation IV systems for demonstration and deployment.

Finally, in the context of sustainable development, a new balance is being sought between economic factors, environmental quality, and social responsibility. Generation IV advanced reactor systems will have to address societal and public perception issues such as concerns about safety, the open or closed fuel cycle, and whether environmentally acceptable wastes disposed in geological repositories can include long-lived actinide materials, in order for future nuclear power systems to meet the social objectives of sustainable development.

THE PATH FORWARD

Because the future nuclear energy market is a world market, Generation IV technology will be a world product. As such, the U.S. Department of Energy Office of Nuclear Energy, Science and Technology, under the leadership of William D. Magwood IV, is organizing a broad international dialog related to the requirements and attributes of the next generation of reactor technology. An international Generation IV working group consisting of senior government and technical personnel have begun to meet to discuss common goals, interests, and to establish bilateral and multilateral relationships and agreements that will allow the next generation of technology to be developed through joint R&D programs.

THE ANS ROLE

The American Nuclear Society, consistent with its mission and goals to be the recognized leaders in the advancement of nuclear science and technology and to be an active contributor to nuclear policy issues, is active in the planning and execution of the Generation IV strategy. ANS officers and members are key participants in U.S. DOE planning and international working groups and forums dealing with Generation IV. ANS, as a respected professional society, can organize and facilitate technical forums for government, industry, the R&D and education community, and international leaders to discuss and debate the global issues important to nuclear energy and to support the formation of consensus and actions that foster a healthy future for the technology. ANS sponsors workshops, technical sessions and topical meetings for the presentation of technical papers related to Generation IV and its technologies. The ANS Officers engage in regular meetings with senior officials from the U.S. Government in Washington DC in order to provide them with technical information with which to make sound policy decisions. Finally, ANS adopted a Resolution late last year advocating the design, construction and operation of a Generation IV nuclear power plant in the near term.

CONCLUSIONS AND IMPLICATIONS FOR THE FUTURE

The economic, operations, and safety performance of nuclear power in the U.S. and around the world is very good. This provides a solid foundation for us to envision a sustainable future for nuclear power that is very bright so long as we can respond to the economic, safety, nuclear waste, proliferation-resistant, and infrastructure challenges. These are challenges worthy of our best efforts.

Pergamon

www.elsevier.com/locate/pnucene

Progress in Nuclear Energy, Vol. 40, No. 3–4, pp. 309–325, 2002

Printed in Great Britain
0149-1970/02/$ - see front matter

PII: S0149-1970(02)00024-0

REQUIREMENTS FOR NUCLEAR ENERGY IN THE 21st CENTURY

NUCLEAR ENERGY AS A SUSTAINABLE ENERGY SOURCE

G. KESSLER

formerly Forschungszentrum Karlsruhe
P.O. Box 3640, 76021 Karlsruhe, Germany

ABSTRACT

Nuclear energy must compete against other energy technologies in the 21st century. It must be economical and it must be proven that it fulfills the conditions for sutainability. This means that the requirements of

- no short term depletion of resources
- extremely low emission of noxious or radioactive substances to the environment
- extremely low release of radioactivity from a nuclear plant in case of the most severe accidents

and

- the present very long term problem of high active waste must be transformed into a few hundred years problem through destruction of plutonium, transmutation of the minor actinides and the most important very long lived fission products. © 2002 Published by Elsevier Science Ltd.

1. INTRODUCTION

In 1999 there were 433 operating nuclear power plants providing about 350 GWe and producing about 16% of the electrical power worldwide. In the USA and Canada 118 reactors, in Western Europe 150, in Eastern Europe and the former Soviet-Union 68 and in the Far East 90 reactors were operated. Germany had 19 reactors with 22 GWe (7% of the world wide nuclear power capacity) in operation. In addition 36 nuclear power plants were under construction, 34 of them in Eastern Europe and Asia. As a capital intensive and advanced technology about 83% of the nuclear capacity was concentrated in highly industrialized countries. Some countries, e.g. France have a 75% share of nuclear power for their total electricity production. Most of the operating nuclear reactors are Light Water Reactors (LWRs), published by Nuclear Technology Review (2000).

Up to 1999, about 220 000 tons (Germany 8000 tons) of spent nuclear fuel have been discharged from civil nuclear reactors. Each year this value rises by about 9500 tons. About 1400 tons of plutonium have been generated worldwide (Germany 83 tons of plutonium) in the spent fuel until 1999. Part of this plutonium has been extracted by chemical reprocessing of spent fuel. But the major part of the plutonium is still contained in spent fuel intermediate storage. The spent fuel storage capacity world wide is roughly 100.000 tons. Additional storage capacity is under construction, published by Nuclear Technology Review (2000).

Plutonium recycling in LWRs is being practised in Germany, Belgium, France, Switzerland and in Japan. The Mixed Oxide (MOX) fuel production capacity worldwide was about 220 tons per year in 1999, published by Nuclear Technology Review (2000). For 2010 an increase of this capacity up to 550 tons of MOX fuel is envisaged. In Germany, Belgium, France, Switzerland and Japan a number of LWRs have been granted already the licence to operate with MOX fuel elements, published by Nuclear Technology Review (2000).

Radioactive waste disposal is one of todays dominant public acceptance issues. Nevertheless more than 100 disposal facilities for low level radioactive waste (LLW) have been built worldwide. Several countries are currently engaged in studies for deep disposal of highly radioactive waste (HAW) (Belgium, Canada, Finland, France, Germany, India, Japan, Sweden, Switzerland). However, the opening of civil HAW repositories is still more than one or two decades away, published by Nuclear Technology Review (2000).

2. FUTURE CONDITIONS (DESIGN AND OPERATING REQUIREMENTS) FOR NUCLEAR ENERGY TECHNOLOGY

Nuclear energy has to compete with a number of other energy systems already now. Most of these energy systems will still be present for the next few decades (gas, oil) or even for the next centuries (coal and the renewable energies: photovoltaics, wind power, biomass). Some of the renewable energy systems (photovoltaics, wind power, biomass) are striving to enter into the energy market now.

It is therefore necessary to compare nuclear energy with all these competing energy systems along the criteria of

- economical competitiveness
- sustainability according to the definition of the Brundtland Commission of the World Commission for Environment and Development (WCED) of the United Nations, published by Voss (1999).

Sustainability of energy systems according to the Brundtland Commission can be cast into the requirements, published by Voss (1999):

- no short time depletion of resources
- extremely low emission of noxious or radioactive substances which could jeopardize the world climate, the environment (atmosphere, water, biosphere) or the health of the human population
- extremely low risk of the energy system for the population.

From the last two requirements one can ideally derive the requirements for nuclear energy also defined by Y. Fujii-e, published by Fujii-e (1994):

- almost zero release of radioactivity from the nuclear power plant and from the plants of the fuel cycle during normal operation
- extremely low release of radioactivity from the nuclear power plant during e.g. a core melt down accident. No emergency measures shall be necessary outside of the plant in such cases. (This requirement was cast into an amendment of the German Nuclear Law in 1994).
- The present issue of a very long term high active waste disposal problem should be transformed into a few hundred years problem through destruction of the plutonium, transmutation of the minor actinides and of the most important very long lived fission products.

In the following sections it will be analysed to which extent and how these above conditions can be satisfied by nuclear energy.

3. ECONOMICAL COMPETITIVENESS OF NUCLEAR ENERGY

Light Water Reactors (LWRs) do compete well on the energy market already now, despite of their relatively high investment costs. The capital requirements of a nuclear plant are typically 60% of the electricity generation cost, whereas the fuel cost are only about 15%. This is opposite to fossil plants which have much lower capital costs but much higher fuel cost contributions, e.g. 45% for coal and 70% for natural gas. Gas combined cycle plants and coal plants are the main competitors of nuclear plants now. But especially gas combined cycle plants which can be constructed presently in about 1/3 of the construction time of nuclear plants are vulnerable to rising gas prizes which are usually coupled to the oil prize. Fig. 1 shows present energy production costs in ¢/kWh for fossil energy technologies (coal, oil, natural gas), hydropower, nuclear fission and fusion reactors and other renewable energy technologies, e.g. solar (photovoltaics), wind and biomass. The data are based on the most recent studies of the European Community, published by Kröger (1999) and Voss (1999) and complemented by expected future fission and fusion breeder data. For wind and solar power also the present subsidized values for feeding electricity into the electrical grid in Germany are indicated. Future fission breeders or plutonium and actinide burner reactors will have higher (20-30%) electricity production costs than present LWRs. Also future fusion reactors might have higher electricity production costs than fission reactors, published by Kessler (2000), if their technical feasibility can be realized.

In addition to electricity production costs Fig. 1 contains also cost contributions based on an assessment for additional costs accounting for health and ecological damage by noxious gases, toxic pollutants and particular greenhouse gas emissions (GHG), published by Nuclear Technology Review (2000), Kröger (1999), Kessler (2000). Determining the cost impact of these external costs, is not simple and – of course – not yet accounted for in the energy market at present. However, a comprehensive Extern-E project of the European Commission (EC), published by Kröger (1999), Voss (1999) developed full chain methodologies to place a monetary value on health and climate impact so that the varying effects on the electricity generation costs can be seen. Coal and oil have the highest external costs. Fission reactors have only small external costs although their impact of emissions of radioactivity was fully accounted for. Wind power is presently subsidized by about 9 ¢/kWh and photovoltaics is subsidized by about 49 ¢/kWh (marked by * in Fig. 1) in Germany and in some other countries.

Fig. 1 shows clearly that nuclear fission power, natural gas and hydropower (large plants) are the most economic energy technologies followed by oil and coal. The often expressed hope that the renewable energy systems (photovoltaics, wind and biomass) will soon enter the energy market with a considerable share and competitive energy prizes will be very hard to become reality.

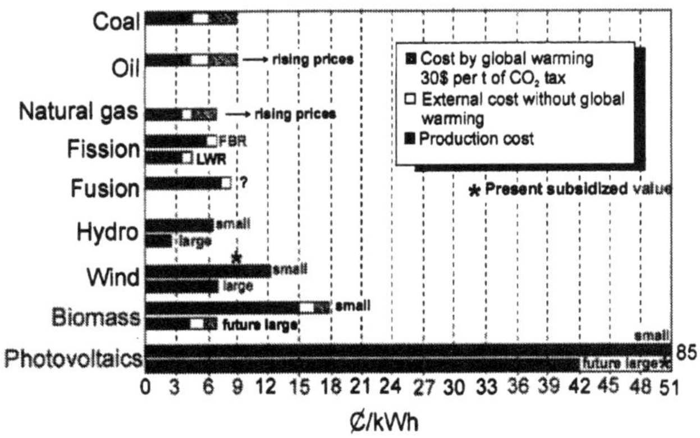

Fig. 1: Cost of electricity production for different energy systems, published by Nuclear Technology Review (2000), by Kröger (1996), by Voss (1999), complemented by fission and fusion breeder data.
*indicates present subsidized costs for wind and solar power in Germany

Table 1: Timely availability of different energy resources, published by Kessler (2000), by Nakicenovic (1998) and by Häfele (1991

Energy system	Energy availability
natural gas oil	50-100 years
hard coal lignite	several hundred years
solar wind hydro	timely unlimited
nuclear: fission	LWR-technology 50-100 years Breeders U-238, Th-232 many 1000 years
nuclear: spallation	Breeder U-238, Th-232 many 1000 years
nuclear: fusion	D-T cycle with Li many 1000 years D-D cycle unlimited

4. NUCLEAR ENERGY AS A SUSTAINABLE ENERGY SYSTEM

It will now be analyzed how nuclear energy can satisfy the criteria of sustainability as they were defined in Section 2.

4.1 Depletion of primary energy resources (published by Nakicenovic (1998), and Häfele (1991)

In Table 1 the energy resources of nuclear energy are compared with those for fossil energy and renewable energies. It is a well accepted fact now, that the world fossil energy reserves and resources have been underestimated in the 1970's (after the first oil crisis 1974). It can be hardly predicted now when fossil energy resources will be really depleted. Table 1 shows a presently widely accepted view over the timely availability of the different energy reserves. Coal and lignite will be available for at least several 100 years. Oil reserves might become depleted after the end of the first half of this century. But then oil shales and tar sands might come into the picture, but at higher prices. Hydropower can hardly be more extended in the future. Nuclear fission energy is presently based on the availability of the naturally occurring U-235 in natural uranium (0.7% enrichment of U-235 in natural uranium). But in a few decades Fast neutron breeders or spallation breeders, which utilize the 100 times more abundant U-238 and the Th-232, could take over. The technical feasibility of fast neutron breeders was already demonstrated in the USA, Russia, Europe and Japan, published by Häfele (1977).

Fusion breeders on the basis of the D-T cycle will also be timely limited by their breeding material lithium. But there will be enough lithium available for many 1000 years, published by Häfele (1977). However, the technical feasibility of fusion breeders must still be demonstrated, published by Kessler (2000).

4.2 Emission of noxious gases

CO_2 emissions are the main problem of present fossil technologies as this is considered to lead to global warming (Fig. 2). New energy technologies like wind power, fission and fusion reactors produce only very little CO_2 per energy unit generated. However, they are not totally CO_2 emission free, since in the course of their plant construction, carbon based energy technologies (e.g. production of steel, copper etc) must be used which emit CO_2, published by Kröger (1999) and White (1998). The highest CO_2 emissions are attached to the carbon based energy technologies. Photovoltaics have a surprisingly high CO_2 emission, since the production of solar cells (steel and copper etc.) preceding construction of a photovoltaic plant produces CO_2, and also SO_2, NO_X, published by Kröger (1999).

Similarly the emission of SO_2 and NO_X is high for fossil energy systems, published by Kröger (1999). Coal as an energy system emits 750 kg/GWhe SO_2 and 730 kg/GWhe NO_X. Gas emits only about 30% SO_2 and 67% NO_X whereas photovoltaic generates also about 30% SO_2 and 36% NO_X of that of coal. Wind power, hydropower and fission or fusion emit only about 10% and 5% respectively of these emission values for coal, published by Voss (1999).

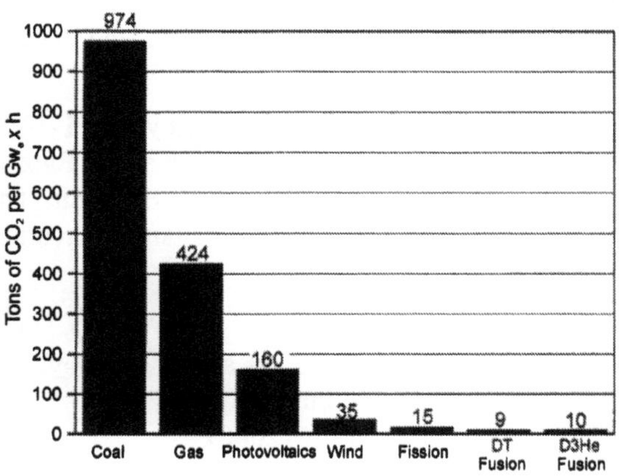

Fig. 2: CO_2 emission of different energy systems published by Kröger (1999), Voss (1999) and White (1998)

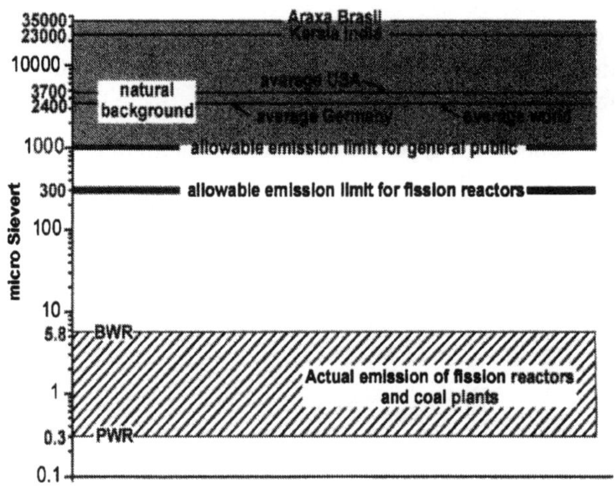

Fig. 3: Yearly radiation exposure from fission reactors and coal plants compared to yearly natural background radiation and medical exposure

4.3 Emission of radioactivity during normal operation

Nuclear fission reactors emit small amounts of radioactivity during their normal operation. The maximum allowable emission is set to 300 μSv to the air and water environment by regulatory bodies in Germany. Experience and governmental control show that present LWRs emit only 1-2% of this allowable limit (Fig. 3). This corresponds to less than the deviations of the naturally occurring radiation at different location on Earth.

Coal fired plants emit as much radioactivity as nuclear fission reactors, published by Halbritter (1982), since coal contains the daughter products of uranium and thorium in ppm contents and a 1 GWe coal plant has a throughput of about 2,6 million tons of coal per year.

4.4 Risk of nuclear fission reactors

Fig. 4 shows risk data for coal, oil and gas collected by insurance companies. These data include the frequency of occurrence and the respective damage in form of fatalities as a consequence of severe accidents, published by Frischknecht (1998). Besides from data for coal, oil, and gas it includes also the results of risk studies for LWRs. It shows that coal and oil are responsible for severe accidents with up to more than 1000 fatalities. The most severe accidents were: 3000 fatalities in a coal mine (Fushan, China 1931), 3000 fatalities during oil transport (crash of oil tanker with ferry boat, Philippines, 1989), 600 fatalities with liquid gas accident (Ufa, Russia, 1989) and 2500 fatalities after crash of a dam of a hydroplant (Machu II, India, 1979).

Also for nuclear fission reactors (LWRs) consequences of up to more than several 1000 early fatalities have been estimated, published by Bayer (1981). However, this would happen only with frequencies of occurrence which are by 3 orders of magnitude lower than for coal, oil and gas. Data on other renewable energies (wind, photovoltaics) are not available yet, published by Breitung (2001).

From Fig. 4 one might take the position that core melt down accidents in nuclear fission reactors can have similar severe accidental damages (fatalities) to the population as other energy systems (coal, hydro), but at much lower probabilities of occurrence. One must, however, admit that there is one essential difference between nuclear energy and other energy systems. In case of a core melt accident with very low frequency of occurence – as it happened at Chernobyl – large areas of land can be contaminated by radioactivity, published by Egorov (2000).

Future fusion reactors will release several 100 gr of tritium to the environment as a consequence of a severe accident. This would lead only to a radiation exposure of 10-100 mSv which is well below the 250 mSv limit set by US-Federal Regulations in 10 CFR 100, published by Voss (1999). This would not lead to early fatalities.

In a Chernobyl type accident these contaminated areas must be evacuated for several hundred years. Table 2 shows that the Chernobyl accident caused a few 1000 km^2 of contaminated land, published by Egorov (2000). Similar results could be also obtained from risk studies (see Fig. 5), published by Ehrhardt. The financial damage would be by orders of magnitude higher than e.g. for oil spills in case of tanker havaries, pipeline breaks etc.

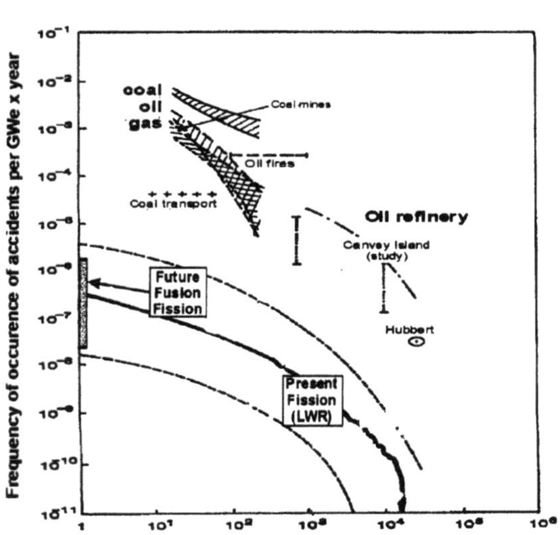

Fig.4: Frequency of occurrence of severe accident and number of early fatalities for different energy systems, published by Frischknecht (1998) and Bayer (1981) complemented by data for future nuclear fission and fusion reactors

Table 2: Cs^{137} contaminated areas after Chernobyl accident [in 1000 km^2] published by Egorov (2000)

Contamination level Ci/m^2	Resettlement possible 5-15	Evacuation necessary 15-40	>40
European Russia	5,7	2,6	0,3
Ukraine	3,2	0,9	0,6
Belarussia	10,2	4,2	2,2
Total:	19,1 km^2	7,2	3,1

In addition one should consider that nuclear fission reactors, e.g. in Europe, are built and operated in highly populated areas.

Therefore, nuclear fission reactor systems to be built and operated in the 21. century should be designed such that large scale contaminations causing evacuation cannot occur any more (amendment of German Nuclear law 1994). The argument that such severe accidents and following land contaminations can only occur at extremely low probabilities might not be sufficient for nuclear fission energy to be classified as a sustainable energy source.

5. FUTURE REQUIREMENTS FOR THE SAFETY DESIGN OF NUCLEAR FISSION REACTORS

Future nuclear fission reactors must be designed such that in case of a core melt accident, there will be no large scale radioactivity release to the outside environment. This means that the outer containment must remain intact and leaktight in case of a severe melt down accident. Many probabilistic safety and risk analyses have shown that there are essentially a few accident sequences which are responsible for containment failure after a core melt accident. If one can prove that these accident sequences cannot destroy the leak tightness of the outer containment one will have fulfilled the above requirement. The Research Center Karlsruhe (FZK) and the French Commissariat à l'Energie Atomique (CEA) are performing research along these lines since about 1990. The essential accident sequences, e.g. in a Pressurized Water Reactor (PWR), leading to outer containment failure and the corresponding actual research results are:

- the steam explosion after contact of molten core masses with remaining water in the lower part of the pressure vessel: FZK proved by theoretical analysis and experiments that as a consequence of a large scale steam explosion the upper part of the pressure vessel (cover and bolts) will not fail and therefore not jeopardize the outer containment leaktightness, published by Struwe (1999).
- the hydrogen deflagration to detonation (DDT): In the course of a core melt accident the cooling of the 1000 °C hot zircaloy claddings of the fuel elements by steam will generate about 1500 kg of hydrogen after a $Zr\text{-}H_2O$ chemical reaction. The large masses of hydrogen can ignite within the containment and generate high pressure spikes during a DDT event. FZK proved that even in the case of large scale hydrogen DDTs and of detonations the leaktightness of the outer containment – if properly designed – will not be jeopardized, published by Breitung (2001). In addition hydrogen catalytic recombinators can decrease the amount of hydrogen after the core melt accident.

  In case of loss of offsite power (LOOP) followed by a failure of the steam generators, core melt at high pressure can be initiated. GRS, published by Löffler (2001), proved that in the sequence of this accident chain, the primary piping will fail and the primary cooling system will depressurize prior to a melt through of the core at high pressure. In that case only a core melt through the lower part of the pressure vessel will occur at low pressure (several bars). The leaktightness of the containment will in such a case not be endangered.
- So called by-pass accidents can be avoided by proper design of the cooling circuits and the containment compartments.
- Core melt through the bottom of the containment can be avoided by a core catcher and proper cooling of the molten core masses as proven by FZK, published by Alsmeyer (2000).

The German-French design of the European Pressurized Water Reactor (EPR), published by Fabian (1999) is required by the respective Safety Commissions to be designed along the above outlined rules, published by Technical Guidelines (2000). The present probabilistic safety concept applied for fission reactors is only allowed for the optimization of the safety systems, not for the outside consequences of severe core melt accidents. The German future Boiling Water Reactor Design SWR-1000 shall follow the same rules as required for EPR, published by Brettschuh (1998). Some small reactor designs, e.g. the German pebble bed modular (PBMR) gas cooled reactor can – in principle – also fulfill the same requirements, published by Lohnert (1990). This safety concept relies on a strong negative temperature coefficient such that, in case of loss of offsite power and cooling, the maximum temperature of the fuel (ceramic coated fuel particles) will not exceed 1600 °C and the fuel particles will not release the accumulated fission products. The afterheat can then be released through radiation to outside steam generator modules. Also future liquid metal cooled fast breeder reactors, e.g. IFR (ANL), published by Chang (1989) or the advanced HLMC reactor, published by Spencer (2000) can be designed to fulfill the above requirements. Similar aims are followed by the Generation IV initiative of the department of energy (DOE), USA, published by Generation IV Workshop and in Russia, published by Nuclear Power in the 21st Century (2000).

According to the amendment of the German Nuclear Law of 1994, published Artikelgesetz (1994) any future reactors must be built such that emergency actions (evacuation of the population etc.) outside of the nuclear plant shall not become necessary in case of core melt accidents.

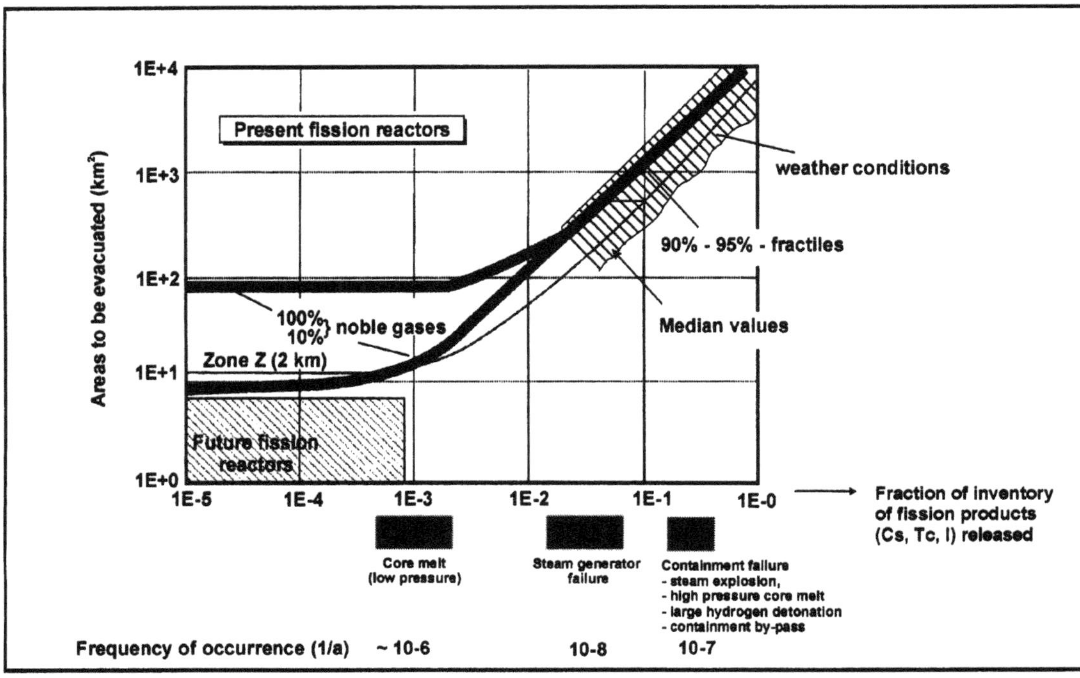

Fig. 5: Evacuation areas (km²) as a function of different internal severe accident chains (radioactivity releases) and frequencies of occurrences for present and future fission reactors

Consequently, for future fission reactors – according to the new R&D-results, e.g. for PWRs – the above discussed essential severe accident consequences (large releases of radioactivity caused by internal events) will not exist anymore. Then, land contamination and the necessity for evacuation of the population outside of the nuclear plant will not occur anymore (Fig. 5). Higher contamination will be restricted to the plant site itself. Also the number of early and late fatalities as displayed in Fig. 4 will shrink to a minimum value (indicated only qualitatively) on the left side of Fig. 4 (similar to fusion reactors, which would release essentially tritium), as the radiation exposure outside of the reactor plant remains at a low level (determined by the requirement of no evacuation) which does not cause early fatalities.

Such future reactor designs will fulfill all conditions for sustainability as they were defined in section 2. This holds not only for normal operation, but also in case of severe core melt accidents.

For the facilities of the closed fuel cycle (UO_2 and MOX fuel fabrication plant, reprocessing and HAW vitrification plant) the design conditions to be satisfied are less severe since the radioactivity inventory is lower and the fuel is neither at a high temperature nor is the cooling system under high pressure, published by Kessler (1983) and Koelzer (1989).

6. WASTE STREAMS FROM ENERGY PRODUCTION

Present energy systems produce different amounts of waste which are partly toxic or as in the case of nuclear energy technologies also radioactive. Table 3 shows the amounts of waste in t/GWe·h or in m^3/GWe·h, published by Kröger (1999)

While coal as an energy source produces with 200 t/GWe·h by far the most (partly toxic) waste, oil, gas and wind generate only little of this waste category. Surprisingly photovoltaics produce also toxic waste (As, Cd, Pb). The nonradioactive waste associated to fission reactors is generated during their construction and decomissioning period. Low active waste is building up during the operation of the fission reactors and during operation of fuel cycle facilities. Highly radioactive waste contains the fission products and the higher actinides of the spent fuel of fission reactors, published by Kröger (1999). Fusion reactors will generate only low radioative waste. Coal ashes contain K^{40}, uranium and thorium as well as their daughter isotopes. They must be classified as low radioactive waste, published by Halbritter (1982). Similarly, oil and gas processing pipings can approach levels of radioactivity which are classified as low level radioactive waste, published by OECD-NEA (1999).

Table 3: Waste generated by different energy systems, published by Kröger (1999) (slightly complemented by data for coal, oil, fission and fusion energy)

	Nonradioactive Waste t/GWe·h	Low radioactive m^3/GWe·h	High radioactive m^3/GWe·h
Coal	200	10-30 (coal ashes)	-----
Oil	7	oil pipings	-----
Gas	5	gas pipings	-----
Fission	6	$7 \cdot 10^{-2}$	$5 \cdot 10^{-4}$
Fusion		$3{,}5 \cdot 10^{-1}$	-----
Hydro	20	-----	-----
Photovoltaics	30	-----	-----
Wind	5	-----	-----

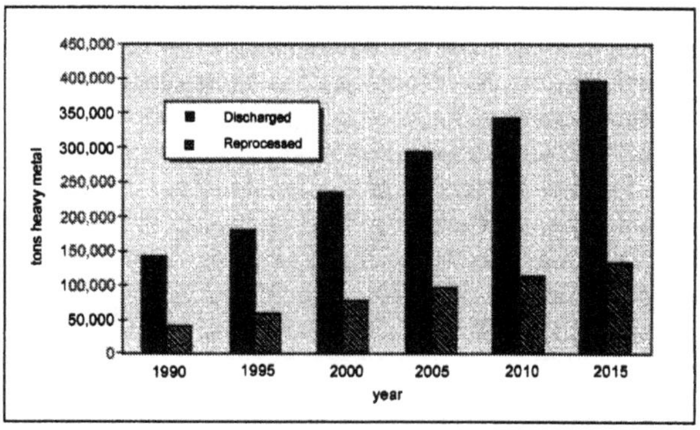

Fig. 6: Worldwide spent fuel discharged and reprocessed, published by Nuclear Technology Review (2000)

6.1 Spent fuel and radioactive waste from fission reactors

Two spent fuel management paths are being pursued at present. On the one side it is the closed fuel cycle with reprocessing of the spent fuel elements (after intermediate storage) and recycling of the plutonium and uranium. The high active waste containing fission products and the minor actinides are vitrified and after further storage of some 40-50 years being deposited (final deep storage) in a waste repository. On the other side it is the Once Through Cycle with direct spent fuel disposal. The spent fuel elements – after 40-50 years storage – are further conditioned and shall also deposited in a repository (deep storage).

The cumulative spent fuel discharged worldwide (Fig. 6) over the past 4 decades amounts to some 220 000 tons by the end of 1999, with 75 000 tons (about 1/3) already reprocessed. As already stated in section 1, that amount of spent fuel contains about 1400 tons of plutonium. Part of this plutonium after reprocessing and MOX-fabrication was already recycled once in LWRs. Some 145 000 tons of spent fuel are still stored at or away from the reactor site, published by Nuclear Technology Review (2000).

Direct spent fuel disposal was promoted by the USA since 1982 mainly for nonproliferation reasons, published by USA Nuclear Waste Policy (1982) and INFCE (1980). Some countries e.g. Sweden, Finland, partly Germany and Switzerland follow this concept. The nuclear industry of these countries hopes that the direct spent fuel disposal concept has economical advantages. However, some of the decision parameters have changed over the past few years and will continue to change in the future:

- the economical advantage (about 15% lower fuel cycle cost) of direct spent UO_2 fuel disposal could fade out, as present contractual costs for reprocessing and MOX fabrication still have a potential for lower costs in the future. In addition one must consider that the fuel cycle costs are only 1/6 of the electricity production costs (section 3).
- the quantities of waste from reprocessing have been reduced enormously by special developments at the French reprocessing plant LaHague. The quantities of waste from direct spent fuel disposal and reprocessing are about equal now, published by OECD (2000).
- recent OECD studies show that the impact of radiation exposure of direct spent fuel disposal and of the closed fuel cycle are almost equal for workers (operational data). For the public the difference is only between 1,6 (for the direct spent fuel disposal) and 2,6 man Sv/GWe·year collected dose at 500 years (for the reprocessing case), published by OECD (2000).
- if the present 1400 tons of plutonium following the direct spent fuel disposal concept will be put in deep geological storage the plutonium will accumulate in the repository (plutonium mine). This holds the more for all future spent fuel elements to be deposited in the repository. Contrary to vitrified HAW glass the conditioned spent fuel elements cannot be released from IAEA safeguards, published by Gmelin (2000). The repository with spent fuel elements (one ton UO_2 spent fuel element contains about 10 kg of plutonium, one ton MOX spent fuel element contains about 25 k g of plutonium) will have to be surveyed and inspected for all times, published by IAEA (1997) and Gmelin (2000). The physical protection of the spent fuel elements by radiation of the fission products decays away after about 300 years, published by National Academy of Sciences (1995).
- Plutonium contributes considerably to the long term radiotoxicity of the waste up to about 10^5 years, published by Review of Modern Physics (1978).
- the closed fuel cycle with recycling of the plutonium in MOX fuel elements has the potential of burning the plutonium (and later also the minor actinides). In a one time recycling strategy – followed at present – one can only burn about 1/3 of the plutonium. However, with multirecycling essentially all plutonium can be burnt (destroyed), published by C.H.M. Broeders (1997).

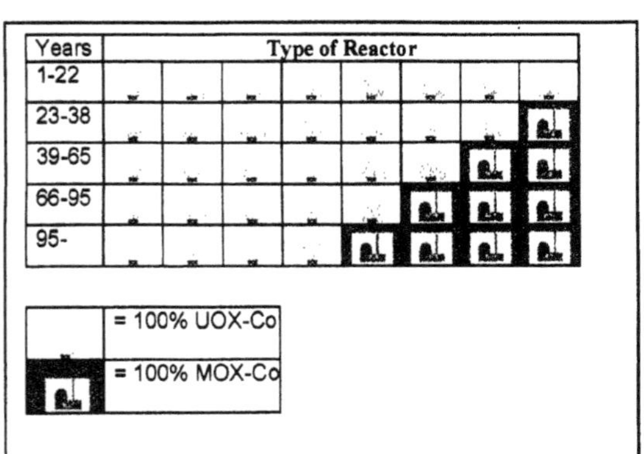

Fig. 7: Pool of 8 UO_2 LWRs which are subsequently replaced by MOX-LWRs according to plutonium availability

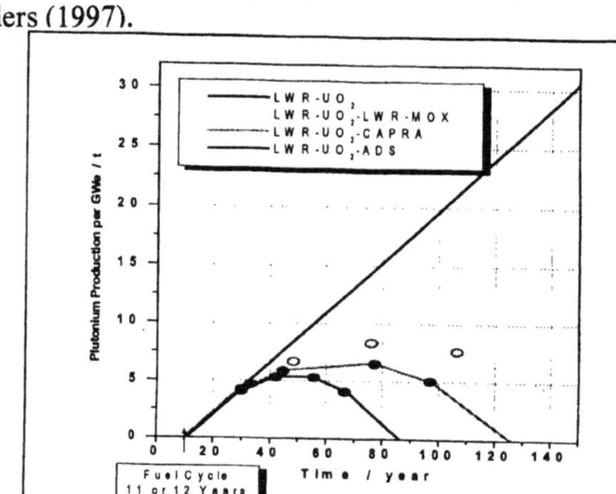

Fig. 8: Plutonium in tons per GW_{el} for different scenarios as a function of time (year)

6.2 Plutonium destruction by multirecycling

Both for nonproliferation reasons (no accumulation in waste repository) and because of its high contribution to the radiotoxicity of spent fuel, plutonium should be destroyed with first priority. Recent studies in Germany show that special multirecycling schemes for plutonium in MOX fuel of reactors (LWRs,), fast reactors (FRs) or accelerator driven systems (ADS) lead to a complete destruction (except the about 0,1% losses going the HAW) of the generated plutonium, published by Kessler (2001). Conditions are that the plutonium which is discharged from burner reactors in spent MOX fuel elements is always blended (mixed) during reprocessing with the plutonium coming from spent UO_2 fuel elements of UO_2 fuelled reactors (UOX). In addition there must always be enough plutonium available to feed the special burner reactors, e.g. full MOX core LWRs or full MOX core fast reactors (FRs) the core design of which is optimized for plutonium burning (these FRs are named CAPRA in Fig. 8).

A LWR completely filled with MOX fuel can incinerate about 420 kg Pu/GWe·year. The LWR, however, because of its thermal neutron spectrum, also builds up relatively high amounts of americium and curium. Therefore the multirecycling scheme with LWR MOX-burners should only be followed for a certain time period. Fast neutron reactors (FNR CAPRA), because of their fast neutron spectrum, have much better destruction properties. They can burn about 570 kg Pu/GWe·year. Accelerator Driven Subcritical Systems (ADS) can burn about 600-700 kg/GWe·year, but have in addition lower core inventories which makes them more efficient for plutonium burning, published by Kessler (2001), C.H.M. Broeders (1997), Bowman (1999), Wade (2000)and C.H.M. Broeders (2000).

Fig. 7 shows a scheme of UO_2 and MOX LWRs, in which always a new MOX fuelled reactor is started when enough plutonium is available. Fig. 8 shows as a function of time the plutonium per GW_{el} in tons for such a scenario which starts with a pool of 8 UO_2 LWRs. Depending upon the plutonium availability UO_2 reactors are – after some time – replaced by burner reactors (LWR-MOX-burner, FNRs (CAPRA) or ADS).

In Fig. 8 the upper straight line shows the amount of plutonium generated as a function of time for a direct spent fuel disposal strategy with UO_2-LWRs. The rising amounts of plutonium will accumulate in the repository. The lower curve shows as a function of time the amount of plutonium in tons per GWe for the above described multireycling case (Fig. 7) with reprocessing and disposal of the HAW to the repository. The difference between both curves is the amount of plutonium burnt or not coming into existence because UO_2-LWRs are replaced slowly by LWR-MOX burner reactors, FNRs (CAPRA) or ADS which do not produce but burn plutonium. In a real case one would start with LWR (MOX)-Burner reactors and then, after several decades, switch over to FNRs (CAPRA) and to ADS reactors.

The time periods to burn the plutonium are in the range of 100 years. In reality one would not burn all the plutonium but rather go to some kind of an intermediate equilibrium level. The burn or destruction capabilities of different burners can certainly be still more optimized in the future by using e.g. uranium free fuel etc. or thorium and seed-blanket strategies, published by Galperin (2000).

6.3 Transmutation of minor actinides

The transmutation of the minor actinides – aside from the destruction of plutonium – is the second most important task to lower the long term radioactivity of the high active waste. Recent studies (e.g. the Japanese Omega program, the French Spin program) show that also the minor actinides can be transmuted and incinerated in reactors or ADS. Table 4 shows the transmutation capabilities of different reactor systems, published by OECD (1999), C.H.M. Broeders (1997), Baetslé (2000), OECD (1999), Bowman (1999) and Wade (2000). Only reactors with a fast neutron spectrum (FNRs and ADS) have good transmutation capabilities for minor actinides. Transmutation of Np and Am in LWRs is very problematic as these minor actinides would be transformed mainly into Pu-238 and Cm-244. This would increase the radiotoxicity.

Table 4: Transmutation capabilities of reactor systems in kg/GW_{el}

	PWR	FR (CAPRA)	ADS
Plutonium	420	570	580
Neptunium	very proble-matic	140	800
Americium		80	400

Table 5: Separation efficiencies of different advanced reprocessing schemes for partitioning of minor actinides in laboratory tests

	Neptunium	Americium/Curium
DIDPA	99,95%	99,99%
TRPO CYANEX 301	99,95%	99,95%
DIAMEX/SANEX/ SESAME	99,9%	>99%

Two alternative reprocessing schemes are followed in research at present:

- aqueous reprocessing schemes with ceramic fuel refabrication (mainly Europe, Japan), published by Madic (2000)
- pyroprocessing schemes (electrorefining) with metallic fuel (mainly USA), published by Chang (1989), Bowmann (1999), Wade (2000). But Europe and Japan will also investigate pyroreprocessing schemes in the future.

With aqueous reprocessing schemes plutonium and the different minor actinides (Np, Am, Cm) can be separated (partitioned) with high separation efficiency (Table 5). If the refabricated actinide fuel is loaded separately into the reactor core, high incineration rates (Table 4) can be achieved, published by OECD (1999), Baetslé (2000), OECD (1999), Bouquin (1996), Salvatores (1999).

With pyroreprocessing schemes uranium and fission products are separated from the transuranic waste (Pu, Np, Am, Cm). The latter are refabricated in Zirconium based metallic fuel elements, published by Chang (1989), Wade (2000), Mc Pheaters (1994). The incineration rates in ADS are about 750kg (Pu, Np, Am, Cm) per GWe·year, published by Wade (2000), Mc Pheaters (1994).

The feasibility of partitioning has been demonstrated in several laboratory tests but its verification in pilot test facilities and in larger technical facilities still will need another several decades, published by OECD (1999). Separation of neptunium is much easier than the separation of americium/curium. The latter must first be separated from the Lanthans and finally americium and curium must be separated.

Recycling of curium seems only feasible with the pyroprocessing scheme and metallic fuel element fabriction, because of the very high α-, γ-radiation and because of its neutrons radiation, published by Wade (2000), Mc Pheaters(1994). In the aqueous processing scheme followed by nitride fuel fabrication, this does not seem to be feasible. One option would be interim storage of curium for about 150 years and recycling of the remaining Pu-240/Cm-245 mixture.

First experience for the fabrication of special fuel pins containing high contents of Am (20%) and Np using the SOL-GEL process is available. One complete LWR-MOX fuel element containing 5% Np was already fabricated in France, published by OECD (1999).

Fig. 9 shows a possible 22 GWe scenario (Germany) adapted from, published by Salvatores (1999), with LWR-UO_2-Reactors, LWR-MOX-Burner reactors, FR-MOX-Burner reactors (CAPRA) and ADS-Burner which can incinerate all plutonium, all minor actinides and some long lived fission products. The reprocessing losses of 0,1% Pu and 1% minor actinides will go into the vitrified HAW. The scenario describes an equilibrium state. The inventory remaining at the end of the fission reactor episode (or age) in energy technologies is not taken into account in that consideration.

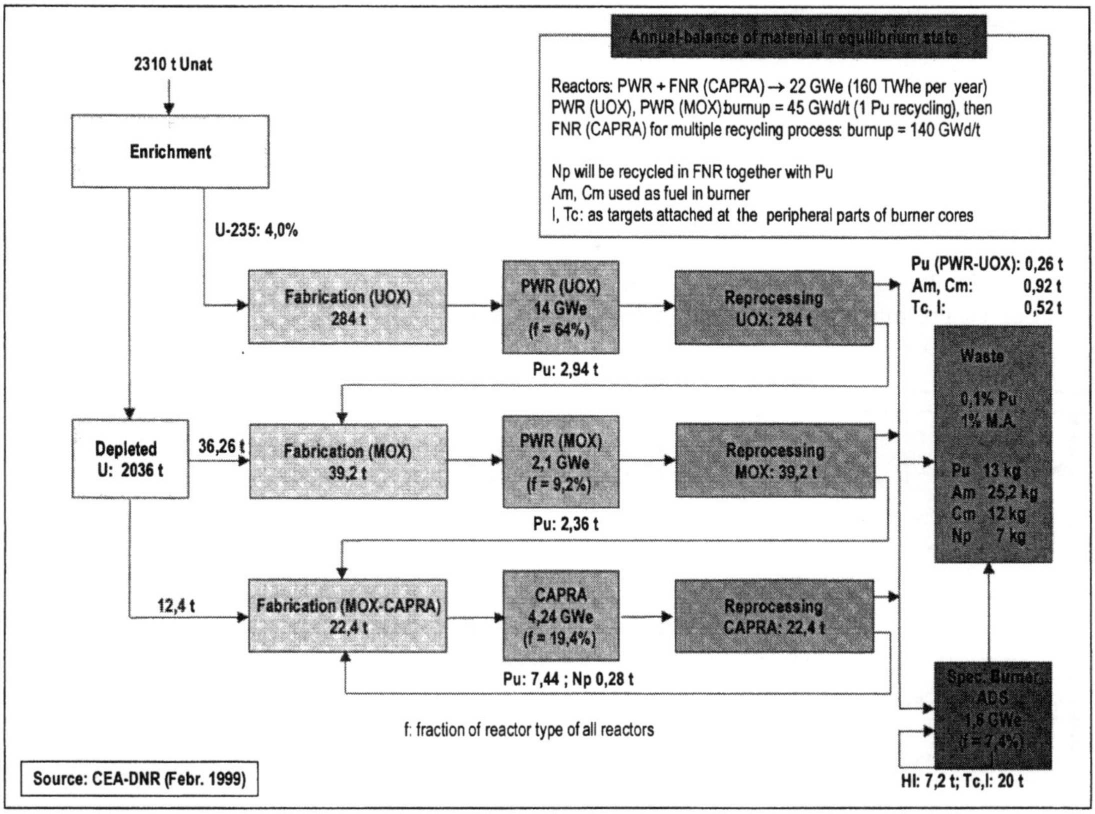

Fig.9: 22 GWe scenario (example Germany) in which Pu, minor actinides (Np, Am, Cm) and 2 fission products, I-129, Tc-99, are transmuted or fissioned (adapted from, published by Salvatores (1999) (HI corresponds to the heavy isotopes Pu + minor actinides)

6.4 Transmutation of long lived fission products

The majority of radioactive fission products will have decayed after about 800 years. From the point of view of reduction of radiotoxicity the transmutation of the longer living fission products is of much less interest (about 3 orders of magnitude smaller) as for the minor actinides. But some of the fission products are very mobile in certain geological environments. One way to tackle the high mobility of some of the fission products could be to immobilize them by appropriate waste conditioning and special engineered barriers around the vitrified waste in the repository. Very long lived fission products which contribute to the very long term radiological effects of the HAW are Tc-99, Cs-133, I-129, Zr-93 etc..

I-129 which is a dominant isotope in both the radiological effects of reprocessing effluents and of spent fuel in certain geological formations can be recovered from the dissolver off-gas during reprocessing of the spent fuel. Tc-99 can also be separated from the TBP-streams during reprocessing. Tc-99 and I-129 are single isotopic species and can be transmuted by neutron capture into nonradioactive Ru-100 and Xe-130. The net transmutation of Tc-99 in fast reactors is about 110 $kg/GW_{el} \cdot a$ and for I-129 about 18 $kg/GW_{el} \cdot a$, published by Bouquin (1996).

Cs can also be separated from HAW but it is not a single isotopic species. It is a mixture of Cs-135 (long lived), Cs-137 (short lived) and Cs-133 (stable). Transmutation is not considered feasible without using additional enrichment techniques, e.g. laser enrichment techniques. Other radiologically important fission products like Se-79, Zr-93, Pd-107, Sn-126 also must be addressed.

Some of the fission products (Se-79) exist either only in very small concentrations in the liquid HAW or they appear also with nonradioactive stable isotopes (Zr-93, Sn-126). Activation products like C-14 and Cl-36 must probably also be considered. They can lead to an accumulation in the biosphere and play an important radiotoxicological role.

6.5 The merit of separation efficiencies and transmutation

Fig. 10 shows a comparison of the radiotoxicity of HAW for different efficiencies of partitioning and transmutation for U, Pu and minor actinides (MA). As for the radiotoxicity the curve for the 99.9% transmutation of U, Pu, minor actinides falls below the radiotoxicity of natural uranium after about 800 years.

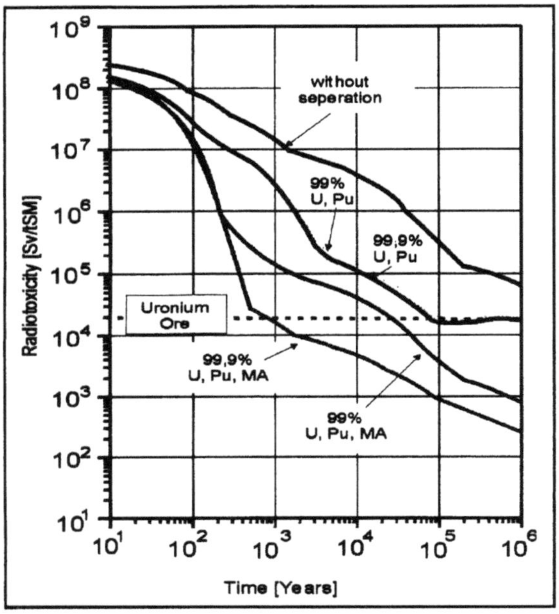

Fig. 10: Merit of partitioning and transmutation of U, Pu and minor actinides (MA)

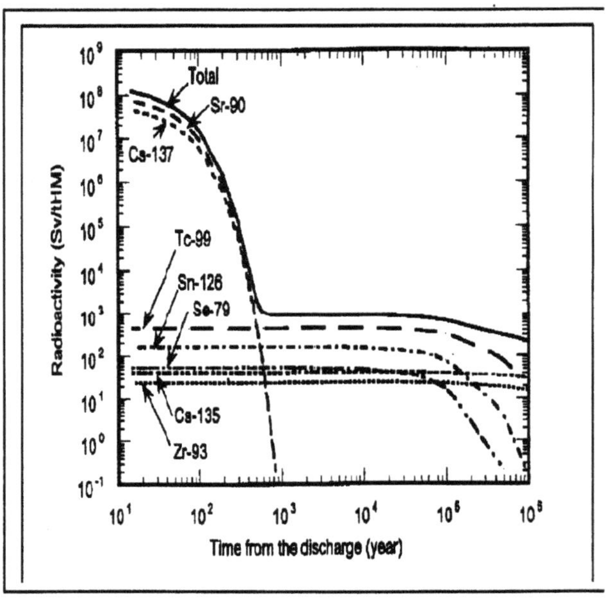

Fig. 11:Radiotoxicity of long lived fission products

Fig. 10 demonstrates clearly the difference between direct spent fuel disposal and recycling and transmutation of U, Pu, and minor actinides. Much R&D work still has to be performed until these goals (99.9% incineration of U, Pu, MA) of Fig. 10 can be attained.

Fig. 11 shows the radiotoxicity of the long lived fission products. Their contribution to the radiotoxicity attains about the same values as the 99,9% U, Pu, MA case (Fig. 10) after about 10^5 years.

The need for transmutation of the different long lived fissile products eventually depends on the

- chosen disposal concept in geological formation (salt, rock, clay)
- the encapsulation concept of the HAW (waste conditioning concept and engineered barrier)
- the sitedependent mobility of the fission product and its radiological significance

Which long lived fission product shall be included in a partitioning and transmutation concept can only be decided on the basis of the waste encapsulation and disposal (geological formation) concept, on the long term safety concept and on radiological consequence calculations which must fit to the radiological standards for long term exposure. Research is still going on in this area.

CONCLUSION

It has been shown that nuclear energy in the 21st century can clearly meet the requirements for a sustainable energy source. It also will remain an economical energy source which can compete with all other remaining or uprising renewable energy technologies.

It must however satisfy the design conditions for extremely low emission of radioactivity especially in case of severe accidents so that no emergency measures outside of the plant become necessary. Recent developments for the future of the fuel cycle show that new separation techniques for the actinides and long lived fission products (partitioning) and their transmutation in dedicated reactor facilities can modify the present waste disposal concepts drastically and satisfy the conditions for sustainability.

REFERENCES

Alsmeyer, H., et al. (2000): Controlling and cooling core melts outside the pressure vessel, *Nucl. Eng. & Design, Vol. 202, 2-3.*

Amendment of German Atomic Energy Act (Artikelgesetz 1994).

Baetslé, L.H. (2000), *Partitioning and transmutation in a strategic perspective*, EUR 19143 EN.

Bayer, A., Heuser, F.W. (1981), Basic aspects and results of the German Risk Study, *Nuclear Safety, 22, 695.*

Bouquin, B., Le futur du retraitement, *Clefs No. 33, Automne 1996*Breitung, W., Krieg, R., *Hydrogen detonations and the response of the outer containment*, to be published, FZKA (in 2001)

Bowman, C.D. (1999), Sustained nuclear energy without weapons or reprocessing using accelerator driven systems, *Journal de Physique IV, France 9, 1999.*

Brettschuh, W., Wagner, K. (1998), Das Sicherheitskonzept des SWR-1000, Tagungsbericht an KTG, Königswinter.

Broeders, C.H.M., et al. (1997), Recent Neutron Physics Investigations on the incineration of Plutonium and Other Transurania Elements. In E. Merz, C.E. Waltar (eds.): Advanced Nuclear Systems Consuming excess Plutonium, Kluwer Academic Publishers, p. 27-42.

Broeders, C.H.M., Broeders, I. (2000), Neutron physics analysis of accelerator driven subcritical assemblies, *Nuclear Eng. & Design, Vol. 202 (2-3).*

Chang, Y. (1989), The Integrated Fast Reactor, *Nucl. Technology, 88.*

Ehrhardt, J., Qu, J., Strahlenschutzkommission, SSK Band 25, 123.

Egorov, N.N., et al.. (2000), The radiation legacy of the Soviet Nuclear Complex. IASA, EARTHSCAN Publications Ltd., London.

Fabian, H.U., Teichel, H. (1999), Der Europäische Druckwasserreaktor (EPR): Stand und Ausblick, Deutsches Atomforum, Inforum-Verlag.

Frischknecht, S., et al. (1998), *Project GaBE, Comprehensive assessment of energy systems, Severe accidents in the energy sector.* PSI Bericht Nr. 98-16, ISS 1019-0643.

Fujii-e, Y.: Nuclear Science and Technology in the 21st century. Proc. Int. Symp. Global Environment Nuclear Energy Systems, TIT, Susono, Japan, 1994.

Galperin, A., et al. (2000), A Pressurized Water Reactor Plutonium Incinerator Based on Thorium Fuel and Seed Blanked Assembly Geometry , *Nuclear Technology, Nov. 2000, 214.*

Generation IV Workshop, http://www.anlw.anl.gov/gen4/

Gmelin, W. et al. (2000). *Euratom safeguards for the Direct Disposal of Spent Fuel in Geological Repositories*, EUR 19143 EN.

Häfele, W. et al.: *Fusion and Fast Breeder Reactors, International Institute for Applied System Analysis* (July 1977).

Häfele, W. (1991), *Energy in a Finite World.* IASA, Laxenburg

Halbritter, G. et al. (1982), *Beitrag zu einer vergleichenden Umweltbelastungsanalyse am Beispiel der Strahlenexposition beim Einsatz von Kohle und Kernenergie zur Stromerzeugung,* KfK 3266.

INFCE-Summary volume and report of INFCE working groups, IAEA, Vienna, 1980.

Kessler, G. (1983), *Nuclear Fission Reactors*, Springer Verlag.

Koelzer, W. (1989), Plutonium, KfK 4516.

Kessler, G. (2000), Potential of Mirror Systems as Future Fusion Power Reactors, *Trans. of Fusion Technology, Vol. 39.*

Kessler, G. et al. (2001), Moderne Strategien zur Beseitigung von Plutonium, *atw, 46. Jg, 2001, Heft 1.*

Kröger, W., Hirschberg, S., Auf dem Weg zur Nachhaltigkeit: Die wesentlichen Ergebnisse für die Kernenergie, Paul Scherrer Institut, Villigen, Schweiz, 1999.

Kyong W. Han et al. (1997). Radioactive waste disposal: Global experience and challenges, Special Report, IAEA Bulletin 39/1.

Ledermann, P. (1994), Operating UP-3. Three years of experience. *Nuclear Engineering International.*

Löffler, H. et al.: *High pressure core melt and its consequences*, to be published, GRS (in 2001)

Lohnert, G.H. (1990), Technical design features and essential safety-related properties of the HTR-Module. *Nucl. Eng. & Design, 121, 259.*

Madic, Ch., et al. (2000), *New Partitioning Techniques for Minor Actinides*, EUR 19149 EN.

Mc Pheeters et al. (1994), *Pyroprocessing oxide spent nuclear fuels for efficient disposal*, ANL/CMT-84354, CONF-941207-31.

Nakicenovic, H., Grübler, A., McDonald, A. (1998), Global Energy Perspectives. Cambridge University Press.

National Academy of Sciences: Management and Disposition of Excess Weapons Plutonium: Reactor Related Options, National Academy Press, Washington DC, USA, 1995.

Nuclear Power in the 21st Century, *Nuclear News, November 2000.*

Nuclear Technology Review, IAEA General Conference GC (44)/9, Vienna, August 2000.

Radiological Impacts of Spent Nuclear Fuel Management Options, OECD 2000, August 2000.

Report of the Advisory Group Meeting on Safeguards for Final Disposal of Spent Fuel in Geological Repositories, IAEA-AGM 995, Vienna, 1997.

Review of Modern Physics, Vol. 50 No. 1, Part I, Am. Phys. Society, Jan. 1978.

Salvatores, M., *Role of ADS in the Back-end of the Fuel Cycle Strategies, lecture given at FZK*, 2 Febr. 1999.

Spencer, B.W. et al.: An advanced Modular HLMC Reactor concept featuring economy, safety and proliferation resistance, 8th International Conference on Nuclear Engineering, April 2-6, 2000, Baltimore, USA.

Status and assessment report on actinide and fission product partitioning and transmutation. OECD-NEA, 1999.

Struwe, D., et al. (1999), *Consequence evaluation of in-vessel fuel coolant interactions in EPR*, FZKA 6316.

Technical Guidelines for the Design and Construction of the Next Generation of Nuclear Power Plants with Pressurized Water Reactors, adopted between the GPR/German experts plenary meetings, held in Paris on October 19 and 26, 2000.

USA Nuclear Waste Policy Act (1982).

Voss,A.: Nachhaltige Entwicklung ohne Kernenergie, Deutsches Atomforum, Inforum Verlag, 1999.

Wade, D., et al, *ATW neutronic design studies,* Susono Seminar, Japan, Nov.6-9, 2000.

White, S., Kulcinski, G.L. (1998), *Energy payback Ratios and Lifetime CO_2 emissions from wind generated electricity in the Midwest.* UWFDM-1093.

Pergamon
www.elsevier.com/locate/pnucene

Progress in Nuclear Energy, Vol. 40, No. 3–4, pp. 327–333, 2002

0149-1970/02/$ - see front matter

PII: S0149-1970(02)00025-2

THE ROLE OF NUCLEAR ENERGY IN 21st CENTURY FOR SUSTAINABLE DEVELOPMENT IN KOREA

BYONG WHI LEE

Professor Emeritus
Department of Nuclear Engineering
Korea Advanced Institute of Science and Technology
Taejon 305-701, Korea
Fax.+82-2-544-0477, Tel.-544-0479
E-mail: bwlee@cais.kaist.ac.kr

ABSTRACT

With the growing international consensus concerning the harmful health and environmental impact of fossil energy use, there is on the political level recognition of nuclear power's potential role in delivering large quantities of energy without releasing common environmental pollutants and greenhouse gases. The energy consumption in Korea has greatly increased with rapid economic growth and industrialization since 1970. The average annual growth rate was 8% in 1970s and more than 10% in 1980s and 1990s except during the '98 financial crisis.

Due to the lack of domestic energy resource bases, the rapid economic growth and industrialization has to be supported by the imported energy. Thus, the imported energy dependence of total energy supply has rapidly increased from 47.5% in 1970 to 97.5% in 1997. The fossil fuels share of energy consumption grew up to 88.2% in 1997. This resulted to CO2 emission of 140 million ton-C, which account for 1.8% of world greenhouse gases emission. (MOCIE, 2000) Because of rapid industrialization, Korea has relatively higher energy intensive industries compared to most of the developed countries with 3.1 ton-C/capita and 0.49 ton-C/million Won.

Thus, energy policy is being focused on the improvement of energy efficiency and optimum energy mix for the reduction of GHGs. At present, 16 units of Nuclear Power Plant are in operation, 6 units under construction. The nuclear share of electric power generation was 43% last year. This share will be increased up to more than 50% by 2015. In order to meet voluntary GHGs emission target, a drastic switching to non-carbon energy bases would be necessary.

1.INTRODUCTION

Since the Beginning of the Peaceful Uses of Atom and up until the 1st Oil Crisis, the nuclear energy was considered to be the energy for the future and the most promising alternative energy to oil. However, after TMI and Chernobyl accidents, the public perception has been changed rather negative in many western industrialized countries due to possible proliferation and health risks. Thus, nuclear energy is at a cross-road and its future rather uncertain. Nevertheless, Nuclear Power is a substantial contributor to world electricity production, supplying more than 16% of total world electricity. With current trends and projection in the East Asia, regional energy supply will grow and could significantly increase mainly in developing countries such as Korea, China, and India. In view of such large increase in energy requirements and the limitation of fossil fuel resource in particular oil and gas supply for the 21st Century, an issue is how to maintain the stable future supply of clean energy in support of rapidly increasing large electricity demand for sustainable development in the region.

However, the Kyoto protocol calls for Green House Gas (GHG) emission reduction to prevent further Climate Change for sustainable development. This implies that the future energy supply mix would likely to depend upon non-carbon energy base such as Hydrogen, Nuclear Energy, and Renewable. Concerns about global climate change are steadily growing, leading the industrialized country to agree to limit their GHG emissions and focus on sustainability. The question is where will all this environmentally clean energy or electricity come from? Except for Nuclear or HydroPower, no other source of energy is likely to meet the expected large scale electric power demands, satisfy the GHG emission concerns and can be economically viable. Renewable energy sources are important, but are expected to be producing only 2 to 5% of world electricity by the year 2020 due to low energy density.

With the growing international consensus on the harmful health and environmental impact of energy use, there is on the political level recognition of nuclear energy's potential role in delivering large amount of energy without releasing common environmental pollutants and GHG. With its remarkably low fuel requirements, nuclear power and heat supply can contribute to meeting national and global goals for sustainable development.

In spite of rapid economic growth and industrialization since 1970 in Korea, the annual electric power consumption per capita is still at the 5,000 kWh levels, which is only about 2/3 of Japan and 2/5 of USA. Thus, the electric power demand is likely to continue with strong growth into 21st century. Due to lack of the indigenous energy resource base, the vigorous economic growth in the past had to be supported by the imported energy. 98% of total energy consumption was imported last year while only 50% in '71. Fossil fuels contributed 88.2% of total and close to 140 Mt-c emissions, which accounts for 1.8% of world GHG emission in '97. Korea has relatively higher energy intensive industry, with 3.1 t-c/capita and 0.49 t-c/m-won, than that of industrialized developed countries.

Current high oil price will impact heavily on our economy. The oil import from Middle East this year would be expected to amount to about 36 billion US dollars. For imported energy from overseas, the nuclear fuel has clear advantages over any other fuel because of distinct high energy density per unit weight at stable price level, no GHG emission, and improving energy security with sustainable long-term supply at affordable price.

2.NEED FOR NUCLEAR ENERGY

At the COP-5 held in Bonn, Germany last year, Korea expressed its willingness to the non-legally binding voluntary participation for GHG emission reduction to prevent global warming, and introduced nationwide a policy on energy conservation. Efficiency and reduction of GHG emission had been stepped up further. Also, Korea is seeking bilateral and regional cooperation in order to materialize specific international cooperation regarding the climate change. The only realistic way to meet the Kyoto protocol by the target date would be the wider use of Nuclear Energy such as Nuclear Power, Desalination, District Heating, Switching to non-carbon energy base such as Hydrogen Production for the transportation sector, etc.

Korean commitment on Nuclear Power is motivated by a number of considerations:

- Keeping nuclear power as a major base load option available to meet rapidly rising electric demand,
- Reducing energy import dependence,
- Improving air quality and reducing GHGs emission,
- Easing energy supply logistics, and
- Benefiting from the technological spin-offs of high technology.

The planned rapid deployment of nuclear power encounter several problems to be overcome by the viable and workable international cooperation either through International Atomic Energy Agency or Nuclear Energy Agency/OECD so as to be nuclear energy the sustainable alternative to fossil fuel in particular Oil in future, such as

- Improve Economic Competitiveness,
- Correct Public Misconception,
- Strengthening Safety Culture,
- Improving Effectiveness of Spent Fuel and High Level Radioactive Waste Management, and
- Proliferation Resistant.

The higher energy demand will result not only from very fast economic development, but also from faster growth of population in developing countries. In order to achieve sustainable development in the 21st century, the wise choice of optimum energy mix should be selected in harmony with the environment. The evolution of energy use has been toward sources of higher energy density, notably to sources, which yield more energy than others per weight of fuel. They are less bulky and easier to transport and store resulting in better economy. Therefore, Nuclear Energy is attractive to improve the energy security of an importing country. At present 16 nuclear power units of 13.7Gwe comprising 8 PWR, 4 Korea Standard Nuclear Power Plant (KSNP), and 4 CANDU units are in operation. 4 KSNP of 4 GWe are under construction. Also under Korea peninsular Energy Development Organization (KEDO) project, 2 KSNP units of 2 GWe are under construction in Keumho, North Korea. In accordance with the fifth long term electric power development program (MOCIE, 2000), total installed nuclear power capacity is expected to reach 26.5 GWe by 2015. This will represent 33.2% of total generating capacity and 49% of total electricity generation. The target for Korea Next Generation Reactor (KNGR) of 1.4 GWe Korea Standard Evolutionary APWR is set around 2010 to commence commercial operation. The major improvement KNGR compared with KSNP are larger capacity, longer design life, more safety margin, double containment, severe accident mitigation features, and passive safety system to improve safety and economics.

The International Energy Agency (IAE) assumes that roughly 75-80% of the energy needs of Asia will be met by an approximate doubling of coal and oil use by 2015. Uses of electricity is projected to grow by a factor of two to three, with nuclear energy making significant contribution to meeting the needs for power in a number of Asian countries. Providing this needed power without worsening already serious environmental degradation represents the major challenges. Because of the financial difficulty in a number of the countries in this region, there will be some slow down of Nuclear Power Plant construction for the time being. However, it will pick up again in future since no good alternative option except Nuclear Energy can meet the large demand.

3. ROLE OF NUCLEAR ENERGY

In accordance with the Mid and Long-Term Nuclear R&D Program (MOST, 2000) up to 2006 initiated by Ministry of Science and Technology from 1992 in accordance with the revised Atomic Energy Law, this program is to be reviewed and revised every five years of implementation. The major projects of 1997 revised program are System-integrated Modular Advanced ReacTor (SMART), Korea Advanced Liquid Metal reactor (KALIMER), and Hybrid Power Extraction Reactor (HYPER) for reactor technology development and DUPIC, KNGR fuel for nuclear fuel technology development. Reactor and fuel cycle technology development is as shown in Fig.1.

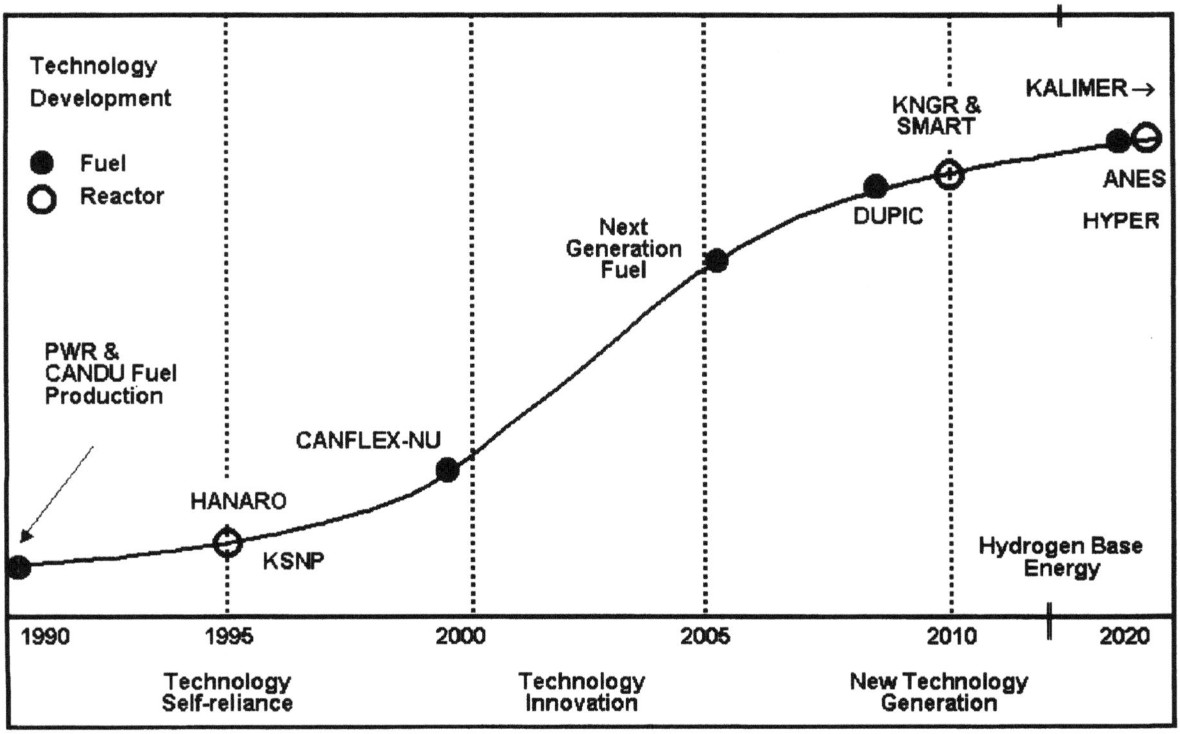

Fig.1. Technology Development Trend

Recent "Long Term Outlook of Oil" by US Energy Information Administration (USDOE/Energy Information Administration, 2000) projected that the world peak production of oil would take place in the first quarter of 21st century - i.e. 2004, 2010, 2015, or 2022 on different scenarios. The recent high oil price would likely persist until next year and stabilize at about the 20-25 US dollars/ barrel level until the world peak production reaches. Afterwards the oil price escalates steadily. Then, there would be certainly fuel switching to gas to take place. Energy intensive industries are clearly large emitters with many undesirable pollutants and wastes. The unconstrained open market use of oil and natural gas continue to grow, and hence lead to increased GHGs emission. In order to meet the Kyoto Protocol target, it would be more likely to switch to the non-carbon base fuel such as Hydrogen, or Nuclear Energy. Although Hydrogen use is not necessarily only for transportation, it is transportation that provides the largest synergism for the production of hydrogen and GHG emission free electricity.

3.1 <u>Nuclear power for Hydrogen production</u> (Duffey et al., 1999)

The extensive US study states that "As the ultimate fuel, hydrogen, once established, will provide a single transition to a stable alternative fuel vehicle and infrastructure---a transition to hydrogen vehicle---would last for the foreseeable future". Nuclear power source and hydrogen are uniquely synergistic in reducing potential cost, electricity generation emissions and end-use transportation emissions. The basic concept is that the reactor is used to generate electricity for the grid and hydrogen production by electrolysis of water.

By-products would be heavy water, oxygen and process heat. Where hydrogen production is centralized, the very economical production of heavy water is produced as a by-product during the electrolysis of water into hydrogen and oxygen. Heavy water is an essential and expensive moderator and coolant of CANDU reactor. Since nuclear power plant is operating as base load plant, the excess electric power generated during deep night can be used for the electrolysis of water to produce hydrogen and heavy water. This would be able to ease the need for pump storage power plant for peak load management. Thus, there is a very substantial additional synergy, which significantly reduces capital cost of CANDU reactor, in the application of nuclear power to the hydrogen production.

3.2 KNGR in DUPIC cycle

In order to improve economic competitiveness, KNGR in Tandem Fuel Cycle with CANDU namely DUPIC fuel would be able to improve not only economics, but also fresh Uranium feed saving. Korean Reactor Strategy is based on PWR as main and CANDU as supplementary. For an equilibrium system, in which the PWR to CANDU reactor ratio of electrical generation is such that the spent PWR fuel provides all the fuel needs of the CANDU reactors, the DUPIC fuel cycle could reduce the natural uranium requirements by about 25% compared with direct once-through cycle, in which CANDU reactors are fueled with natural uranium and PWR reactors are fueled with enriched uranium. Approximately 40 to 60% more energy can be derived from PWR spent fuel by burning it again as DUPIC fuel. This results in a three-fold reduction in quantities of spent fuel arising per unit electricity generation, compared with once-through cycle. Thus DUPIC may offer a good option for reducing the spent fuel management problem. This reduction of spent fuel is very attractive because it can also reduce the handling, storage and transportation activities prior to final disposal. Also DUPIC is a much simpler process than conventional wet chemical reprocessing. Hence, DUPIC fuel processing and fabrication would be less expensive than reprocessing and MOX fuel fabrication. In parallel with DUPIC, extended high burn-up fuel to KSNP and KNGR would further improve availability and economic competitiveness. Therefore, good combination of Extended High Burn-Up fuel in KNGR and DUPIC fuel in CANDU would not only improve the economic competitiveness and availability, but social acceptance by minimizing spent fuel and radioactive waste discharge for sustainable development also.

3.3 SMART for dual purpose applications

Small and medium reactors are attractive because of the unique features: such as better suitability for smaller electric grids as well as non electric applications, lower investment cost and risk, smaller components and equipment to facilitate higher design flexibility as well as adoption of passive system simplification to achieve a high degree of safety and modularization, shop fabrication and assembly further improving quality assurance and reliability of Nuclear Steam Supply System (NSSS), and better adaptability to mass production to turn production cost down.

The shortage of potable water supply is expected to be acute in Korea early 21st century. Rapid build up of heavy industrial complex since '70s and the past several years long drought in the south east region of Korea gave rise to the shortage of potable water supply. Thus feasibility of desalination in this region was studied. For the seawater desalination and cogeneration of electricity, a design concept of dual purpose SMART with a rated thermal power of 330MWt was developed by Korea Atomic Energy Research Institute. A concept of an integrated nuclear desalination plant coupled with SMART to produce 40,000 m^3/day of potable water with MED process and generate about 90Mwe of electricity was also established. It was found from the preliminary safety analysis that the key safety parameters of the limiting design base events do not violate the safety limit. Also, preliminary economic evaluation shows that SMART desalination plant can produce potable water comparable to or more economical than the similar size fossil fired plant in terms of production cost. If this project is successfully implemented, this plant would be the first dual-purpose cogeneration of desalination and power in Korea.

3.4 HYPER for Transmutation of Long-Lived Nuclides (Chang, 2000)

HYPER

HYPER (Hybrid Power Extraction Reactor) has been developed for the transmutation of nuclear waste and energy production through the transmutation process. One of the major features of the HYPER system is the sub-critical operation that needs an accelerator for the production of seed neutrons. The sub-criticality of HYPER can provide many attractive features that can not be obtained in a critical reactor; for instance, (1) the safety of the system is enhanced considerably by removing the possibility of reactivity induced accident and simplifying the power control mechanism. (2) the flexibility of the fuel (or targets to be incinerated) enables the system to have high transmutation capability and to have higher proliferation resistance, and (3) the surplus neutrons by the spallation increase the transmutation capability. The HYPER program is within the framework of the national mid and long-term nuclear research program. KAERI is aiming at developing the system concept and a type of roadmap by 2001 and complete the conceptual design of HYPER system by 2006.

Transmutation of Long-Lived Nuclides

The long half-lives of some radionuclides ultimately cause the risks of environmental contamination due to the possible release of disposed radionuclides into the environment by a certain reason. One of the feasible approaches to cope with this problem is nuclear transmutation because it could basically eliminate the existence of long-lived radionuclides from the wastes. A few concepts for the transmutation of long-lived radionuclides are being considered at present in several countries. One of prospective technologies for the transmutation is to utilize an Accelerator-Driven System (ADS), namely, a hybrid system composed of a proton accelerator and the sub-critical reactor, HYPER.

3.5 Assuring operational safety of nuclear facilities and Plant Life Management (MOCIE & KEPCO, 2000)

In order to enhance societal acceptance and assure nuclear safety, the first periodic safety review (PSR) of Kori Unit 1 PWR plant was performed last May in accordance with the IAEA Convention on Nuclear Safety. This is to evaluate and improve safety of operating Nuclear Power Plant periodically and systematically. PSR may be established as a useful evaluation system for continuous operation beyond design lifetime.

As a part of the Plant Life Management (PLiM) Program, the first phase of Lifetime study for Kori Unit 1 was made during 1993 - 1996. The conclusion was that the continuous operation beyond 40 years was feasible in the technical and economical aspects. The pressurized thermal shock evaluation of Kori Unit 1 reactor pressure vessel was made during 1996-1999. The evaluation concluded that the reactor pressure vessel failure frequency would be less than the regulatory guide criterion for 60 years. The second phase of Lifetime study for Kori Unit 1 is in progress for the detailed life evaluation and aging management. PLiM first phase of Lifetime study for CANDU Unit 1 is initiated for the feasibility of Wolsong Unit 1.

3.6 Plant Life Extension

Since the replacement of steam generator with Inconel 690 U tube was made for Kori Unit 1 in 1998, the PLiM and PSR may establish a good basis for licensing renewal beyond the design lifetime. With such life extension, the economic competitiveness of nuclear power would certainly further improve with some back fitting. The updated licensing base, and current licensing back fitting of the same type of nuclear power plant in advanced country is expected to be applied for Kori Unit 1 licensing renewal process.

4. CONCLUDING REMARKS

Thanks to the healthy electric power demand growth prospect, the Nuclear Power and Fuel Cycle R & D can be focused on the definite target with clear policy objective and supporting Mid and Long Term R and D programs.

To improve economic competitiveness is the most important realistic issue for the utility and nuclear industry. This can be overcome by the good combination of new reactor and fuel cycle development such as KNGR with Extended High Burn-Up fuel in tandem with advanced CANDU with DUPIC fuel. Also, life extended aging PWR or KSNP in tandem with advanced CANDU with DUPIC CANFLEX fuel. Such synergistic effect would be not only for economics, but minimizing fresh uranium feed as well as the spent fuel and radwaste discharge. Further reduction of the minor actinide in the radwaste can be achieved when the Korea Advanced LIquid MEtal Reactor (KALIMER) of actinide burner or HYPER become operational in future. However, the Energy Security is the most important energy policy issue for highly import energy dependent country like Korea. Thus, nuclear energy is not only the viable option for Korean sustainable development, but also essential for national security in 21st century particularly during an escalated high international oil price level. In order to improve efficiency of utilities and nuclear industry, the effective restructuring is in progress to maintain credibility by continuous managerial innovation. Under volatile global energy market situation, the efforts are concentrated on how to satisfy customer needs with better assurance of stable electricity supply for sustainable development.

For correcting some public misconception, good and timely correct information of nuclear energy uses have to be made available to the public particularly those who would be nearby resident of nuclear facility. The basis for improving public perception is the correct education. In order to win public support for nuclear energy, plutonium phobia should be corrected. The proven correct information have to be widely disseminated through educational materials of primary, secondary, high school, and university level where ever appropriate for correct education of public and the next young generation in particular.

Proliferation risk can be reduced somewhat by technical fix or strengthening IAEA Safe Guard System. At the moment, START I and implementation of START II to follow are making good progress between two major nuclear weapon states. Thus proliferation risk would be much reduced in future. Hopefully Nuclear Weapon Free World would come in 21st Century.

Since the capability and resources available to single nation is very much limited in this rapidly changing age of multitude Challenges for sustainable development in 21st century, the key to the success is viable international cooperation in harmony with environment.

REFERENCES

MOCIE (2000), *Energy Statistics*, Ministry of Commerce, Industry and Energy, Republic of Korea.

MOCIE (2000), *The 5th Long Term Electric Power Development Program*, Ministry of Commerce, Industry and Energy, Republic of Korea.

MOST (1997), *The National Mid and Long Term R&D Plan for Nuclear Energy*, Ministry of Science and Technology, Republic of Korea.

USDOE/EIA (2000), *Long Term Outlook of Oil*, US Department of Energy, Energy Information Administration.

Duffey, R.B., Hancox, W.T., Miller, A.I. and Pendergast, D.R. (1999), *Hydrogen and Nuclear Energy : Building Non-Carbon Bridges To The Future*, 9th Canadian Hydrogen Conference, Vancouver, BC, February 7-10, 1999.

Chang, In-Soon (2000), *The Mid and Long Term Nuclear R&D Program in Korea*, Plenary Sessions, PBNC 2000: Proceedings of 12th PBNC, Seoul Korea, October 29 – November 2, 2000.

MOCIE and KEPCO (2000), *White Paper on Nuclear Power Generation*, Ministry of Commerce, Industry and Energy & Korea Electric Power Corporation, Republic of Korea.

Pergamon
www.elsevier.com/locate/pnucene

Progress in Nuclear Energy, Vol. 40, No. 3–4, pp. 335–342, 2002

Printed in Great Britain
0149-1970/02/$ - see front matter

PII: S0149-1970(02)00026-4

Japan's Program for Back-end of Nuclear Fuel Cycle - Flexibility is the Key

Atsuyuki Suzuki
Professor of Nuclear Engineering
Department of Quantum Engineering and Systems Science
University of Tokyo

Introduction

Looking to the coming century, it would not be an overstatement that the greatest interest in the nuclear community is the management of spent fuel from nuclear power stations, namely, measures for back-end of fuel cycle.

Some people think spent fuel should be disposed of directly, while others think that such fuel has high value as energy, and should be re-used .

The value of spent fuel as an energy is actually a serious issue, because plutonium is a substance that easily becomes a source of international argument. So the decision between direct disposal and reprocessing supercedes a company's level or an industrial level, and is frequently debated as an issue of national policy.

In case of the direct disposal of spent fuel, the methods for final disposal of the wastes must be decided quickly. In fact, in the U.S., the Department of Energy (DOE) decided on the candidate site for disposal, and has been proceeding with extensive surveys as to its appropriateness as final repository construction. However, since there has been opposition at the state government, it has been forced to delay its plan. In Sweden as well, the plan has fallen behind the schedule.

Nevertheless, the tenacious and persistent efforts have paid off, and we have got the impression that these nations have gradually begun to make a certain progress with their programs, even though it is not very visible. At this point, however, in all of these situations, we do not see things proceeding according to the original plans. Rather than one choice out of two alternatives, it seems that the main force driving these activities might be seen as a mixture of the two or somewhere in between the two.

In other words, if we take the case of Sweden, before the final disposal is conducted, we see them proposing the set-up of a sufficient temporary storage period together with a prior experimental burial, and the proposal is with a philosophy of flexible measure, theoretically not neglecting possible future reuse of the material as an energy resource. The flexibility has also been observed in the U.S. program as well. Now, the DOE is proposing the institutional period of about 300 years or more before the closure of repository at Yucca Mountain. This simply suggests that the direct disposal policy would not be completed under the current generation but left with the far future generation. So, my understanding is that the Yucca Mountain project might no longer follow the original plan but be modified with more flexibility. And I think this is a good modification for realization of the disposal program.

Another good news is that at the end of March 1999, the U.S. began actual operations of the waste disposal facility called WIPP (Waste Isolation Pilot Plant). The WIPP is for the disposal of radioactive wastes from military uses, and the fact that the wastes are buried deep underground, indicates some similarities with civil high-level radioactive waste in means and method of final disposal. It took approximately 15 years to gain local public acceptance, but the fact that the world's first deep underground repository for nuclear waste has begun operations, is so significant.

A similar thing is happening in countries leaning toward reprocessing policy. As with direct disposal of spent fuel, the final disposal for reprocessing HLW is a difficult problem.

France has spent 10 years for deliberations in the Congress, and recently, has engaged an intensive investigation for locations to serve as candidates for final disposal facilities. This is a great advance in their plan.

Meanwhile, in Japan as well, recently, there have been developments deserving attention in the area of the disposal of this HLW.

Technical Report on HLW Disposal

First of all, the Japan Nuclear Cycle Development Institute, last year published a comprehensive report on research and development to date for HLW disposal program, that is in accordance with the Atomic Energy Commission directives. One of the things Japan has been paying more and more attention to is the transparency, and the JNC report, which is called "The Year 2000 Report", was peer reviewed by OECD/NEA experts group before review by the AEC that has been completed very recently.

In Japan, because the geology is varied, specific identification of the geological environment is difficult. In technical discussions, the geological environment must be conceived widely. Discussions on the possibilities including a wide range and variety of geology, from hard-rock granite to softer sedimentary rock, are features of the report.

It is concluded: "even considering the geological features of our country, which lies in an area of crustal movement, there is a wide distribution of geological environments where deep underground final disposal is technically possible." In drawing this type of conclusion, basically the two types of natural phenomena were considered; long-term natural phenomena such as earthquakes, fault movements, and volcanic activity phenomena, which are relatively localized, are prone to arise violently. The upheaval, depression, and climatic and seawater movements are, on the other hand, broadly distributed phenomena categorized as slow-moving or inactive. In the report, they examined the extent of possible extrapolation from data concerning these phenomena's past histories of activity over a given period of time. The conclusion was that a hypothetical time span of ten thousand years was possible.

A number of reports of this type have been published in the past overseas. Some countries have already decided on candidate sites for disposal facilities like Finland as well as the U.S. Yet the fact that Japan, late though it may be, has been able to prepare such a technical report, is of great significance.

Blueprint for Implementation of HLW Disposal System

Related to the planning for disposal of HLW, the governmental organizations have shared the responsibilities. The AEC makes recommendations for the government in terms of general policy, the NSC is now preparing the safety regulation, the MITI made the proposal for legislation of the law which is specifically dedicated to the implementing body which is in change of the actual business. The law was passed in the National Diet, in May, 2000. And of course, the JNC is responsible for R & D. There is a common movement between the organizations. That is the greatest efforts particularly dedicated to public perceptions.

High-level Radioactive Waste Disposal Law

After enormously intensive and extensive discussions we had these several years in the government and industries, the Diet approved the specific law proposed by the MITI. The legislation of the law has two purposes: establishment of the implementing body and necessary capitalization of disposal program.

In other words, in order to proceed with HLW disposal program, provisions are being put forth as government proposals in the legal system, with the goals of clarifying responsibilities both on a national level and for nuclear power plant operators (electric utilities), obtaining firm assurances for mandatory submission of disposal costs, and establishing an entity to perform the disposal.

The scope covered by these disposal expenses will be itemized, beginning with the technology development of site evaluation, the disposal operations, and the monitoring conducted after the closure of facility.

Also, the capacity of the disposal facility was supposed to be determined by the total number of HLW vitrified waste canisters to be accepted, and considering the economies of scale, be set at the smallest scale not significantly being dependent on the unit cost for disposal; namely 40,000 canisters.

Disposal facility also was supposed to designed under the assumption that the disposal schedule would be determined by the body in the year 2000, with operations beginning from the year 2035, to the closure in the year 2095, and then for the next 300 years, continue monitoring, and then enacting measures after the shut-down.

Under these assumptions, the estimates for the disposal expenses was set at about 3 trillion yen.

Concerning the actual calculation of the costs to be submitted, an appropriate discount of this estimate given above is set, and the results of a trial calculations of the unit cost for disposal with a discount in the range of 0-4%, the amount per unit of electricity generated by nuclear power came to roughly 0.1 yen/kWh to 0.3 yen/kWh.

Developing the Technology to Partition and Transmute Long-lived Nuclides

Before disposing of materials with latent high-level risks, such as HLW, is it not technically possible to change them into safer ones? This type of question, or expectation, is frequently expressed by the general public.

Actually, there are a few specialists who do feel that it is possible. This is not seen only in Japan, but also in foreign countries as well. The OMEGA (Option Making Extra Gains of Actinides) Project, which has been advocated for about the past 10 years in Japan, is one such concept, a sort of extinguishing of the HLW radioactivity through nuclide transformation. The OMEGA Project has easily gained many advocates among non-specialists, out of a feeling of expectation of enormous benefits in the event that the goal of the project achieved.

A couple of years ago, the OMEGA Project had a check and review carried out by the Atomic Energy Commission. Their findings are summarized as follows: Quote!

"The technology here, that of altering material at the atomic level, belongs to a field that differs from the everyday chemical reaction, and the possibilities it offers is not limited simply with nuclear transmutation. Also, when developing this technology, there are problems which would be difficult to solve solely with existing technology, and we look forward that solving these problems will lead to the breakthroughs to new technology."

"Internationally, touched off by Japan's OMEGA Project, a great number of plans of the same type have been started in Europe, mainly France, and the U.S. As one part of our contribution to the international community, in the future our country will, while conducting appropriate evaluation of the research and development, hope to play an important role in this field."

In France, they have been proceeding with their own version of the OMEGA Project, as vital and important research that can serve as a substitute proposition, in order to gain the public understanding on their final disposal program for HLW. Also, in the U.S., looking toward the next century, in order to regain their leadership in the field of nuclear energy development, the U.S. itself sees the need to engage in the development of new technology, and is trying to go ahead with their nuclear energy development for the next century. The projects, named "Nuclear Energy Research Initiatives" or "Generation IV", are what may be called the "American-style nuclear non-proliferation type OMEGA Project".

One of the characteristics of the above mentioned AEC check and review changed Japanese words for "partitioning" and "transmutation", and this change indicates a greater expression of reality.

The former Japanese terms for "Partitioning" and "Transmutation" refer to separating the entire radioactivity, and treating it to eliminate completely, which gives the impression of actualizing an concept. Their new Japanese words pull this concept as close as possible to reality.

The tendency to exaggerate the results of research and development because it is easy to obtain the endorsement of ordinary people is at times seen among scientific communities, and we should welcome this AEC recommendation on change in thinking, in the sense that it raises a goal that has an actual reality, and rejects such temptations to exaggerate.

Another aspect which is noticeable in the recent Japan's situation in connection with partitioning and transmutation is the fast reactor option. Now, it seems to me there is a wide and increased understanding that fast reactions are useful for further burning of minor actinides as well as plutonium. And the understanding would probably lead to the public acceptance that the Monju could be restart its operation in the near future.

Strategic Interim Storage of Spent Fuel

Japan is basically proceeding with choice of reprocessing-recycling of spent fuel. This way of thinking does not change, yet in terms of the actual amounts and timing for reprocessing, the policy that we should be more flexible in our thinking has been clearly brought forth in the past several years.

This is because, for the amount of spent fuel produced at nuclear power stations, our country's reprocessing capacity is inadequate, and this fact is becoming clearer with the passage of time.

In various countries, there is no generic problem with the management of spent fuels. However, whether it is disposed directly, or reprocessed and recycled, the point held in common is that the result is the unavoidable problem of the final disposal of HLW. Needless to say, any countries can not avoid taking tenacious endeavors to resolve the issue with HLW disposal.

Where we actually are right now, is that various countries are in some senses repeating by trial and error to solve this very same problem. In particular, what ought to be remembered is that gaining the understanding of the people is something that takes time. From this point of view, the need for intermediate storage of spent fuel is clear.

In Japan as well, we must push forward steadily with those plans.

In that case, an important point of view, I believe, is that spent fuel should be dealt with as one type of unutilized resource. At this point in time, it is not necessary to decide it is a waste. In particular, for countries like Japan, vulnerable in the area of energy security, considering future uncertainties, rather than disposing of it as waste, storing it as resources for the future can be seen as a more enlightened policy.

Concluding remarks

In fields such as policies on back-end of nuclear fuel cycle, where nothing will develop without the understanding of the people, we must, as a result, rely on our last resort, the democratic decision-making process. This process is, in other words, a political problem. It requires legal provisions, and when in some cases it becomes something that must be discussed amongst international politicians, it is vital that there be responsible interaction at the governmental or Diet level.

As in some countries where this fuel cycle problem has been raised in the government and in the legislatures, recently the discussions on this topic have become extremely active on the National Diet in Japan as well. We expect to steadily actualizing a Japanese style of policy to resolve these problems, by making good use of Japan's social and cultural tradition.

Nevertheless, there are a number of roles that the technological community should play in the process. One of the important aspects is transparency, where we have two-fold meanings. The decision-making process within the community must be transparent and technical measures to improve the transparency must be developed with continuous efforts. Information technologies would be obviously of great help. This is particularly true in the use of plutonium, for instance.

Then, flexibility is also very crucial for planning the future back-end of fuel cycle. It is a responsibility of technological community to show the public that we have at least two options technically available. A binary decision which we used to take, either direct disposal or reprocessing, might be too simple and not flexible enough. There is another alternative between the two.

Last but not least, perhaps what is the most important role that only the technical community could play is to overcome such a challenging issue as back-end of fuel cycle with technological innovations. In that context, there are a variety of areas that are worth while to challenge. There is no doubt this meeting would provide tremendous contributions in exploring the technical possibilities.

Pergamon

www.elsevier.com/locate/pnucene

Progress in Nuclear Energy, Vol. 40, No. 3–4, pp. 343–348, 2002

Printed in Great Britain
0149-1970/02/$ - see front matter

PII: S0149-1970(02)00027-6

RESEARCH AND DEVELOPMENT IN JAPAN ON LONG-LIVED NUCLIDE PARTITIONING AND TRANSMUTATION TECHNOLOGY

SANAE AOKI

DIRECTOR FOR PLANNING, RADIOACTIVE WASTE POLICY DIVISION ,

SCIENCE AND TECHNOGY AGENCY

2-2-1, Kasumigaseki , Chiyoda-ku ,Tokyo, 100-8966, JAPAN

Measures to treat and dispose of radioactive waste are one of the most important issues in the development and application of nuclear energy. The Atomic Energy Commission has carefully considered how to classify radioactive waste properly and how to dispose of it according to these classifications. For high-level radioactive waste (HLW), efforts towards implementing disposal have been progressing steadily.

At the same time, recognizing the nature of the radioactive nuclides contained in HLW, the aim since the early days of nuclear energy has been to develop technology either to separate useful elements and nuclides in order to reuse them effectively, or to transmute long-lived nuclides into short-lived or stable – i.e., non-radioactive – forms by irradiation.

In Japan, reference to P&T technology for long-lived and other nuclides first appeared in the Long-Term Program for Nuclear Research, Development and Utilization (or "long-term nuclear program") back in 1972. That program noted the need for research and development in order to ensure effective processing of radioactive waste.

Under that program, in 1988, the Atomic Energy Commission's Advisory Committee on Radioactive Waste Measures issued a report entitled, "Long-Term Program for Research and Development on Nuclide Partitioning and Transmutation Technology."[1] This can be considered to have been the first systematic R&D program on P&T technology in Japan. It presented a plan for R&D that ran from 1988 to 2000 and was divided into two phases: Phase I, covering the first four to nine years, which included evaluation of various concepts and R&D on key technologies; and Phase II, covering the next four to nine years, which included engineering experiments on

[1] Called the "OMEGA Program" – an acronym for Options Making Extra Gains from Actnides and Fission Products.

key technologies and demonstrations.

The Atomic Energy Commission's Advisory Committee on Nuclear Fuel Cycle Back-End Policy investigated and considered matters concerning P&T technology for long-lived and other nuclides, based on the evaluation schedule stated in the long-term nuclear program issued in 1994.

The Committee issued a report entitled, "Research and Development in Japan on Long-Lived Nuclide Partitioning and Transmutation Technology" in March, 2000. The outline of this report is as follows.

1. RESULTS TO DATE AND ANALYSES OF CURRENT STATUS

1.1 Elements Subject to P&T

○ Long-lived (i.e., long half-life). In particular:
- High radiotoxity due to the emission of α rays
- Fast migration through geological formations via underground water, when disposed of geologically.

○ Relatively short-lived and heat-generating, producing most of the heat in HLW.

○ Rare and useful elements.

1.2 Results to Date and Analyses of Current Status

1.2.1 The Partitioning Process

JAERI is developing a four-group partitioning process, in which elements in concentrated HLLW are separated into four groups: MAs, Tc-platinum group metals, Sr-Cs, and others. Basic experiments using simulated high-level liquid waste (HLLW) helped to establish the concept. Test runs with both simulated and actual HLLW on a scale 1/1000 that of an actual plant confirmed the expected capabilities of group partitioning.

JNC is developing an improved PUREX process – an advanced version of the conventional reprocessing process – to recover Np. It is also developing a CMPO-TRUEX process to separate MAs from highly-concentrated HLLW and an electrolytic extraction method to separate Tc-platinum group and other elements from aqueous reprocessing solutions.

CRIEPI is developing a reductive-extraction process using molten chlorides and liquid metal solvents. Basic data were obtained for the behavior of elements in this type of molten-salt-and-liquid-metal system. Experiments on the separation recovery of TRUs were carried out using some 10 milligrams of TRUs and some 100 grams of chlorides.

1.2.2 The Transmutation Cycle

JNC and CRIEPI are studying transmutation technology using fast breeder reactors (FBRs). This concept is centered on the use of fast reactors for power generation. A shift from the conventional light-water reactor cycle to an FBR cycle is adopted, with power generation and the transmutation of MAs and other elements carried out simultaneously by the FBR. In contrast, JAERI's concept is the "double strata fuel cycle," where a dedicated system for transmuting MAs, such as an actinide burner fast reactor (ABR) or an accelerator-driven subcritical system (ADS) is at the center of the transmutation cycle, allowing a commercial power generation cycle and a transmutation cycle to be developed and optimized independently for their individual purposes.

(1) The Fuel Production Process

JAERI is developing MA-nitride fuel. In addition, burn-up tests of U-Pu mixed nitride fuel produced on a trial basis were carried out.

JNC is developing fuel in which Np and/or Am is added to the MOX fuel. For the addition of Np to MOX fuel, a vibro-packing process is being developed in a joint international research effort. In addition, irradiation experiments were carried out at the experimental fast reactor "Joyo".

As a part of an international joint research effort, CRIEPI is developing metallic fuel – an MA-containing U-Pu-Zr ternary alloy. Fuel pins have been made on a trial basis and physical characteristics and other basic data have been obtained.

(2) The Transmutation Process

JAERI is developing concepts for ADSs and ABRs. Nuclear data on MA nuclides were obtained through international cooperation. In the development of a proton accelerator for the ADS, major new components were developed.

JNC is developing a MOX-fueled FBR. Nuclear data on MA nuclides were obtained and evaluated via nuclear reactors and accelerators. Design studies were carried out on acceptable amounts of MAs and rare-earth elements, and on actual fuel loading.

CRIEPI is developing the concept of metallic-fueled FBRs. Regarding nuclear data on MA nuclides, analysis programs for MA transmutation were developed and analyses were carried out.

(3) Fuel Processing

JAERI is developing a pyrochemical process similar to dry reprocessing. Molten-salt electrolysis of U-nitride, Np-nitride and Pu-nitride on a gram scale were carried out to confirm that transuranic metals can be recovered.

JNC is considering the same method as for the partitioning process.

CRIEPI is developing molten-salt electrorefining and reductive extraction similar to the partitioning process. Feasibility of the conversion of oxides of spent fuel to metal, and of the spent-salt treatment process, are still to be confirmed.

1.2.3 Conclusion

R&D at the three research institutes has resulted in establishment of processes for P&T technology with the expected performance. The aims of Phase I R&D have thus been achieved. R&D in Phase II has experienced some delays, the primary reasons being that Japan is redefining its entire FBR program, and facilities to handle MAs and other materials have yet to be constructed. In carrying out further R&D, it is important to promote cooperation with domestic and foreign organizations in order that experimental facilities – including those for engineering experiments – can be used efficiently.

1.3 Technical Issues

The implementation of experiments to demonstrate processes using actual HLLW is an issue common to the three organizations. Also common to JAERI and the JNC, which are developing aqueous partitioning processes, are development of a method to more efficiently separate MAs and rare-earth elements, and technologies to reduce the volume of secondary waste. At CRIEPI, where dry partitioning is being developed, the main technical issues in the handling of molten chlorides are material development and molten-salt transport technology. Common issues to the three organizations are preparation of a database on fuel irradiation behavior for performance analysis, and development of fuel production technology.

2. EFFECTS AND SIGNIFICANCE OF PARTITIONING AND TRANSMUTATION TECHNOLOGY

2.1. Radioactive Inventory in Waste

It is a social imperative to control, as far as possible, the generation of hazardous waste produced by industrial activities. Reduction of long-term radioactive inventory through the removal of long-lived nuclides from HLW by P&T technology helps to meet this requirement.

2.2. Effects on Geological Disposal

2.2.1 Long-Term Safety

① Effects on the underground water migration scenario
Maximum dose can be reduced by about two orders of magnitude by separating and removing 99% of the Cs-135, and 99% or more of the Np and Am, which are parent nuclides of Th-229.

② Effects on the human intrusion scenario
Risks can be reduced by two orders of magnitude by separating and removing 99-99.99% of the actinide elements.

2.2.2 Impact on Geological Disposal Site Design

Approximately two-thirds of the heat from HLW is generated by Cs and Sr, and separation and removal of these elements would shorten the required storage period and reduce the size of the site. In addition, disposal site design could be rationalized by, for example, disposing of the HLW in one large cavity rather than in several smaller ones.

2.3 Effective Use of Resources

Among the materials in HLW, some can be used effectively as resources. For example Platinum group elementsare widely used as catalysts to reduce nitrogen oxides in vehicle exhaust gases,. With HLW having inherent heating properties, solidified, thermally stable packages could be used as heat sources. Nuclear fission is being considered as a means for the transmutation of TRUs, and the energy generated in this process, too, could be used for power generation.

2.4 Reducing the Amounts of MAs and LLFPs, and the Times Required

Even if P&T technology is employed, some MAs and LLFPs will remain in the waste, and geological disposal of such waste will eventually be necessary. If transmutation is carried out with same energy outputs , the reduction in the amounts of target nuclides as well as the times required to achieve this reduction should be the same, regardless of the fuel type, reactor type or processing method used.

2.5 Generation of Secondary Waste

Compared with reprocessing, the volume of secondary waste is lower because P&T technology deals with the very limited quantities of HLLW generated by reprocessing.

2.6 Short-Term Increase in Radiation Dose

When P&T technology is employed in the nuclear fuel cycle, exposure dose could increase in the short term as new processes and facilities are added and as the volume of MAs and LLFPs to be dealt with in such processes increases. It will be possible, however, to keep exposure dose for workers and the public below the statutory standard, and as low as reasonably possible, by measures such as enhanced shielding.

2.7 Economy

R&D to date has focused on basic studies to obtain fundamental data for designing processes and systems, and no one is in a position to make a reliable cost estimate for the P&T technology.

3. FUTURE RESEARCH AND DEVELOPEMENT

Modern society demands maximum control of hazardous waste produced by industry, as well as recycling to preserve resources and protect the environment. It goes without saying that the nuclear industry, too, must take all effective measures. P&T technology applied to long-lived nuclides can be useful in, for example, reducing the long-term radioactive inventory in nuclear waste, and it is appropriate that R&D should be carried out on an ongoing basis.

3.1 P&T Technology and the Nuclear Fuel Cycle

P&T technology cannot be separated from the nuclear fuel cycle. The question of how and in what part of the nuclear fuel cycle P&T technology should be incorporated in order to optimize the cycle, should be considered. The purpose of R&D is to suggest scenarios for introducing P&T systems into the nuclear fuel cycle, and to develop designs and establish key technologies for such systems. Keeping the totality of the fuel cycle in mind, it is necessary to correctly evaluate issues of economy, energy security, and reduction of the radioactive inventory in waste, and to analyze the trade-offs among those factors.

3.2 Research on System Design and Development of Key Technologies

R&D for P&T technology consists of research on system design with the aim of introducing this technology into the nuclear fuel cycle, and development of the key technologies necessary to realize the system. Development of key technologies takes time and should be carried out in a progressive manner, while at the same time being properly integrated with research on system design. R&D should be carried out on a wide range of systems and technologies so as to extend the options for the nuclear fuel cycle. In this regard, it is important to enhance the available basic data and actively provide information so that R&D results can be reflected in discussions on the future of the nuclear fuel cycle.

3.3 Current Implementation Issues

P&T technology based on the use of power-generating fast reactors, and P&T technology based on the double-strata fuel-cycle concept, on which R&D is being carried out in Japan, have characteristics of their own. These two concepts also provide new options for the fuel cycle and it is therefore appropriate at this stage to continue the development of both. The target is to study scenarios, including a possible blending of these two concepts, in order to introduce a feasible P&T technology system into the nuclear fuel cycle, and to establish the necessary technologies. System design and P&T introduction scenarios will continue to be studied. According to the scenarios thus defined, and based on the results of R&D to date, small-scale tests to demonstrate the feasibility of a series of processes will then be conducted. Following this, systems whose feasibility has been successfully demonstrated will be subjected to engineering tests in order to obtain data on their safety. It will be important at this stage to manage the R&D under a system of checks and reviews, and to update the scenarios on a regular basis.

3.4 Implementing Systems

Although there are differences in the concepts, reactor types and systems, many common issues exist in R&D on P&T technology. JAERI, JNC and CRIEPI should work together in an effort to resolve those common issues, strengthening their cooperation through the sharing of their R&D results. At the same time, it is important to carry out the R&D effectively by, for example, cooperating with other domestic or international organizations including universities and using their existing facilities.

3.5 R&D Schedule and Evaluation

As previously described, P&T technology is inseparable from the nuclear fuel cycle, and it is therefore appropriate to conduct R&D in this area on a time schedule compatible with nuclear fuel cycle R&D. At present, investigations and research on FBRs and related aspects of the nuclear fuel cycle are being carried out. In the process, studies on useful technologies and candidates for commercial use, will be integrated and evaluated by about 2005, and R&D scenarios for taking these technologies into commercial use will be reviewed. Thus, around the year 2005 is deemed to be an appropriate time to reconsider all R&D scenarios, including the use of FBRs for power generation, and the double-strata fuel cycle. Thereafter, progress, results and R&D policy will be checked and reviewed every five years or so. Evaluations of P&T technology system concepts, and reviews of introduction scenarios, should also be conducted.

Pergamon
www.elsevier.com/locate/pnucene

Progress in Nuclear Energy, Vol. 40, No. 3–4, pp. 349–356, 2002

Printed in Great Britain
0149-1970/02/$ - see front matter

PII: S0149-1970(02)00028-8

R&D Activities based on Fast Reactor Cycle Technologies for Transmutation of TRU and LLFP by JNC

KIYOTO AIZAWA

Japan Nuclear Cycle Development Institute
4-49, Muramatsu, Tokai-mura, Naka-gun, Ibaraki, 319-1184, JAPAN
Tel: +81-29-282-1122/FAX: +81-29-282-4904/E-mail: aizawa@hq.jnc.go.jp

Abstract

Research and development(R&D) activities on partitioning and transmutation of trans-uranium nuclides (TRU) and LLFP and future R&D program in JNC were summarized. Feasibility design studies have been conducting to investigate the characteristics of a fast reactor core with TRU and LLFP transmutation. It was reconfirmed that the fast reactor has a strong potential for transmuting TRU and LLFP, effectively. R&D for establishing partitioning process of TRU apart from the high-level radioactive wastes have been carried out. By several counter-current runs of the TRUEX process using highly active raffinates, a process flow sheet capable of selective partitioning of actinides and fission products was newly developed. JNC has settled a new R&D program concerning partitioning and transmutation of long-lived radioactive waste based on recommendation of check & review for OMEGA program performed by the Ad Hoc Committee under the Atomic Energy Commission of Japan (AEC). The R&D program is composed of the design studies and development of element technologies (fabrication, irradiation) in the "Feasibility Studies" on commercialized fast reactor system and the basic studies with experiments (nuclear data, reactor physics, fuel property, etc.) to establish database and analytical tools for the TRU- and LLFP- containing fuel and core design.

1. INTRODUCTION

Some of trans-uranium nuclides (TRU: Np, Am, Cm) and long-lived fission products (LLFP: I-129, Tc-99, etc.) contained in residual waste produced by reprocessing have extremely long-term radiotoxicity. The issue of radioactive waste generation, as one of the crucial issues of fission energy use, has to be improved in the scenario for the handling of the long-lived radiotoxicity.

One of the excellent features of fast reactors is its good neutron economy. Utilizing the excess of neutrons enables us to construct flexible nuclear cycle system including reactors and fuel cycle facilities such that they breed or burn plutonium in consideration of plutonium balance, transmute TRU and LLFP for reducing radiotoxicity and enhance safety and non-proliferation features.

JNC has been performing the R&D on Partitioning and Transmutation (P&T) technologies in the framework of the development of the Advanced Fuel Recycle Systems combined fast reactors. This paper describes the R&D activities on P&T of TRU and LLFP and future R&D program in JNC.

2. R&D ACTIVITES ON P&T TECHNOLOGIES

2.1 Partitioning Technologies for TRU and LLFP

A dual solvent extraction (SX) system has been proposed for enhanced governance of TRU in the spent fuel for the coming back-end fuel cycle. In the JNC's scheme, actinides (U, Pu, Np) are separated from dissolver solution by the PUREX process. Residue of them and all of TRU (Am and Cm) are additionally recovered from the high level liquid waste (HLLW) by the TRUEX process using a CMPO (O ϕ D[IB]CMPO : Octyl [phenyl] -N, N-diisobutyl carbamoylmethyl phosphine oxide) - TBP mixed extractant. By such TRU recycling, high radioactive toxicity sustained in the ultra long terms of the vitrified HLLW is expected to be shortened to permit the human being's supervision.

2.1.1 TRU Partitioning by TRUEX System

Among more than 20 partitioning methods, CMPO-TRUEX method was primarily nominated by JNC so far, because it meet well with several partitioning requirements. Major advantage on CMPO is defined that it will not require additional dilution of HLLW due to its high TRU extractability still in highly concentrated nitric acid media, hence that will be helpful to design simpler reprocessing process.

Actual R&Ds have been pursued at the hot lab. in the CPF (JNC's Chemical Processing Facility in Tokai-mura) using real HLLW (HAR from PUREX tests with "Joyo" FR spent fuels, and concentrated HLLW from commercial reprocessing at "TRP" with LWR spent fuels). Five TRUEX runs and two TRU/Ln separation runs were carried out so far. Major lessons from the results of the hot runs are summarized as follows[(1),(2)] ;

(i) TRU-nuclides were successfully separated from highly active raffinate by improved TRUEX flowsheet with SF(Separation Factor)$>10^3$,where they were fractionally stripped depended with their valences by contacting with a series of salt-free reagents.

(ii) Separation of Am from light Ln^{3+} (La~Nd) was achieved by "SETFICS(Solvent Extraction for Trivalent f-elements Intra-group Separation in CMPO-complexant System) " method using DTPA with SF more than several tens. About 80% of Ln was separated from Am.

2.1.2 FPs Partitioning by Electrolytic Extraction System

Incentives on FPs partitioning in the spent fuel are summarized that (i) partitioning of LLFP (Tc-99, Pd-107, Se-79, etc) for transmutation, (ii) ion partitioning (Ru-106, Tc-99) to increase total DFs of the SX process, (iii) elemental recovery of rare metals (Ru, Rh, Pd, Tc,

Se, Te) as potential resources in future. Among ca. 35 fission products in the spent fuel, some rare metals including platinoids are characterized by their nobler redox (reduction and oxidization) potentials for M^{n+}/M couples in the acidic media. In situ aqueous electrolytic extraction (EE) method(3) therefore can separate those metal ions and has an advantage not to generate secondary radioactive wastes (*i.e.*, salt-free).

2.2 Transmutation Technologies for TRU and LLFP

2.2.1 Design studies on TRU Transmutation in a Fast Reactor(4)

Feasibility studies have been performed to investigate the basic characteristics (transmutation rate, burnup reactivity, Doppler coefficient, sodium void reactivity, maximum linear heat rate, etc.) of a fast reactor core for TRU transmutation. The following items were considered:

(i)Study on loading method of TRU in the core (homogeneous, heterogeneous, hybrid), blanket, etc.,

(ii)Study on the maximum permissible amount of rare earth (RE) nuclides and

(iii)Effect of TRU recycling on core characteristics and fuel cycle system.

The main results of the studies are summarized as follows.

(1) Study on TRU Loading method

The TRU transmutation in a fast reactor core has no serious drawbacks in terms of core performance, provided that the homogeneous loading method can be employed with a small ratio of TRU to fuel (~5wt%). Since a 1000 MWe-class LWR produces about 26 kg of TRU per year, a fast reactor with 5wt % TRU loading can transmute the TRU mass from six LWRs .

(2) Study on the permissible RE level in homogeneously-loaded TRU fuel

The homogeneous loading of TRU and RE has no serious effects on the reactor core performance, provided that the amounts of TRU and RE in the fuel are less than 5 and 10wt% respectively.

(3) Effect of TRU recycling on core characteristics and fuel cycle system.

The recycling of TRU in a fast reactor is feasible from neutronic and thermal-hydraulic points of view. However, during multi-recycling the Np fraction is significantly reduced compared to the unirradiated fuel, and the fraction of Cm is greatly increased because of neutron capture of Am. The accumulation of Cm as a result of the TRU recycling will bring about some problems concerning fuel handling and reprocessing, because of the increases in both the decay heat and the neutron emission rate from Cm-244.

2.2.2 Design studies on FP Transmutation in a Fast Reactor(5)

A moderated target subassembly was used for FP transmutation. The subassembly consists of moderator pins and FP target pins distributed between the moderator pins. The moderated target subassemblies were loaded in the radial blanket or shield region of a fast reactor. A new concept of duplex pellets was also examined : a moderator annulus surrounding a Tc-99 adopted to get a better Tc-99 transmutation performance.

The transmutation rate of Tc-99 in the new target subassembly is higher than that in the subassembly consisting of separate $ZrH_{1.7}$ moderator pins and Tc-99 target pins. A maximum Tc-99 transmutation rate of about 10%/year was obtained by using the new target subassembly loaded in the blanket region of the fast reactor.

Several calculations were performed to determine the I-129 transmutation performance. Iodine-129 was loaded as NaI. The transmutation rate of I-129 was 5.2% and the transmuted amount was 18kg per year. The amount of I-129 produced by a 1000MWe class PWR is about 5.0kg, so the transmuted amount of I-129 was equal to the output from 3 PWRs.

2.2.3 Measurements of TRU, RE and FP cross sections

Fission cross section ratios of minor actinide nuclides (Np-237, Am-241 and Am-243) relative to U-235 in the fast neutron energy region have been measured to evaluate the accuracy of TRU nuclear data, using a back-to-back (BTB) fission chambers at YAYOI fast neutron source reactor at Tokyo university.

Making use of BTB fission chambers and a lead slowing-down spectrometer coupled to a 46MeV electron linear accelerator at Kyoto university, the fission cross sections of Np-237, Am-241, Am-242m and Am-243 have been measured relative to that for $^{235}U(n,f)$ reaction in the energy range from 0.1 eV to 10 keV. Each of the fission cross sections in the JENDL-3.2 and ENDF/B-VI libraries was compared with the measurement.

As a part of TRU nuclear data evaluation, the analysis of irradiated Np-237 sample in JOYO has been performed. Additional irradiation test of Np, Am and Cm samples in JOYO was performed.

Measurements of keV-neutron capture cross sections of RE nuclides (Sm-147, Sm-148, Sm-150, Ce-140, Pr-141, Eu-153, Nd-143 and Nd-145) have been performed to evaluate the accuracy of the nuclear data libraries using the 3-MV Pelletron accelerator of the Research Laboratory for Nuclear Reactors at the Tokyo Institute of Technology.

JNC has been measuring the thermal neutron capture cross sections and resonance integral of radioactive fission products(6),(7). The thermal neutron capture cross section and resonance integral have been measured for Sr-90, Cs-137, Tc-99, I-129, Cs-135, etc.

2.2.4 Fabrication of MOX Fuel containing TRU

As for fabrication and investigation of irradiation behavior of MOX containing TRU, the systematic program has been planned in JNC. Two fabrication methods, pellet and vibro-packing have been studied for neptunium-based fuel pins. The pellet type Np-based fuel will be fabricated at Tokai Works of JNC, and the fabrication of Np-based fuel by vibro-packing method will be performed at PSI in collaboration with JNC. For Am-based fuels, the Alpha-Gamma Facility (AGF) at O-arai Engineering Center of JNC has already been reformed to fabricate MOX fuel pins containing Am at first and then containing Am and Np. The irradiation tests of Np- and Am-contained MOX fuels are planned in JOYO.

3. R&D PROGRAM ON P&T TECHNOLOGIES

The Ad Hoc Committee under AEC has checked and reviewed concerning P&T technology for long-lived radioactive waste, and issued the report in March 31st, 2000. In this report, the effect and significance of related technologies was analyzed, and based on that, recommendations on how to promote R&D for P&T in the future were summarized as follows.

Present society demands maximum control of toxic waste produced by industrial activities from the viewpoint of the global environmental protection and recycling to preserve resources.

It goes without saying that the nuclear industry, too, must take all effective measures. P&T technology for long-lived radioactive waste can be useful in, for example, reducing the long-term radioactive inventory in waste, and it is appropriate that R&D be implemented steadily.

P&T technology is strongly related with the nuclear fuel cycle technology. The purpose of R&D for P&T is to suggest scenarios for introducing P&T technology systems into the nuclear fuel cycle, and to perform design studies for such systems and establish necessary key technologies. For the purpose, it is necessary to accurately evaluate issues of economy, energy security, and reduction of the radioactive inventory in waste, keeping consistency of the fuel cycle in mind, and to analyze the trade-offs among those factors.

R&D for P&T technology consists of system design study from the point of view of introducing it into the nuclear fuel cycle, and development of necessary key technologies in order to realize the system. R&D for P&T should be carried out on a wide range of systems and technologies so as to enlarge options for the nuclear fuel cycle.

The P&T system using commercialized fast reactors by JNC and CRIEPI and the double strata system by JAERI, R&D on which is being carried out in Japan, have characteristics that can add diversity to options for the nuclear fuel cycle. It is thus appropriate to carry out technological development of both for the time being. Hereafter, system design and P&T introduction scenarios will be studied. According to such scenarios, basic experiments to demonstrate the feasibility of a series of processes will be conducted, and engineering experiments with the systems whose feasibility is successfully demonstrated will be implemented in order to obtain data on, for example, the safety of such systems.

At present, the "Feasibility Studies" on commercialized fast reactor system including related fuel cycle system are being implemented. In those studies, R&D scenarios toward commercialization of fast reactor system will be reviewed by about 2005. Thus, around the year 2005 is deemed to be the time to reconsider all R&D scenarios for P&T. Thereafter, progress, results and how to carry out R&D will be checked and reviewed every five years or so. Evaluations of P&T technology system concepts, and reviews of introduction scenarios, should also be conducted.

The AEC recently started soliciting comments on the final draft of the "Long-term Program for Research, Development & Utilization of Nuclear Energy" from the general public. This draft program endorses to pursue solution to reduce environmental burden by TRU burning and LLFP transmutation technologies.

JNC has settled a new R&D program concerning P&T of long-lived radioactive materials based on the recommendations of the report issued by the Ad Hoc Committee. The R&D program is composed of the studies as part of the "Feasibility Studies" on commercialized fast reactor systems including related fuel cycle and the basic studies. The scheme of the relationship among "Feasibility Studies", basic studies and international and domestic collaboration is shown in Fig.1.

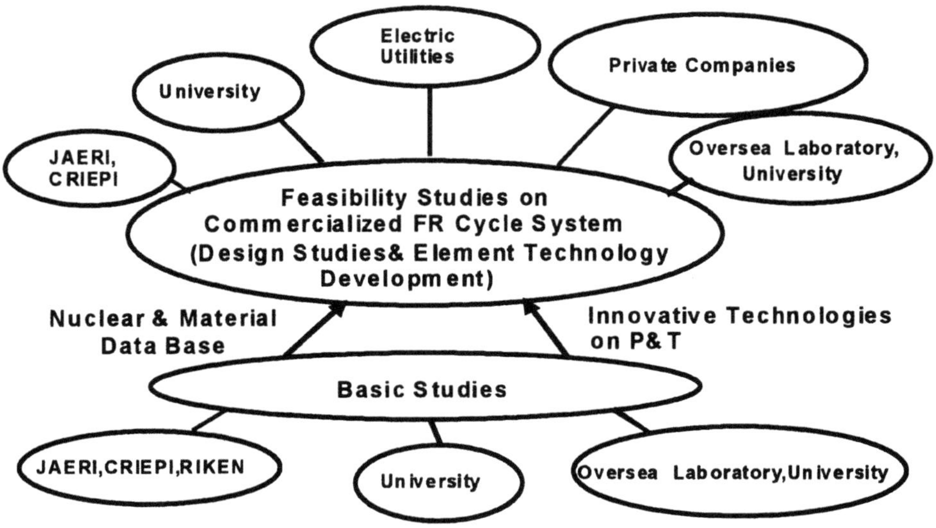

Fig.1 Scheme of the relationship among "Feasibility Studies", basic studies and international and domestic collaboration

In the "Feasibility Studies", target nuclides to be transmuted are selected based on element separation for the purpose of achieving economical recycling technologies. There are plutonium, neptunium, americium and curium as TRU and I, Cs, Tc and Sr as fission products. The design studies and development of element technologies (element separation, TRU-contained fuel fabrication, irradiation, etc.) will be performed in the studies.

In the basic studies, R&D for P&T will be carried out on a wide range of systems and technologies. Experiments and evaluation for nuclear data, reactor physics, and fuel physical property. will be performed to establish database and analytical tools for the TRU- and LLFP-contained fuel and core design. In addition, new separation technologies including isotope separation for Se-79,Zr-93,Sn-126,etc.will be studied, and conceptual design studies to increase the transmutation efficiency will be also implemented. The results of the basic studies will be transferred to the "Feasibility Studies".

The total scope of the R&D program for P&T is shown in Fig. 2. The schedule of R&D for P&T is shown in Fig. 3. In this studies, JNC will actively promote international collaboration from the viewpoint of the most effective use of available resources.

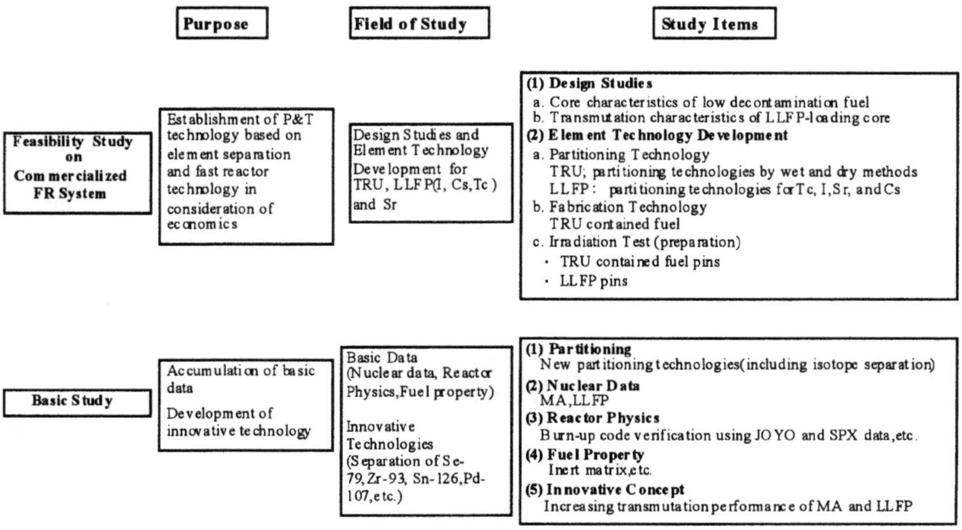

Fig. 2 Total scope of the R&D program for P&T in JNC

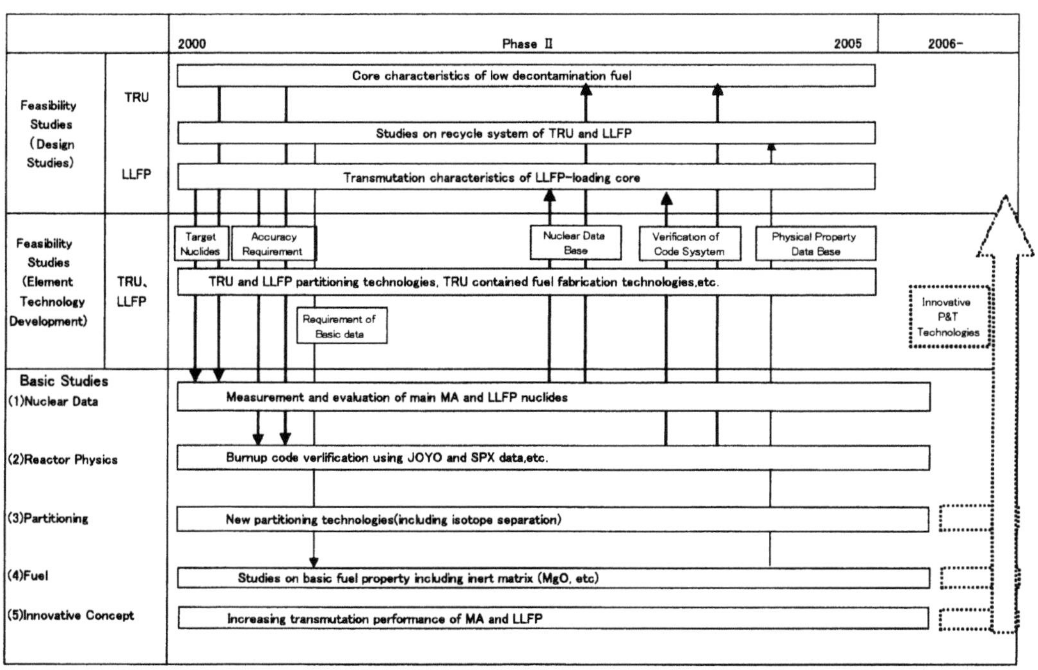

Fig. 3 Schedule of the R&D program for P&T in JNC

REFERENCES

(1)M.Ozawa, et al., En hancing actinide separation by consolidated PUREX and TRUEX process intensified by salt-free requisite, Global'95 Proc.1(1995)585.

(2)Y.Koma, et al., Trivalent f-element intragroup separation by solvent extraction with CMPO-complexant system, J.Nucle.Sci.Eng.35(2)(1998)130.

(3)Y.Sano, et al., Electrolytic extraction of fission noble metals for waste minimizing in advanced actinide separation system, RECOD'98 Proc.3 (1998)717.

(4) T. Wakabayashi , K.Takahashi and T. Yanagisawa, "Feasibility Studies on Plutonium and Minor Actinide Burning in Fast Reactors", Nuclear Technology ,Vol.118, p14-25, Apr.1997.

(5)T. Wakabayashi and N. Higano, "Study on MA and FP Transmutation in Fast Reactors", Progress in Nuclear Energy,Vol.32, No.3/4, p259-264 , 1998.

(6)H. Harada et al., "Measurement of Thermal Neutron Cross Section and Resonance Integral of the ^{99}Tc (n, γ)^{100}Tc Reaction", J. Nucl. Sci. Technol.,32 , 395(1995).

(7)S. Nakamura et al., "Measurement of Thermal Neutron Cross Section and Resonance Integral of the ^{129}I(n, γ)^{130}I Reaction", J. Nucl. Sci. Technol.,33 , 283(1996).

Pergamon
www.elsevier.com/locate/pnucene

Progress in Nuclear Energy, Vol. 40, No. 3–4, pp. 357–364, 2002

0149-1970/02/$ - see front matter

PII: S0149-1970(02)00029-X

U.S. ACCELERATOR-DRIVEN TRANSMUTATION OF WASTE (ATW) PROGRAM OBJECTIVES, AND TECHNOLOGY DEVELOPMENT EFFORTS

GREGORY J. VAN TUYLE (1)
DEBORAH R. BENNETT (1)
JOHN W. HERCZEG (2)
EDWARD D. ARTHUR (1)
DAVID J. HILL (3)
PHILLIP J. FINCK (3)

(1) Los Alamos National Laboratory, P.O. Box 1663, H816 Los Alamos, NM 87544
(2) U.S. Department of Energy, 1000 Independence Ave. SW, Washington, DC 20503
(3) Argonne National Laboratory, Argonne, IL 60439

ABSTRACT

The U.S. Program to evaluate accelerator-driven systems for transmuting problematic, long-lived nuclear waste stream components was initiated during fiscal year 2000, based largely on the Accelerator-driven Transmutation of Waste (ATW) Technology Development Road Map developed during 1999. The Road Map (DOE/RW-0519) effort provided a long-range plan, involving technology development, demonstration, and deployment, as well as a recommended initial effort to evaluate the technology options for five or six years. This paper summarizes the ATW Research and Development Plan currently in draft form. Technology Readiness Levels (TRLs), which are based in part on a system used by the U.S. National Aeronautics and Space Administration in determining levels of flight readiness, was developed for use in assessing and advancing technologies relevant to waste transmutation. Based on TRLs and other considerations, the Program is screening technology options and prioritizing the long-term research and development effort. A top-level schedule illustrates the efforts planned to advance the important technology options in preparation for integrated system tests.

Presented at the International Seminar on Advanced Nuclear Energy Systems Toward Zero Release of Radioactive Wastes, Susono Seminar House, Shizuoka, Japan, Nov. 6-9, 2000.

1 INTRODUCTION

The Accelerator Transmutation of Waste (ATW) R&D Plan [Ref. 1] provides a structured framework for identifying and prioritizing activities leading to technically justifiable proof-of-performance testing of ATW. The Plan builds from the decision objectives specified for ATW, utilizes informational input from the ATW Roadmap and programmatic System Point Design efforts, and employs the knowledge and expertise provided by professionals familiar with ATW technologies. With the firm intent of understanding what, why, and when information is needed, including critical interfaces, the Plan then develops a progressive strategy for developing ATW technologies with the use of a Technology Readiness Level (TRL) scale.

The TRL approach is first used to develop a comprehensive, yet generic, list of experimental, analytical, and trade critical to developing ATW technologies. Technology-specific and concept-specific aspects are then laid over the generic mapping to readiness levels. Prioritization criteria for (1) reducing technical uncertainty, (2) providing information to decision points, and (3) leveraging international collaborations are then applied to focus analytical, experimental, and trade activities.

This process clearly identified R&D activities that will be critical in the near term to the success of the ATW program, with a clear path for development in the longer term leading to proof-of-performance testing. A three-pronged R&D strategy provides for (1) resolving critical performance objectives through integrated system trade studies, (2) performing design assessments to resolve phenomenological and performance sensitivities, and (3) performing critical-path technology developments in separations, fuels, long lived fission product (LLFP) targets, nuclear data and codes, and materials.

The R&D Plan used input from "A Roadmap for Developing Accelerator Transmutation of Waste Technology," the Roadmap's companion documents [Refs. 2–8] and from System Point Design Information developed by programmatic staff [Ref. 9]. The System Point Design information generally assessed various target/blanket configurations based on lead-bismuth eutectic (LBE), sodium, gas, and molten salt. System Point Design Information can be found in the "Compendium of Initial System Point Designs for Accelerator Transmutation of Radioactive Waste."

Three primary assumptions were designated during preparation of the R&D Plan:

- That R&D is the integration of analytical plus experimental activities plus trade studies across interfaces;
- That the R&D Plan must include what is needed, why it is needed, when is it needed, who are the interfaces, and what is the needed data fidelity; and
- That the R&D Plan can lead to integrated, proof-of-performance testing in approximately 10 years.

A classic definition of research and development might include only experimental activities necessary for developing a given technology. However, accelerator transmutation of waste is a complex mixture of developing and combining *multiple* technologies, all at various states of readiness. It was, therefore, assumed that the R&D Plan would work to include and integrate analytical efforts in the form of design assessments, experimental activities in the form of technology development, and trade studies across interfaces that are critical to integrating the system together.

2 FUNCTIONAL BASELINE

A functional baseline was used to help categorize the multitude of phenomena, processes, components, and subsystems within a generic accelerator-driven transmutation system. As depicted in Fig. 1, the Functional Baseline follows the overall system process. It is assumed that each element has a set of performance parameters, each element influences the performance of other elements (e.g., fuel performance affects blanket performance), and all elements are integrated to define the overall system performance and transmutation efficiency.

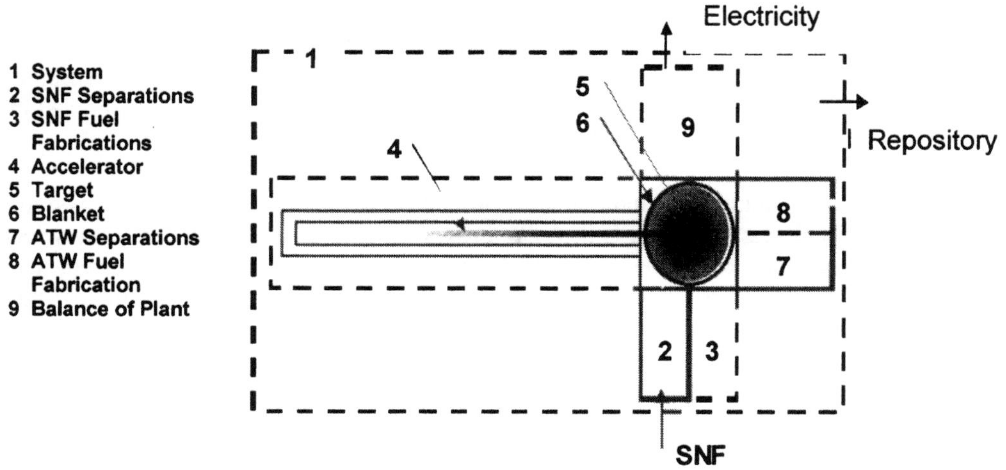

Fig. 1. ATW Functional Baseline Elements

3 TECHNOLOGY READINESS LEVEL

The TRL scale was derived from a NASA model for assessing readiness of space exploration technologies during development. The transmutation TRL scale provides a framework for identifying the analyses, experimentation, and trade studies necessary to demonstrate phenomena, assessing component and subsystem performance, and integrating the overall system. The TRL scale generally increases in component complexity and integration (coupon to component to subsystem to system), and increases in test environment prototypically (laboratory, relevant, prototypic).

TABLE 1. TECHNOLOGY READINESS LEVEL DEFINITIONS

TRL Level	Stage	Definition
1	Research	Basic principles observed and formulated
2	Research	Technology concepts and/or applications formulated
3	Proof of Phenomena/ Principle	Analytical and experimental demonstration of critical function and/or proof of concept
4	Technology Selections	Component and/or breadboard validation in a laboratory environment
5	Development	Component and/or breadboard validation in a relevant environment
6	Proof of Concept	System/subsystem model or prototype demonstration in relevant environment
7	Proof of Performance	System prototype demonstration in prototypic environment
8	Prototype Demonstration	Actual system completed and qualified through test and demonstration
9	Full Implementation	Actual system proven through successful mission operations

For purposes of the R&D Plan, the integration of the TRL activities (analytical and experimental) should be sufficient to demonstrate the phenomenology and performance, at relevant scales, of an integrated system: likewise, it is assumed that technologies should achieve TRL3-4 within two years, TRL5 within five years to rely on fuel feasibility assessments, TRL6 within 10 years for orderly progression into integrated proof-of-performance testing, and TRL7 within 15 years. The scale was truncated at the completion of the TRL7 because it was assumed that progression into TRL8 and TRL9 would be indicative of a prototype activity and full transmutation implementation, respectively.

A partial Summary of Generic Transmutation Technology Readiness Levels is provided in Table 2. Included is the overall system, the separations for light-water reactor (LWR) spent nuclear fuel, the fuel fabrication based on the LWR fuel, and separations for the ATW spent fuel. A more complete table, including other major subsystems such as the spallation target and nuclear assembly, are included in the full R&D Plan [Ref. 1].

The next step involved completing a technology-specific table that included all technology options under consideration. During fiscal year 2000, the U.S. ATW Program was actively considering many technology options, including several that could be considered second-generation systems. Current technology readiness and the steps necessary to advance the technology readiness to Level 6 within 10 years were considered. In some cases, differences in the time and effort to development a given technology determines whether it would be considered a near-term technology option or a second-generation technology option. Although an important part of the planning process, these tables are quite lengthy and appear only in the appendices to Ref. 1

4 TECHNOLOGY SCREENING AND PRIORITIZATIONS

Formal completion of the technology options screening process requires a documentation effort that has not been completed. Similarly, the process of formally designating a primary set of technology options and perhaps one or two alternate approaches has not been completed. However, to proceed to a 10-year R&D plan, it was necessary to make some assumptions as to the outcome of those choices.

It was assumed that some of the technology options would be categorized as second-generation options. In some cases, technology development activities unrelated to waste transmutation may advance those technologies within a few decades to a degree in which they may evolve into near-term options. For example, the fusion community's efforts on tokomacs may advance that technology to become a viable neutron source (replacing accelerator-driven spallation neutron sources) in the future.

It was also recognized that some systems suffered disadvantages in terms of system efficiencies or performance. For example, the process of producing neutrons using electron accelerators is less efficient than the process using proton accelerators. In addition, some spallation source materials, for example, would be much less efficient in producing useful neutrons than lead or tungsten.

The technology options that are thought most viable over the near term include linear accelerators; separations based on aqueous or pyrometallurgical processing; spallation targets based on tungsten or lead-bismuth; and nuclear assemblies based on metallic, nitride, or pyrolytic carbon fuels and cooled using sodium or helium to assure a fast neutronic spectrum. A leading set of technology options would most likely include sodium-cooled metallic fuels, lead-bismuth spallation targets, and pyrometallurgical processing—at least for the ATW spent fuel. An interesting set of alternate technologies would include gas coolants and nitride fuels.

It is emphasized that this description of technology choices does not necessarily correspond to the formal screening process and technology prioritization outcome that will be documented shortly. However, it should be close enough to warrant the discussion that follows.

5 A 10-YEAR PLAN FOR TECHNOLOGY TESTING

A top-level schedule for ATW proof-of-performance testing is provided in Fig. 2. It includes the 10-year effort to advance the TRLs to Level 6, in preparation for integrated system tests. It also includes a top-level view of the integrated proof-of-performance testing envisioned once an appropriate facility becomes available. TRLs for the various technologies are indicated in small numbers within boxes just above the schedule bars. Currently, they are typically at Levels 2 and 3, but can be advanced to Level 6 by 2010, assuming key test facilities are available in the specified time period.

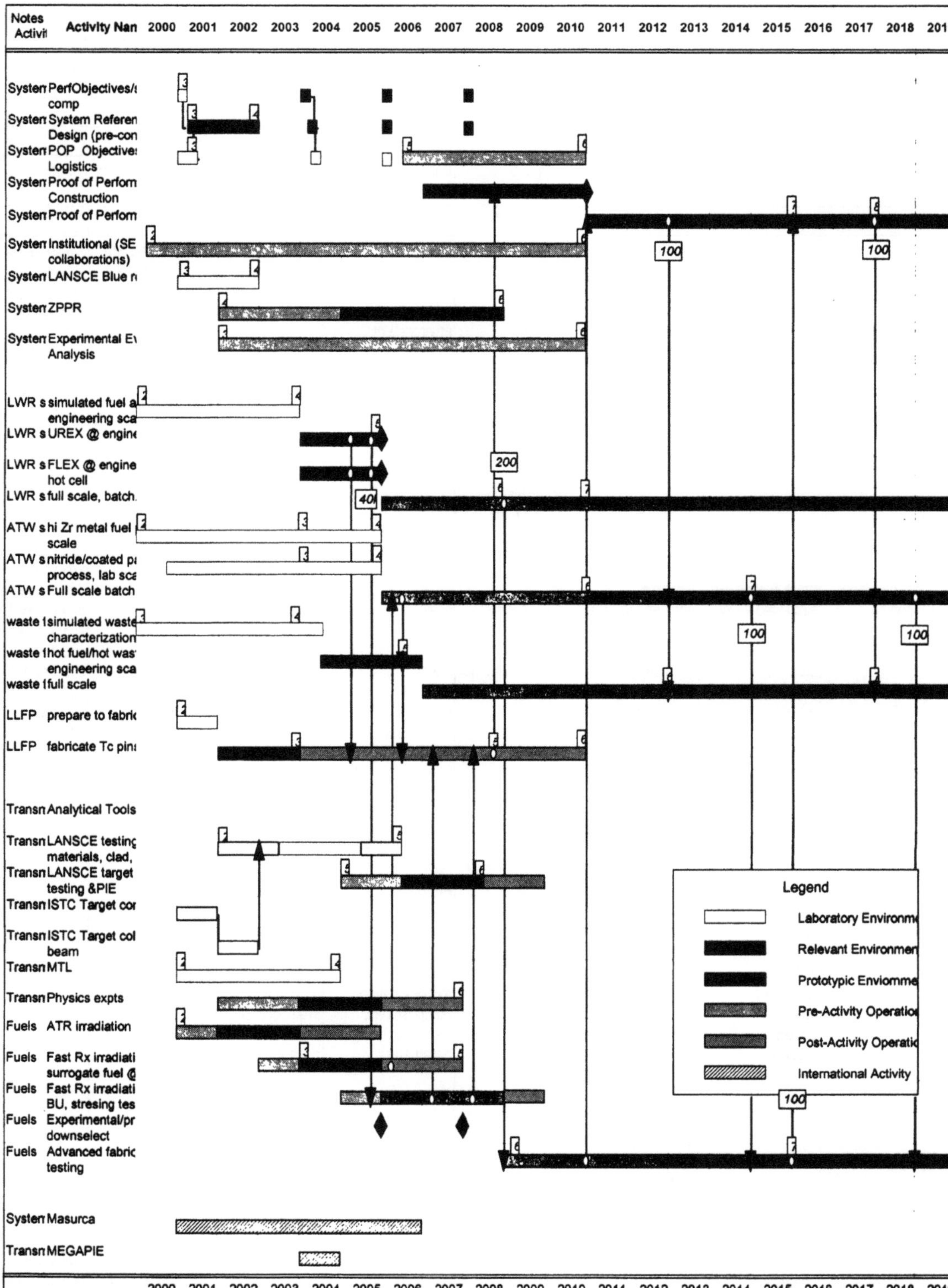

Figure 2. Top-level Schedule to Advance ATW Technology Readiness to Level 6 in 10 Years, then Level 7 Using Integrated Facility

TABLE 2. SUMMARY OF GENERIC TRANSMUTATION TECHNOLOGY READINESS LEVELS

Transmutation Technology Readiness Level Scale								
1	2	3	4	5	6	7	8	9
Completion by...Year 2				Year 5	Year 10	Year 15		
Proof of Principle			Proof of Concept			Proof of Performance		
Technology Selection			Final Process Selections and Integration		Full-Scale Integrated Testing		Full-Scale Demonstration	
1	2	3	4	5	6	7	8	9
Basic principles observed and formulated	Technology concepts and/or applications formulated	Analytical and experimental demonstration of critical function and/or proof of concept	Component and/or breadboard validation in a laboratory environment	Component and/or breadboard validation in a relevant environment	System/subsystem model or prototype demonstration in relevant environment	System prototype demonstration in prototypic environment	Actual system completed and qualified through test and demonstration	Actual system proven through successful mission operations
System								
Establish performance objectives	Assess phenomena sensitivities	Assess component sensitivities	Optimize component performance	Integrate components into subsystems and assess subsystem sensitivities	Optimize subsystem performance	Integrate subsystems into system and assess system sensitivities	Optimize system performance	
LWR Separations								
Basic concept development for different fuel types; identification of process options; close interaction with Fuels Working Group	Flowsheet development scoping experiments on process options; simulated ATW spent fuel types; preliminary systems engineering	Bench-scale batch testing with simulated ATW spent fuel; confirmation of process chemistry; preliminary equipment design concept testing; reagent selection; systems engineering	Unite preparations testing at engineering scale with simulated ATW spent fuel, actinide, and nonradioactive FP content; process validation; design concept testing materials balance flowsheet confirmation; model development.	Unit operations testing at engineering scale; actual ATW irradiation fuel, existing hot cells; process validation; design concept validation	Unit operations testing with full-scale equipment, actual ATW spent fuel, and existing hot cells (not collocated); process and equipment evaluation; validation; model verification; control systems checkout	Integrated system cold shakedown testing; new facility; full-scale equipment; simulated fuel	Demonstration with full-scale equipment, integrated system, actual ATW spent fuel; short (~3 months) periods of sustained operation	Full-scale demonstration at 4-t/y level; actual ATW spent fuel; sustained operation for two-year minimum
LWR Fuel Fabrication								
Technical review leading to technical options	Proposed fuel candidates based on #1	Lab-scale experimentation and characterization, including fabrication process development; property measurement and ex-pile tests	Establish proof of concept. Fabrication parameters for processes from #3; design parameters and features established; irradiation test samples fabricated to QA requirements; irradiation testing (proof of concept); performance phenomena identified	Irradiation testing under nominal prototypic conditions	Prototypic pin/rod/compact irradiation in prototypic environment; representative compositions; prototypic design parameters	Fabrication and irradiation testing of fuel derived from SNF and from received ATW fuel; irradiation in representative environment; prototypic design; prototypic fabrication processes; representative compositions		
ATW Separations								
Basic concept development	Flowsheet development; scoping experiments. w/ simulated ATW blanket spent fuel	Confirmation of process chemistry; bench-scale testing w/ simulated ATW blanket fuel	Process validation; systems engineering and design; unit operation testing w/ simulated ATW blanket fuel	Process validation through unit operation testing at engineering scale with irradiated ATW blanket fuel	Process and equipment final validation; unit operation testing at full scale w/ irradiated ATW blanket spent fuel	Integrated system cold shakedown testing; simulated ATW blanket fuel	Full-scale demonstration with full-scale equipment; irradiated ATW blanket fuel; hot shakedown	Full-scale demonstration at 4 t/y level with irradiated ATW blanket fuel; sustained operations

In the systems area, the effort is partly experimental, partly design, and partly evaluation. The experiments are to focus on using accelerator-driven spallation targets coupled with subcritical assemblies to test the behavior of such accelerator-driven systems (although the theory is well-known, there have been few relevant tests of such coupled systems). The design effort will be focused on establishing and maintaining a reference system design in order to provide a baseline for the experimental efforts. The evaluation efforts will address institutional, safety, nonproliferation, and other issues.

In the separations area, much of the work is confirmation of processes adapted for waste transmutation purposes and the eventual scale-up of these processes to full-scale batches. An exception is work on the so-called "FLEX" Process, which provides a potential alternative to using aqueous separations to remove the oxygen from spent LWR fuel. Another exception is the work on separations for nitride or pyrolytic carbon fuels, where a few years of lab-scale testing is anticipated before programmatic decision-making.

Waste forms are to provide long-term isolation for ATW waste stream materials. Metallic and ceramic waste forms are anticipated, corresponding to steps in the separations processes, and additional waste forms may be developed. Samples can be fabricated based on separations processes using appropriate simulants. Testing of waste forms can go on *ad infinitum*, but it begins with laboratory efforts to characterize the waste form.

Work on LLFP targets in Europe suggests technetium metal and sodium-iodide are promising options. Test rods will be fabricated and irradiated in appropriate flux environments. As it is highly desirable to irradiate these rods to high burnups, an extensive campaign is anticipated in which a few rods are withdrawn periodically and examined.

In the transmuter (spallation target and subcritical assembly), both experimental and analytical efforts are anticipated. For the spallation target, lead-bismuth has long been considered an attractive option if certain technology issues can be resolved. The scheduled program addresses the remaining open issues by irradiating candidate materials with lead-bismuth in the LANSCE proton beam, while simultaneously advancing alternate target options that are based on tungsten. Fuel testing introduces the greatest programmatic uncertainty, given the dwindling number of sodium-cooled fast-spectrum reactors in the world. The Fast Flux Test Facility, Phenix, MONJU, and BOR-60 are all candidates to provide an appropriate fuel-testing environment, but the logistics associated with fuel transport and refabrication range from moderate to highly challenging. It is in the fuels testing area that international collaboration is almost a programmatic imperative. Analytical efforts related to the transmuter are significant, as design analyses require fairly sophisticated simulations. Fortunately, useful computer models are already available, so the tasks are primarily related to upgrades (including data bases).

Two international tests are identified as being of particular interest. The Masurca facility provides for testing subcritical assemblies and their response to a neutron source-driver. Megapie is a lead-bismuth target to be tested at Paul Scherrer Institute in Switzerland. There are certainly other international testing efforts of great interest to the U.S. Program, even if they haven't been identified in Fig. 2.

6 INTEGRATED PROOF OF PERFORMANCE TESTING

There are presently no facilities capable of performing integrated testing of an accelerator-driven spallation target and subcritical multiplication system. However, there are visions for such a facility in France and elsewhere, and such a facility has now been postulated as part of the U.S. Program. Such a facility would have two primary objectives: provide a demonstration of integrated system performance and an appropriate combination of high-energy protons, spallation neutrons, and fission neutrons (hard spectrum) for testing fuels and materials. The schedule includes the assumption of such a facility becoming available around 2010, so the schedule thereafter is based largely on the existence of such an environment. However, the short time frame for the completion of such a facility may be in conflict with some of the fuels throughput indicated in the schedule. The implication is that larger fuel throughput may require concurrent irradiation in fast reactors, or may be delayed until the facility can be upgraded to higher power. Assuming the integrated test facility is designed and built with sufficient flexibility, the ultimate upgrade option could lead directly to a full-scale demonstration of an ATW system.

7 REFERENCES

1 "Accelerator-Driven Transmutation of Waste (ATW): Research and Development Plan," (September, 2000), LA-UR-00-5336.
2 "A Roadmap for Developing Accelerator Transmutation of Waste (ATW) Technology—A Report to Congress," (October, 1999), DOE/RW-0519.
3 "A Roadmap for Developing ATW Technology: System Scenarios and Integration," (September, 1999), ANL 99/16.
4 "A Roadmap for Developing ATW Technology: Accelerator Technology," (September, 1999), LA-UR-99-3225.
5 "A Roadmap for Developing ATW Technology: Target-Blanket Technology," (September, 1999), LA-UR-99-3022.
6 "A Roadmap for Developing ATW Technology: Separations and Waste Forms Technology," (September, 1999), ANL-99/15.
7 "A Roadmap for Developing ATW Technology: Performance Assessment of a Geological Repository Containing High-Level Waste and Waste from Accelerator Transmutation of Waste (ATW)," (September, 1999), PNNL-13020.
8 "A Roadmap for Developing ATW Technology: Estimated Cost of an ATW System," (October, 1999), PNNL-13018.
9 "Compendium of Initial System Point Designs for Accelerator Transmutation of Radioactive Waste," (February 2000), LA-UR-01-1817.

Pergamon
www.elsevier.com/locate/pnucene

Progress in Nuclear Energy, Vol. 40, No. 3–4, pp. 365–374, 2002

Printed in Great Britain
0149-1970/02/$ - see front matter

PII: S0149-1970(02)00030-6

MULTI-COMPONENT SELF-CONSISTENT NUCLEAR ENERGY SYSTEM FOR SUSTAINABLE GROWTH

MASAKI SAITO

Research Laboratory for Nuclear Reactors; Tokyo Institute of Technology
2-12-1, O-okayama, Meguro-ku, Tokyo 152, Japan
Tel: +81-3-5734-3074 Fax: 03-5734-2959
Email: msaitoh@nr.titech.ac.jp

ABSTRACT

Environmental harmonization of nuclear energy technology is considered as an absolutely necessary condition in its future successful development for peaceful use. Establishment of Self-Consistent Nuclear Energy System, that simultaneously meets four requirements - energy production, fuel production, burning of radionuclides and safety, strongly relies on the neutron excess generation. Implementation of external non-fission based neutron sources into fission energy system would open the possibility of approaching the Multi-Component Self-Consistent Nuclear Energy System (MC-SCNES) with unlimited fuel resources and zero radioactivity release. This provides strong evidence that nuclear energy could be considered as a base for the future sustainable growth in long perspective.

1. INTRODUCTION

It is broadly recognized that existing nuclear reactors represent the first stage of nuclear energy development. Along with demonstrating the possibility to safely utilize nuclear energy for peaceful purposes it brought in public worries about nuclear wastes that have been customary relating to fission products (FP), minor actinides (MA) and Plutonium. The waste terminology, however, seems to be confusing since it mixes the nuclides inconvenient in and useless for energy generation. Recent trends, such as, for example, desirable increase in fuel burnup and nuclear proliferation issues might take the advantage of TRU involvement in fuel cycle, needless to say about the efficiency of uranium use. For this reason it is prudent to reserve the term "waste" for the radioactive FP whose transmutation into stable nuclides is an obvious way to the wasteless nuclear technology.

The main bulk of the studies on FP transmutation could be well categorized in two groups reflecting historically the following general trends. At first the main attention was given to the neutronics of various particular transmuters like fission reactors (Taube, 1976), accelerator-driven (Takahashi *et al*, 1995), fusion

(Parish *et al*, 1980) and even such an exotic as muon-catalyzed fusion installations (Kase *et al*, 1990). Notwithstanding the diversity of transmuters proposed, they stressed the achievable transmutation rate compared to natural decay. With rare exceptions, the isotope separation of the target nuclides was assumed in these studies.

The second category of the studies relates to establishing the global nuclear energy system oriented for the long-term stable and safe operation with simultaneous breeding of nuclear fuel and transmutation of radionuclides. This approach was formulated as an ultimate goal for nuclear development (Fujii-e *et al*, 1992) in the form of four universal principles of Self-Consistent Nuclear Energy System that implies:

- energy generation based on fission energy recovery with maximum efficiency,
- fuel breeding to provide inexhaustible fuel supply,
- burning of radionuclides produced in the system,
- system safety relied on simple logic which is to be intelligible both to experts and public.

It is worth to single out some other concepts of harmonized nuclear systems brought out in the past decade known as an Equilibrium Nuclear Society (Sekimoto *et al*, 1991), Radiologically Clean Nuclear Power (Salvatores *et al*, 1994) and the concept of Radiation Equivalency (Adamov *et al*, 1993). All these concepts deal with such an entity as equilibrium state. The Radiological Equivalency addresses the local radiation equilibrium and envisages the disposal of radioactive nuclides after some period of post-reactor management in the places of uranium excavation. Artificial transmutation is not required at all if the system is based on PWR with open-through cycle, and, possibly only ^{99}Tc and ^{129}I are the candidates for transmutation in the case of complete use of uranium excavated. However, this approach of nuclear harmonization for the states with no uranium resources will be obviously at the cost of uranium exporting countries, that shifts the accent from the scientific to political feasibility, unless the sea water uranium is used.

With the exception of Radiological Equivalency concept, to avoid deep geological storage is considered as the most desirable option in approaching the environmental harmonization. The entity of equilibrium is translated to the state when the concentration of each nuclide involved in the fuel cycle is saturated. Thus, the main goal attributed to FP transmutation turns to keeping their inventory at the level that is manageable within the reactor fuel cycle. As a result, the focus in FP transmutation studies was shifted to providing the neutron excess to cover transmutation requirements and to analysis of equilibrium mass balance. So far, the most studies have dealt with critical Fast Reactor (FR) (Sekimoto *et al*, 1991; Fujii-e *et al*, 1992) or hybrid Accelerator-Driven System (ADS) with fast spectrum (Salvatores *et al*, 1994; Rubbia *et al*, 1997) as leading candidates to harmonize nuclear power. Post irradiation fuel management heavily relies on element separation and, to incinerate FP, envisages their isotope separation.

The present paper gives a review of the current studies on nuclear harmonization. The logic of the paper follows the pathway of the approach to self-consistency. The first part is devoted to featuring the mono-component SCNES based on fast reactors. The rest of the paper reveals an important advantage of external neutron sources that make the harmonized nuclear energy system inherently multicomponent. The structure of MC-SCNES includes spallation and fusion neutron generating elements which serve to improving the quantity and quality of neutrons in the system. The main message to public from the side of this project is that that nuclear energy could be considered as a base for the future sustainable growth in coming millennium and far beyond.

2. SELF-CONSISTENT NUCLEAR ENERGY SYSTEM BASED ON FISSION TECHNOLOGY

To advertise the inner potential of fission technology logically and historically was the first step toward nuclear harmonization. Below considered are the main area of activities, motivations and achievements on this way following the four principles of self-consistency listed in the previous section. To begin with, transuranium isotopes which are important byproducts of energy production are not treated as a waste. Rather than this, they are considered as a fuel that helps to completely use the excavated uranium. Another corner stone is artificial transmutation of fission products to avoid deep geological storage. It makes an excess neutron generation to be the key element of nuclear harmonization. It is defined as a number of neutrons left after all the inevitable neutron consumption in the course of reactor operation. The advantage of fast neutrons is based upon two main facts of neutronics:

- the harder the spectrum the larger is the number of neutrons released as a result of fission process (the same fissile material is assumed for thermal and fast spectra);
- the harder the spectrum the less is parasitic absorption rate in structural materials.

These phenomena have been already found their application in fast breeder reactors. So, it has been a natural further step to extend this idea to transmutation of radionuclides in SCNES (Fujii-e *et al*, 1992), fuel breeding being referred to as an inevitable consumption also. Table 1 gives an illustration of neutron balance in typical fast and thermal reactors. One can conclude that, in contrast to thermal reactor, there is neutron excess in fast reactors (0.72 n/f) that might be destined to FP transmutation. Neutron requirements for FP transmutation are determined by number of FP isotopes to be transmuted and necessity to provide FP isotope separation. The last two issues govern the parasitic neutron capture in transmutation process, because most of the discharged FP have small capture cross-sections and parasitic capture in the FP bearing matrix or in stable isotopes of the same element might be significant. Illustrative is the estimation of neutron requirements in a metal-fuel FR for non-separation option done in (Ohsaki and Fujii-e, 1996) sketched in Fig.1 as a function of FP half-lives, that is being assumed as a selection criterion to identify candidates for transmutation. For 29 FP with half life more than one year (^{79}Se, ^{85}Kr, ^{90}Sr, ^{93}Zr, ^{93m}Nb, ^{94}Nb, ^{99}Tc, ^{106}Ru, ^{102}Rh, ^{107}Pd, ^{108m}Ag, ^{109}Cd, ^{113m}Cd, ^{121m}Sn, ^{126}Sn, ^{125}Sb, ^{129}I, ^{134}Cs, ^{135}Cs, ^{137}Cs, ^{146}Pm, ^{147}Pm, ^{151}Sm, ^{152}Eu, ^{154}Eu, ^{155}Eu, ^{158}Tb, ^{166m}Hm, ^{171}Tb) in equilibrium state it requires neutron consumption in FP as high as 6.78 n/f, total mass being 121 tones per GW_{el}. Neutron requirements are reduced with respect to increase in half-lives of selected FP, however exceed the neutron excess available in FR assumed in (Ohsaki and Fujii-e, 1996) as high as 0.58 n/f.

Table 1. Normalized to one fission neutron balance in fast and thermal reactors*

	Fast	Thermal	Remarks
Neutron generation	2.9	2.9	
Neutron captures:			
Energy generation	1	1	Fission chain
Capture by fissile	0.13	0.4	Capture by fissile
Fuel breeding	0.85	1.3	Breeding ratio 1
Other absorption	0.2	0.25	Coolant, Structure
Total captures :	2.18	2.95	
Neutron excess	>0	<0	

*Data from (Shimizu, A., and Fujii-e,Y, 1995)

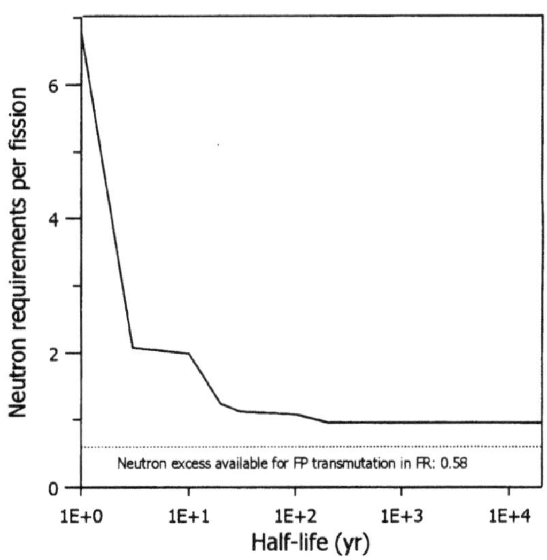

Fig.1.Neutron requirements to transmute fission products in a fast reactor as a function of their half-lives assumed as a selection criterion; derived from the data presented in (Ohsaki and Fujii-e, 1996)

For the case of isotope separation, confinement of all FP with half-life longer than 1 yr requires only 0.25 n/f which is markedly less than neutron excess produced and total inventory at equilibrium state becomes 6.2 tones per GW_{el}, so, it was concluded that, provided FP isotope separation, there is a possibility to confine these FP in a reactor fuel cycle (Akatsuka, *et al*, 1995). Various separation technologies: atomic vapor laser isotope separation (AVLIS), plasma separation process (PSP) and gas centrifugation method were assessed for twenty-one elements (I, Pd, Cs, Zr, Tc, Sn, Se, Sm, Sr, Pm, Ru, Kr, Eu, Sb, Nb, Ho, Cd, Tb, Ag, Tm, Rh). Analysis of energy balance in SCNES to run these technologies reveals the fact that large-scale isotope separation in SCNES is feasible (Ezoubtchenko, *et al*, 1998). Nuclear data performance is another essential issue in FP transmutation study. Measurement of nuclear cross-sections is ongoing by using experimental equipment available at Research Laboratory for Nuclear Research of Tokyo Institute of Technology. In combination with evaluated data presented in nuclear data libraries this serves to improve the reliability of the transmutation efficiency estimated for SCNES (Igashira and Ohsaki, 2000).

Safety issues of SCNES are well exemplified by promoting the concepts of self-controllability (Ninokata, *et al*, 2000) and self-terminability (Endo, *et al*, 2000). The first makes focus on inherent and passive mechanism to withstand against anticipated transient without scram in liquid metal FR. The second stresses Controlled Material Relocation (CMR) to provide for neutronic shut down of a reactor by removing the molten fuel out of the core before critical pool formation. These concepts are transparent to be understood by general public which is very much important in a democratic society for adopting the large scale technology.

Whatever complicated set of burnup equations was considered in (Ohsaki and Fujii-e, 1996), this approach remains to be rather idealistic, since it goes through the simple treatment of sigma-fi product, without taking into account degradation of neutronics environment (flux and spectra) that follows the introduction of the FP in the real reactor core. Nevertheless, the conclusion of scientific feasibility to approach self-consistency is, of course, important and valuable. It gives a philosophical base for analysis of real core configurations and fuel cycles toward waste transmutation. In the framework of SCNES three main options are under consideration: MOX (Fujimura, *et al*, 2000), metal (Fujita, *et al*, 2000) and nitride (Ikeda, *et al*, 2000) fuel cycles. The list of the isotopes included in the core analysis was shortened to nuclides with half-lives more than 30 years, focusing transmutation of no more than the following 7 long-lived isotopes ^{79}Se, ^{126}Sn, ^{99}Tc, ^{93}Zr, ^{135}Cs, ^{107}Pd, and ^{129}I on condition of their isotope separation. Close inspection of the nuclides given by (Arie *et al*, 1998) excluded the long-lived ^{135}Cs from species to be transmuted because of the largest equilibrium inventory even on condition of isotope separation and smallest atomic density in bearing matrix. It is destined in elemental form together with ^{137}Cs and ^{90}Sr to serve as in-reactor shielding.

It is worth to note that isotope separation was assumed even at the stage of feasibility assessment of the SCNES concept. More over, SCNES appears as a monocomponent structure based upon advanced reactors. Being oriented for the ultimate goal of nuclear development it denies the long experience gained on energy production with PWR. All this stimulated the scientists belonging to this project to have a close look at the neutron excess generation with non-fission neutron sources that is the subject of the next item.

3. MULTI-COMPONENT NUCLEAR ENERGY SYSTEM

The structure of MC-SCNES is sketched in Fig2. An essential feature is that MC-SCNES gives rooms for each particular nuclear technology: fission (represented by both thermal and fast reactors), fusion and accelerator technology. They are not competitors but are combined in a way to realize their unique advantageous features for getting the credit for the system as a whole.

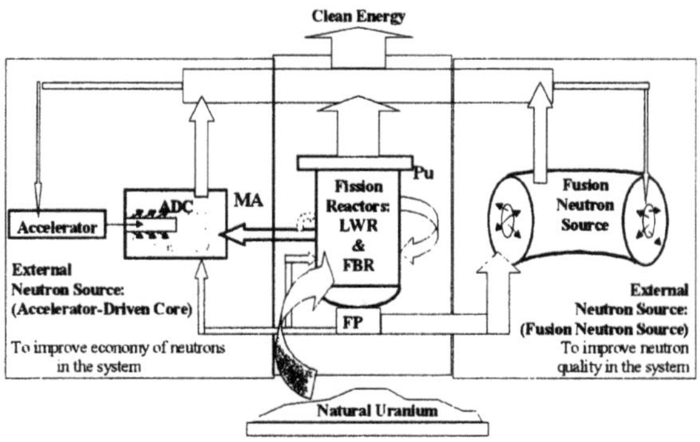

Fig.2. Multi-Component Self-Consistent Nuclear Energy System

Fission reactors are considered as the main energy and fuel (the last is mostly for FR) producers while advanced accelerator and fusion technologies reveal themselves as a neutron sources destined to improve quality and quantity of the neutron population that is especially important for waste transmutation. Below are presented some specific issues of MS-SCNES.

3.1 Advanced fuel for MC-SCNES

Anticipated spreading of nuclear power over developing countries created a new trend with treatment of the MA instead of using them like fuel in FR. In reactor with thermal and epi-thermal spectra, MA could help to create isotopic radiation barrier against uncontrolled proliferation for Pu and radiation barrier for themselves (by increased neutron and gamma radiation of transplutonium). The key point here is accumulation of ^{238}Pu ($T_{1/2}$=87 years, α-decay) whose decay heat is rather high- 570W/kg that makes plutonium with increased fraction of ^{238}Pu difficult to use in weapon manufacturing.

What important is that in a soft spectrum both parents of ^{238}Pu in transition chain ^{237}Np and ^{241}Am are strong absorbers and their transmutation through capture reaction just releases the reactivity giving birth to more fissionable ^{238}Pu, simultaneously creating the radiation barrier. Figure 3 exemplifies typical criticality behavior for UOX fuel doped with ^{237}Np (Nikitin, *et al*, 1999). To keep initial appropriate criticality at increased neptunium fraction requires enhanced enrichment, so, it serves to storing well enough reactivity to achieve burnup values in the vicinity of 100 GWd/t that is three times of the standard PWR fuel. Even higher burnup might be achieved through substituting light water with heavy one and tightening the fuel lattice: 200 GWd/tHM (see Fig.3). Plutonium isotopic vector might be well illustrated by one given for U-Np-Pu fuel performance in a small PWR with burnup of 140 GWd/tHM (Nikitin, *et al*, 2000). At the beginning of cycle it contains 91.5% U, 3.5% Np and 5% Pu, the fraction of ^{238}Pu in plutonium mixture being 55%.

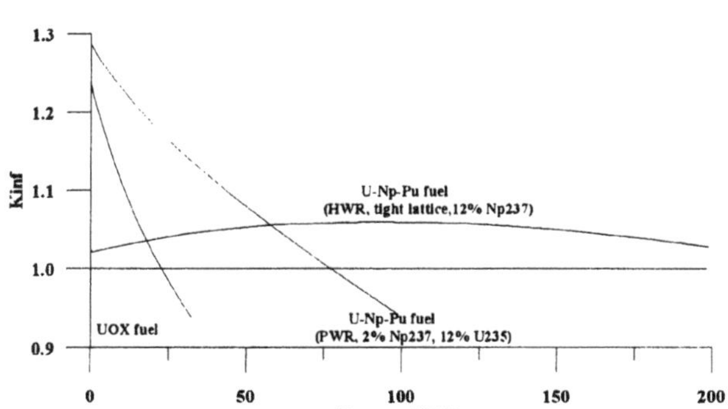

Fig.3 K_{inf} as a function of burn-up for UOX fuel and U-Np-Pu fuels (from: Nikitin, *et al*, 1999)

High proliferation resistant properties along with high burnup give the right to appeal against killing the MA through fission. Rather than this, MA might be treated as useful species important for fuel cycle management. The question on their increased generation by application of some advanced technologies has been already put forward (Shmelev, et al, 1995).

3.2 Neutron excess generation in MC-SCNES

Neutron production, wherever it might occur, is inevitably accompanied by energy generation. Regardless of the possibility to approach positive energy balance in a single neutron producing facility, if the dominant energy producer is a fission reactor, it is desirable to minimize any energy which is associated with neutron generation outside reactor. That is the key issue of the methodology developed by (Saito, *et al*, 1998). Energy might be consumed in accelerator to drive spallation neutron generation, for example, or be released in the form of kinetic energy, like in a fusion neutron source; in any case, it should be treated as an additional burden to deal with. In our approach we take into account nuclear processes involved in neutron production rather than single installations. From this viewpoint, neutron source can be determined as a combination of two types of the processes. The first is a fission process associated with neutron source (to feed accelerator or to support plasma performance in case of negative energy balance). The second type includes all the possible reactions inherent in neutron production which are distinguished from the ordinary fission: spallation, fusion, stripping. Neutrons, thus produced, initially should cover necessary neutron consumption inside the neutron source itself: both fuel self-sustaining and elimination of FP, if any. So the meaning of 'neutron excess' stands for 'clean neutrons', which are free from the burden of radionuclides and are to be used to cover the transmutation requirements of fission reactions destined to net power output from the system.

As a basic system parameter we make use the following energy utilization factor:

$$EUF = \frac{E_f^{net}}{E_{th}} \tag{1}$$

where: E_{th} - total thermal power generated **in** the nuclear energy system, E_f^{net} - net fission power produced **by** the system. This equation may be rewritten as

$$EUF = \frac{1}{1+\delta} \tag{2}$$

where δ - system configuration - ratio of the power associated with neutron source to the net power:

$$\delta = \frac{e_f N_f^{NS} + e_i N^{NS}}{e_f N_f^{net}} \tag{3}$$

herein superscripts 'NS' and 'net' relate to reactions associated with neutron source and the net energy production, correspondingly, e_i - energy release per one inherent reaction (energy release in spallation target of accelerator, fusion energy etc.); e_f - fission energy; N_i^j - reaction rates. From consideration of neutron balance in the system it appears that δ is governed by transmutation requirements of the net energy production (n_{tr}^{net}) and neutron source characteristics:

$$\delta = ENC\frac{n_{tr}^{net}}{e_f} \tag{4}$$

where complex denoted as ENC incorporates all the terms related to the neutron source:

$$ENC = \frac{e_i^{NS} + e_f \dfrac{N_f^{NS}}{N^{NS}}}{S^{NS} - n_{br}^{NS} - n_c^{NS} - n_l^{NS} - n_i^{NS} - n_{tr}^{NS}\dfrac{N_f^{NS}}{N^{NS}}} \tag{5}$$

where S^{NS} -neutron source strength - number of neutrons generated per one inherent reaction; n_{br}^{NS} - breeding requirements, if any; n_i^{NS} - neutron consumption in transmutation of radionuclides different from FP; n_c^{NS} - parasitic capture; n_l^{NS} - useless neutron leakage; n_{tr}^{NS} - neutron requirements to transmute FP from fissions dedicated to neutron source. Structure of the nuclear energy system with energy and neutron flows is illustrated by **Fig.4**. Denominator in Eq(5) stands for the excess neutron generation in the neutron source and numerator - for energy inevitably accompanying this process. The literal meaning of this ratio is energy burden of one 'clean neutron'. This can be treated as an excess neutron cost (ENC). It is useful combination of characteristics to estimate through

Eq.(5) and Eq.(2) the integrated effect of neutron source on the nuclear energy system as a whole. Analysis of various neutron sources has been done in (Saito *et al*, 1998 and Artisyuk, *et al*, 1998). In Fig.5 energy utilization factor is shown as a function of ENC. For quantitative example neutron requirement of 0.7 n/f were chosen that significantly larger than one needed for FP transmutation in FR on condition of their isotope separation (Ohsaki and Fujii-e, 1996). Fusion Neutron Source (**FNS**) with ITER-like parameters (Q-value is equal to 15) being introduced in nuclear energy system can bring it to the self-consistency with minimum burden of energy production associated with innovative fusion technology. Especially it is contrasted with ADS with traditional lead target.

The important conclusion that might be drawn from the ENC analysis is that FNS has a potential to bring into self-consistency the nuclear energy system with large fraction of PWR thus giving the honorable place to the mature and technologically proved energy producing technology.

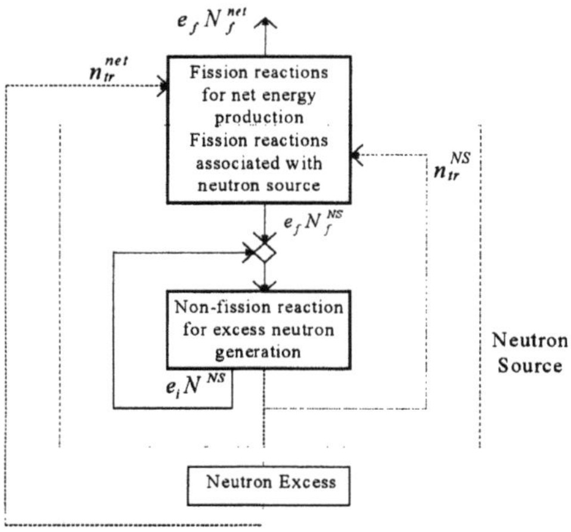

Fig.4 Structure of nuclear energy system with external neutron source (Saito, *et al*, 1998).

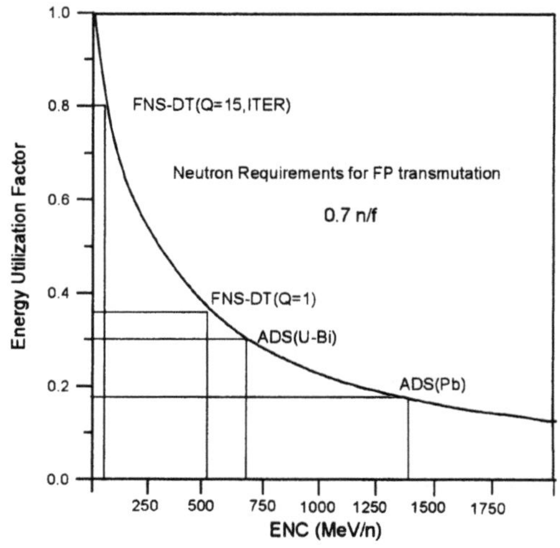

Fig.5 Energy utilization factor as a function of excess neutron cost (Saito, *et al*, 1998).

3.3 Transmutation efficiency in a MC-SCNES

As it was mentioned in the previous section, non-fission neutron producing reactions give some advantage in neutron excess generation compared to fission. In addition to that, FNS has flexibility in creating neutron environment (flux and spectrum). Shown in Fig.6 is the neutron spectrum in a transmutation zone of the FNS which is very well thermalized due to a large fraction of carbon inserted in the zone. It appears to be even more softer than in standard thermal PWR reactor. This, of course, increases significantly cross-section of transmutation reaction and all the characteristics of transmutation efficiency listed in Table 2 (Saito, et al, 1999).

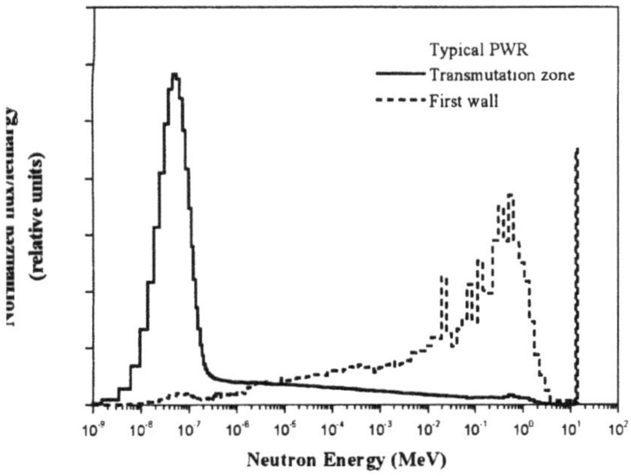

Fig.6 Neutron spectra in FNS compared from PWR (Saito, et al, 2000,a).

Table 2. ^{135}Cs transmutation characteristics in typical neutronics environments

	Fast*	Thermal*	FNS**
Neutron Spectra	(0.2MeV)	(1eV)	
Capture cross-section (b)	0.07	1.1	3.12
Neutron flux (n/cm^2 s)	10^{14}	10^{15}	4.75×10^{14}
Mean life-time (yr)	240	450	21.4

*Data from (Salvatores, *et al*, 1998)
**First Wall Loading 1 MW/m^2

Advantage of transmutation environment in FNS is instructive to compare with that in FR. The feature is a mean life-time under irradiation. That means a long time to generate equilibrium inventory. Usually it is much longer than life-time of a reactor itself. This casts a shade on their transmutation potential. It is true that FR does give the neutron excess that could be used for FP transmutation, but it is hardly believed that the society gets ready to invest in the transmutation that accomplishes its goal within centuries. A meaningful example is transmutation of ^{135}Cs in a highly moderated high-flux blanket of FNS with ITER-like First Wall Load that gives time to reach equilibrium about 20 years, neutron excess being enough to avoid its isotopic separation (Saito *et al*, 1999). Incorporation of FNS in nuclear infrastructure would facilitate attaining the self-consistency within time period comparable with the life-time of FNS. Preliminary scoping studies indicate that fraction of energy accompanying FP transmutation in FNS does not exceed 5% of the fission energy accompanying their accumulation (Saito *et al*, 1999). Similar optimistic results were obtained for other key isotopes like Zr (Takibaev, *et al*, 2000) and Pd without isotope separation (Apse *et al*, 2000).

Accelerator-driven transmutation might find its place in nuclear infrastructure also. Specially designed ADS would provide transmutation environments difficult to achieve in critical fission and fusion driven facilities. The example is transmutation of ^{244}Cm proposed in (Arisyuk *et al*,1999). As for the FP, ADS with no fissionable material in the blanket is capable of creating neutronics necessary to effectively transmute ^{126}Sn whose capture cross-section is very small.

3.4 Nuclear data requirements for MC-SCNES

Neutron production in accelerator-driven systems originates some radioactive species that are rather unique for nuclear scientists in both fission and fusion fields. So-called spallation products (SP) accumulated in primary neutron producing targets are among those often neglected in analysis of radiological hazard. There is not enough information to qualify SP among which one can reasonably expect nuclides scarcely presented in the waste stream of fission and fusion technologies. This mainly relates to some rare earth (RE) elements with atom numbers between 145-155. Samarium, Gadolinium and Dysprosium are known to have isotopes unstable to alpha decay with half-lives ranging from several decades (^{148}Gd, $T_{1/2}$ =74.6 yr) to million years (^{146}Sm, $T_{1/2}$=1.03×10^8 yr). Being alpha emitters, they would give much higher specific toxicity (in a contrast to beta emitting long-lived FP), and because of their small mass number they cannot be fissioned in the same manner as transuranics. These nuclides give a significant contribution into overall toxicity of spallation targets (Saito, *et al*, 2000 b). Shown in Fig.7 and 8 are examples of toxicity analysis done for lead, tungsten and tin targets (Stankovsky, *et al*, 2000). This seems to be an important factor to make a design choice of accelerator-driven neutron source.

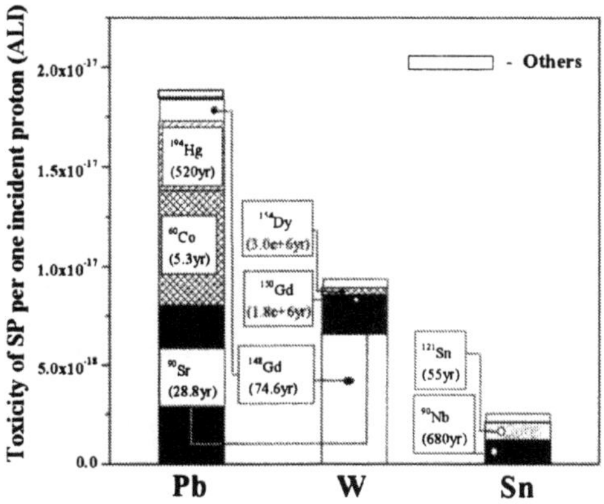

Fig.7 Toxicity generated in various targets by one incident 1.6 GeV proton

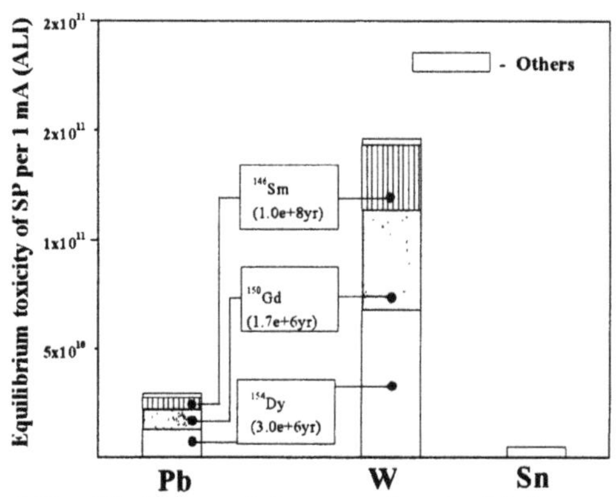

Fig.8 Equilibrium toxicity associated with beam current of 1 mA, proton energy - 1.6 GeV

4. CONCLUSIONS

To realize harmonization of nuclear technology with human environment requires for the nuclear energy system to be neutronically self-consistent, to demonstrate effective transmutation of dangerous radionuclides and to be safely protected against uncontrolled proliferation of fission materials. All these requirements are satisfied with the help of external neutron sources driven by accelerator and fusion. It reveals that SCNES is an inherently multi-component system.

REFERENCES

Adamov, E., *et al* (1993), Attainment of Radiation Equivalency in Nuclear Power Radioactive Product Management, *Nuc. Technol., 104, 233.*

Akatsuka, H., *et al* (1995), Scientific Feasibility of Incineration in SCNES, *Progr. Nucl. Energy., 29, 477.*

Apse, V., *et al* (2000), Transmutation of Elemental Fission Products in Fusion Facilities with Tritium Self-Sustainability, *International Seminar on Advanced Nuclear Energy Systems Toward Zero Release of Radioactive Wastes*, Susono, Shizuoka, Japan, Nov.6-9, S 8-10.

Arie, K., *et al* (1998) Long-Lived FP Burning Based on the Actinide Recycle Metal Fuel Core, *Progr.Nucl. Energy 32, 665.*

Artisyuk, V., *et al* (1998), Potential of Neutron Excess Generation by DD-fusion in SCNES, *Progr.Nucl. Energy 32, 721.*

Artisyuk V., *et al* (1999), ^{244}Cm Transmutation in Accelerator-Driven System, *J. Nucl.Sci. Techn. 36, 1135.*

Ezoubtchenko, A., Akatsuka, H., Suzuki, M.(1998), Isotope Separation for Self-Consistent Nuclear Energy System, *Prog. Progr.Nucl. Energy 32,729.*

Endo, H., *et al* (2000), Elimination of Criticality Potential for the Self-Consistent Nuclear Energy System, *International Seminar on Advanced Nuclear Energy Systems Toward Zero Release of Radioactive Wastes*, Susono, Shizuoka, Japan, Nov.6-9, S 8-4.

Fujii-e, Y., *et al* (1992), An Approach to Self-Consistent Nuclear Energy System, *Trans. Amer. Nucl. Soc., 66, 342.*

Fujimura,K., *et al* (2000), Feasibility Study of Large MOX Fueled FBR Core Aimed at the Self-Consistent Nuclear Energy System, *Progr. Nucl. Energy 37, 177.*

Fujita, R., *et al* (2000), An Application of Metal Fuel Cycle Technology Toward Self-Consistent Nuclear Energy System (SCNES) Concept, *Prog. Nucl. Energy 37, 169.*

Igashira, M., Ohsaki, T.(2000), Neutron Economy and Nuclear Data for Transmutation of Long-Lived Fission Products, *International Seminar on Advanced Nuclear Energy Systems Toward Zero Release of Radioactive Wastes*, Susono, Shizuoka, Japan, Nov.6-9, S 8-1

Ikeda, K., *et al* (2000), Feasibility Study of Nitride Fuel Core and Recycle System Toward Self-Consistent Nuclear Energy System, *Progr. Nucl. Energy 32, 673.*

Kase, T., *et al* . (1990), Transmutation of Cs-137 Using Muon Catalized Fusion, *Muon Catalyzed Fusion, 5/6, 521.*

Ninokata, H., *et al* (2000), Evaluation of Self-Conrollability of LMFRs Against ATWS, *International Seminar on Advanced Nuclear Energy Systems Toward Zero Release of Radioactive Wastes*, Susono, Shizuoka, Japan, Nov.6-9, S 8-3

Nikitin, K, *et al* (1999), An Approach to Long-Life PWR Core with Advanced U-Np-Pu Fuel, *Ann. Nucl. Energy, 26 1021.*

Nikitin, K, *et al* (2000), Long-Life Water Cooled Small Reactor with U-Np-Pu Fuel, *10-th Int. Conf. on Emerging Nuclear Energy Systems*, Petten, The Netherlands, Sept.24-28, p.458.

Ohsaki, T. and Fujii-e, Y. (1996), Fall Meeting of the Atomic Energy Society of Japan, A-56.

Parish, T.A., Davidson, J.W.(1980), Reduction in the Toxicity of Fission Product Wastes through Transmutation with Deuterium-Tritium Fusion Neutrons, *Nucl.Technol., 47, 324.*

Rubbia, C, *et al* (1997), CERN-Group Conceptual Design of a Fast Neutron Operated High Power Energy Amplifier, *In the Status Report Accelerator Driven Systems: Energy Generation and Transmutation of Nuclear Waste*, IAEA-TECDOC-985.

Saito M., *et al*, (1998) Contribution of External Neutron Sources in Excess Neutron Generation for SCNES, *Progr.Nucl. Energy, 32, 697.*

Saito M., *et al*, (1999), Transmutation of Elemental Cesium. *Int. Conf. on Future Nuclear Systems GLOBAL'99*, Wyoming, Aug.30-Sept.2.

Saito, M., *et al* (2000,a), Fusion-Driven Transmutation of Fission Product Cesium in its Elemental Form, *J. Nucl.Sci.Techn., 37, 1024.*

Saito, M., *et al* (2000,b), Hazard of Long-Lived Spallation Products, *IAEA Workshop on Advanced Nuclear Reactor and Fuel Cycle Concepts*, USA, Argonne National Laboratory, Nov.28,-Dec.1.

Salvatores, M., *et al* (1994), A Global Physics Approach to Transmutation of Radioactive Nuclei, *Nucl. Sci. Eng., 116, 1.*

Salvatores, M., *et al* (1998), The Transmutation of Long-Lived Fission Products by Neutron Irradiation. *Nucl.Sci.Eng.130, 39.*

Sekimoto, H. and Takagi, N.(1991), Preliminary Study of Future Society in nuclear quasi-equilibrium, *J. Nucl Sci.Techn, 28, 941.*

Shimizu, A., Fujii-e,Y.(1995), Self-Consistent Nuclear Energy Systems, *Prog. Nuclear Energy* ***29,25***.

Shmelev, A, *et al* (1994), A Conceptual Study of Proliferation Resistant Fuel Cycles for LWR, *Nuclear Energetics 4, 34 (1995) in Russ.*

Stankovsky, A., *et al* (2000), Accumulation and Transmutation of Sapallation Products in Accelerator-Driven Core, *10-th Int. Conf. on Emerging Nuclear Energy Systems*, Petten, The Netherlands, Sept.24-28, p.410.

Takahashi, H. (1995), Transmutation of High-Level waste by a Charged-Particle Accelerator, *Nucl. Technol. 111, 149.*

Taube, M. (1976), The Transmutation of Sr-90 and Cs-137 in a High-Flux Fast Reactor with a Thermalized Central Region, *Nucl.Sci. Eng, 61, 212.*

Pergamon
www.elsevier.com/locate/pnucene

Progress in Nuclear Energy, Vol. 40, No. 3–4, pp. 375–402, 2002

Printed in Great Britain
0149-1970/02/$ - see front matter

PII: S0149-1970(02)00031-8

TRANSMUTATION : ISSUES, INNOVATIVE OPTIONS AND PERSPECTIVES

Massimo SALVATORES

Nuclear Reactor Directorate
Bldg. 707 CEA CADARACHE
13108 Saint Paul Lez Durance Cedex

Abstract

Since more than a decade, transmutation studies have been again a topic of wide interest and it is legitimate to ask at this point, what is known, what has been done and what should be done.

Since the motivations of national programs are often different, due to a different context, we will take for granted that transmutation is generally seen as an option for the back-end of the fuel cycle in order to reduce the burden of potential geological storages of radioactive wastes (whatever their nature).

Finally, we also acknowledge the fact that some highly respected scientists have at several occasions during this decade expressed their doubts about the value of the transmutation option. A typical example is the position expressed by Pigford and Rasmussen, reporting the results of a study for the US National Research Council (IAEA Tec DOC 990, 2/12/1997).

1. INTRODUCTION

To give a state-of the art of the transmutation studies, one could make use of international publications or proceedings of the specialized conferences that have been mushrooming in this field. Of course a significant example is the OECD-NEA state of the art report published in 1999 : "Actinide and Fission Product Partitioning and Transmutation. Status and Assessment Report".

We will have a more modest goal and hence, we will not give credit to many outstanding contributions in this field.

In fact, we will limit our analysis to a few points that we consider of special relevance. Successively, we will review some ongoing research and experimental validation studies, in order to provide a list of relevant expected results, which could have impact to shape (or re-shape) future programs.

In this perspective, we will also indicate which are in our opinion the "missing" experiments or studies, the absence of which could jeopardize the process of decision-making.

This analysis will of course be made mostly from a "European point of view".

The nuclear energy "environment" is a changing one, and it is of interest to review some activities / studies / concept proposals, which are not strictly speaking in the "transmutation" domain, but which could have an impact on the conclusions which could be drawn on the potential role of transmutation.

Finally we will attempt to summarize a list of open questions. Of course, this paper does not deal with chemistry issues, but rather with reactor and fuel cycle technology, since the physics of transmutation is today well understood (see for example Ref. 1).

2. WHERE ARE WE ?

Transmutation is of course an R and D endeavour. The potential "customers" of such R and D can be found more in the society at large and its political representative bodies more than in industry. By the way, it is not evident that even a fundamental feature of transmutation (i.e. the need to reprocess the fuel) is clearly understood in that context.

Moreover, there is an inherent difficulty to evaluate the real status of the research in the transmutation field with respect to the potential utilisation of the technology in a reasonable time horizon within stated performance and cost criteria.

In view of this difficulty, we have preferred to single out some specific scientific results, which could characterize our present understanding of transmutation and its use as an option for the fuel cycle.

2.1. The IFR concept and the homogeneous recycling

The IFR concept (Ref. 2) is still the most outstanding example of an "inherently transmuting" concept in the so-called "homogeneous" recycling mode (see figure 1 and Ref. 3). The IFR concept can be seen as an energy producing system capable to recycle Pu and minor actinides (MA), to reach equilibrium, both stabilizing the Pu and MA mass flows, and sending to the wastes only a very small fraction of the radiotoxic isotopes. This fraction is of the order of 0.1% or less, according to the announced performances of the pyrochemical process involved, which has still to be demonstrated at large scale in the frame of the transmutation application.

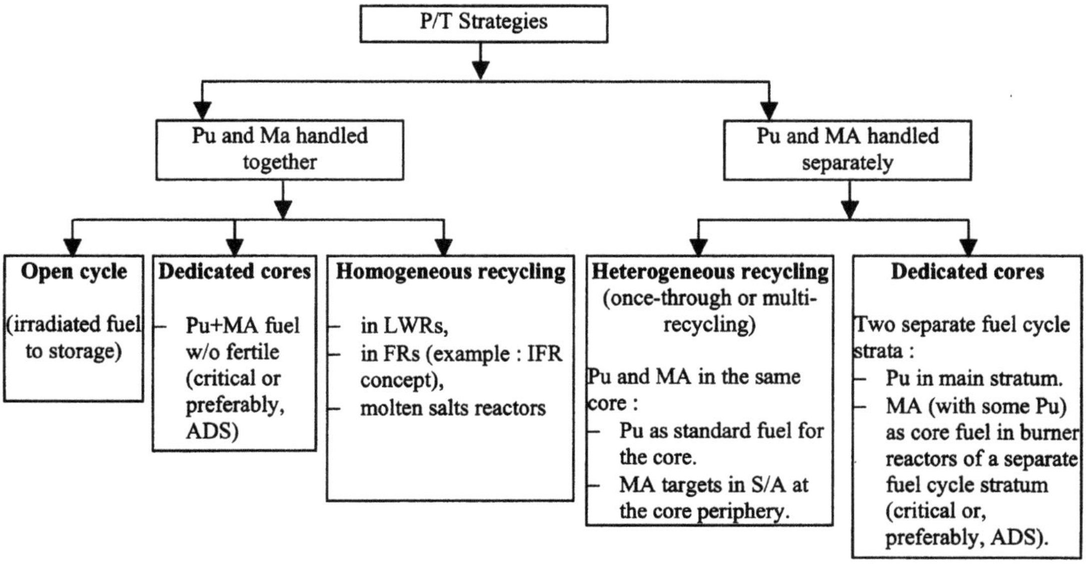

Remark : If LLFP management is required, they can in principle be handled in the different scenarios as targets to be irradiated at the periphery of the different core types.

Figure 1 : Pu and MA management in P/T strategies

The appealing aspects of the IFR concept in the frame of transmutation are :

- the concept is mainly designed to produce energy, making an optimised use of resources and using a robust reactor and fuel cycle layout.
- The fuel cycle does not imply the separation of Pu and MA.
- The concept can accommodate in principle several options in terms of reactor size, reactor coolant, waste-forms, etc...

In general, the homogeneous recycling has equivalent performances for whatever the type of fuel in the fast reactor. In fact, if the losses at reprocessing are assumed to be of the order of 0.1 %, the homogeneous recycling allows to reach a reduction of the potential radiotoxicity with respect to the open cycle scenario of a factor of 200 and more, and this over all the time scale ($10^2 \rightarrow 10^6$ years), (Ref. 4). This reduction is such that the radiotoxicity in a deep geological storage becomes comparable to that of the initial uranium ore, after less than a thousand years. However, the consequences on the fuel cycle have to be taken into account (see table 1), and their impact evaluated. With respect to this last point, it has to be mentioned that recent studies performed at CEA-France, allow to envisage also a multirecycle of both Pu and MA in PWRs. If a specific core assembly design can in principle allow acceptable core performances, the impact on the fuel cycle (e.g. at fuel fabrication) is much stronger of what indicated in table 1 for the homogeneous recycle in a FR, and probably not acceptable.

Table 1
Consequences on the fuel cycle of MA recycling in FR (a)
Variation expressed as ratio with respect to the corresponding values for the reference case : PWR - MOX Pu content 12%, taken as 1

		PWR - MOX (12% Pu content)	FR : Multi-recycling of Pu, Am, Cm in the homogeneous mode	FR : Multi-recycling of Pu, Once - through irradiation of Am + Cm targets (d)
Fuel Fabrication	Activity	1	0.1 (c)	0.1
	Heat due to :			
	α	1	0.5	1.7
	β	1	0.2	0.4
	γ	1	1.5	8.7
	Neutron source	1	30	104
Spent fuel Reprocessing (b)	Activity	1	0.1	0.10
	Heat due to :			
	α	1	0.2	0.11
	β	1	0.5	0.09
	γ	1	0.2	0.06
	Neutron source	1	0.4	245 (e)

(a) Oxide fuel, EFR type
(b) 5 years cooling time
(c) This value means that, when fabricating fuel as obtained at equilibrium during Pu+Cm+Am multirecycling in a FR, the activity is 10 times less than the activity of the MOX fuel (Pu content : 12 %) for a PWR.
(d) Heterogeneous recycling total fission rate : 90%
(e) Effect due mainly to ^{252}Cf

2.2. Heterogeneous recycling and its potential limitations

An option has been explored, mainly in Europe and in particular at CEA in France (Ref. 5) and at JNC in Japan (Ref. 6), to perform the transmutation of MA in the form of targets to be loaded in critical cores of a "standard" type. The mode of recycling has been called "heterogeneous" (see fig. 1), the potential advantage being to concentrate in a specific fuel cycle the handling of a reduced inventory of MA (separated from Plutonium). The major obstacles to that approach are :

- The very high irradiation doses needed to fission a significant amount of MA (which implies very high damage rates).
- The need to separate Am and Cm from Pu and to keep them (Am and Cm) together, in order to reach high values (~ 30) for the radio-toxicity reduction (Ref. 3).
- The need to load the MA targets in a very large fraction (~ 30 ÷ 50 %) of the reactor park,-possibly made of fast reactors, due to their favourable characteristics for this mode of recycling (high fluxes, which can be easily tailored in energy to increase fission rates).
- Consequences on the power distributions and their evolution with time.

In any case, the consequences on the fuel cycle are relevant, if one wants to reach a factor of radiotoxicity reduction of ~ 30 ÷ 40 (see table 1 and Ref. 7).

2.3. Dedicated systems

Making again reference to the scheme of fig. 1, a possible approach to keep the MA fuel cycle and the transmutation technology separated from the electricity production, is the one which calls for the use of "dedicated" cores, where the fuel is heavily (> 30%) loaded with MA, the rest being, e.g., Plutonium (the ratio Pu/(Pu+MA) being < 0.3). Work performed at JAERI in Japan (Ref. 8) and in Europe (Ref. 9), has shown that critical "dedicated" cores can have difficulties, related to the safety parameters degradation, due in particular to a very low delayed neutron fraction (< 0.2% $\Delta k/k$) according to the fraction of Pu (and its isotopic vector) in the core, and to a reduced Doppler effect. This characteristic has indirectly helped to promote the Accelerator-Driven Sub-critical Systems (ADS) and the so-called "double strata" fuel cycle concept (Ref. 8).

It should be said, however, that a convincing, scientifically based comparison of a critical and a corresponding (i.e. same fuel, same power, same coolant) sub-critical core is still lacking, in particular in terms of safety performances for a well defined range of MA/Pu ratios and for a variety of Pu isotopic vectors. As far as transmutation performances, obviously both critical and sub-critical systems (loaded with the same fuel and for the same power) are equivalent. The "dedicated" cores approach implies that ~ 3 - 5 % of a power park (in other term a "support" ratio of ≈ 20) is needed to stabilize MA and Pu mass flows (see for example paragraph 3.3.3).

The "dedicated" core approach has been a powerful incentive to support ADS research studies.
It is also worth to notice, that the concept of a "dedicated" core applies both to a strategy of reduction mainly of MA or of reduction of Pu <u>and</u> MA (see fig. 1). In fact, Pu and MA can be kept together and their inventory reduced drastically, using a fuel without "fertile" (i.e. Uranium) support. In this case (where the Pu/(Pu + MA) ratio is of the order of 0.7 ÷ 0.8), ADS can still offer a valuable core concept.

2.4. Long Lived Fission Products (LLFP)

The transmutation of LLFP has been always a controversial issue. A better understanding of the list of potential candidates for transmutation has been achieved (Ref. 10). However, there is no doubt that the task of transmutation, even if limited to Tc-99 and I-129 (Cs-135 being out of question) is a formidable one, both in terms of fuel cycle and of neutron availability requirements. In fact, to transmute Tc-99 and I-129, the nuclear power park should provide 0.1 ÷ 0.2 neutrons/fission just for that task and, moreover, the relatively slow pace of transmutation even in the most optimised neutron environment (i.e. high flux and well-tailored thermalized spectrum), implies the presence in the fuel cycle at equilibrium of very large amount of these fission products (Ref. 11). The use of ADS can slightly help, in terms of neutron availability, but the decrease of the support ratio (unless I-129 isotopic separation is envisaged), makes all the scenarios of LLFP transmutation very unlikely.

3. WHAT IS BEING DONE ?

For the four areas indicated above, that we have chosen as relevant to assess the potential of the transmutation technology, it is important to review ongoing research and experimental validation studies, in order to evaluate the time horizon for their possible implementation.

3.1. The homogeneous recycling and the IFR concept

Unfortunately, after a period of intensive development, the IFR program has been stopped. No significant activities have been left alive on the metal fuel development and the pyroprocess, with the exception of activities started at CRIEPI (Japan) and now extended to TUI - Karlsruhe - These new activities concern the fuel reprocessing too (see Ref. 12). In particular, an experiment (METAPHIX) is planned, in order to irradiate metal fuel pins, loaded with MA and rare earths (RE) (Ref. 13 and fig. 2).

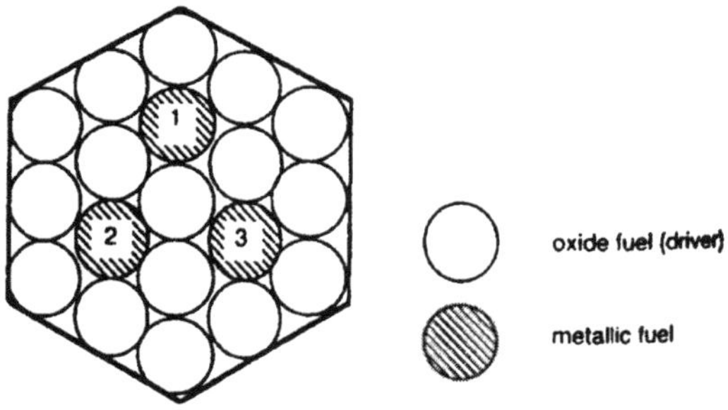

Figure 2 : Arrangement of fuel pins in a rig for the METAPHIX experiment (CRIEPI - TUI) (from ref. 13)

Nine metallic fuel pins are prepared for the METAPHIX irradiation study : three pins of UPuZr, three pins of UPuZrMA2%RE2%, three pins of UPuZrMA5%, and UPuZrMA5%RE5%. They are planned to be inserted in the positions 1, 2 and 3, respectively, in the rig. Three rigs consisting of three sample metallic fuel pins and sixteen driver oxide pins will be prepared. Three rigs correspond to three different values of the burnup : 1.5, 5. and > 10 %, respectively.

As far as homogeneous recycling in standard oxide fuels, some experimental knowledge has been obtained with the SUPERFACT experiment (Ref. 14). More will come from experimental programs conceived at JNC and which should take place in JOYO beyond 2003. On the contrary, no experience exists on MA-loaded oxide fuels in standard light water reactors.

Finally, it has to be noted that the present revival of interest for pyroprocessing techniques, has been largely motivated by the relevance of these techniques to handle "hot" fuels, like those which are foreseen for transmutation. It has to be noticed that a modest but significant program, PYROREP, has been launched as a European Union contract for the 5th Frame Work Program.

3.2. Heterogeneous recycling

Apart from conceptual studies at JNC and CEA, experimental activities have been launched in Europe (e.g. the EFTTRA collaboration) and some useful indications have been gathered (Ref. 15). The EFTTRA-T4 and T4-bis experiments concern ^{241}Am, at a 12% volume fraction, inside a matrix of $MgAl_2O_4$, for a maximum fission rate of 28%. The swelling due to the decay of the Cm-242 produced by neutron capture, has been relevant, and triggered further research on the form of inert matrix / actinide fabrication (e.g. micro-dispersion versus macro-dispersion). It is worth to notice that experiments performed up to now, did not cover the presence of ^{243}Am and the presence of Cm.

Further experiments are planned, and in particular the ECRIX experiment, which should take place in PHENIX and the CAMIX and COCHIX experiments (J.C. Garnier, CEA, Private communication), also planned in PHENIX. The CAMIX experiment will provide informations on "micro-dispersion" of a $(Am, Zr, Y)O_{2-x}$ compound in MgO and COCHIX informations on the same compound "macro-dispersed" in MgO or $(Zr_{0.6}\,Y_{0.4})O_{1.8}$. All 3 experiments are planned to reach a fission rate equivalent to 30 at %.

A significant global experiment is presently worked-out in the frame of the collaboration between MINATOM (Russia) and CEA (France) with the participation of FZK and TUI-Karlsruhe, as partners of CEA. In this experiment (AMBOINE), Am targets $Am\,O_2 + UO_2$ and $Am\,O_2 + Mg\,O$ should be fabricated at RIAR according to the VIPAC process. These targets should be irradiated in BOR-60 and reprocessed by pyroprocess after irradiation again at RIAR, providing in that way a full validation of the whole fabrication - irradiation - reprocessing cycle for CERCER targets (S. Pillon, CEA - Private communication).

3.3. Dedicated systems

3.3.1. Fuels for dedicated systems

For both critical and sub-critical dedicated cores, the major issue in the path towards feasibility demonstration, is the fuel development. Many candidates have been considered (see for example fig. 3), but limited experimental work has been done, in order to characterize the basic properties of these potential fuels, their fabrication processes and their behaviour under irradiation.

Metal fuels

- Need to improve thermal properties ⇒ add non-fissile metal with high melting point (e.g. Zr) ⇒ Pu-MA-Zr alloy
- However : mutual solubility of Np and Zr ?

Oxide fuels

Mixed transmutation oxides as a logical extension of MOX.
However : smaller margin to melting (low thermal conductivity)

Nitride fuels

- Good thermal behaviour
- However : need to enrich in ^{15}N
- Lower stability against decomposition at high temperatures

Composite fuels : the role of Zr

Ad-hoc "tailoring"
- MgO + (Zr, An) O_{2-X} (CERAMIC - CERAMIC)
- Zr + (Zr, An) O_{2-X} (CERAMIC - METALLIC)
- Zr + (An, Zr) alloy (METAL - METAL)

However : fabrication can be difficult (also : size and distribution of the disperse actinide phase).

Coated particle fuels

Special form of composite fuels. However in the case of fast spectra, little is known on potential candidates (TiN ?)

⇒ A generic problem : the high He production under irradiation.

Figure 3 : Dedicated Pu + MA fuels
(adapted from Ref. 16)

A common feature for these fuels is to be fertile-free, or, at least, "U-free", since Th is sometimes considered as an acceptable support, in particular for strategies that promote the replacement of the U-cycle with the Th-cycle.

Well structured programs for fuel development are missing in practically all the major transmutation programs. A significant exception is the JAERI program, focussed on nitride fuels.

The CONFIRM project, sponsored by the European Union 5th Frame Work Program is also devoted to nitride fuels ((Pu, Zr) N and (Am, Zr) N). The project aims to the fabrication, characterisation and irradiation of these fuels, and addresses also the issue of ^{15}N enrichment.

Finally, we recall the initiative of the European Technical Working Group (TWG) on ADS, chaired by professor Rubbia, which has set up an ad-hoc task force an Fuel Fabrication and Processing, in order to produce a state of the art report and an agreed work plan in the frame of a road-mapping towards ADS deployment (Ref. 16).

As far as reprocessing, the situation is, obviously, not fully satisfactory either. The use of dedicated fuels imposes their reprocessing in all envisageable schemes. Again, the program proposed by JAERI, includes laboratory scale experiments of reprocessing, together with a flow diagram of a process to enrich in ^{15}N the fuel (Ref. 17).

The present European efforts are reviewed in Ref. 16 and Ref. 12. A program for pyrochemistry development is also being set up in France at CEA. The rationale for it can be found in the report : "Assessment of Pyrochemical Processes for Separation / Transmutation Strategies : Proposed Areas of Research "CEA/PG - DRRV/Dir/00-92, March 2000".

3.3.2. ADS systems

A special case is the research activity in the ADS domain.

In the last two years, relevant initiatives have taken place. The ATW Road-mapping in the US (Ref. 18), should give rise to a focussed program in very near future. The joint KEK-JAERI project, has given a place to ADS development in Japan, in the frame of a multipurpose facility (Ref. 19).

In France, the GEDEON program (Ref. 20) has gathered a large community of physicists around the basic physics items of research for ADS (nuclear data, spallation physics, subcritical core neutronics, materials, but also pyrochemistry, molten salts, thorium and system studies).

In Europe the Technical Working Group mentioned above has been established. In that frame two concepts for ADS are being studied (see fig. 4) and a rationale is emerging for a "step-by-step" validation and demonstration of the ADS concept (see fig. 5) and its waste transmutation potential, in the frame of a specific road-mapping, which is being finalized at present. Few comments will be made in what follows on some ongoing experimental steps, like the MEGAPIE project and the MUSE program. Finally, the European Union is supporting a number of projects, in the frame of the 5th and Framework Program.

The issues and programs related to high power proton accelerators (HPPA), although essential, will not be dealt with here. We only remind the R and D work that has been initiated on the topic of "accelerator reliability" and "beam trips", which have been the subject of two NEA-Nuclear Science Committee workshops (MITO 1998, Aix-en-Provence 1999).

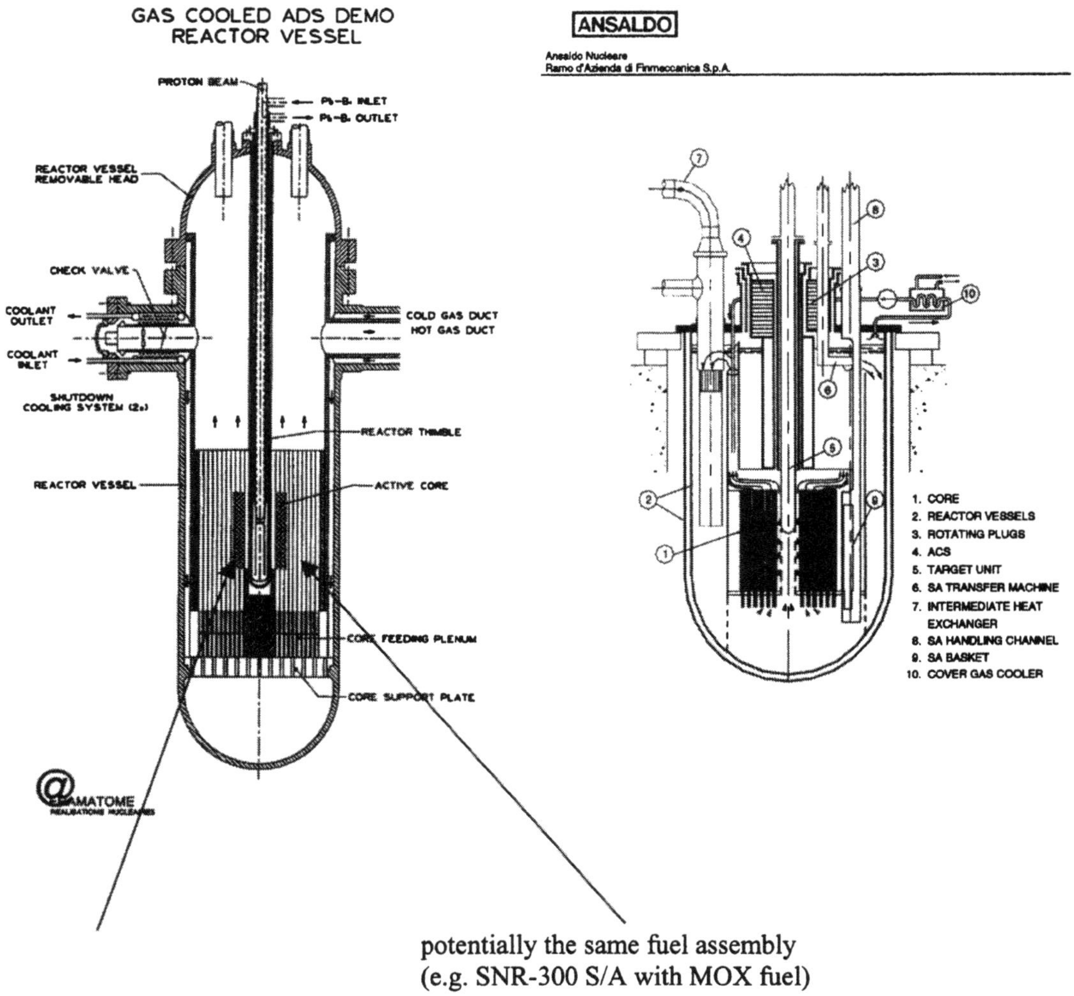

Figure 4 : Sketch of ADS, liquid metal cooled (right) and gas cooled (left) (not to scale), representative of the European EADS proposals (ANSALDO and FRAMATOME)

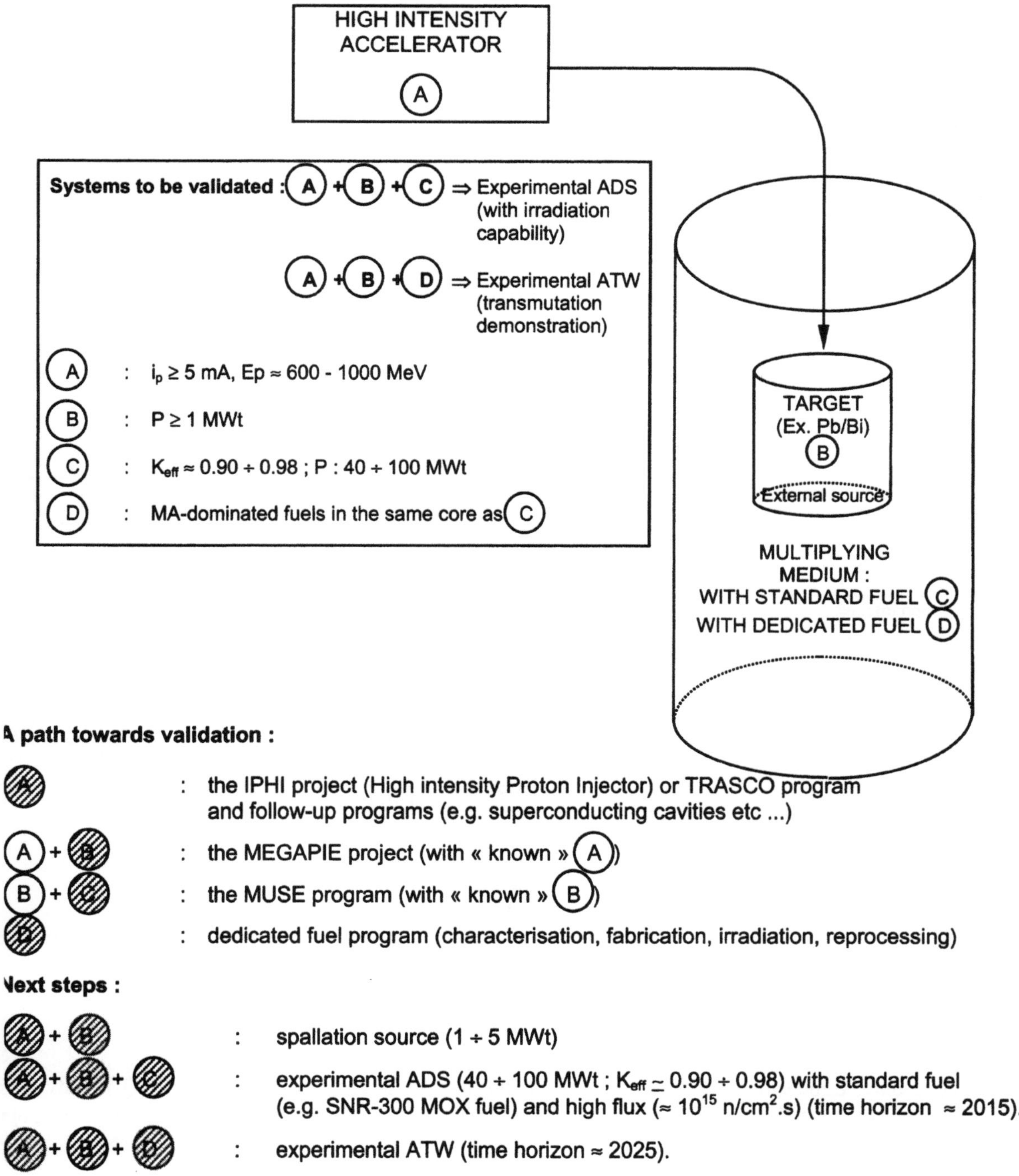

Figure 5 : A step-by-step approach to the validation and demonstration of the ADS concept

3.3.2.1. The MEGAPIE project (Ref. 40)

MEGAPIE is an international (CEA, PSI, FZK, CNRS France, ENEA, SCK-Mol. JAERI will join soon) experiment to be carried out in the SINQ target location at the Paul Scherrer Institute in Switzerland and aims at demonstrating the safe operation of a liquid metal target at a beam power in the region of 1 MW. The minimum design service life will be 1 year (6000 mAh).

The target material will be the PbBi eutectic mixture. Existing facilities and equipment at PSI will be used to the largest possible extent. In fact, the MEGAPIE target will be used in the existing target block of SINQ.

A vertical cut through this target block and parts of the proton beam line is shown in fig. 6.

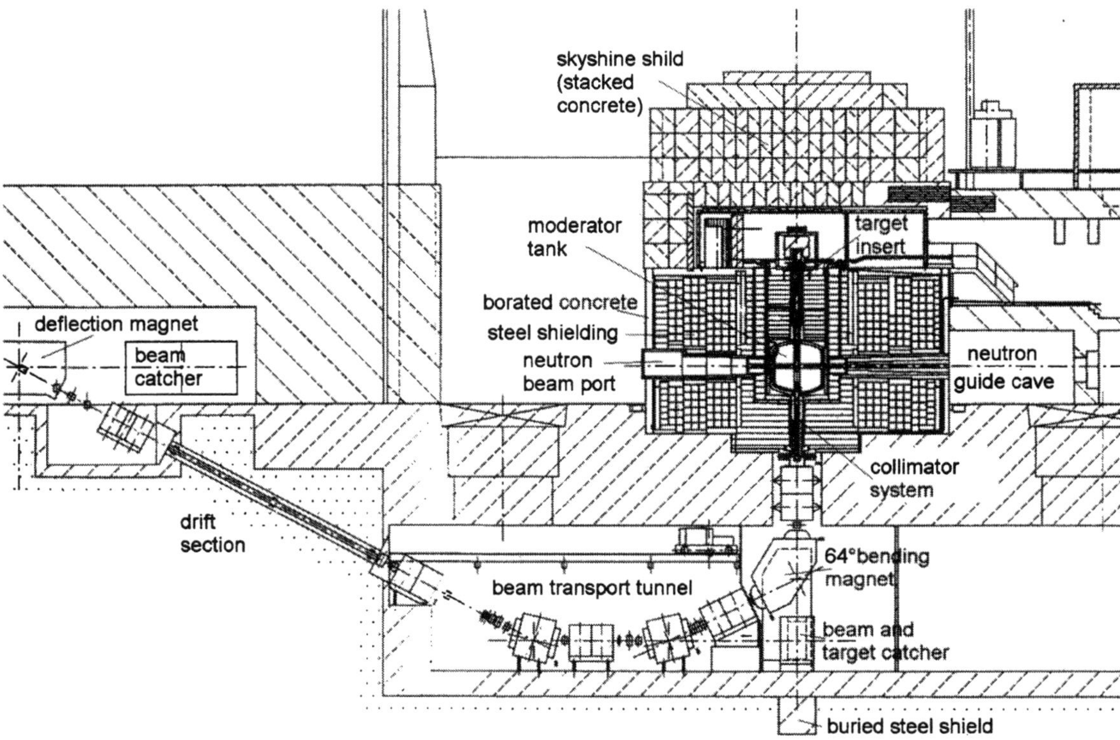

Figure 6 : Vertical cut through the target block and part of the proton beam transport line of SINQ

The target's outer dimensions must be such that it fits into the target position of the SINQ facility, the existing target exchange flask including its contamination protection devices and the existing target storage positions.

The target will be designed for 1MW of beam power at a proton energy of 575 MeV, i.e. a total beam current of i_p : 1.74 mA.

It is also important to realise that the stability of beam delivery cannot be guaranteed at all times. The MEGAPIE heat removal system must be able to cope with frequent short beam trips and occasional unstable operation, up to days long shutdown periods.

A sketch of the MEGAPIE spallation target is given in figure 7.

The major objectives of the MEGAPIE Initiative are :

- full feasibility demonstration of a spallation target system,
- evaluation of radiation and damage effects of structures and beam window in a realistic spallation spectrum,
- effectiveness of the window cooling under realistic conditions,
- liquid metal / metal interactions under radiation and stress,
- post irradiation examinations (PIE),
- demonstration of decommissioning.

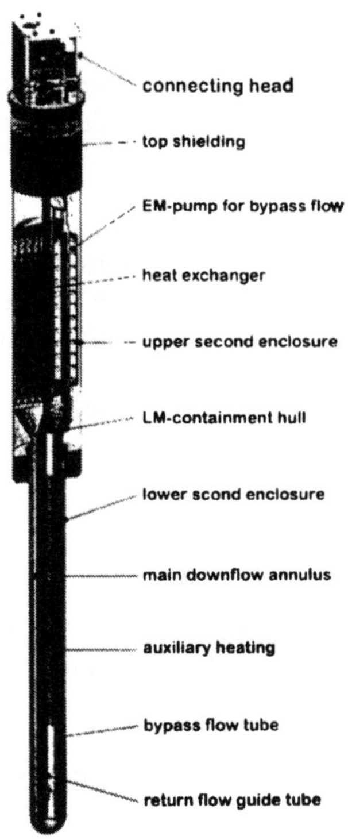

Figure 7 : Sketch of the 1 MW exploratory liquid lead-bismuth spallation target MEGAPIE

It has to be reminded that two European Union contracts, established in the frame of the 5th FWP (Ref. 21), SPIRE (material irradiation) and TECLA (physico-chemical properties of lead alloys : corrosion...), provide a relevant R and D back-up to the MEGAPIE project. Moreover, experimental laboratories have been launched in support of these activities (like the KALLA laboratory in FZK-Karlsruhe) or are re-oriented (like the ENEA laboratory in Brasimone : the CIRCE loop).

3.3.2.2. The MUSE experiments

The MUSE experiments, launched in 1995 (Ref. 22), provide a simulation of the neutronics of a source-driven sub-critical system, using the physics characteristics of the separation of the effects due to the presence of an external neutron source from the effects of the neutron multiplication. In fact for a wide range of sub-criticality values (e.g. Keff : 0.9 ÷ 0.99) the space dependence of the energy distribution of the source neutrons is quickly (in approximately one mean free path) replaced by the fission-dominated neutron energy distribution.

In practice, external known neutron sources have been introduced at the centre of a sub-critical configuration in the MASURCA reactor. The more recent of these experiments is made of a deuton accelerator and a target (deuterium or tritium) at the centre of a configuration, where actual target materials (like lead) are loaded, to provide the neutron diffusion representative of an actual spallation target (see fig. 8 and ref. 23). The neutrons issued from (d,d) and (d,t) reactions provide a reasonable simulation of the spallation neutrons, in terms of energy distribution (see fig. 9).

Static (e.g. flux distributions, spectrum indexes, importance of the source neutrons) and kinetic parameters (e.g. time dependence of neutron population, effective delayed neutron fraction, with appropriate weighting, etc ...) have been or will be measured (see e.g. ref. 23). Subcriticality itself, is measured with static or dynamic techniques.

Finally, the proposed MUSE-4 experiment start-up procedure i.e. 1) critical configuration with accelerator hole but no beam, 2) sub-critical configuration with accelerator hole but no beam, 3) same, but with beam on, allows to establish a precise reactivity scale in step 1, which can be used both to calibrate eventual control rods and to measure in a standard way (e.g. with the Modified Source Multiplication, MSM, method) the level of sub-criticality of steps 2 and 3.
The MUSE-4 experiments, described elsewhere (Ref. 23), are also partly supported by a European Union contract for the 5th Framework Program.

3.3.2.3. Streamlining basic physics experiments

The ADS research development has also motivated a significant number of experimental activities in the field of spallation physics and nuclear data measurement and evaluation (mainly actinides and LLFP) in general countries.

A major experiment takes place at GSI-Darmstadt (Germany), defined in order to gather much needed informations on spallation product yields and distributions in (A, Z) (Ref. 24)..

If present uncertainties in nuclear data allow to make reasonable pre-conceptual design assessments, future detailed studies will require more accurate data, with drastically reduced uncertainties. The relevant sensitivity studies have started (see ref. 25), but they have not yet tackled in a satisfactory way the problem of the accuracy needs in the intermediate (i.e. 20 MeV ≤ E ≤ 200 MeV) energy range.

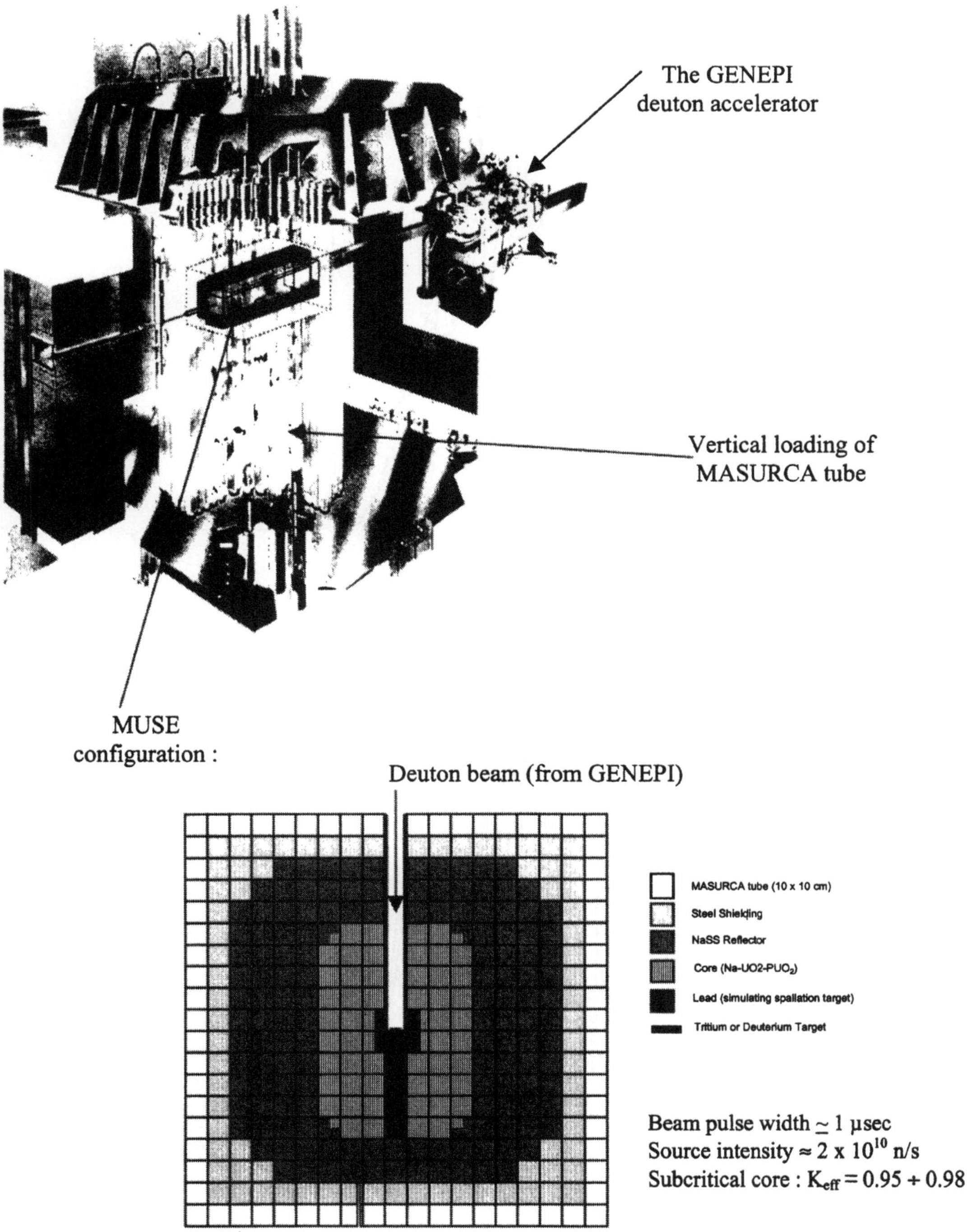

A subcritical MUSE-4 configuration

Figure 8 : The MASURCA installation for the MUSE program

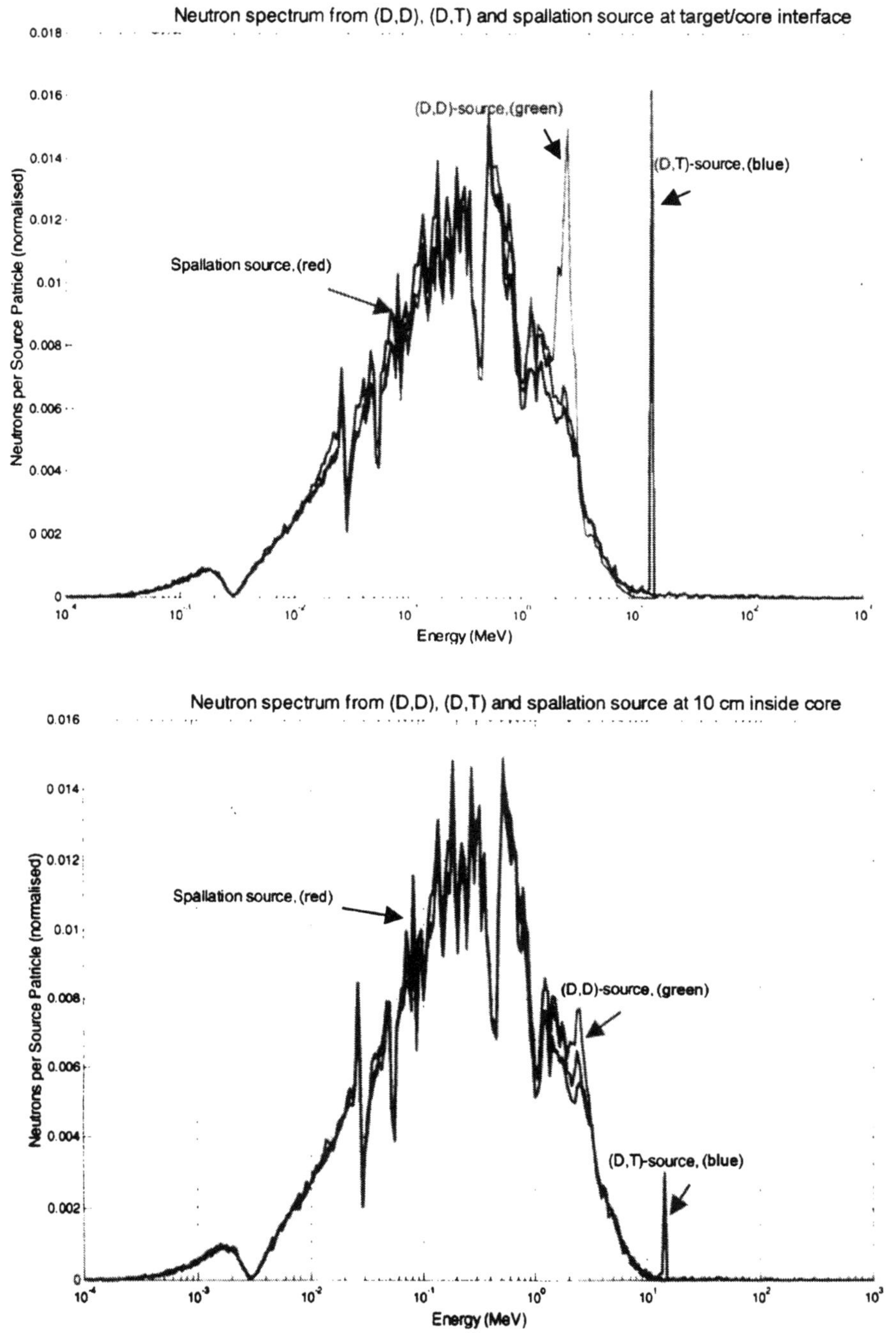

Figure 9 : Comparison of the neutron spectra obtained with (D,D) and (D,T) neutron sources (as in MUSE experiments), with the reference spallation source, in the same configuration (P. Seltborg, private communication)

3.3.3 Scenarios studies

Scenario studies have allowed during this decade to get a global picture of the transmutation potential, mass flows at equilibrium, and consequences on the power park structure.

In the illustration of fig. 10 (ref. 26), the same type of ADS is used in order to transmute MA (double strata approach), or Pu+MA (double component type of power park, ref. 27).

The fractions of ADS in the park at equilibrium are shown (respectively 3.4 % and 16 %), and also the MA and Pu yearly mass flows, including total losses towards a deep geological storage.

3.4. LLFP

In this area, after the performance of irradiation experiments on Tc-99 and Iodine (Ref. 28), not much is being done apart from conceptual studies, that underline the need to use high fluxes and thermalized spectra, like in the so-called "Leakage-with-Slowing Down (L.S.D.) approach (Ref. 11).

Projects related to the transmutation of ^{90}Sr and ^{137}Cs have finally been abandoned everywhere.

Cs transmutation is more and more considered as non realistic (even with isotopic separation). Other activation products have been mentioned as candidates for transmutation, but the inherent difficulty of high neutron-consuming processes has discouraged further experimental programs.

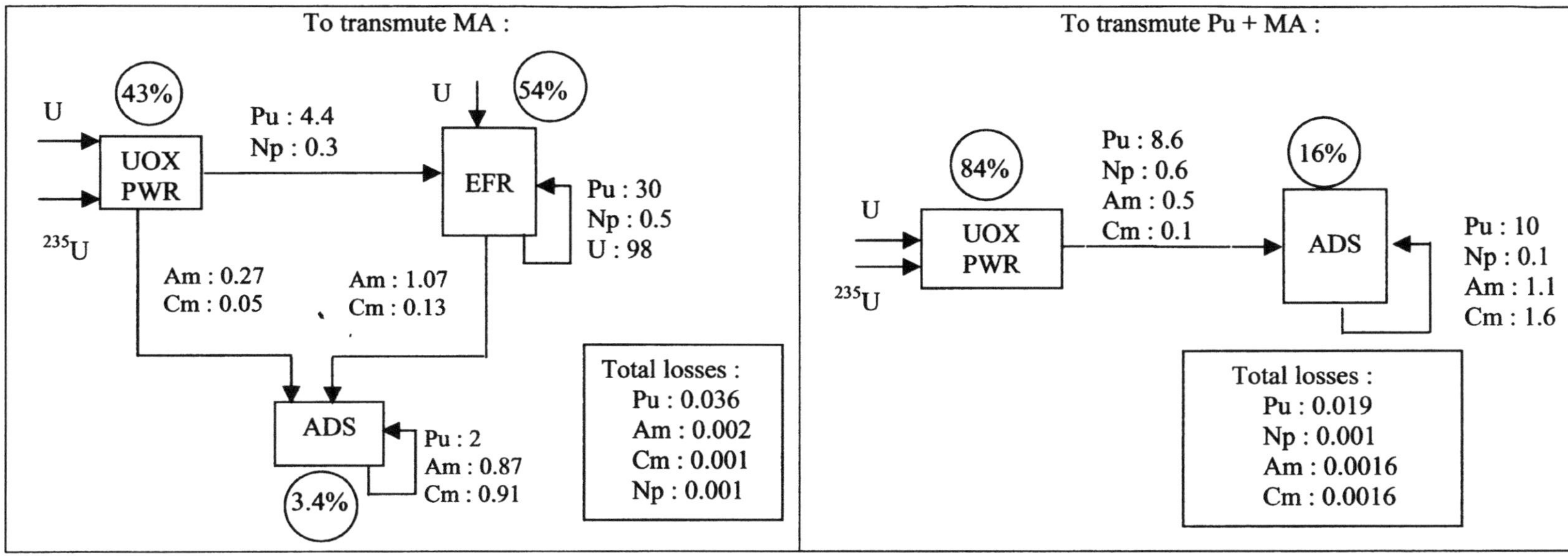

(%) : fraction in the power park

UOX - PWR : 4.9% ^{235}U enrichment. BU : 60 GWd/t

Cooling time : 5 years

Ageing before irradiation : 2 years

Losses to the wastes : U, Pu, MA = 0.1%

Figure 10 : Scenarios at equilibrium for a 60 GWe Park - Mass flows / year (t) (only TRU)

Double strata : to transmute MA (separated from Pu)

Double component : to transmute Pu + MA (non separated)

Same type of subcritical ADS (gas-cooled, particle-fuel)

Power : 1500 MWt

Initial sub-criticality : $k_{eff} = 0.98$ ($i_p \approx 17$ mA, $E_p = 1$ GeV)

4. MAJOR MISSING POINTS

The analysis of the ongoing activities and of their relevance allows to attempt an indication of areas where further efforts are needed, in order to consolidate the present knowledge and to provide elements to judge feasibility.

a. It seems that what is probably still needed with a high priority is a (re)assessment of the criteria to judge the performance of transmutation systems. Transmutation as a waste management option, is indissolubly related to a constant or expanded use of nuclear energy and its impact should be evaluated on the full fuel cycle (cost and licensing of new installations, doses to the workers, secondary wastes, acceptability, ...).

b. Experimentation about fuels is a priority. No concept can be considered seriously, if the appropriate fuels are not defined, which means characterized, fabricated, irradiated and reprocessed. Now, very limited facilities are available to deal with MA fuel fabrication and reprocessing (wet or dry routes) and their work load is already very demanding. Moreover, the problem of Cm has been up to now somewhat "forgotten" and, on the contrary, it can be crucial to define an optimum transmutation strategy or even to identify potential "show stoppers".
   To start with compounds and fuel basic properties assessment should be a priority. An international coordination and share of work should be envisaged.

   Also, irradiation tools with fast neutrons will be dramatically reduced in the coming years with the remarkable exception of the JOYO reactor (at least up to ~ 2015). Again, an international initiative could be envisaged to harmonize programs and to allow the best use of existing resources.

   Reprocessing of irradiated fuels should be foreseen as an essential step of any program on fuels, for homogeneous recycling of fuels, heterogeneous recycling of targets, or dedicated fuels. The priority, in the opinion of the author is with fuels for homogeneous recycling and with dedicated fuels, if the double strata, (or the "double component") approach is accepted.

c. In the case of heterogeneous recycling, the feasibility of a fission rate > 90% should be experimentally verified, and that demonstration should be made in the case of a target containing both Am and Cm. In fact, multi-recycling of targets should be avoided.

d. It is more and more evident that transmutation studies could not necessarily require the separation of individual MA and MA from Pu. Quite the opposite option seems to be more attractive (Ref. 3). In that respect, partitioning / conditioning strategies (Ref. 29) could represent an option to be investigated, and which could justify partitioning by itself.

e In the case of ADS, an Experimental ADS (XADS) realisation at the 2015 horizon, is a need, in order to prove the technology at a significant scale. The priority is the engineering concept (i.e. component coupling, control, reliability and licensing) validation. The double role of an XADS as a facility to validate the concept but also able to provide the appropriate fast neutron field for advanced fuel irradiation at high damage rates, is a strong point to be made. The present status and the necessary research in the accelerator field is not reviewed here, although this is an essential issue since the expected performances of the "dedicated" high power accelerators are very demanding in terms of reliability and availability.

f. LLFP transmutation research, if at all needed, should concentrate on a realistic approach for I-129 handling, if any. Reliable target materials and high transmutation rates should be the priority goals in this field. Since one -through transmutation is hard to be envisaged, the recovery of I in the irradiated target and its reprocessing, should also be the object of research.

5. A CHANGING ENVIRONMENT

The research in transmutation experienced a revival in the mid-eighties, essentially in the context of waste management within programs which gave a definite value to Plutonium and which implied the reprocessing of irradiated fuel.

The transmutation approach was successively identified in some countries with the approach to Pu elimination (both weapon and civil Pu) (see fig. 11). Reactor physics problems were indeed very similar. In that way, the interests of the two previously separated communities (i.e. Pu = resource versus Pu = liability), were somewhat federated in particular in terms of fuels development and their reprocessing by pyroprocesses.

The third step in the evolution of the transmutation approach, is underway at present, since the objectives of a "Generation IV" or, in general, the objectives of a future nuclear power development (beyond the horizon 2030 - 2050) are being globally rediscussed.

Transmutation and waste minimisation are then part of the potential criteria to define future energy systems (reactor plus fuel cycle).

In this changing environment, it can be useful to single out some concepts or research areas, which can have impact on the future of transmutation studies.

5.1. Evolutionary reactor concepts

A few well worked-out reactor concepts have emerged in the last few years, which, besides attractive safety and economics characteristics, have a potential to be "inherent" MA transmuters in the homogeneous recycling mode.

Besides the IFR concept, often quoted as a paradigm in the present paper and the Energy Amplifier proposed by C. Rubbia, we can remind the BREST lead-cooled fast reactor concept developed in Russia (Ref. 30), the CAPRA reactor in France (Ref. 31), the SCR (Super-critical Water Cooled) concept, developed in Japan (Ref. 32), but also the APA concept (Ref. 33), despite the fact that it concerns mostly an innovative assembly design.

In particular, the nitride fuel foreseen for the BREST reactor, favours the MA transmutation by neutron spectrum hardening. Using a pyrochemical process, it is possible to envisage for this fuel, by multirecycling, the transmutation of the actinide produced during irradiation. This mode, close to the one indicated for the IFR concept, has the same advantages and of course similar drawbacks, in particular due to the build-up of spontaneous fission neutron emitters (Cm isotopes, Cf isotopes, see paragraph 2.1 and table 1).

1st phase (mid-80's to beginning of 90's)

The revival : an answer to
- a "national" challenge (OMEGA Program)
- a parliament law (SPIN Program)

Key words : Nuclear power should develop (in a better public acceptability environment). The spent fuel is reprocessed, the use of FR is foreseen
⇒ The "Camelot" period of the Transmutation community !

2nd phase (mid-90's to end of the 90's)

The "Plutonio-clasts" join the transmutation community (and vice versa)
Pu versus MA : it's a question of proportions in the fuel recipe !
The ADS emerges as an "ecumenical" tool (in the meantime, SUPERPHENIX is shut down...)

3rd phase (the new millennium)

The "holistic" approach : the Pandora box opens.
The future competition requires to re-invent a nuclear reactor and an associated fuel cycle with redefined constraints :
- economy,
- safety,
- non-proliferation,
- simplified fuel cycle (for acceptability and economy)
- minimised wastes (for acceptability)

The "transmutation" community joins "GENERATION IV" !

Figure 11 : TRANSMUTATION
the evolution of a concept

Finally, interest in gas-cooled fast reactors has been renewed, in particular to keep open the fast reactor (FR) option, due to FR flexibility with respect to resources utilisation and their potential for waste minimisation. In view of the "political" opposition to Na as coolant in some countries, the gas cooling is being revisited.

The potential of any fast reactor, whatever the fuel type (oxide, nitride, metal) and whatever the coolant (Ref. 34), indicates that the priority for GCFRs is to design a viable reactor (in terms of safety) with a realistic fuel form (e.g. particle, avoiding as far as possible graphite) for which no firm candidate has been proposed up to now.

5.2. Molten salts

Molten salt reactors, besides their specific interest as energy producing systems, have also a number of perceived advantages for transmutation, in particular :

- high burnup potential (up to 600 MWd-t^{-1}) limited only by absorption due to fission products, minimizing the quantity of fuel to be reprocessed (a few litters of salt per day).
- Actinide losses minimized in the ultimate wasteform (0.01-1% of the actinide inventory).

Molten salts characteristics result also in increased flexibility :

- continuous input of purified salt fuel and output of irradiated salt: by adjusting the TRU concentrations, reactivity can be controlled without the use of poison or fertile material.
- Long-lived fission products (Zr, I, Tc, Cs, ...) can be added directly to the salt with no detrimental effect on its physicochemical properties.
- In the case of Thorium cycle, fertile Thorium can be added to the salt to fabricate ^{233}U rather than TRUs ; the ^{233}U can be "quickly" extracted without ^{232}U.

Recent studies on molten salts concepts (Ref. 35, 36) point out the application to MA and Pu elimination, but also indicate the way towards improved fuel cycle and waste management scenarios (e.g. the TASSE system, Ref. 37 and figs. 12-13).

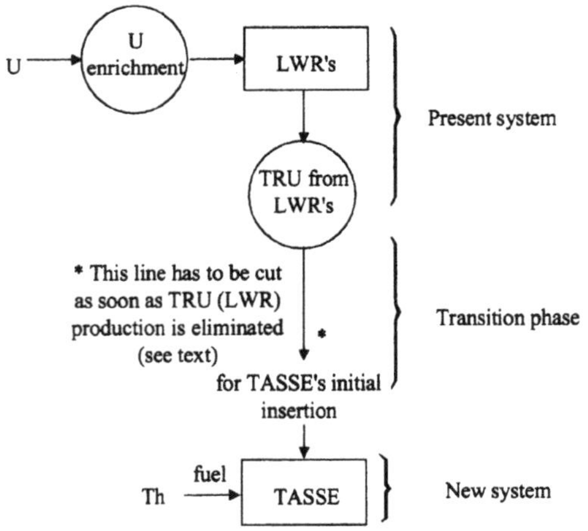

Figure 12 :
From LWRS AND U-cycle to TASSE with natural Th feed (ref. 37)

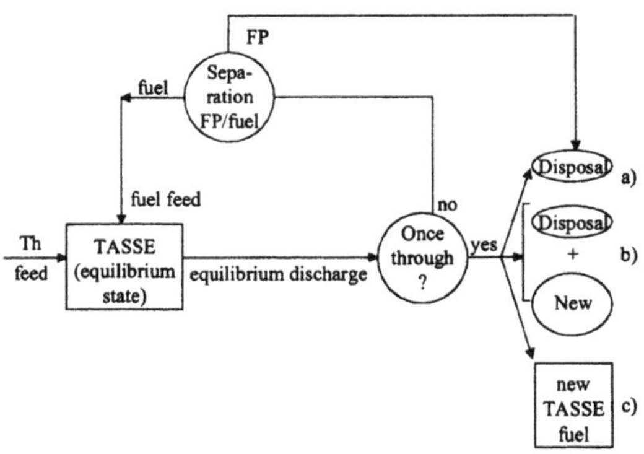

Figure 13 :
Fuel cycle options for TASSE at equilibrium state (ref. 37)

5.3. The Thorium cycle

A recent study of the European Union (in the frame of a contract for the completed 4th FWP, to be continued in the 5th FWP) has addressed the issue of "Thorium as a waste management option" (Ref. 38).

The objective of the work was a re-assessment of Thorium cycles in the context of limitation of nuclear waste production and prospects for waste burning. The aim was to obtain a review of the major steps of the fuel cycle, focusing to the waste aspect. A restriction was made regarding reactor ty_ ; : PWR, FR and ADS.

The final report of that study shows that there are important advantages of Thorium cycles with respect to the waste issue, that we will quote in detail from ref. 38 :

- Long-lived radio-toxicity of mining waste is expected to be relatively small, which leads to more manageable waste as compared to the uranium case.
- Fabrication of Th/Pu-MOX fuels is comparable with U/Pu-MOX fabrication methods as long as fresh Th, fresh U and recycled Pu are used. Recycling of U bred from Th, however, needs remote handling and reprocessing techniques specific to Thorium.
- The use of Thorium in PWRs always requires make-up fuel and therefore a self-sustaining mode is impossible in such a reactor.
- To reduce the radio-toxicity of PWR waste in an once-through mode, one has to avoid U-238 and therefore use Thorium together with make-up fuel like U-233 or highly-enriched U-235. Advantages in terms of waste radio-toxicity are seen during the first 10,000 years of disposal. Recycling gives a further reduction of radio-toxicity up to 10,000 to 50,000 years of disposal.
- The long-term residual risk of directly disposed fuel in a Thorium matrix is still not known very well, but there are indications on improved performance. Further experimental work is needed to clarify this point.
- Th-assisted Pu burning, using a Th/Pu-MOX type of fuel in a PWR, is an attractive option with respect to mass reduction of Pu.
- Fast neutron reactors and accelerator-driven systems offer both (with similar characteristics) the possibility of a closed Th cycle without make-up fuel, except to start the cycle, reducing mining needs and radiological risks. Full recycling of actinides gives impressively low radio-toxicity results for the wastes over a long period of disposal, starting after the bulk of fission products has decayed.

Non-proliferation concerns are also treated in the report.

The interest of the Th cycle should justify a number of experimental developments, in particular :

- reprocessing and fabrication techniques of Th fuels could be extended from laboratory scale to industrial scale and further optimised,
- co-extraction of actinides by pyrochemistry in molten salts, aiming at losses of the order of 0.1% should be demonstrated,
- nuclear data and physico-chemical data should be established or improved,
- some simple irradiation experiments should be foresee, (as it is the case in the new Thorium project for the European Union 5th Frame Work Program).

Finally, the considerations of ref. 37, further enhance the potential of Thorium, if powerful accelerators are used.

5.4. Multipurpose neutron source installations

Recently, the "transmutation" community has become involved in the discussions around "multipurpose" facilities, based on a high power proton accelerator which provides neutrons, by spallation on one (or several) target(s), for different applications.

The most known example is of course the joint KEK-JAERI project (Ref. 19). A new initiative is under study in Europe. The ADS experimental installation could be one of the "potential" customer of the neutrons (as it is in Japan), and, consequently, the transmutation community could be interested both to the possibility to demonstrate the concept, and to irradiate the dedicated fuels and targets needed to assess feasibility (see paragraph 4).

However, the need to single out a "leading" customer, can somewhat jeopardize the performance allowable for the "lesser" customers. A typical example, is the debate on the pulsed or continuous mode of operation of the high intensity proton accelerator.

6. HEALTHY CRITICISM

In the reference quoted at the begin of this report, Rasmussen and Pigford express their doubts about the value of P/T with arguments that should be carefully taken into account still today. Three of their arguments seem of particular relevance, besides the economical and institutional issues, which, although of fundamental importance, have to be adapted to each specific situation :

1. The total inventory of untransmuted radioactivity in the reactor and fuel cycle must also (besides what is sent to the repository as losses at reprocessing) be considered as a potential waste and it takes centuries to reduce it.

2. In the search for an adequate measure of performance, the repository "intrusion" scenario, is claimed to be the most affected by P/T.
   However, if one considers "intrusion" in a repository, why not to consider "intrusion" in the installations of the fuel cycle, where most of the inventory is kept !

3. P/T will give rise to significant amounts of new secondary wastes;

As far as arguments 1 and 2, it is clear (and should be always made clear in front of any type of audience), that P/T strategies are definitely associated to an (expanded) use of nuclear energy, with fuel (re)processing and relevant investments in new facilities. However, this (expanded) use of nuclear energy can be made acceptable to the public by the very fact that the burden to repositories is reduced by P/T, and that potentially physical means to eliminate all nuclear materials are provided by the same P/T technologies, even if they should be operated for long periods of time.

Finally, the problem of secondary wastes, and the more general problem of the impact of P/T on the fuel cycle installations, often mentioned in the present report has to be carefully quantified, in particular in terms again of social acceptability.

In conclusions, arguments against P/T can be seen simply as arguments in favour of a simplified fuel cycle, and not necessarily in favour of the once-through cycle based on Uranium utilisation.

7. CONCLUSIONS AND PERSPECTIVES

Transmutation of wastes has been revisited in the last decade and, although no spectacular breakthrough has been made, a number of significant results have been obtained.
Besides the relevant results in the aqueous chemical separation process domain (which have not been reviewed here), one can quote :

- understanding of the physics of transmutation and of the "neutron availability" concept,
- understanding of the role of innovative fuels (including molten salts and particle fuels) to improve the characteristics of the fuel cycle and to minimise wastes,
- understanding, in that context, of the potential of pyrochemical processes both for fuel fabrication and for irradiated fuel reprocessing,
- understanding of the role of ADS to handle Pu and MA, but also to provide an option for an extended use of the Thorium cycle,
- understanding of the role of fast neutron spectra and their flexibility. In this frame, the discussion around the coolants for FR would benefit from a better international agreement on pro and cons of the different options.

Since fuels play a central role in all scenarios of waste minimisation and nuclear power development, an international share of efforts around nitrides, oxides and metals should be organised in order to insure an optimum use of resources in the few existing laboratories to handle very active fuels. In that frame, the availability of irradiation facilities, in particular able to provide fast spectra (and high damage rates) is a key point and a major concern.

No convincing case can be made in favour of transmutation, without the full experimental demonstration of its feasibility. Experiments are then needed and the relevant installations should be kept available, with enough experienced teams. Besides the case of the installations for fuel characterisation, fabrication and irradiation, often mentioned in this report, installations related to basic physics (nuclear data and neutronics) will remain vital for all scenarios of development.

In the field of ADS, the development of high power proton accelerators and the construction of a 60 ÷ 100 MWt Experimental facility, at a realistic but not too far away, time horizon, seem to be necessary in order not to loose credibility.

International initiatives should be upgraded and, besides the very valuable information exchange goal, should address the practical share of work in key fields and should help to focus on some most promising concepts, promoting joint experiments and avoiding dispersion of efforts. In this respect, a coordinated activity on pyrochemical processing is strongly suggested.

Finally, the results of the new study of OECD-NEA presently underway will certainly help to better understand and to agree on the relative merits of two of the major options (i.e. critical fast reactors and ADS) for waste transmutation (Ref. 39).

8. ACKNOWLEDGMENTS

The author acknowledges the relevance of many results provided by M. Delpech sometimes before publication, and the fruitful discussions with him, I. Slessarev, J.P. Schapira and A. Zaetta.

9. REFERENCES

1. M. Salvatores et al. "A Global Physics Approach to Transmutation of Radioactive Nuclei", Nucl. Sci. Eng. 116, 1 (1994). See also several papers by Sekimoto and co-workers, e.g. Nucl. Sci. Eng. 135, 84 (2000).

2. "The Technology of the Integral Fast Reactor and its Associated Fuel Cycle "Progress in Nucl. Energy 31, (1997).

3. M. Salvatores "Transmutation and Innovative Options for the Back-end of the Fuel Cycle "Proc. GLOBAL'99 Conference (1999).

4. M. Delpech et al. "The Am and Cm transmutation. Physics and feasibility" Ibidem.

5. J.L. Klosterman et al. "Strategies for the Transmutation of Am" Proc. GLOBAL'97 Conference (1997).

6. M. Ozawa and T. Wakabayashi "Status on Nuclear Waste Separation and Transmutation Technologies in JNC" Ibidem.

7. M. Delpech, to be published.

8. T. Mukayama et al. "Partitioning and Transmutation program OMEGA at JAERI", Proc. GLOBAL'95 Conf. (1995).

9. M. Salvatores et al., Nucl. Inst. and Methods A414 (1998) 5-20.

10. M. Salvatores et al. "The Transmutation of Long-Lived Fission Products by Neutron Irradiation". Nucl. Sci; Eng. 130, 309 (1998). Also H. Sekimoto and K. Kanai, Ann. Nucl. Energy 25, 793 (1998).

11. M. Salvatores "Advanced Options for the back-end of the fuel cycle", Proc. 5th Int. Exchange Meeting on P/T, Mol, Nov. 1998.

12. J.P. Glatz et al. "Demonstration of Pyrometallurgical Processing for Metal Fuel and HLLW". To be presented at the 6th Inf. Exch. Meeting on P/T, Madrid, Dec. 2000.

13. M. Kurata et al. "Fabrication of U-Pu-Zr metallic fuel containing minor actinides" Proc. GLOBAL'97 conf. (1997).

14. J.F. Babelot, N. Chauvin "SUPERFACT Experiment", Tec. Note JRC-ITU-TN 99/03, 1999.

15. R.J. Konings et al. "The EFTTRA-T4 Experiment on Am Transmutation" EUR-19138 EN (2000).

16. R.J. Konings et al. "Fuel and Fuel Processing sub-group of the European TWG. State of the art report". To be published.

17. Y. Arai and T. Ogawa "Research on Nitride Fuel and Pyrochemical Process for MA Transmutation", to be presented at the 6th Inf. Exch. Meeting on P/T, Madrid, Dec. 2000.

18. A Roadmap for Developing ATW Technology" DOE/RW-0519 (1999).

19. "The Joint Project for High. Intensity Proton Accelerators" JAERI-Tech 99-056, KEK Report 99-4 (August 1999).

20. M. Salvatores, J.P. Schapira, H. Mouney "French Programs for Advanced Waste Management Options" Proc. 2nd Int. Conf. on Accelerator Driven Technologies, Kalmar (1996).

21. M. Hugon and V.P. Bhatnagar "Partitioning and Transmutation in the EURATOM RTD 5th Framework Programme", to be presented at the 6th Inf. Exch. Meeting on P/T, Madrid, Dec. 2000.

22. M. Salvatores et al. "MUSE-1 : A First Experiment to Validate the Physics of Subcritical Multiplying Systems" Proc. 2nd Int. Conf. on Accelerator Driven Technologies, Kalmar (1996).

23. R. Soule et al. "The MUSE Experiments for Subcritical Neutronics Validation", to be presented at the 6th Inf. Exch. Meeting on P/T, Madrid, Dec. 2000.

24. J. Benlliure et al. "New data and MC simulations on Residue Production in Spallation Reactions Relevant for the Design of ADS", to be presented at the 6th Inf. Exch. Meeting on P/T, Madrid, Dec. 2000.

25. G. Palmiotti et al. "Uncertainty Assessment for Accelerator Driven Systems" Proc. GLOBAL'99 (1999).

26. M. Delpech, private communication.

27. M. Salvatores et al. "Role of Accelerator-Driven Systems in Waste Incineration Scenarios" Proc. GLOBAL'97 (1997).

28. R.J. Konings et al. "Transmutation of Technetium in the Petten HFR", Nucl. Sci. Eng. 128, 70 (1998). See also : J. Nucl. Mat. 254, 122 (1998), J. Nucl. Mat 274, 336 (1999) and J. Nucl. Mat. 244, 16 (1997).

29. J. Boussier et al. "Assessment of Nuclear Power Scenarios. Allowing for Matrix Behaviour in Radiological Impact Modelling of Disposal Scenarios", to be presented at 6th Inf. Exch. Meeting on P/T, Madrid, Dec. 2000.

30. E. Adamov et al. "Conceptual Design of Lead. Cooled Fast Reactor" Proc. ARS94, vol. 1, p.509, Pittsburgh (1994).

31. A. Languille et al. "The CAPRA core studies. The Oxide Reference Option". Proc. Global'95 (1995).

32. Y. Oka et al., Nucl. Technology 103, 295 (1993). Also Nucl. Technology, 109, 1 (1995) and Proc. GLOBAL'95 (1995).

33. A. Puill, J. Bergeron "Advanced Plutonium fuel Assembly", Nucl. Technology 119 (1997).

34. S. Oki "Comparative study for MA Transmutation in Various Fast Reactors Core Concepts", to be presented at the 6th Inf. Exch. Meeting on P/T, Madrid, Dec. 2000.

35. J. Vergnes et al. "Limiting Pu and MA inventory. The AMSTER concept ». Proc. GLOBAL'99 (1999).

36. C. Bowman, "The physics design of ADS", Proc. GLOBAL'95 (1995).

37. M. Salvatores et al. "Review and proposals about the role of ADS in Nuclear Power", to be published in Progress of Nucl. Energy (2000)

38. H. Gruppelaar, J.P. Schapira "Thorium as a Waste Management Option" EUR-19142 EN (2000).

39. On going OECD-NEA system study on the comparison of FR and ADS for Transmutation, under the coordination of P. Wydler and L. Vandendurpel.

40. M. Salvatores, G. Bauer, G.Heusener "The MEGAPIE Initiative. Executive Outline and Status as per November 1999", to be published.

Pergamon

www.elsevier.com/locate/pnucene

Progress in Nuclear Energy, Vol. 40, No. 3–4, pp. 403–413, 2002

Printed in Great Britain
0149-1970/02/$ - see front matter

PII: S0149-1970(02)00032-X

PARTITIONING AND TRANSMUTATION STUDIES AT JAERI BOTH UNDER OMEGA PROGRAM AND HIGH-INTENSITY PROTON ACCELERATOR PROJECT*

T. MUKAIYAMA, H. TAKANO, T. OGAWA, T. TAKIZUKA
M. MIZUMOTO

JAPAN ATOMIC ENERGY RESEARCH INSTITUTE

Tokai-mura, Ibaraki-ken 319-1195, Japan

ABSTRACT

At the Japan Atomic Energy Research Institute (JAERI), active and comprehensive studies on partitioning and transmutation (P&T) of long-lived nuclear waste from the reprocessing processes of spent fuel has been carried out under the OMEGA program. Studies at JAERI include a design study of dedicated transmutation systems both of an MA burner fast reactor (ABR) and an accelerator-driven subcritical system (ADS), and the development of a high intensity proton accelerator as well as the development of partitioning process, nitride fuel fabrication/dry separation process technologies and nuclear data studies.

During the course of studies, JAERI developed the concept of the double-strata fuel cycle, in which a dedicated system is used for transmutation. Comparing the various transmutation systems, such as thermal neutron spectrum or fast neutron spectrum systems, power reactors or dedicated systems, from the viewpoints of reactor physics, nuclear fuel cycle and socio-technical issues, it was concluded that the ADS is the best option for transmutation of minor actinide(MA). JAERI, therefore, decided to concentrate its R&D efforts on the development of ADS and related technologies.

One of the goals of R&D is to provide a basis for designing demonstration facilities of ADS, aqueous partitioning process and nitride fuel fabrication and dry separation technologies. As the initial step toward this purpose, the construction of an ADS experimental facility is planned under the High-Intensity Proton Accelerator Project which JAERI and the High Energy Accelerator Research Organization (KEK) are jointly proposing since 1998.

The paper discusses the some of the results of P&T studies and the outline of the High-Intensity Proton Accelerator Project under which ADS experimental facility will be constructed.

* The paper is based on the paper by Mukaiyama and *et al.* 2000

1. INTRODUCTION

JAERI has been developing technologies for P&T of long-lived nuclides in high-level radioactive liquid waste (HLLW) since mid 1970's. In 1988, the Atomic Energy Commission started a long-term R&D program on P&T technology, called the "OMEGA" program. In Japan, since plutonium is to be used as a nuclear fuel material, the objective of P&T is to remove and transmute mainly MA and long-lived fission product (LLFP) in high-level radioactive waste (HLW) associated with the nuclear power generation.

JAERI developed the concept of double-strata fuel cycle, in which P&T is carried out in a dedicated and small-scale fuel cycle for P&T attached to the commercial fuel cycle. The R&D activities at JAERI cover the whole area of reactor and fuel cycle technologies related to P&T and high power proton accelerator for ADS with the intention of realizing the double-strata fuel cycle concept.

Through the system study of transmutation, it became clear that a dedicated system can more effectively and efficiently transmute MA than fast reactors for electricity generation, and that an ADS is better than dedicated critical reactors for MA transmutation from the safety viewpoints. Based on these results, JAERI decided to concentrate its future transmutation study on the development of ADS and related technologies.

JAERI and KEK were jointly proposing the High-Intensity Proton Accelerator Project to pursue frontier science in life science, material science, particle physics, nuclear physics and ADS technology, using a new high power proton accelerator. The construction of an accelerator facility complex at the JAERI Tokai site was accepted to start in 2001.

With regard to ADS, JAERI has been carrying out the development both under the OMEGA program and the High-Intensity Proton Accelerator Project; the former covers the general R&D activities on P&T, the latter includes, as a part of the project, the construction of an ADS experimental facility.

2. OMEGA PROGRAM AT JAERI

2.1 Double strata fuel cycle

JAERI developed the concept of the double strata fuel cycle (Murata and Kuroi, 1983, Murata and Mukaiyama, 1983), which consists of a power reactor fuel cycle and a P&T cycle (Fig. 1). The important features of the double strata fuel cycle are that the P&T

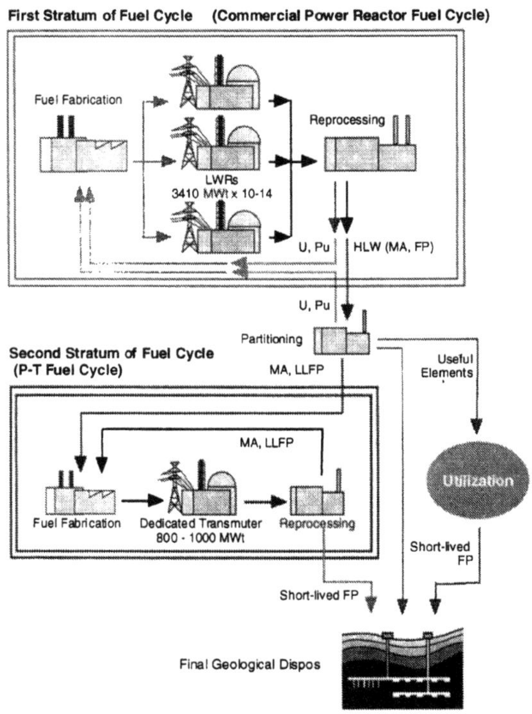

Fig. 1 Double strata fuel cycle with P&T

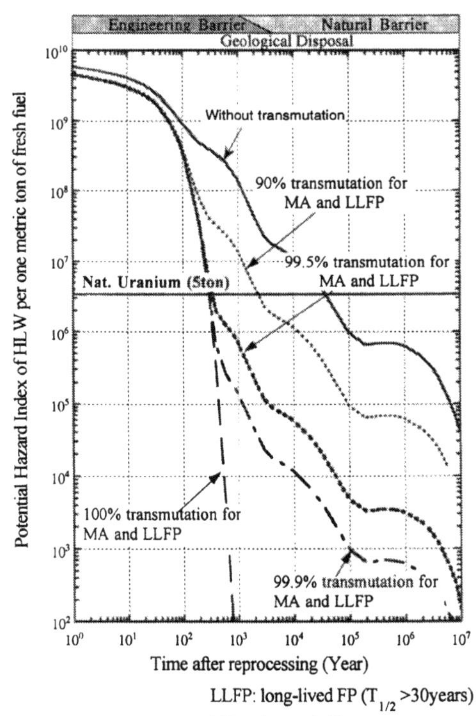

Fig. 2 Reduction of radiological toxicity of long-lived nuclide by transmutation

cycle is independent from the power reactor fuel cycle, and since the P&T cycle handles only the elements recovered from HLLW of the commercial fuel cycle, the heavy metal throughput in the cycle is significantly smaller than that in the commercial fuel cycle. These features will result in compact and economical P&T cycle facilities and in effective partitioning and transmutation. Further, the troublesome MA from the fuel handling viewpoints can be confined in one small P&T cycle.

The goal of development is 99.5% removal and transmutation of MA, Tc-99 in HLLW and I-129 from the PUREX reprocessing. The combined loss as low as 0.5% MA in these two processes can be technically achieved in case of the double strata fuel cycle scheme where only small quantity of MA is handled in a dedicated system. In Fig. 2, the effect of P&T is illustrated for the several cases of P&T efficiencies.

JAERI's activities cover the following areas: (a) development of the aqueous partitioning process, (b) design study of dedicated transmutation systems, (c) development of an high power proton accelerator, (d) development of nitride fuel cycle technologies, and (e) basic research on nuclear data and fuel property data.

2.2 Dedicated transmutation systems

The design study of MA transmutation systems at JAERI has been concentrated on designing dedicated transmutation systems that have very hard neutron spectra and a high neutron flux for effective and efficient transmutation of MA.

In order to clarify the advantages of dedicated transmutation systems over transmutation using power reactors, the reactor physics and ex-core fuel handling characteristics were compared. The generation of much heavier nuclides such as Cf in power reactors, namely light water reactor (LWR) and FBR, is significantly larger than in a dedicated fast neutron spectrum system (Mukaiyama., 1994).

The guidelines for designing dedicated transmutation systems are as follows:

- MA as the major fuel material,
- Very hard core averaged neutron spectrum for direct fission of MA,
- Very high neutron flux for high transmutation rate.

In the design study, nitride was selected as fuel material because of its good thermal properties and compatibility with pyrochemical reprocessing. Two types of dedicated transmutation systems have been designed, one is an actinide burner fast reactor (ABR) and the other is an ADS.

2.2.1 Actinide Burner Reactor (ABR)

In the early study, a sodium-cooled MA alloy fuel ABR (M-ABR) and a helium-cooled MA nitride particle fuel ABR (P-ABR) were designed. In the M-ABR, the reactor core is composed of two zones; the inner core of Np-(Pu)-Zr alloy fuel and the outer core of AmCm-Pu-Y alloy fuel. Two alloy.systems were needed in the design based on the theoretical prediction that similar to U-rare earths, Np and Am are not mutually soluble and the melting point of MA metal is rather low. To improve the low melting point of MA metal, Zr and Y were used as thermal diluents. In these alloys, Pu was mixed in the fuel in order to reduce a critical mass and to compensate for the significant reactivity gain during MA fuel irradiation, which is caused as a result of conversion of Np-237 to Pu-238 and Am-241 to Am-242m. In the P-ABR design, fuel material is MA nitride (Mukaiyama and *et al.*, 1988). In Fig. 3, the innovative concept of particle fuel is shown.

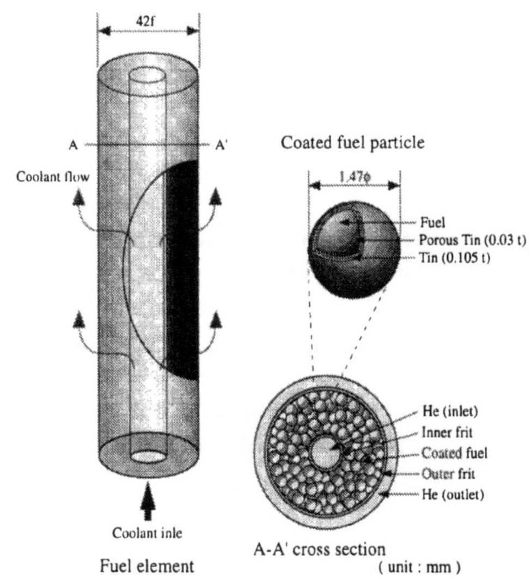

Fig. 3 Coated particle fuel concept for He-cooled ABR

Disadvantages of these designs are the very small effective delayed neutron fractions, around 0.17%, which are smaller than one-half of that of a typical metallic fuel

FBR, and the coolant-void coefficient was very large positive of M-ABR.

To improve the safety characteristics, the design of the original ABRs was modified. In a dedicated MA burner reactor design, the content of U-238 in fuel shall be minimal in order to avoid further generation of MA because U-238 is a mother material of MA. Due to the very hard neutron spectrum and the absence of U-238, the effective delayed neutron fraction in an ABR becomes very small even if Pu exists in the fuel. To increase β_{eff}, 90% enriched uranium was added in the later design. To avoid the large positive sodium-void, liquid lead was adopted as coolant material. Because of its good thermal properties and compatibility with pyrochemical reprocessing, the nitride was selected as fuel material in place of alloy fuel.

Two types of ABR were designed, namely a lead-cooled pin fuel ABR and a helium-cooled coated particle fuel ABR. The fuel material in these ABRs is the nitride mixture of 65% MA and 35% highly enriched uranium.

For a nitride fuel, 99.9% N-15 enriched nitrogen is used to avoid the generation of C-14 which is a long-lived nuclide. ABRs designed have very hard neutron spectra with a core-averaged energy around 720 keV. The amount of MA fissioned in an ABR is about 250 kg/y, which is the amount of MA generated in about ten LWR units yearly (Mukaiyama and *et al.*, 1993).

Under the present Nuclear Non-proliferation Treaty, the use of highly enriched uranium in a civilian reactor is practically impossible. For this reason, we have concluded that ADS is the best option for MA transmutation. The other advantages of ADS are the flexibility in subcritical core designing and the large margin for criticality accidents.

2.2.2 Conceptual design study of Accelerator driven subcritical system (ADS) for transmutation

1) Early design study

The study aims at designing an ADS, which can burn MA from more than10 units of LWR per year. This calls for ADS with about 800MW thermal power.

In the very early design study, a system was designed for MA transmutation by direct use of spallation reactions. The advantage of this concept is less production of long-lived nuclides than fission process. However, this concept was abandoned because of the bad energy balance: even with a proton accelerator with 1.GeV, 300mA, only MA from four LWRs can be transmuted yearly,.

An ADS with use of fuel material as spallation target material was designed. The fuel in the target area generates significant power both by spallation and fission and this causes a lower core averaged power density due to high power peaking (the peaking factor is around 5). It was concluded that a spallation target material should be different from fissionable material. A sodium cooled ADS was designed following designs of a contemporary liquid metal fast breeder reactor (LMFBR) by Takizuka and *et al.* (1991). With a 1.5GeV, 33mA proton beam, the subcritical core having an effective neutron multiplication factor of around 0.95 produces 820MW thermal power. The net transmutation rate is approximately 10%/y. One third of electric power is supplied to the accelerator operation.

One of the major disadvantages of the sodium-cooled system is the large positive void reactivity, even for ADS.

Molten-salt and molten-alloy ADS concepts were studied because of their possibility of continuous on-line charging of MA and removing of reaction products by Kato and *et al.* (1992), and Katsuta and *et al.* (1993). However, because of material compatibility problems and the safety concern of reduced defense-in-depth, the design study on fluid-fuel systems was discontinued.

2) Lead-bismuth cooled ADS : Reference ADS design

Based on the early design study, we selected a reference model of

Table 1 Major system parameters of the Pb-Bi cooled ADS

Target /Coolant	Pb-Bi
Fuel	MA/Pu mononitride
Fission Thermal Power	800 MW
Core Height/Radius	1000/1200 mm
k_{eff} initial/max./min.	0.95/0.95/0.93
Power Density max./ave.	310/180 MW/m^3
MA/Pu Inventory	2500/1660 kg
Coolant Temperature in/out	330/430 °C
Coolant Velocity max.	2 m/s
Transmutation Rate	500 kg/cycle (20 %/cycle)

ADS. The components of the reference ADS are 1) a superconducting (SC) proton linac, 2) a lead-bismuth spallation target and 3) a subcritical core with a fast neutron spectrum of, using nitride fuel and lead-bismuth coolant.

Technology of lead-bismuth is less mature than that of sodium though Russia has experience of developing lead-bismuth cooled reactors for submarine propulsion and 80 reactor•years of operation experience.

A conceptual drawing of a lead-bismuth cooled ADS is shown in Fig. 4. An accelerator injects proton beam through a beam window into the spallation target region located at the center of the subcritical core. Fuel material of the subcritical core is actinide mono-nitride. The operating temperature range of the primary lead-bismuth is from 330 to 430 °C, which is about 100 °C lower than those for commercial LMFBR designs. The lower operating temperature is expected to mitigate the material corrosion/erosion problem in lead-bismuth.

The major system parameters designed by Tsujimoto and *et al.* (1998) are summarized in Table 1. In this design, the system consists of a 1.5GeV, 14mA proton accelerator and the 800MWt subcritical core with an effective neutron multiplication factor of 0.95. The transmutation rate is approximately 250 kg/y when the load factor is 80%. An electric output of 280 MW is obtained with the plant thermal efficiency of 35%.

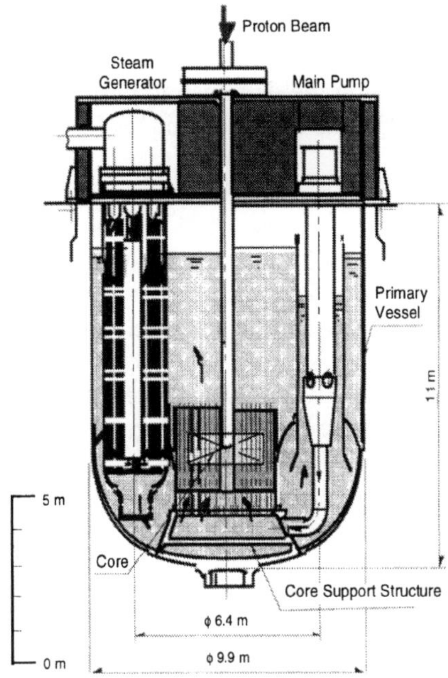

Fig. 4 Concept of Pb-Bi cooled ADS

2.3 Partitioning process

An aqueous partitioning process was developed for separating elements in HLLW into four groups: namely, transuranium elements (TRU), Tc-platinum group metal, Sr-Cs, and others. The process consists of three steps: (a) DIDPA(Di-isodecylphoshoric acid; an organic extractant) extraction process separating TRU and lanthanides, (b) separation of Tc-platinum group metal by denitration with formic acid or reductive adsorption with active carbon, and (c) adsorption of Sr and Cs with inorganic ion exchangers. In a laboratory-scale test at the Nuclear Fuel Cycle Safety Engineering Research Facility (NUCEF) of JAERI, the recovery of more than 99.9% was demonstrated for all actinides (Morita and *et al.*, 1999).

2.4 Nitride fuel fabrication and pyrochemical process

In the double-strata fuel cycle concept (Fig. 1), MA nitride fuel may be used and processed by a molten-salts electrorefining technique for MA transmutation with ADS. The higher thermal conductivity compared with the oxides gives a greater safety margin in the fuel performance, while permitting higher power rating in the fuel elements. These advantages can be utilised for more efficient burning of MA. Another important advantage is the mutual miscibility among the actinide mono-nitrides, in contrast to the metallic fuels

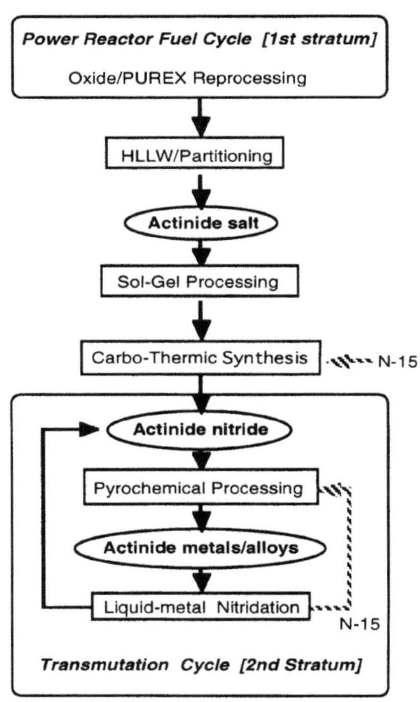

Fig. 5 Process of nitride fuel treatment

where the mutual solubility between lighter actinides (U-Np) and heavier trivalent actinides (Am-Cm) is limited (Ogawa and *et al.*, 1997). The mutual miscibility of actinide mono-nitrides has been demonstrated recently by the fabrication of (Cm, Pu)N, while Cm and Pu are almost immiscible as solid metals. The possibility of applying the pyroprocess to the MA nitrides is being studied. In the reference pyroprocess flowsheet, the irradiated nitride fuels will be submitted to the molten-salts electrorefining to remove the fission products. The process of nitride fuel treatment is illustrated in Fig. 5.

Following the demonstration of anodic dissolution of UN in the LiCl-KCl eutectic melt (Kobayashi and *et al.*, 1995), the same techniques have been tried on PuN and NpN (Shirai and *et al.*, 2000, Arai and et *al.*, 1999).

One of the drawbacks of using nitride fuels is the cost of highly N-15 enriched nitrogen. The cost accompanying the use of N-15 may become prohibitive, unless one can recover and recycle it in the fuel cycle. While N-15 may be diluted with natural nitrogen in ordinary aqueous processes, it may be readily recovered in a dry process such as molten-salts electrorefining. This point has been demonstrated by dissolving NdN and DyN, which are regarded as surrogates of MA nitrides (Kobayashi and *et al.*, 1999). A crude estimate showed that the enriched nitrogen price for 99.9% enrichment of N-15 could be readily accommodated when 90% of the initial enriched nitrogen is recovered during reprocessing. When the 99.9% enriched nitrogen is used, the generation of C-14 in the nitride fuel cycle is similar to that of the oxide fuel cycle.

The new facility for TRU high temperature chemistry where gram-quantities of Am can be handled will become available at JAERI in the fiscal year 2002.

2.5 High power proton accelerator development

Development of accelerator technology has been significant during the last decade so that very efficient and high intensity accelerators can be designed and built for various energy related applications including ADS.

The present typical concepts of ADS apply a high-intensity proton linac with proton energy around 1GeV and beam currents of several tens mA to a few hundred mA. For those high intensity accelerators, low beam loss and very high efficiency are important. Optimization should be made with the complex combination of RF frequency, aperture size, focusing, injection condition to maintain the low beam loss, high efficiency and low cost. SC linac will be an appropriate system for ADS applications where high efficiency is important.

JAERI have long experiences in construction and operation of linacs including an electron linac, a SC free electron laser (FEL) and a heavy ion linac. The technologies for the high power beam handling and radio frequency (RF) source have been also accumulated in the field of nuclear fusion such as neutral beam injection and RF plasma heating. Based on these accelerator-related technologies, the design study was made to confirm the technical feasibility to accelerate a high peak current with high duty operation from the beam dynamics point of view.

R&D for a high-intensity proton linac was started in 1991 to develop a linac with 1.5GeV, peak current 100mA, duty factor 10% for pulse mode operation for spallation neutron source at the initial stage, then continuous wave (CW) mode

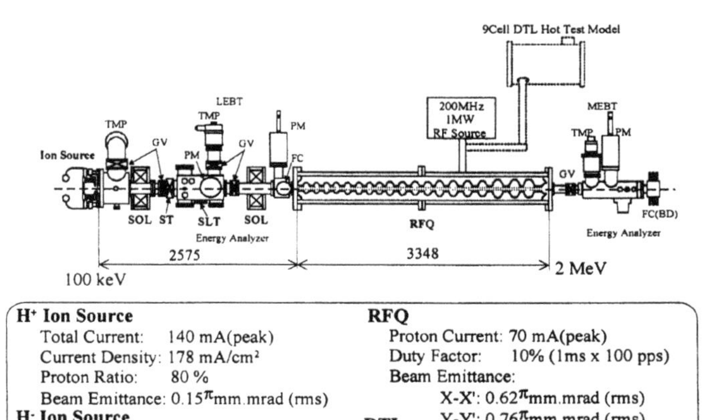

Fig. 6 Results of Low Energy Part of Accelerator development

operation for ADS experiments at the second stage (Mizumoto and *et al.*, 1998).

In the newly proposed JAERI/KEK High-Intensity Proton Accelerator Project, the accelerator complex consists of a 600MeV linac, a 3GeV rapid cycling synchrotron and a 50GeV synchrotron. In Phase 1 of this Project, the 3GeV synchrotron will deliver a beam power of 1MW and it will be upgraded to deliver 5MW in Phase 2. A SC linac will be used for energy ranging from 400 MeV to 600MeV for ADS experiments (Yamazaki and Mizumoto, 1999).

With a high intensity linac for ADS application, there are several technical issues to be solved. Those are low beam loss, high efficiency and high reliability. Various optimizations have been carried out with complex combination of RF frequency, aperture size, focusing, injection conditions and so on to meet those requirements.

The R&D results of the low energy part of the accelerator are summarized in Fig. 6. The RFQ is designed to accelerate 100mA peak current of protons to 2MeV with a duty factor of 10% (Hasegawa and *et al.*, 1997). A DTL hot test model with 9 cells for mock-up of the first part of the DTL. Duty factors of 20% and 50% were achieved for field gradients of 2MV/m and 1.5MV/m, respectively. As a high energy part of DTL, a separated type of DTL (SDTL) has been studied. SDTL is an attractive option for the high duty operation in the energy region of 50-100MeV.

In the SC linac part, the value of β (velocity of proton relative to that of light) gradually changes from 0.43 to 0.92 corresponding to an energy change from 100MeV to 1.5GeV. The main concern is the mechanical strength of a cavity under the vacuum load for the low β region (β<0.7). Mechanical structure calculations were made to determine cavity shape and electromagnetic parameters.

In order to determine the layout of the SC accelerating structure, a SC linac composed of several different β sections was studied. In the design study, one cryomodule (an accelerating unit cooled by liquid helium) was equipped with two cavities, each of which had 7 cells. One doublet-focusing magnet was laid between two cryomodules. The maximum peak surface field was 16MV/m. Two sets of single niobium SC test cavity of β=0.5 (proton energy E_p=145MeV) were fabricated and so-called vertical tests were conducted to examine the RF and mechanical properties. The maximum surface peak fields of 24MV/m at 4.2K and 44MV/m at 2.1K have been successfully achieved for the second cavity. A single cell cavity of β=0.89 (E_p=1.1GeV) was also fabricated. The maximum surface peak fields of 29.5MV/m at 4.2K and 51MV/m at 2.1K were achieved. These test results satisfied the specification of the conceptual layout of the SC linac (Ouchi and *et al.*, 1998). 5-cell cavities were also fabricated as shown in Fig.14. The maximum surface peak fields of 18MV/m and 20MV/m for β=0.5 and β=0.89 cavities, respectively, were achieved at the preliminary vertical tests.

No basic road blocks to the construction of a high intensity linac for ADS applications have been identified but more study of space charge dominated beam transports is needed to ensure low beam loss. The large beam intensities require further detailed tests of all accelerator components to ensure the long term stable operations.

3. ADS DEVELOPMENT UNDER THE HIGH-INTENSITY PROTON ACCELERATOR PROJECT

A high power proton accelerator for ADS application has been developed under the OMEGA Program since 1991. In the course of development, JAERI recognized that the neutron scattering community had been desiring a neutron source with a flux at least two order of magnitude stronger than that of existing sources. The neutron is the ideal probe to study basic microscopic structure and dynamics of materials at the atomic level. JAERI has been one of major neutron suppliers in Japan with its research reactors, and has been asked to continue to be such a supplier in the future. Especially, neutron scattering is now exciting a lot of interest among biologists for probing large organic molecules. A limiting factor for neutron scattering research is the intensity of neutron beams. Under these circumstances, JAERI started the Neutron Science Project in 1996 (Mukaiyama and *et al.*, 1997).

3.1 Neutron Science Project of JAERI

The objective of the project was to construct a multi-purpose high power proton accelerator with proton energy of 1.5GeV, proton beam power of 6 to 8 MW, and a research facility complex for basic sciences and ADS development. The major facilities are a SC proton linac, a 5MW spallation neutron source station allowing neutron pulses for neutron scattering research, and research facilities for ADS experiments with a CW proton beam, neutron physics, materials irradiation, and radioactive isotopes (RI) beam production for exotic nuclei science.

3.2 The Joint Project for High-Intensity Proton Accelerator of JAERI and KEK

Since the autumn of 1998, JAERI and KEK have been proposing to pursue frontier science in materials science, life science, nuclear/particle physics, and ADS technology, using a high beam power new proton accelerator. Previously, KEK proposed the Japan Hadron Facility (JHF) that covers a broad range of science from high-energy physics to materials science, by using the primary and various secondary beams from the high power proton accelerator. Its accelerator complex would consist of a 200-MeV proton linac, a 3GeV booster proton ring and a 50GeV main proton synchrotron ring (JHF Project Office, 1997).

The present new joint plan is based on these two past proposals. It is also proposed that accelerator facility complex of this Joint Project be constructed at the JAERI site. The Joint Project has been accepted for facility construction starting from FY2001 (The Joint Project Team of JAERI and KEK, 1999).

The Joint Project has two phases. Phase 1 accelerator complex with a pulse mode would consist of:

- 400MeV normal-conducting linac, used as an injector to 3GeV ring,
- 400 to 600 MeV SC linac, used for ADS experiments,
- 3GeV, 333μA (1MW) synchrotron ring,, used for an spallation neutron source and as an injector to 50GeV main ring, and
- 50GeV, 15 μA (0.75 MW) synchrotron ring,

In addition, an accelerator upgrade to a beam power of 5MW in the low GeV energy region is proposed as Phase 2 project of the present proposal. Fig. 7 shows the basic configuration of the proposed Phase 1 accelerator complex.

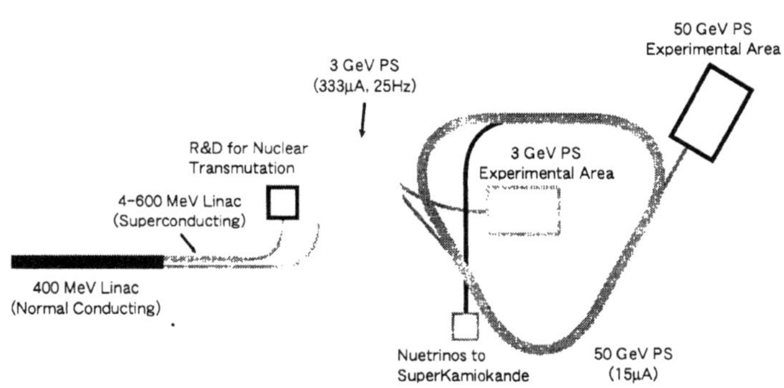

Fig. 7 Accelerator & experimental facilities of High-Intensity Proton Accelerator Project

The ADS experiments will be carried out using proton beam from 600MeV SC linac.

The 3GeV ring is designed to provide a beam power of 1MW spallation neutron source for structural biology, materials sciences and condensed matter physics. In addition to neutrons, muon beams are used for the study of μSR (muon spin rotation/relaxation) and other materials sciences. The 3GeV ring would also be used as a booster synchrotron for the 50GeV main ring. With the 50GeV proton synchrotron, nuclear/particle physics experiments using kaon beams, antiproton beams hyperon beams and primary proton beams are planned. Experiments on kaon rare decays, an experiment on neutrino oscillation from ν_μ to ν_τ using the Super-Kamiokande detector, which is located 295km away from JAERI Tokai site, will also be carried out.

The beam power is sufficiently high for all the above sciences. However, a higher beam power will enrich, in particular, neutron sciences. Thus, we plan in Phase 2 to upgrade the accelerator to 5MW for spallation neutron source. The construction of Phase 1 was accepted to start in 2001 and it will complete in 2006.

3.3 ADS development

Technical issues to be solved for ADS development include a high power accelerator, materials for target, beam window, coolant and fuel, system control, and safety.

Technical challenges can be classified under three categories: namely, 1) problems unique to ADS, 2) technical issues common to ordinary nuclear systems and 3) demonstration of ADS.

For category 1), the demonstration of the principle of the ADS is of primary importance, showing that a subcritical system can be driven safely by an accelerator accompanied with significant fission energy. The technical issues to be solved are the development of a high intensity and highly reliable proton accelerator, a beam window that can withstand high differential pressure, high heat generation, high thermal stress, and high fluxes of protons and neutrons. The principle of the ADS is to be demonstrated stepwise using very low beam power at the initial stage, then using higher beam power depending on the development progress. To solve these problems and to demonstrate the principle of the ADS, an experimental facility will be constructed under the High-Intensity Proton Accelerator Project.

For category 2), the experience and the knowledge accumulated during the FBR development are variable resources for the development of subcritical core technology. Some of the basic reactor physics experiments will be carried out to obtain a basis for designing a subcritical core and for reactivity related safety parameters using the existing fast critical assembly FCA of JAERI. The development of the partitioning process, nitride fuel fabrication/pyrochemical treatment of ADS spent fuel, nuclear and fuel property databases are being carried out under the OMEGA program.

The final goal of the development is to demonstrate the ADS technologies as a whole, including partitioning of HLLW, nitride fuel fabrication, ADS spent fuel processing and waste management. In the initial demonstration of ADS, fuel material of a subcritical core will be UO_2 and fission power will be around 40-50MW. MA transmutation will be demonstrated using MA nitride fuel pins at the beginning, later using a subassembly of MA nitride fuel. In the final demonstration, fuel material will be MA nitride and fission power of a subcritical core will be around 80-100MW, which is about one tenth of a full-scale ADS plant.

The scenario for the ADS development is shown in Fig. 8. The development will proceed in a stepwise manner. The first and second phases of the experimental program for ADS are planned within the framework of the High-Intensity Proton Accelerator Project.

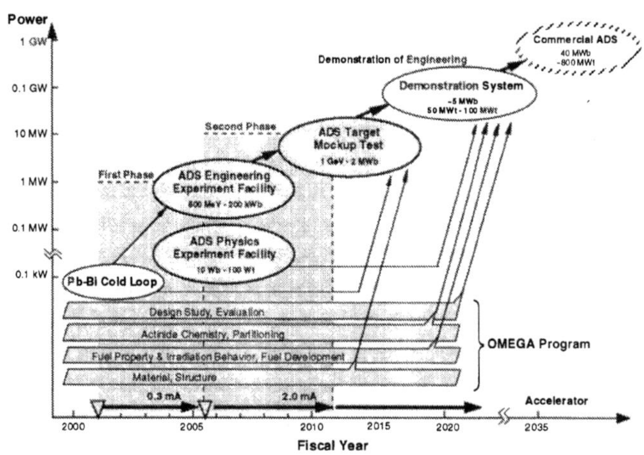

Fig. 8 Development scenario of ADS both under OMEGA & High-Intensity Proton Accelerator Project

3.4 ADS experimental facility

Under the High Intensity Proton Accelerator Project, two ADS experimental facilities are being designed, one for reactor physics of a subcritical core and the other for thermal-hydraulic and material investigations of a beam window and a lead-bismuth target. The former is the ADS Physics Experimental Facility "PEF," and the latter is the ADS Engineering Experimental Facility "EEF "(The Joint Project Team of JAERI and KEK, 1999, Ikeda and *et al.*, 2000).

3.4.1 Physics Experimental Facility "PEF"

Prior to Phase 1 experiments, zero power subcritical reactor physics experiments are scheduled at the fast critical assembly FCA of JAERI using a Cf-252 neutron source and later a 14MeV D-T pulsed neutron source. Data from the FCA experiments will be used in designing the PEF facility and planning the experimental program. The structure of the PEF is based on that of the FCA facility. With this flexible structure, experiments can be carried out for various core layouts of fast neutron spectrum. Fuel material of the PEF is 20% enriched uranium oxide and the effective neutron multiplication factor is in the range of 0.90-0.98.

In Phase 1, reactor physics experiments of subcritical cores and demonstration of the principle of ADS are to be performed. In Phase 2, proton energy will be increased to 1.0GeV by upgrading the linac. However, the proton beam current at the PEF will remain to be 10W due to the same limited heat removal capacity as in Phase 1 experiments.

3.4.2 Engineering Experimental Facility "EEF"

Lead-bismuth eutectic is the primary candidate for target/coolant material in the ADS design study. The major objective of EEF is to accumulate data for the design of a spallation target system. In Phase 1, the 600-MeV linac would deliver a 0.3mA proton beam (200 kW) to EEF. Radiation damage data of window materials for high intensity protons and neutrons is of critical importance in the ADS development.

In Phase 2, the beam power to the EEF will be increased to utmost 2MW. The 2MW target will be a mockup of a full-scale ADS target. The main goals of Phase-2 experiments are to verify the target and window design concept, to verify the reliability and safety, and to demonstrate the structural integrity of the spallation target and beam window. The upgrade will extend the coverage and volume of irradiation experiments.

SUMMARY

JAERI has been conducting a comprehensive R&D for P&T under the OMEGA program. The work includes a design study of dedicated transmutation systems both of an MA burner fast critical reactor ABR and ADS, and the development of a high intensity proton linac as well as the development of partitioning process, nitride fuel fabrication/dry process technology and nuclear data studies.

The concept of the double-strata fuel cycle was developed, in which a dedicated system is used for transmutation. Comparing the various transmutation concepts from the viewpoints of reactor physics and nuclear fuel cycle, the ADS was concluded to be the best option for transmutation of MA. JAERI has been concentrating its R&D efforts on the development of ADS and related technologies.

An aqueous partitioning process was developed and in a laboratory-scale test, the recovery of more than 99.9% was demonstrated for all actinides.

MA Nitride fuel fabrication and a pyrochemical technology for separation were demonstrated in a laboratory-scale. Recovery of enriched nitrogen is the key technology of nitride fuel from the cost consideration and it has been demonstrated by dissolving NdN and DyN, which are regarded as surrogates of MA nitrides.

For the ADS application, a high-intensity proton linac with proton energy around 1GeV and beam currents of several tens mA has been developed. No basic roadblocks to the construction of such a high intensity linac have been identified so far but further detailed tests of all accelerator components to ensure the long-term stable operations are required.

Construction of the ADS experimental facility is planned under the High-Intensity Proton Accelerator Project of JAERI and KEK. The Project aims at the construction of a multi-purpose proton accelerator for neutron scattering, particle and nuclear physics as well as ADS experiments. The construction of the facilities was accepted in 2000 with some delay in construction of ADS experimental facilities.

R&D on partitioning and transmutation attracts young scientists and engineers. These fresh brains and a diversity of thinking may bring a major breakthrough in nuclear technology for the future.

REFERENCES

Arai Y., Iwai T., Nakajima K., Sirai O. and Suzuki Y. (1999), Experimental Research on Nitride Fuel Cycle in

JAERI, *International Conference on Future Nuclear Systems, Global'99*, Jackson Hole, 29 August-3 September.

Hasegawa K., Mizumoto M., Ito N., Kusano J., Oguri H., Touchi Y. and Murata H. (1997), Development of a High Intensity RFQ at JAERI, *Jour. Nucl. Sci. Technol.* 34, 622.

Ikeda Y., Oigawa H. and Sasa T. (2000), *private communication.*

Joint Project Team of JAERI and KEK (1999), *The Joint Project for High-intensity Proton Accelerators*, JAERI-Tech 99-056 (KEK Report99-4, JHF-99-3).

JHF Project Office (1997), *Proposal for Japan Hadron Facility*, KEK Report 97-3, JHF-97-1.

Kobayashi F., Ogawa T., Akabori M. and Kato Y. (1995), Anodic Dissolution of Uranium Mononitride in Lithium Chloride-Potassium Chloride Eutectic Melt, *J. Am. Ceram. Soc.*, 78, 2279.

Kobayashi F., Ogawa T., Takano M., Akabori M. and Itoh A. (1999), Dissolution of Metal Nitrides in LiCl-KCl Eutectic Melts, *International Conference on Future Nuclear Systems, Global'99*, Jackson Hole, 29 August- 3 September.

Mizumoto M., Kusano J., Hasegawa K., Ouchi N., Oguri H., Kinsho M., Chishiro E., Tomisawa T., Touchi Y., Ikegami M., Honda Y., Mukugi K., Ino H., Akaoka N. and Kaneko H., (1998), A High Intensity Proton Linac Development for JAERI Neutron Science Project, *LINAC98*, Chicago, 24-28August.

Morita Y., Yamaguchi I., Fujiwara T., Koizumi H. and Kubota M. (1999), The First Test of 4-group Partitioning Process with Real High-level Liquid Waste, *Forth International Conference on Future Nuclear Systems, GLOBAL'99*, Jackson Hole, 29 August-3 September.

Mukaiyama T., Takano H., Takizuka T., Ogawa T. and Osakabe M. (1988), Conceptual Study of Actinide Burner Reactors, *International Reactor Physics Conference,* Vol. IV, p.369, Jackson Hole, 18-22 September.

Mukaiyama T., Ogawa T. and Gunji Y. (1993), Minor Actinide Burner Reactor and Influence of Transmutation on Fuel Cycle Facilities, *Technical Committee Meeting on Safety and Environmental Aspects of Partitioning and Transmutation*, Vienna, p. 105, 29 November-2 December (IAEA-TECDOC-783, 1995).

Mukaiyama T. (1994), Importance of Double Strata Fuel Cycle for Minor Actinide Transmutation, *3rd OECD/NEA Information Exchange Meeting on P&T*, p.30, Cadarache, 12-14 December.

Mukaiyama T., Oyama Y., Mizumoto M., Takizuka T. and Hino R. (1997), Neutron Science Project at JAERI, *Topical Meeting on Nuclear Applications of Accelerator Technology*, p.398, Albuquerque.

Mukaiyama T., Takizuka T., Mizumoto M., Ikeda Y., Ogawa T., Hasegawa A., Takada H. and Takano H. (2000), Review of research and development of Accelerator-driven System in Japan for transmutation of long-lived nuclides, *submitted to Progress in Nuclear Energy, ADS special edition.*

Murata H. and Kuroi H. (1983), A proposed Concept for Actinide-Waste Transmutation, *in Nuclear Technologies in Sustainable Energy System*, Springer-Verlag.

Murata H. and Mukaiyama T. (1983), Fission Reactor Studies in View of Reactor Waste Programmes, *Atomkernenergi-Kerntech*, 45, 23-28.

Ogawa T., Akabori M., Kobayashi F. and Haire R.G. (1997), Thermochemical Modeling of Actinide Alloys related to Advanced Fuel Cycles, *J. Nucl. Mater.* 247, 215.

Ouchi N., Kusano J., Akaoka N., Fechner B., Hasegawa K., Takeuchi S., Mizumoto M., Saito K., Noguchi S., Ono M., Kako E. and Inoue H., (1998), R&D Activities for Superconducting Proton Linac at JAERI, *APAC98*, Tsukuba, 23-27 March.

Shirai O., Iwai T., Shiozawa K., Suzuki Y., Sakamura Y. and Inoue T. (2000), Electrolysis of Plutonium Nitride in LiCl-KCl Eutectic Melts, *J. Nucl. Mater.* 277, 226.

Takizuka T., Takada H., Kanno I, Nishida T., Akabori M. and Kaneko Y. (1991), Conceptual Design of Transmutation Plant, *Specialist Meeting on Accelerator-Driven Transmutation Technology for Radwaste and Other Applications*, p. 707, Stockholm, 24-28 June.

Tsujimoto K., Sasa T., Takano H., Takizuka T., Mukaiyama T., Hayano M., Yokobori H. and Kamishima Y. (1998), Conceptual Study of the Lead-Bismuth Cooled Accelerator-Driven Transmutation System, *Second International Topical Meeting on Nuclear Applications of Accelerator Technology (AccApp '98)*, Gatlinburg, 20-23 September.

Yamazaki Y. and Mizumoto M. (1999), Accelerator Complex for the Joint Project of KEK/JHF and JAERI/NSP, *Particle Accelerator Conference*, New York, 29 March– 2 April.

Pergamon
www.elsevier.com/locate/pnucene

Progress in Nuclear Energy, Vol. 40, No. 3–4, pp. 415–422, 2002

Printed in Great Britain
0149-1970/02/$ - see front matter

PII: S0149-1970(02)00033-1

RELATIONSHIP BETWEEN DEPOSITED NUCLIDE INVENTORY AND HLW REPOSITORY PERFORMANCE

JOONHONG AHN

Department of Nuclear Engineering
University of California, Berkeley
Berkeley, CA 94720-1730

Abstract

The assertion of this paper is that the insensitivity of predicted repository performance to P-T efforts arises primarily due to the concentration-based measure for repository performance, and due to over-simplification in repository performance models. A mass-based performance assessment model and measures are proposed. Sensitivity of the mass-based measure is demonstrated by numerical results for Np-237.

1. INTRODUCTION

Partitioning-and-transmutation (P-T) systems designed to maximize the rate of transmutation of long-lived nuclides could be a remedy for reducing the difficulties of geologic disposal. However, previous analyses for the impacts of P-T application on the performance of geologic disposal show that P-T systems are not justifiable due to their high costs as they provide only modest improvements in radiological safety (Croff, *et al.* (1980), IAEA (1982), Pigford (1990), Ramspott, *et al.* (1992), Ahn (1998), Lowenthal, *et al.* (1999), and Ahn (1999a)). This conclusion is based on the observations (1) that due to low solubilities and high sorption distribution coefficients of actinides in geologic formations, decrease in the masses of actinides in the waste does not linearly decrease the actinides' radiological impact, (2) that if the recovered uranium is sent to the repository as waste, then much of the claimed benefit of P-T to waste management has been reduced, and (3) that repository performance is already good enough without P-T efforts. In this paper, discussions are given for (1) only.

The assertion of this paper is that the insensitivity of predicted repository performance to fuel cycle schemes, related to observation (1), results from the following two reasons.

- Repository performance is currently measured by exposure dose rates determined by concentrations of radionuclides in groundwater, whereas effects of P-T schemes manifest as reductions in masses of radionuclides contained in the waste streams, and
- Previous performance assessment models have been over-simplified, or tuned to a specific combination of waste types and geologic conditions, so that mechanisms of radionuclide retention in the repository may not be properly represented, which may be important if the combination is changed.

These points are further discussed in Section 2 by reviewing existing performance assessment models.

To better understand the impacts of P-T systems on geologic disposal, we need to develop a model, into which relevant parameters are incorporated. With such a model, new measures for repository performance could be proposed. New measures are expected to express effects of P-T system parameters on repository performance more sensitively.

In Section 3, a new repository performance model is established, with which the masses of a radionuclide retained in the repository region and those existing in the far-field region are obtained. These can be utilized as the basis of mass-based repository performance measures. In the model, shrinking source conditions, solubility-limited release, multiple waste canister configuration, and multiple-member decay chains are incorporated.

In Section 4, it is investigated if these newly-proposed, mass-based measures would actually be sensitive to differences in wastes. It is confirmed by numerical results with the model that the conventional performance measure based on the radionuclide concentration in groundwater is insensitive to the difference in the initial radionuclide mass in the repository, due to solubility-limited release. The mass-based measure proposed in this study, on the other hand, could differentiate the cases with different initial masses. To illustrate, ^{237}Np is selected in this study. Neptunium-237 is considered to be an important contributor to the radiological hazard (Ahn, 1999a, Ahn, 1999b).

2. PREVIOUS PERFORMANCE ASSESSMENT MODELS

Previous studies for effects of reduced radionuclide mass on repository performance have been performed with existing repository performance models developed specifically for a certain geologic condition and waste configuration. Because wastes after P-T application could be substantially different from LWR spent fuel or HLW, the repository design most suitable for P-T wastes could also be different. However, because existing repository performance models have not been developed to accommodate variety of waste types, in previous P-T effect studies, to observe the effect of P-T application, the mass of radionuclides to be disposed of in the repository was decreased by a certain factor with the same repository configuration assumed. For example, the recent ATW Roadmap study by US DOE (1999) was done with the repository performance assessment models developed for TSPA-VA (US DOE, 1998) for the proposed Yucca Mountain Repository (YMR) with the radionuclide inventories reduced by a factor of 1000.

Many performance models were developed for single-canister configuration. The Engineered-Barrier System (EBS) is assumed for the waste canister. Waste matrix alteration, radionuclide release from the waste matrix, and subsequent radionuclide transport in the EBS is analyzed. The outcome of this EBS analysis is the mass release rate of a radionuclide from the EBS into the surrounding host rock, which is used as the inlet boundary condition for the far-field transport analysis. The concentration of the radionuclide evaluated by the far-field transport analysis at a certain point downstream from the repository in the far field is used as the inlet boundary condition for the transport analysis in the biosphere, which finally gives the performance measure such as the exposure dose rate (Sv/yr). Thus, the radionuclide concentration in water in the far field is the basis for the performance assessment.

For such assessment models, effect of multiple canisters is not important. If there are multiple similar canisters in the repository with similar EBS, the cross-section of water stream into which radionuclides are released from EBS increases proportionally as the number of canisters increases. The difference in water-stream cross-section does not affect the radionuclide concentration in the far field significantly, if the mass release from individual canisters is similar.

Another important mechanism that is usually incorporated into the performance model is the solubility-limited release from the waste matrix for hardly-soluble actinides.

What would happen to the assessment results if the radionuclide mass initially placed in the repository is reduced under the same repository condition? Because the release is limited by radionuclide solubility in groundwater, the radionuclide mass does not affect the release from EBS into the far field unless the mass reduction is so significant that the release is not limited by the solubility. If the number of waste canisters is kept the same before and after the P-T application, the cross-section of water stream is the same, resulting in the same radionuclide concentration in the far field even if the initial mass is reduced. This has been observed in the previous studies on P-T effects. Three points can be made for this observation.

- The repository configuration would affect the release of radionuclides, which existing models do not reflect properly. For example, waste after P-T application could be characterized by low concentration and large volume. Rather than disposing such waste in many small canisters placed in a two-dimensional array, a large, silo-type storage might suit better. Then, the cross-section of water stream sweeping the repository could be substantially smaller. Even with the same solubility limited release, the radionuclide concentration in water leaving the repository could be substantially different for different repository configurations, if overlooked mechanisms, such as interference with other waste canisters, are incorporated into the model (see Section 4).
- Shortcomings of the concentration-based measure can be pointed out. (1) Concentration continuity in the performance model is often not conserved. To de-couple the EBS model from the far-field transport model, the concentration at the EBS/far-field interface is assumed zero, while the radionuclide mass flux could be conserved at the interface. Resulting from the de-coupling, the cross-sectional area and the water flow rate in the water stream in far field must be determined artificially. The concentration thereafter can only be relatively compared. (2) With a greater cross-sectional area for contaminated water stream, the probability of taking the contaminated water into the biosphere could also be greater. But, this is not incorporated in existing performance models.
- Because effects of P-T schemes manifest as reductions in masses of radionuclides contained in the waste streams, the repository performance should also be measured by quantities based on masses of radionuclides.

3. MASS-BASED PERFORMANCE MEASURE

3.1 Physical Processes and Assumptions

A repository is assumed to be saturated with water. Canisters containing waste matrix are placed in a two dimensional array fashion. The Engineered Barrier System (EBS) consists of the waste canister surrounded by a homogeneously porous buffer-region, where water is at rest. All canisters in the repository are assumed to fail at 1,000 yr after emplacement. Radionuclides are released from failed canisters by waste matrix alteration. The pore water in the EBS region is assumed to be chemically reducing, so that actinide elements are hardly soluble in water. Their concentrations in groundwater in the vicinity of the waste matrix reach their solubilities. Radionuclides are transported by molecular diffusion through the buffer, and then by advection through the near-field rock. Presence of colloids is neglected in this analysis because of assumed colloid filtration effect by the buffer. The hydrological and geochemical conditions are assumed constant in time. The near-field rock is assumed to be homogeneously porous.

In the present model, we consider a compartment, which consists of the waste matrix, the buffer, and the surrounding near-field rock. A row of N compartments in the repository is considered to be included in the same water stream. It is assumed that radionuclides do not escape through the side boundary of each compartment. The side boundaries coincide with the mid-plane between two adjacent rows of waste canisters. If each canister releases radionuclides equally, no mass transport occurs through the mid-planes due to zero concentration gradient perpendicular to the mid-planes.

Uncontaminated water enters the near field rock of the first compartment in the row. The water in the pores of the near field rock is contaminated by radionuclides that are released from the waste matrix and transported through the buffer by diffusion. Contaminated water in the near field rock continues to flow in the near field rock, and contacts with the buffer in the second compartment in the row. The water gets more contaminated by radionuclides released from the second canister. In this fashion, the groundwater flowing in the row through the near-field rock is assumed to be contaminated before it finally flows out of the repository from the near-field rock of the N-th compartment.

The concentration gradient in the buffer is the largest in the first compartment, resulting in the largest mass flux from the buffer. Therefore, the complete depletion of radionuclide from the waste matrix occurs at the earliest time among the waste matrices included in the row. After the first matrix is depleted completely, then the second is exposed to the highest mass flux among the remaining ones. In this fashion, the radionuclide mass in matrix depletes from the upstream side toward the downstream.

3.2 Model

Assuming that volumes of the buffer region and the near-field rock region are constant with time, the mass balance equations for radionuclide k in chain j are:

$$\frac{dM^W_{k,j,n}}{dt} = -\lambda_{k,j} M^W_{k,j,n}(t) + \lambda_{k-1,j} M^W_{k-1,j,n}(t) - q_{k,j,n}(t), 0 < t < T_{e(k,j),n}, \tag{1}$$

$$\frac{dM^B_{k,j,n}}{dt} = -\lambda_{k,j} M^B_{k,j,n}(t) + \lambda_{k-1,j} M^B_{k-1,j,n}(t) + q_{k,j,n}(t) - Q_{k,j,n}(t),\ t > 0, \tag{2}$$

$$\frac{dM^R_{k,j,n}}{dt} = -\lambda_{k,j} M^R_{k,j,n} + \lambda_{k-1,j} M^R_{k-1,j,n} + \phi_{k,j,n-1} - \phi_{k,j,n} + Q_{k,j,n}(t),\ t > 0, \tag{3}$$

$$\frac{dM^F_{k,j}}{dt} = -\lambda_{k,j} M^F_{k,j} + \lambda_{k-1,j} M^F_{k-1,j} + \phi_{k,j,N}(t), t > 0, \tag{4}$$

$$k = 1,2,\ldots,i_j,\ j = 1,2,\ldots,J,\ n = 1,2,\ldots,N,\ \lambda_{0,j} \equiv 0,\ \phi_{k,j,0} \equiv 0,$$

These equations are solved subject to the following initial conditions:

$$M^W_{k,j,n}(0) = \overline{M}_{k,j,n},\ M^B_{k,j,n}(0) = 0,\ M^R_{k,j,n}(0) = 0,\ M^F_{k,j}(0) = 0,\ k = 1,2,\ldots,i_j,\ j = 1,2,\ldots,J, n = 1,2,\ldots,N, \tag{5}$$

where $\overline{M}_{k,j,n}$ [mol] is the initial mass of radionuclide k of chain j in the waste matrix in compartment n.

$M^i_{k,j,n}(t)$ [mol] (i = W, B, and R) is the mass of the k-th member radionuclide in the j-th radioactive decay chain in region i in the n-th compartment. Superscripts W, B, and R stand for waste matrix, buffer and near-field rock, respectively. $\lambda_{k,j}$ [yr^{-1}] is the decay constant of radionuclide k in the j-th chain. i_j is the number of radionuclides included in the j-th chain. J is the number of radioactive decay chains. $T_{e(k,j),n}$ is the time when element e depletes completely from the matrix region in compartment n. $T_{e(k,j),n}$ is determined by calculating numerically $M^W_{k,j,n}(t)$. Subscript $e(k, j)$ denotes that radionuclide k in chain j is an isotope of element e. We call the region covered by the row of N compartments the repository region hereafter. $M^F_{k,j}(t)$ [mol] is the mass of the k-th member radionuclide in the j-th radioactive decay chain in the far-field region, which extends exterior to the repository region.

For the near-field rock region, the third $\phi_{k,j,n-1}$ and the fourth $\phi_{k,j,n}$ terms [mol/yr] on the right side of (3) represent the mass of radionuclide k per unit time flowing in from compartment $n-1$ to compartment n and the mass flowing out of compartment n to compartment $n+1$ by advection. In the present model, it is assumed that the concentration of radionuclide k in chain j is $C_{k,j,n}(t)$ [mol/m^3] homogeneously in the pore water of the near-field rock of compartment n. Then, the mass, $M^R_{k,j,n}(t)$ is expressed as $M^R_{k,j,n}(t) = \varepsilon_p V R_{e(k,j)} C_{k,j,n}(t)$, where ε_p and V [m^3] are the porosity and the volume of the near field rock in a compartment, and $R_{e(k,j)}$ is the sorption retardation factor. These are assumed constant with time, homogeneous inside the near-field rock of a compartment, and independent of compartment index n. With the homogenized concentration $C_{k,j,n}(t)$ defined, we write the mass transfer function between compartments as $\phi_{k,j,n}(t) = FC_{k,j,n}(t)$, where F [m^3/yr] is the volumetric water flow rate through the compartment boundary. F is assumed to be constant with time and unchanged throughout the compartments.

$Q_{k,j,n}(t)$ [mol/yr] is the diffusive mass flux of radionuclide k of chain j at the buffer/rock interface of compartment n. $q_{k,j,n}(t)$ [mol/yr] is the release rate of radionuclide k from the waste matrix into the buffer. It is assumed that all the waste matrices in N compartments dissolve at a constant rate over the leach time, T_L [yr]. In the time period, $0 < t < T_L$, isotopes of element e in the waste matrix are released congruently at the rate, $m_{k,j,n}(t)$. If the dissolution of the waste matrix is so fast that congruently released radionuclides could form precipitates, the release rate $q_{k,j,n}(t)$ is given by the solubility-limited diffusive mass flux, $\psi_{k,j,n}(t)$. Detailed formulation for $m_{k,j,n}(t)$ and $\psi_{k,j,n}(t)$ is given elsewhere (Ahn, 1999b). Thus, $q_{k,j,n}(t)$ is given by

$$q_{k,j,n}(t) = \min\{m_{k,j,n}(t), \psi_{k,j,n}(t)\}, 0 < t < T_L. \tag{6}$$

Once the solubility-limited condition is achieved, then $\psi_{k,j,n}(t)$ is applied until all the isotopes of the element in the waste matrix region are released. It is assumed that the solubility of an element is shared by all the isotopes of that element.

From the downstream side of the row, i.e., from compartment N, radionuclide k is released into the far field. The third term in the right side of (4) represents the mass release rate [mol/yr] of radionuclide k from the down-stream side of the repository to the far field. Because the concentration, $C_{k,j,N}(t)$, is determined by the concentrations of the other upstream compartments, $C_{k,j,n}(t), n = 1,\ldots,N-1$, by (3), $M^F_{k,j}(t)$ determined by (4) includes the details of mass transport in the repository through the term, $\phi_{k,j,N}(t) = FC_{k,j,N}(t)$.

The quantities as prospective performance measures are defined as

$$\mu^{IN}_{k,j}(t) \equiv \frac{\sum_{n=1}^{N}\left\{M^W_{k,j,n}(t) + M^B_{k,j,n}(t) + M^R_{k,j,n}(t)\right\}}{M^o_{k,j,tot}} \text{ and } \mu^{OUT}_{k,j}(t) \equiv \frac{M^F_{k,j}(t)}{M^o_{k,j,tot}}, \text{ where } M^o_{k,j,tot} \equiv \sum_{n=1}^{N} \overline{M}_{k,j,n}. \tag{7}$$

These are the masses of radionuclide k in chain j existing in the region inside the repository and in the region exterior to the repository, normalized by the total mass of that radionuclide that exists initially in the waste matrices. Numerical results for these quantities are shown in the next section.

4. NUMERICAL EXAMPLES AND DISCUSSIONS

Table 1 shows three radionuclides among those included in vitrified HLW (JNC, 2000). The volume V of the near-field rock is assumed to be 9.0 m^3. The water flow rate F is 0.9 m^3/yr. The porosity ε_p is 0.5. With the parameter values assumed, $M^i_{k,j,n}(t)$ (i = W, B, and R) and $M^F_{k,j}(t)$ have been calculated numerically based on the governing mass balance equations (1) through (4). It has been found that radionuclides such as ^{93m}Nb that decay fast compared with the diffusion in the buffer, or those such as ^{241}Am with small solubilities compared with their initial masses in the waste matrix never come out to the near field rock. Np-237 is sufficiently long lived to survive the transport inside the repository.

Table 1: Input Data.

Nuclide	half-life [yr]	$\overline{M}_{k,j,n}$ [mol]$^{(a)}$	Solubility [mol/liter]$^{(b)}$	$R_{e(k,j)}$ $^{(c)}$
Nb-93m	1.36E+01	7.21E-05	1.00E-04	260
Np-237	2.14E+06	3.74E+00	2.00E-08	2600
Am-241	4.32E+02	1.88E-01	2.00E-07	13000

(a)(b): from JNC (2000). (c): calculated with sorption distribution coefficients given in JNC (2000).

Hereafter, we consider transport of ^{237}Np in two different configurations. In one configuration, N compartments are lined up in the direction parallel to the water flow in the near field rock. Compartment 1 is located at the upstream end, while compartment N is on the down-stream end of the row. In the other configuration, N compartments are lined up in the direction perpendicular to the water flow. In this configuration, all compartments are independent with each other, because no water flow is assumed to cross the boundary between adjacent compartments. N is arbitrarily set to be 64 in the examples shown below. The precursors of ^{237}Np are neglected. By preliminary calculations, it was confirmed that the effects of precursors (^{245}Cm and ^{241}Am) are negligible. They decay to ^{237}Np while they are in the waste or in the buffer. Thus, by lumping the initial masses of the precursors on that of ^{237}Np, almost identical results could be obtained. Results shown in Figures 1 to 4 have been obtained for such lumped mass of ^{237}Np.

Figure 1 shows the temporal change of the mass of ^{237}Np in the row of 64 compartments and of the mass in the far field, for the *perpendicular* configuration. In the "no mass reduction" case, the initial mass placed in the repository is $M^{o}_{237,tot} = \sum_{n=1}^{64} \overline{M}_{237,n}$ by (7). Hereafter, the subscript 237 is dropped. For the "1/10 mass reduction" case, the initial mass placed in the repository in 64 canisters is $M^{o}_{tot}/10$. The 1/100 and 1/1000 reduction cases can be defined similarly. For each case, three curves are shown: the mass in the repository region, the mass in the far-field region, and the total mass. These masses are normalized by M^{o}_{tot}. For no reduction case, the mass in the far field increases with time, while the mass in the repository decreases rapidly after 1 million years, and becomes zero at 5.5 million years. The far-field-mass curve coincides with the total-mass curve at that time.

With reduced initial mass, obviously the mass in the repository is reduced by the same factor. The peak value of the mass in the far field is also reduced by a similar factor. The mass in the repository depletes earlier for a greater reduction factor. For the no reduction case, more than 99% of initial ^{237}Np remains in the repository until 1 million year. About 2/3 of initial Np decays in the repository, and only 1/3 accumulates in the far field at 6 million years. With a greater reduction factor, virtually all neptunium is released into the far field. However, because the initial mass is reduced, the mass released into the far field is also reduced by the same factor.

For the no-mass-reduction case, 1/10-, and 1/100-reduction cases, the far-field-mass curve traces the same trajectory until it meets the total-mass curve. This is because the release from the waste matrix is limited by the solubility. The far-field-mass curve for the 1/1000 reduction case is located below the other three curves. It implies that the initial mass is reduced to the extent that the release is not limited by the solubility anymore. This is confirmed by Figure 3.

In Figure 2, same quantities are plotted for the *parallel*-to-flow configuration. In this configuration, 64 compartments are linked sequentially. Mass released in the upstream compartments accumulates as water flows in the row. As mentioned in Section 3.1, release from EBS in downstream compartments becomes smaller, so that the release from the repository is smaller than that from the repository in the perpendicular configuration. This is actually confirmed by comparing Figure 2 with Figure 1. The far-field mass in Figure 2 is smaller than that in Figure 1 for the same initial mass. Due to smaller release from the repository (or longer residence time in the repository), more neptunium decays while it is in the repository. Thus, the mass-in-the-repository curves are shifted to the right, compared with those in Figure 1.

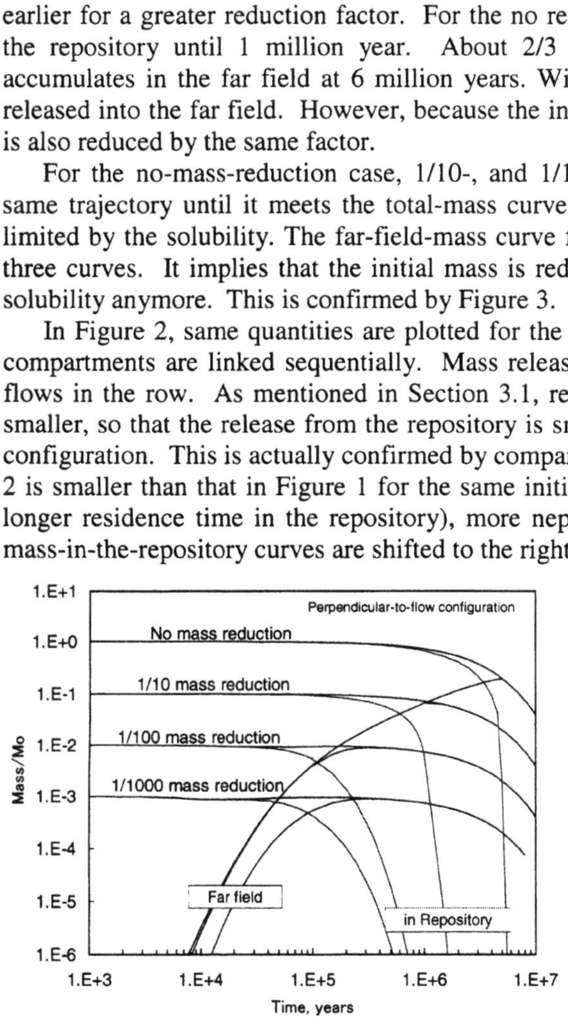

Figure 1: Effect of initial ^{237}Np mass reduction on mass distribution in the repository and in the far field for the perpendicular-to-flow configuration.

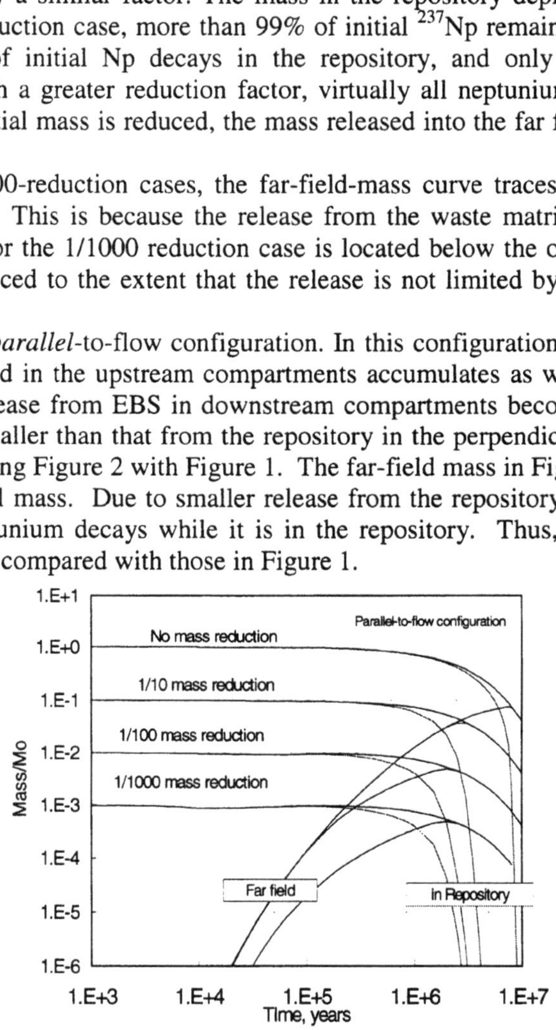

Figure 2: Effect of initial ^{237}Np mass reduction on mass distribution in the repository and in the far field for the parallel-to-flow configuration.

Note that in any case the total-mass curve is an exponentially decaying curve, which is determined by the decay constant of ^{237}Np and the initial mass. It is confirmed by the observation that by summing equations (1) through (4) for all compartments, internal-flux terms such as $q_{k,j,n}(t)$, $Q_{k,j,n}(t)$, and $\phi_{k,j,n}$ cancel, resulting in a simple decay equation.

In Figure 3, the change of the concentration $C_1(t)$ as a function of time is shown for the *perpendicular* configuration. Note that in this configuration, $n = 1$ for all the compartments. For the cases with no, 1/10-, and 1/100-reduction, the concentration is at the same magnitude while the time span decreases as the mass is reduced at a greater factor. This indicates that the release from the waste matrix is limited by the solubility of neptunium. On the other hand, in the 1/1000-reduction case, the concentration is substantially smaller than the other three cases. This is because the release from the waste matrix is controlled by the matrix dissolution (see eq. (6)). The rate of mass release into the far field is obtained by $64FC_1(t)$.

This depicts the situation often observed in previous studies for P-T effects on the repository performance. The concentration depicted in this figure are the inlet boundary condition for the transport model in the far field. Because the magnitude of the concentration in the far field is determined by the inlet boundary concentration, the repository performance would not differ.

In Figure 4, the change of the concentration $C_{64}(t)$ as a function of time is shown for the *parallel* configuration. The rate of mass release into the far field is given by $FC_{64}(t)$. Due to accumulation during the transport through 64 compartments, the concentration at the 64-th compartment is close to the solubility. However, there is a clear concentration decrease by the initial mass reduction. Therefore, if the repository configuration is taken into account in the repository performance model, effects of P-T scheme could have been expressed more sensitively even with the conventional concentration-based performance measure.

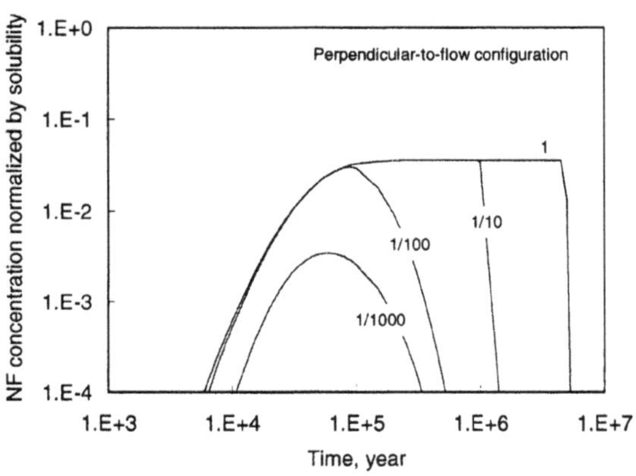

Figure 3: Effect of initial ^{237}Np mass reduction on ^{237}Np concentration in the near field rock for the perpendicular-to-flow configuration. "NF" in the vertical axis stands for the near field.

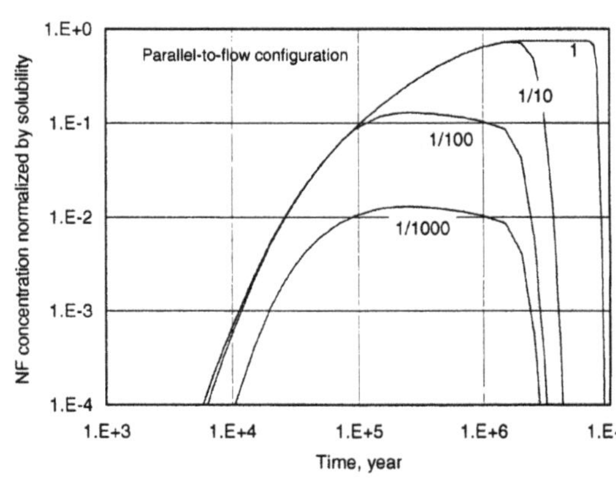

Figure 4: Effect of initial ^{237}Np mass reduction on ^{237}Np concentration in the near field rock for the parallel-to-flow configuration. "NF" in the vertical Axis stands for the near field

5. CONCLUSIONS

A repository performance assessment model has been developed, where repository configuration, including canister arrangement and EBS configuration, is taken into account. Multiple canisters, multiple-member decay chains, solubility-limited release from waste matrix are incorporated into the model as well. With the model, the mass in the repository and the mass in the far field region for a radionuclide can be obtained numerically as a function of time. Observing the mass distributions, the following conclusions can be drawn.

- For long-lived radionuclides, such as ^{237}Np, the repository configuration has significant effects on release from the repository into the far field. Short-lived radionuclides decay out in the buffer.
- The mass of ^{237}Np in the repository is reduced proportionally by the reduction of initial ^{237}Np mass. Smaller mass in the repository is advantageous from the viewpoints of human-intrusion risk and criticality-accident risk.
- The mass in the far field region is also reduced proportionally by the reduction of initial mass, especially at later times. At early times ($<10^5$ yr), the initial mass reduction effect on the far-field mass is not significant.

- The insensitivity of the conventional concentration-based performance assessment results to the initial mass reduction has been reproduced in the perpendicular configuration, which is the same configuration as assumed in the previous performance assessment studies.
- For the parallel configuration, the concentration of neptunium in the water leaving the repository is reduced by the initial mass reduction. Thus, by taking into account canister arrangement and interaction among multiple canisters in the repository, the conventional concentration-based performance measure could also be sensitive to initial mass reduction.

ACKNOWLEDGEMENT

The author is grateful for Mr. D. Kawasaki for his help with the numerical work. The author thanks Prof. P. L. Chambré for stimulating and fruitful discussions.

REFERENCES

Ahn, J. (1998), Integrated Radionuclide Transport Model for an HLW Repository in Water-Saturated Geologic Formations, Nuclear Technology, **121**, 24-39.

Ahn, J. (1999a), Preliminary Assessment of the Effects of ATW System Application on YM Repository Performance, *Global '99, International Conference on Future Nuclear Systems,* August 29-September 3, 1999, Jackson Hole, Wyoming.

Ahn, J. (1999b), Possibility of Safety Improvement for Vitrified HLW Geologic Disposal by Partitioning and Transmutation, *Global '99, International Conference on Future Nuclear Systems,* August 29-September 3, 1999, Jackson Hole, Wyoming.

Croff, A. G., Blomeke, J. O., and B. C. Finney (1980), *Actinide Partitioning-Transmutation Program Final Report*, I. Overall Assessment, ORNL-5566.

International Atomic Energy Agency (1982), *Evaluation of Actinide Partitioning and Transmutation*, Tech Report, 214, Vienna.

Japan Nuclear Cycle Development Institute (2000), *Second Progress Report on Research and Development for the Geological Disposal of HLW in Japan, H12: Project to Establish the Scientific and Technical Basis for HLW Disposal in Japan.*

Lowenthal, M. D., Ahn, J., Chambré, P. L., Greenspan, E., Kastenberg, W. E., Park, B., and Stone, N. (1999), Impacts of Waste Transmutation on Repository Hazards, *Global '99, International Conference on Future Nuclear Systems,* August 29-September 3, 1999, Jackson Hole, Wyoming.

Pigford, T. H. (1990), *Actinide Burning and Waste Disposal*, MIT International Conference on the Next Generation of Nuclear Power Technology, UCB-NE-4176, Univ. of California, Berkeley.

Ramspott, L. D., Choi, J.-S., Halsey, W., Pasternak, A., Cotton, T., Burns, J, McCabe, A, Colglazier, W., and Lee, W. W.-L. (1992), *Impact of New Developments in Partitioning and Transmutation on the Disposal of High-Level Nuclear Waste in a Mined Geologic Repository*, UCRL ID-109203, Lawrence Livermore National Laboratory.

US Department of Energy (1998), *Viability Assessment of a Repository at Yucca Mountain*, DOE/RW-0508.

US Department of Energy (1999), *A Roadmap for Developing Accelerator Transmutation of Waste (ATW) Technology, A Report to Congress*, DOE/RW-0519.

Pergamon
www.elsevier.com/locate/pnucene

Progress in Nuclear Energy, Vol. 40, No. 3–4, pp. 423–430, 2002

0149-1970/02/$ - see front matter

PII: S0149-1970(02)00034-3

SIMPLE ASSIGNMENT OF PARTITIONING & TRANSMUTATION OBJECTIVES TO REDUCE THE OVERALL REPOSITORY IMPACTS DUE TO THERMAL LOAD AND NUCLIDE MIGRATION

OHE, TOSHIAKI

Nuclear Engineering Dept. , Tokai University
1117 Kita-Kaname, Hiratsuka, Kanagawa, 259-1292, Japan
TEL +81-463-58-1211 ext.4144
FAX +81-463-50-2017
e-mail Ohe@keyaki.cc.u-tokai.ac.jp

Partitioning & Transmutation (P&T) technique is considered to reduce the potential radiotoxicity in high-level radioactive waste (HLW) disposal. The goal of the P&T process such as decontamination factors, transmutation rates, are often assigned from the hazard indices (HI) but not fully investigated from a view point of the radiological impact so far. It is pointed out that the removal and reduction of the insoluble nuclide such as Np-237 is insensitive to the total dose for the most probable scenario of the nuclide migration in underground water. This contradiction is caused by the difference of the performance measures used in the safety analysis of disposal (Dose) and P&T design procedures (HI). Simple analyses were thus conducted to assign the practical goal of the P&T to improve the repository performance. (1) As for partitioning, the nuclides were classified into several groups of similar properties of thermal load, migration impact, and chemical behavior. Then optimal nuclide groups were extracted using the cluster analysis of the above-mentioned factors. (2) For transmutation, the limit nuclide inventories corresponding to the 1 ALI/y hazard were calculated by assuming simplified geosphere conditions. The biosphere conditions were neglected to avoid the uncertainty of the parameters but the figures are easily converted to the dose rate afterwards by employing an appropriate biosphere model. The proposed nuclide grouping and the limit inventories are completely independent from process availability of P&T and purely reflect the factors affecting the repository performance. The comparison between the proposal and the existent P&T examples would resolve the contradiction and bring harmonized connection between the HLW disposal and the P&T.

1. INTRODUCTION

The concept of partitioning and transmutation (P&T) has been discussed internationally (OECD,1999,

IAEA,1995) and domestically (Inoue et al., 1991) to overcome the difficulties arising from the long-term stability of the deep geologic disposal. A number of research activities have been carried out so far and the discussions mainly focus on the technical feasibility based on the reactor physics and/or the engineering chemistry. However, it is not clear whether these activities proceed toward the final goal to sweep away the fear of the long-term issue of geologic disposal.

Ultimate zero release of radionuclides out of the fuel cycle is practically impossible if the partial loss in the process is taken into account (Yanama,1996) thus the final disposal of the process wastes is, more or less, not avoidable. This means the consideration is necessary whether the process is available from both points of technical viability and disposal acceptability. However, the latter point has not been fully investigated. The reason of such gap may be caused by the lack of the tight connection among the specialists in the fields of both P&T and the geologic disposal.

This paper is trying to present the practical goal of P&T process from the viewpoint of the deep geologic disposal in order to bridge the gap between P&T and disposal.

2. PARTITIONING OF NUCLIDES

To minimize the environmental impacts, the waste charged into geosphere must be limited as low as possible. If the waste charge is minimum, the resulting environmental change may be minimum. This MINIMUM CHARGE / MINIMUM CHANGE concept is applied in the paper.

Two aspects are considered to make the waste charge minimum; radiological impacts due to migration of potentially hazardous nuclides with long half-lives such as Tc-99 and Np-237, and thermal impact due to heat generation nuclides such as Sr-90 and Cs-37 and.

2.1 Migration impact

Assuming the dispersion coefficient D is approximated by $D=LU_P/10$ (Gelhar et al., 1985), the equations of the peak nuclide concentration in groundwater are simply derived for two ultimate boundary conditions from the analytical solutions (Bear,1979); for instantaneous discharge of soluble nuclide derived by Eq.(1), and for solubility-limited dissolution of insoluble nuclides by Eq.(2),

$$\frac{C_{max}(L)\cdot SL}{M}=\frac{10}{\sqrt{\pi\kappa}}\exp\left[-\frac{(10-\kappa)^2}{4\kappa}-\frac{\kappa}{10}\cdot\frac{\lambda R_f}{U_p}L\right] \qquad \text{Eq.(1)}$$

where, C: nuclide concentration; L: distance from the source; M: nuclide inventory; S: discharge cross-section; λ: decay factor; U_P: pore water velocity

$$\kappa=\left[\sqrt{1+10^2\left(1+\frac{4}{10}\cdot\frac{\lambda R_f}{U_p}L\right)}-1\right]\Big/\left(1+\frac{4}{10}\cdot\frac{\lambda R_f}{U_p}L\right)\cong 10 \quad (\textit{for most cases}).$$

$$\frac{C_{max}(L)}{C_0}=\exp\left\{10\left[\frac{1}{2}-\sqrt{\frac{1}{4}+\frac{1}{10}\left(\frac{\lambda R_f}{U_p}L\right)}\right]\right\} \qquad \text{Eq.(2)}$$

where, C_o: nuclide solubility.

The above equations are characterized by a factor $A(=\lambda R_f L / U_p)$ and similar functions for the two boundary conditions are found as depicted in Fig.1. In particular, the factor A is divided into the nuclide dependent term λR_f and the geologic term L/U_p and the latter is independent of nuclide property thus λR_f (or $T_{1/2}/R_f$) term is used for nuclide grouping.

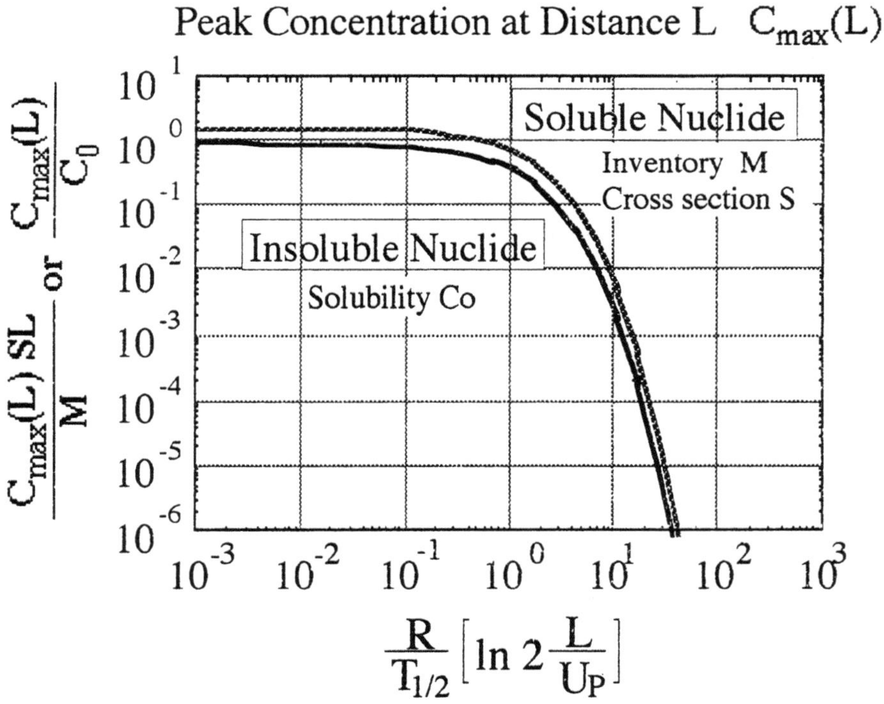

Fig.1 The peak concentration of two ultimate boundary conditions

2.2 Thermal impacts

The temperature increase in the repository due to the decay heat of nuclide was calculated by the heat conduction analysis. For simplicity, the underground repository is treated as a spherical medium and the temperature rise due to a single nuclide is derived by the non-stationary analytic equation (Mufti,1971). The maximum temperature was calculated by the derivative of the non-stationary equation with time and an empirical expression Eq.(3) for the maximum temperature rise was finally derived by the regression analysis,

$$K\frac{\Delta T_{max}}{W_0} \cong \frac{3}{4\pi R}\left[\frac{1}{3} - 0.110\left(R^2\lambda\right)^{0.278}\right] \qquad \text{Eq.(3)}$$

where, ΔT_{max}: maximum temperature rise
K: thermal conductivity of surrounding medium
W_o: initial heat generation in a single waste
R: waste radius.

The equation (3) indicates the maximum temperature rise of the surrounding geologic medium when the spherical heat source with the radius R is placed in the medium of the thermal conductivity K. The general tendency of the temperature rise is a function of $R^2\lambda$ and the thermal impact of each nuclide having initial heat generation W_o can easily be speculated.

2.3 Classification of nuclides with impact factors

Nuclides in a spent fuel were classified into several groups by using the Eq.(1) or Eq.(2) and Eq.(3) from the viewpoints of migration and thermal impacts, respectively. The maximum temperature rise of each nuclide calculated by Eq.(3) with assumptions of 150L/MTU waste volume (R=0.33m) is shown in Fig.2 together with the migration properties $T_{1/2}/R_f$ which was calculated from the appropriate data in the Ref. (JNC 2000). In this figure, three distinct groups are found; heat dominant group, migration dominant group, and others.

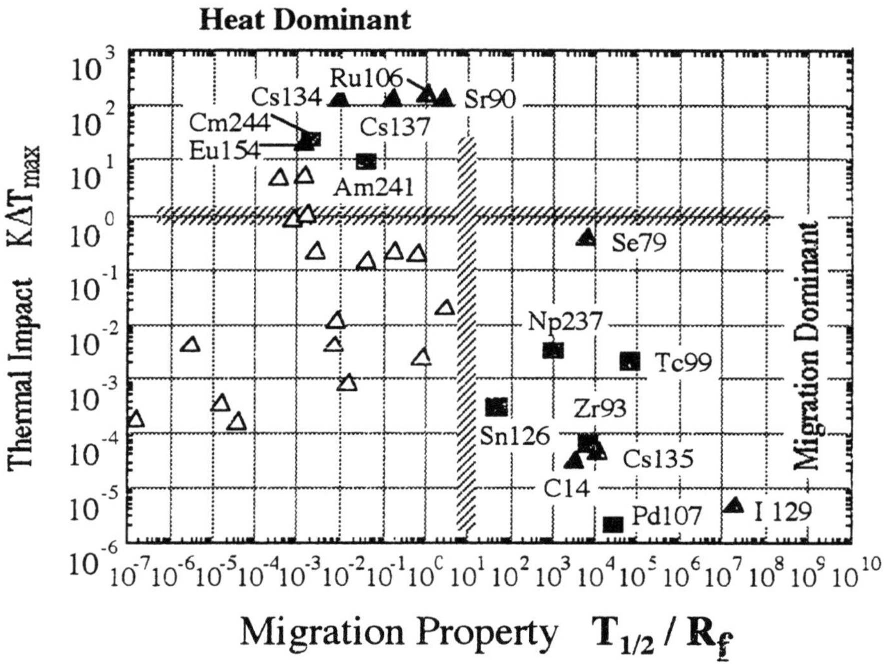

Fig.2 Classification of nuclides by the impact factors of each nuclide

2.4 Cluster analysis for appropriate grouping

Grouping of nuclides was conducted by employing the cluster analysis(Anderberg 1973). In this analysis, the following factors were introduced to identify the properties of nuclide to be concerned:

a. $T_{1/2}/R$: migration impact factor
b. $K\,T_{max}$: thermal impact factor
c. D_{con} : dose conversion factor
d. M : inventory
e. F_{chem} : chemical factor.

The chemical factor F_{chem} was defined as the element property including (1)electronegativity (Pauling, 1960), (2)Gibbs free energy of formation of the corresponding oxide (Phillips,1998, Katz,1986), (3)redox potential of the dominant valence species speculated in undergroundwater (Atkins,1981). The value of the chemical factor F_{chem} was assigned by the principle component analysis (Flury 1988) as a single numeric value for each element. This factor was introduced in the cluster analysis for simply taking into account (1)the degree of complex formation, (2)chemical stability in the waste form, and (3)redox sensitivity in the repository environment, respectively. Five nuclide groups were then extracted from the cluster analysis as seen in Fig.3.

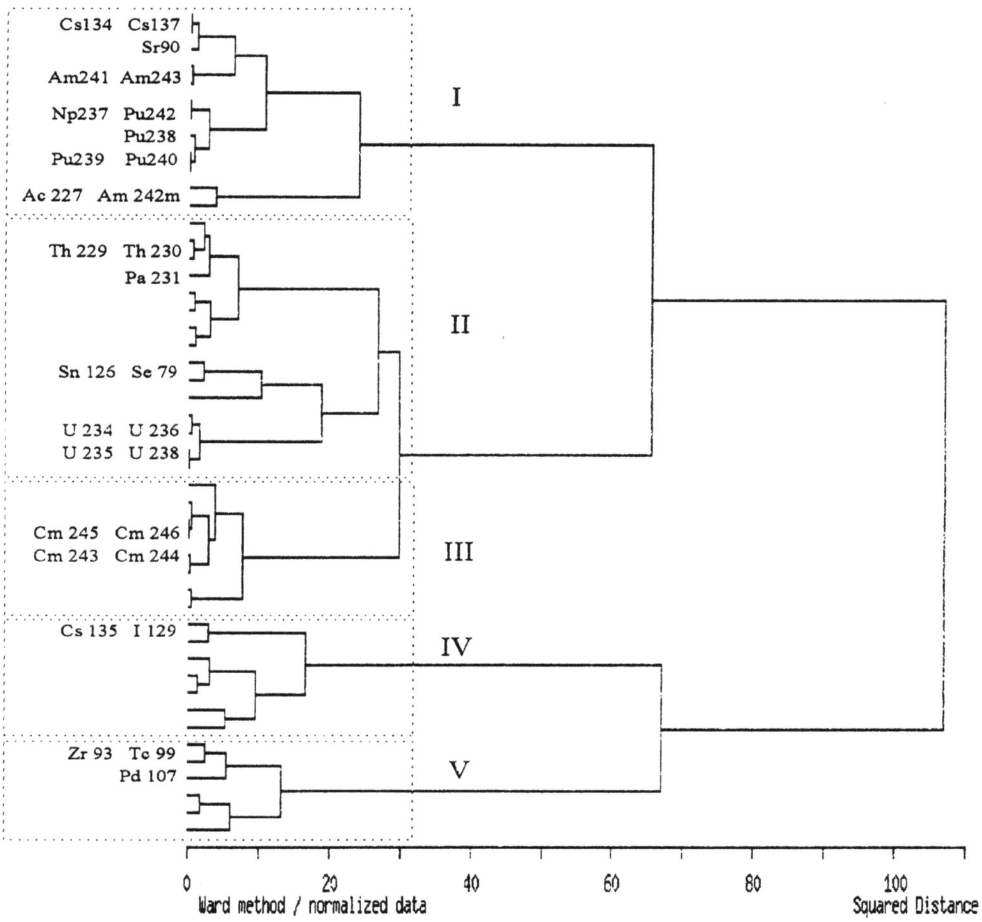

Fig.3 The result of the cluster analysis using Ward method

In practical manner, the partitioning process can be developed by employing elemental separation techniques. The grouping was thus revised by the element basis. The isotope with the maximum migration impact factor $T_{1/2}/R$ was selected as the representative of each element and the grouping was conducted again. The appropriate element groups for partitioning from the viewpoint of underground disposal were extracted as shown in Table 1. The group A is not necessary to be disposed in deep geologic environment after appropriate decay period and another nuclides not shown in Table 1 are to be included in this group A as well. On the contrary, groups B, C, and D require the deep or medium depth geologic disposal (longer migration distance L and/or reducing environment may be required) because of the long-lived nuclides. All decay chain member nuclides are included in group B because the production and accumulation would occur and mix again sooner or later even if complete elemental separation is accomplished.

Table 1 Nuclide grouping extracted by the cluster analysis

Group	Nuclides	Element	Rearrange	Partitioning
I	Cs-134,137 Sr-90	Sr	Sr	A
	Am-241,242,243	Am	Cm, Am, Np, U, Th, Ac, Pa	B (decay chain members)
	Np-237	Np		
	Pu-238,239,240,242	Pu		
II	Ac-229	Ac		
	Th-229,230	Th		
	Pa-231	Pa		
	Se-79 Sn-126	Se, Sn	Se, Sn	C
III	U-234,235,236,238	U		
IV	Cm-243,244,245,246	Cm		
V	Cs-135 I-129	Cs, I	Cs, I	D
	Zr-93 Tc-99 Pd-107	Zr, Tc, Pd	Zr, Tc, Pd	

3. TRANSMUTATION OF ISOTOPES

Transmutation is designed to reduce the hazard index defined as M [Bq]/ALI [Bq/y]. This term however dose not represent the impact of the disposal and contradiction is found due to the different measures of incentives. For example, Cm-245 and Am-241 would have high hazard index value but have not significant migration impact because of its low solubility and relatively short half-life. Thus, alternative index to promote the incentive to reduce the migration impact must be assigned.

3.1 The limit inventory corresponding to 1 ALI annual impact

The limit inventory corresponding to the 1 [ALI] impact was calculated by assuming a very simple migration model. As a preliminary analysis, the performance of the engineered barrier such as waste form, canister, overpack, and buffer materials was pessimistically neglected. The behavior of nuclides in the biosphere was also neglected to avoid the accumulation of uncertainty of the parameters used for calculation. Under these assumptions, the maximum flux at a fixed distance, for example, L=100m was simply calculated by Eq.(4) which is an approximate form (κ=10 and S=1) derived from Eq.(1)

$$U_d C_{\max}(L) S \cong U_d \frac{M_{\lim}}{L}\sqrt{\frac{10}{\pi}}\exp\left(-\frac{\lambda R_f}{U_p}\cdot L\right) \Rightarrow 1\ ALI[Bq/y] \qquad \text{Eq.(4)}$$

where, $R_f = 1 + (1-\varepsilon)K_d\rho/\varepsilon$
K_d : sorption coefficient
ALI : annual limit of intake
ε: porosity of medium
ρ: density of medium
U_p, S, $C_{max}(L)$: see Eq.(1)
U_d, : Darcy velocity
M_{lim}: limit inventory of 1 ALI annual impact.

3.2 Examples of the limit inventory

The preliminary calculation of the limit inventory of 1 ALI annual impact for the long-lived nuclides were conducted by assuming a Japanese repository concept (JNC,2000). The geological medium of the repository is assumed to be granite and wastes correspond to 32000 MTU of PWR spent fuel are disposed into the repository. The calculation results of the limit inventories M_{lim} by applying Eq.(4) are summarized in Table 2. The data indicates the limit inventories correspond to the 1 ALI annual impact for dominant long-lived nuclides are approximately in the range of $10^3 \sim 10^7$ [MBq]. From these figures, requirement for the transmutation to maintain the repository inventories of nuclides less than the limitations may easily be calculated. Two examples are listed in the last two column in Table 2 shown as "elimination rate" which is a ratio of the limit inventory to the initial inventory. One was based on Eq.(4) and the other was on Eq.(1), which is an exact form of Eq.(4). Although large differences between two values were found for Np-237 and Se-79 because of inappropriate assumption (κ =10) for Eq.(4), total inventories enable to transfer into the repository must be eliminated by a factor of $10^4 \sim 10^5$. Good performance of the engineered barrier and biosphere may reduce the requirement and resulting elimination rate may become lower. It must be reminded this value is the response of the inventory to the migration impact and no regulatory limit of the repository inventory can be drawn from this figure. If a larger or smaller impact other than 1 ALI is assumed, the revision of the elimination rate is straightforward because of the linear function of inventory M as depicted in Eqs.(4) or (1).

Table 2 Elimination rate corresponding to 1ALI annual impact*

Nuclides	K_d ** [m3/kg]	ALI [Bq/y]	M_{lim} [MBq]	Elimination Rate*** Eq.(4) applies	Eq.(1) applies
Np-237	1	5.00E+04	1.37E+06	2.4E-03	8.2E-05
Pd-107	0.1	1.00E+09	7.41E+06	4.3E-02	4.1E-02
Se-79	0.01	2.00E+07	1.84E+06	2.8E-03	1.4E-04
Tc-99	0.001	1.00E+08	6.11E+05	3.0E-05	2.9E-05
Zr-93	0.1	1.00E+08	1.84E+06	6.3E-04	5.0E-04
Cs-135	0.05	3.00E+07	2.50E+05	3.6E-04	3.4E-04
I-129	0	5.00E+05	2.80E+03	5.7E-05	7.4E-05

* medium properties : ε =0.01, ρ =2650 [kg/m^3], U_p =1 [m/y], L=100 [m]

** taken from Ref. (JNC,2000)

*** requirement of the reduction of the inventory corresponding to the 32000 MTU PWR spent fuel : burn-up 45000 [MWD/MTU], initial enrichment 4.5[%] (equivalent to the repository designed in the Ref. JNC,2000)

4. CONCLUSION

This paper indicates the simple and practical goal of the partitioning and transmutation from the viewpoint of geologic disposal. A procedure employing the cluster analysis is presented and examined to classify the elements into several groups for partitioning from both thermal and migration impacts. The limited inventory corresponding to the 1 ALI annual migration impact is proposed and calculated as the preliminary objectives of the transmutation.

Although the figures presented here are limited, one would expect the staring value of the discussions of the P&T objectives to reduce the concern about the long-term issue of the high-level radioactive waste disposal. Contradiction still remains among the present study and the existent investigations and further discussions are necessary from both viewpoints of technical feasibility and acceptability.

5. REFFERENCES

OECD/NEA.(1999), *Actinide and Fission Product Partitioning and Transmutation -Status and Assessment Report –* , OECD/NEA.

IAEA,(1995), *Safety and Environmental Aspects of Partitioning and Transmutation of Actinides and Fission Products*, IAEA.

Inoue,T., Sakata,M., Miyashiro,H., Matsumura,T., Sasahara,H. and Yoshiki,N., (1991),Development of Partitioning and transmutation technology for long-lived nuclides, *Nucl.Technol,206, **69***.

Yamana,H and M.Shiotsuki,M.,(1996), Recovery Yield of Nuclides Required for Satisfactory Actinide Recycling, *Radioactive Waste Research, 34, **2(1/2)***.

Gelhar,L.W., et al.,(1985), *A Review of Field-Scale Physical Solute Transport Processes in Saturated and Unsaturated Porous Media*, EA-4190, EPRI, Palo Alto.

Bear,J., (1979), *Hydraulics of Groundwater*, p267-270, McGraw-Hill.

Mufti,I.R.,(1971), Geothermal Aspects of Radioactive Waste Disposal into the Subsurface, *J.Geophys.Res., 8568, **76(35)***.

Anderberg,M.R.,(1973), *Cluster Analysis fro applications*, Academic Press.

Pauling,L.(1960), *The Nature of the Chemical Bond,* Cornell University Press.

Phillips,S.L. et al., (1988), *Thermodynamic Tables for Nuclear Waste Isolation*, NUREG/CR-4864, LBL-22860, Laurence Berkeley Laboratory, Berkeley.

AtkinsP.A. et al, (1998), *Physical Chemistry*, 6th Edition, Oxford.

Katz,J.J, Seaborg,G.T., and Morss,L.S., (1986), *The Chemistry of the Actinide Elements*, 2nd Edition, Chapman and Hall.

Flury,B. and Riedwgl,H.,(1988), *Multivariate Statistics – A practical approach-*, Chapman and Hall.

JNC,(2000), Project Overview Report, H12, Project to Establish Technical Basis for HLW Disposal in Japan, JNC TN1410 2000-001, Japan Nuclear Fuel Cycle Development Institute.

Pergamon
www.elsevier.com/locate/pnucene

Progress in Nuclear Energy, Vol. 40, No. 3–4, pp. 431–439, 2002

Printed in Great Britain
0149-1970/02/$ - see front matter

PII: S0149-1970(02)00035-5

THE ENCAPSULATED NUCLEAR HEAT SOURCE REACTOR FOR LOW-WASTE PROLIFERATION-RESISTANT NUCLEAR ENERGY

E. GREENSPAN and the ENHS Project Team

Department of Nuclear Engineering
University of California, Berkeley, CA 94720
(510) 643-9983; gehud@nuc.berkeley.edu

ABSTRACT

The encapsulated nuclear heat source (ENHS) is a new Pb-Bi cooled modular reactor concept that features a combination of the following useful features that may make nuclear energy more attractive: (1) 20 years of full power operation without refueling. (2) Nearly constant fissile fuel contents and k_{eff}. (3) No on-site refueling and fueling hardware. (4) The ENHS modules are factory manufactured and transported already fueled to the site. (5) No access to neutrons. (6) No mechanical connections between the ENHS module and the energy conversion plant (The ENHS module has the function of a nuclear battery – with 20 years of full power operation at 125 MW_{th}). (7) At end of life, the ENHS module serves as a spent fuel storage cask and, later, as a spent fuel shipping cask. That is, the fuel is locked inside the ENHS from "cradle to grave". (8) 100% natural circulation resulting in passive load following capability and autonomous control. This combination of features offers a highly safe nuclear energy system that is characterized by low waste, high proliferation resistance and high uranium utilization. The low waste and high uranium ore utilization are achieved by recycling the Pu and MA many times using a proliferation-resistant dry process; only fission products are to be extracted between cycles. Spent LWR fuel can provide for the HM make-up. The high level of proliferation resistance is obtained by restricting access to the fuel and neutrons and by eliminating the economic incentive of the client country to invest in sensitive technologies or infrastructure that can be used for clandestine production of strategic nuclear materials.

1. INTRODUCTION

The encapsulated nuclear heat source (ENHS) is a new modular reactor concept (Greenspan and Wade, 1999) that was selected by the 1999 DOE NERI program as a candidate "Generation-IV" reactor. It is a fast- neutron-spectrum reactor cooled by Pb-Bi using natural circulation. It is designed for passive load following, for a high level of passive safety, and for 20 years of operation without refueling. One of the unique features of the ENHS is that the fission-generated heat is transferred from the primary coolant to the secondary coolant across the reactor vessel wall by conduction – providing for an essentially sealed module that is easy to install and replace.

The ENHS modules are factory manufactured and transported already fueled to the site for insertion into the Pb-Bi secondary coolant pool connected to the client's energy conversion plant. The ENHS module has the function of a nuclear battery with a 20 years of full power operation at 125 MW_{th}. At end of life the module is withdrawn and returned to the factory or a spent fuel center and a new module is inserted into the client's energy conversion plant.

The ENHS core is designed to maintain nearly constant k_{eff} throughout its 20 years of operation so that very small excess reactivity need be built into it. This feature, along with negative temperature coefficient, large temperature margin to coolant boiling, and passive decay heat removal makes the ENHS extremely safe. There is no need for active control of reactivity or power level; after the reactor is brought to full power, the compensation for reactivity changes and for power variations is done autonomously.

The ENHS reactor concept offers unique combination of low waste, high proliferation resistance and high uranium utilization. After a brief description of the ENHS reactor concept (Section 2), we discuss the feasibility of multi-recycling for low waste and high uranium ore utilization (Section 3) and follow with a discussion of the unique proliferation-resistant attributes of ENHS-based nuclear energy systems (Section 4). More details about the ENHS concept and its feasibility assessment can be found in publications by Greenspan et al (2000a, 2000b and 2001), Sienicki and Wade (2001) and Saphier et al (2001).

2. BRIEF DESCRIPTION OF THE ENHS

A description of one of many possible embodiments of a reactor concept having a single ENHS module is depicted in Figs. 1, 2 and 3. The reactor consists of nine modules: one ENHS module and eight steam generators. All of these components are inserted into the reactor pool; they are supported by a massive structural platform. There is no mechanical connection between these components; this makes them relatively simple to install and to replace.

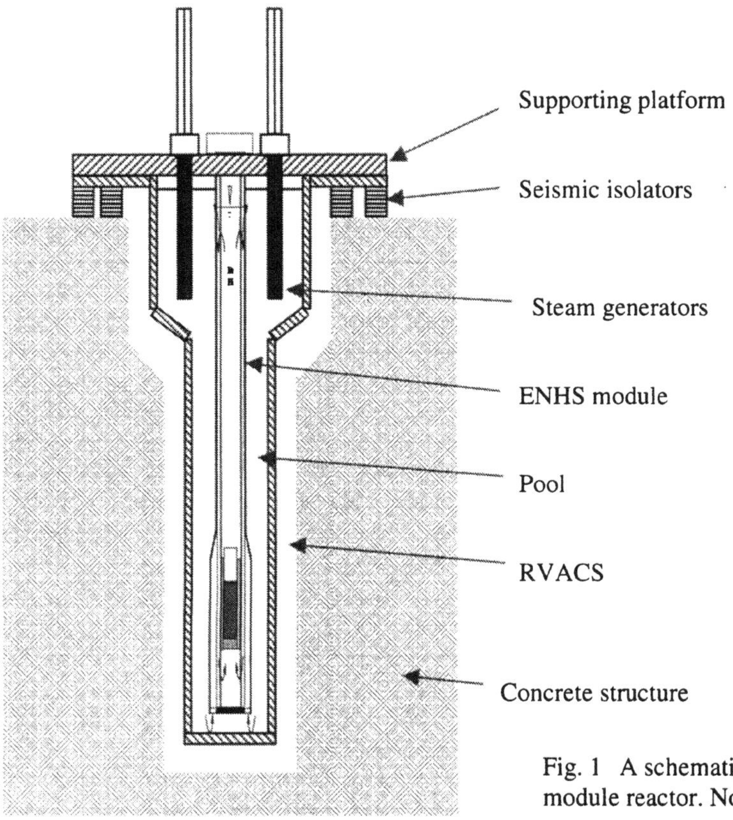

Fig. 1 A schematic vertical view of a single ENHS module reactor. Not to scale.

The ENHS module concept shown in Fig. 2 features 100% natural circulation. There are an inner- and outer structural walls to the ENHS reactor vessel with the Intermediate Heat Exchanger (IHX) in between. The IHX, depicted in Fig. 3, consists of rectangular channels that are connected at their top and bottom to a tube sheet (not shown in Fig. 2). The rectangular channel walls provide the barrier between the primary and the secondary coolants. The primary coolant flows from the riser into the rectangular channels near the top of the vessel through openings in the inner structural wall. The primary coolant flows back into the coolant plenum underneath the core through openings near the bottom of the inner structural wall. The secondary coolant flows from the pool outside of the vessel into the IHX near the bottom of the IHX and flows up between the rectangular primary-coolant channels in the space between the inner structural wall and the outer structural wall. The secondary coolant exits back to the pool near the top of the IHX. Heat is conducted from the primary to the secondary coolants through the walls of the rectangular channels. These walls are designed to be 4 mm thick and the ENHS module dimensions are set to give 100% natural circulation along with a primary-to-secondary temperature drop of 50 °C.

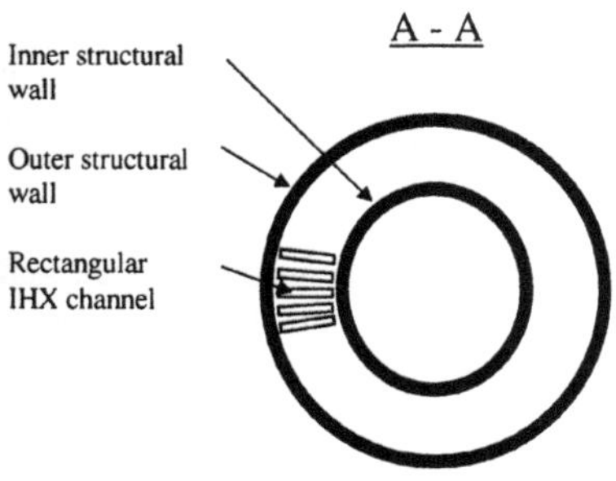

Fig. 3. Schematics of an ENHS IHX made of rectangular channel boxes

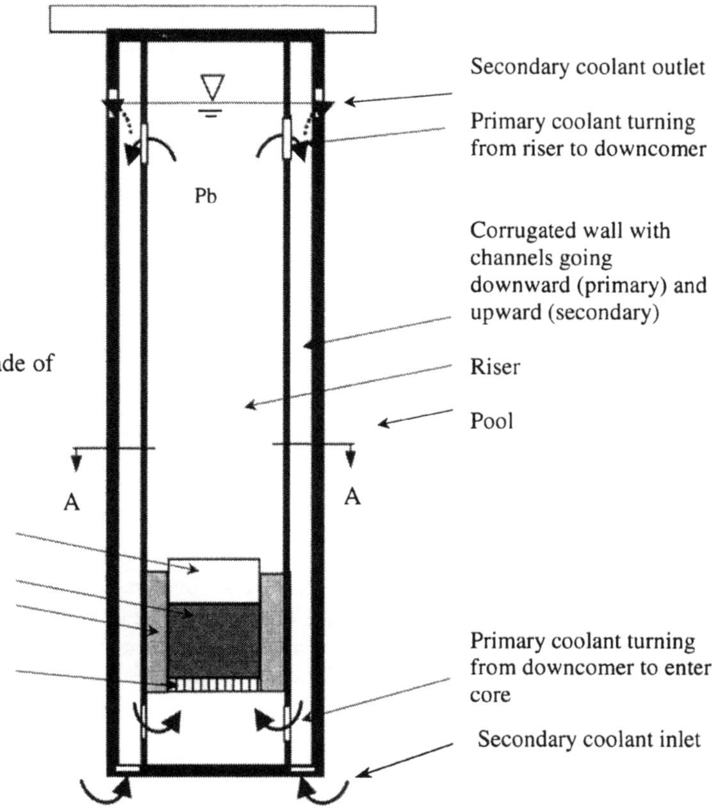

Fig. 2 A schematic vertical cut through an ENHS module having 100% natural circulation. Not to scale.

As presently envisaged, the ENHS will be manufactured and fuelled in the factory and shipped to the site as a sealed unit with solidified Pb-Bi (or Pb) filling the vessel up to the upper level of the fuel rods. Upon insertion to the reactor pool hot Pb-Bi is to be pumped into the ENHS vessel through a small diameter pipe. This hot Pb-Bi, along with the hot Pb-Bi in the pool, will melt the solid Pb-Bi at the lower part of the vessel.

At the end of its life, currently planned to be 20 EFPY, the ENHS will be removed from the reactor pool and stored on site until the decay heat drops to a level such that passive cooling will permit the Pb-Bi to solidify. Before solidification, the coolant will be extracted from the upper part of the ENHS above the level of the fuel rods. The ENHS with the solidified Pb-Bi will serve as a shipping cask.

Currently, the Control and Shutdown Systems (C&SS) used in the 4S reactor (Hattori and Handa, 1989; Hattori and Minato, 1995; Aoki et al, 1995) have been adopted for the ENHS. The C&SS consist of a single neutron absorber assembly located centrally in the core and six reflector drive mechanisms that move six radial reflector segments. The shutdown assembly is securely latched in place during shipping and reactor installation. The shutdown rod includes an electromagnetic latch that does not engage until the startup temperature – 350°C – is achieved. At this point the rod can be withdrawn and held out until permanent shutdown is required. Normal operational shutdowns can be accomplished with the reflectors as needed. The reactor is brought critical by a hydraulic system that moves the reflectors at 1 mm/sec to compensate for the negative temperature coefficient of reactivity. At the full power position, the reflectors are stopped from further upward movement by mechanical stops whose movement is established by high-reliability gear drives. These drives restrict the rate of movement of the reflectors to approximately 1 mm/day. In the 4S reactor this movement is used to compensate for the burn-up reactivity effect. In the ENHS it is anticipated that the power level can be controlled over a wide range simply by using steam pressure and feed-water flow control. In summary, the reactor control should be very simple and fully automated: a combination of axial movement of the radial reflector surrounding the core and passive control by temperature feedback.

Each of the six reflector segments is to consist of approximately 5cm thick tungsten encased in stainless steel. The reflector is as long as the active-fuel region. In the shutdown position the level of tungsten, that is strongly absorbing, coincides with the active fuel level. The highest reactivity-state of the core is obtained with the tungsten withdrawn and Pb-Bi (Pb) surrounds the core. The reactivity worth of each reflector segment is less than half a dollar. Hence, a severe reactivity insertion accident is inconceivable.

The path for decay-heat removal is from the primary coolant through the IHX walls to the secondary coolant and through the pool wall to a passive reactor vessel air cooling system (RVACS). The ENHS is characterized by a very large surface area per MW_{th} of decay heat removal both from the primary to the secondary coolant, and from the secondary coolant to the RVACS. In addition, the ENHS is characterized by a large thermal inertia due to the large inventory of the primary and secondary Pb and the very large margin to coolant boiling. Both features are expected to make the ENHS highly passively safe.

The steam generators are of a tube-in-tube type. They are designed so that the flow rate of steam into the secondary coolant in case of a breech in a steam tube will be small so there can be no significant pressure buildup in the pool. Pressure relief valves are located at the top of the pool.

An alternate ENHS concept being considered (Greenspan et al, 2000a) uses a cover-gas lift-pump that circulates the cover gas from the plenum above the coolant level in the ENHS and injects it into the coolant in the riser through nozzles located at a level above the core. The cover gas bubbles reduce the effective density of coolant in the riser, thus increasing the head for coolant circulation. The circulator would be located above the reactor pool, outside of the ENHS. Thus, no moving mechanical parts and no electrical parts associated with coolant circulation would be located inside the ENHS. More careful analysis is required to verify the feasibility of the lift-pump design.

Table 1 summarizes selected design parameters of two representative ENHS modules while Table 2 compares selected performance characteristics of two representative ENHS cores with those of conventional power reactors.

Table 1 Selected design parameters of representative ENHS modules for $125MW_{th}$

Design parameter	ENHS1	ENHS2
Primary coolant circulation	100% natural	With lift-pump
Average linear heat-rate (W/cm)	60	60
Core height (m)	1.25	1.50
Core diameter (m)	1.98	1.87
Fuel rod diameter (cm)	1.0	1.0
Clad thickness (cm)	0.1	0.1
Lattice (hexagonal) pitch (cm)	1.45	1.50
Overall module height (m)	19.6	10.1
Outer module diameter (m)	3.2	3.3
Weight of Pb in module, when filled (t)	~660	~100
Weight of module structure (t)	~80	~70
Weight of fueled module for shipment (t)	~300	~130
Coolant core inlet/outlet temperature (°C)	400/550	400/550
Coolant flow velocity in core (m/s)	0.51	0.52

3. UNIQUE FEATURES OF THE ENHS

3.1 Long-life high-burnup cores

One goal of the ENHS is to have a long life core without refueling or fuel shuffling. Another goal is to have this long-life core operate with a nearly zero burnup reactivity swing. The design domain for cores that meet these design goals for at least 20 years of full power (EFPY) is defined in Greenspan et al. (2000b). What limits the core

Table 2 Comparison of selected performance characteristics of ENHS versus conventional reactors

Performance characteristics	Reactor type					
	ENHS1	ENHS3[c]	ALMR[a]	PWR	BWR	HWR
Av. q' (W/cm)	60	80	189	176	206	200
Power density (kW/l)	33	51	97	95	56	12
Specific power (W/g)	7	10	38	37	26	20
Fis. Pu inventory (g/kW)	12.5	8.8	2.3			
Peak fuel BU (GWD/tHM)	105[b]	105[b]	140	~50	~40	~10
Fueling intervals (y)	15	15	1.5	1.5	1.5	Cont.
Thermodynamic η (%)	40	40	38(33)	33	33	33

[a] This is a variant of ENHS2 (Table 1) that is designed to have a higher power density
[b] PRISM-S
[c] Limited by radiation damage to clad; otherwise could go to ~200 GWD/tHM

life is the radiation damage to structural materials, assumed to be $4\times10^{+23}$ n/cm^2. The corresponding peak fuel burnup is approximately 105,000 MWD/tHM. The fuel considered for these cores is IFR-like metallic Pu-UZr fuel with $10^{w}/_{o}$ Zr (Hofman et al., 1997). Typical Pu concentration is 11-12 $^{w}/_{o}$ of HM.

Fig. 4 shows the k_{eff} behavior in a typical nearly zero burnup reactivity swing core. The fluctuations in the k_{eff} value are statistical; they are due to use of a Monte Carlo code for these calculations. Detailed discussion of the assumptions and methods used can be found in Greenspan et al. (2000b). By adjusting the lattice p/d ratio and the Pu $^{w}/_{o}$, it is possible to change the slope of k_{eff} vs. BU. The feasibility of designing Pb-cooled cores to have nearly constant k_{eff} is consistent with findings of independent studies by Sekimoto et al. (1995) and by Orlov et al. (1994).

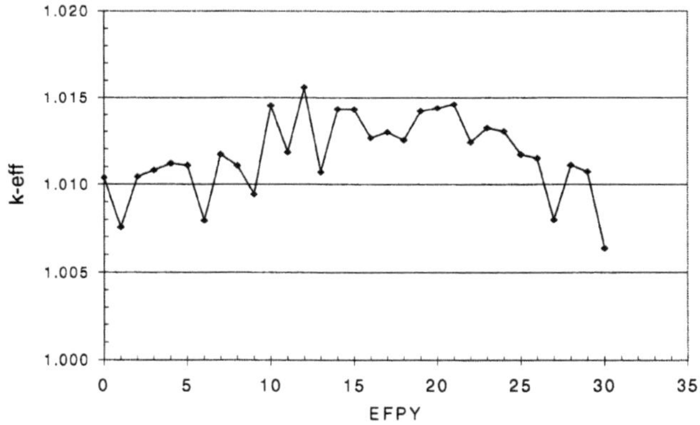

Fig. 4 Evolution of k_{eff} of an ENHS core for 125MW$_{th}$ having 2m height, p/d of 1.30, 11.4 $^{w}/_{o}$Pu and average linear heat rate of 80 w/cm.

3.2 Nearly constant fissile fuel content

Figures 5 and 6 and Table 3 show the change in the fuel isotopic composition with burnup of the same core examined for Fig. 4. What we observe is a slight increase in the concentration of ^{239}Pu and ^{240}Pu and a drop in the concentration of ^{241}Pu and ^{242}Pu. The net effect is a slight increase in the total inventory of Pu – less than 4% increase over 15 EFPY at 80 w/cm average linear heat rate (corresponding to 20 EFPY at 60 w/cm), a slight increase in the concentration of the even isotopes of Pu, and a noticeable decrease in the concentration of ^{237}Np and its precursors – ^{241}Pu, ^{241}Am and ^{245}Cm. The total concentration of the Minor Actinides (MA) keeps building up, while the concentrations of ^{235}U and of ^{238}U keep decreasing almost linearly. The slight increases in the concentrations of Pu and MA is necessary to compensate for the reactivity effect of Fission Products (FP) buildup.

3.3 No on site refueling

The combination of long-life core and nearly constant k_{eff} makes it possible to eliminate on-site refueling altogether. The ENHS module is designed to be factory fueled and to be discarded after 20 EFPY of operation. The fuel is to be enclosed inside the sealed ENHS module from the time the module leaves the factory until the spent module is returned to the factory or to a spent fuel regional or international center. It is envisioned that the ENHS power plants will not even have on-site hardware for refueling.

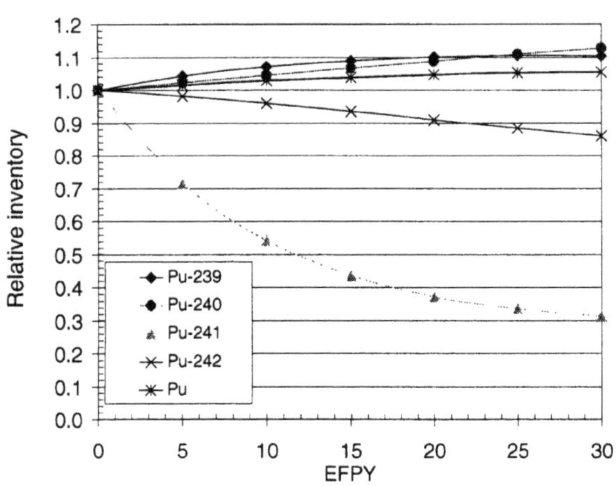

Fig. 5 Change in inventory of Pu isotopes in the core of Fig. 4. BOL weight % of even Pu isotopes is 0.264.

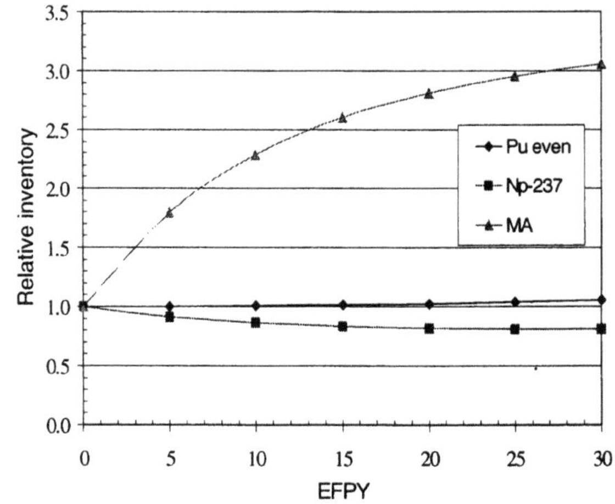

Fig. 6 Change in concentration of even Pu isotopes, of MA and of ^{237}Np and its precursors in the core of Fig. 4.

Table 3 Evolution of inventory (in tons) of fuel isotopes in the core of Fig. 4

	Effective full power years (EFPY)						
Isotope	0	5	10	15	20	25	30
U-234	0	1.15E-05	7.84E-05	0.000228	0.000462	0.000774	0.00115
U-235	0.023261	0.019242	0.015963	0.01329	0.011104	0.009316	0.007856
U-236	0	9.85E-04	1.79E-03	2.45E-03	2.99E-03	3.45E-03	3.82E-03
U-238	11.60731	11.32986	11.05919	10.79629	10.53989	10.28944	10.04469
Np-237	0	1.03E-03	2.02E-03	2.95E-03	3.78E-03	4.52E-03	5.17E-03
Pu-238	0	8.07E-04	2.86E-03	5.39E-03	7.98E-03	1.04E-02	1.27E-02
Pu-239	1.007076	1.049243	1.07846	1.097299	1.108009	1.112101	1.110856
Pu-240	0.325201	0.332238	0.339489	0.34669	0.353678	0.360402	0.366546
Pu-241	0.095912	0.068625	0.051958	0.041839	0.03574	0.032145	0.030096
Pu-242	0.070435	0.069167	0.067579	0.065848	0.064061	0.062287	0.060548
Am-241	0	0.018059	0.02877	0.035067	0.038725	0.040815	0.041998
Am-243	0	2.10E-03	3.88E-03	5.39E-03	6.68E-03	7.76E-03	8.66E-03
Cm-242	0	2.08E-04	3.50E-04	4.36E-04	4.84E-04	5.09E-04	5.26E-04
Cm-243	0	1.69E-06	5.76E-06	1.02E-05	1.42E-05	1.76E-05	2.03E-05
Cm-244	0	1.03E-04	3.52E-04	6.82E-04	1.05E-03	1.42E-03	1.75E-03
Cm-245	0	2.14E-06	1.44E-05	4.09E-05	8.18E-05	1.36E-04	2.00E-04
Cm-246	0	1.29E-08	1.77E-07	7.71E-07	2.11E-06	4.46E-06	8.05E-06

3.4 No opening of the reactor vessel outside factory or spent fuel center

The lack of need for on-site fuel handling, combined with the relatively small number of components inside the ENHS module, enables designing the module in a way that will make it unnecessary to ever open its cover when out of the factory or the spent fuel center. The components inside the module are all robust and will be designed to operate reliably for at least 20 EFPY. Thus we envision the top cover of the ENHS module to be factory welded.

3.5 No blanket and no reflector assemblies in or around core

Even if the top cover of the ENHS module were to be removable on site, it would not be possible to insert targets of fertile material to make fissile material. This is because there are no blanket assemblies in or around the core and no reflector assemblies around the core, as is common to have in sodium-cooled fast reactors. Moreover, we envision the core to be assembled at the factory from the bottom of the ENHS module.

4. LOW WASTE VIA MULTI-RECYCLING WITHOUT SEPARATION OF PLUTONIUM

Consider the ENHS core referred to in Figs. 4, 5 and 6 and in Table 3. As far as reactivity is concerned, the core could have operated at least twice as long. What limits the core life to ~105 GWd/tHM maximum burnup is the radiation damage to the structural materials. Recently Greenspan et al. (1998) proposed to get rid of the radiation-damage effects by partially reprocessing this spent fuel just to extract the volatile and some of the semi-volatile FP and re-fabricate the fuel rods. The make-up fuel, approximately 7% of the fuel loading, can be spent fuel from light water reactors (LWRs) from which the volatile and semi-volatile FP have been removed. Hopefully, this process can be repeated many times. So far we have not carried out the burnup analysis beyond an average discharge burnup of 20%. The technology that appears to be most suitable for the partial reprocessing suggested above is AIROX. Details about the AIROX process were published by Feinroth (1999).

There remains the question of how to come by the first core loading without separation of Pu. One possibility (Greenspan et al. 1998) is to use LWR spent fuel as the feed material and remove from it only part of the uranium and part or all of the FP. For example, if the LWR SF contains 1% Pu and MA, we need to remove approximately 90% of the uranium to make a fuel with 11 to 12 $^{w}/_{o}$ Pu and MA. This can be done in a proliferation resistant way using an IFR-like process. Other processes may be considered as well.

The total Pu and MA inventory in the proposed ENHS-based energy system of a given capacity is fixed. Most of this transuranics (TRU) inventory is well secured inside ENHS modules. The only waste anticipated from this energy system consists of the following: (1) The FP extracted in the dry process. (2) TRU and FP that can not be recovered from the spent fuel clad material. (3) TRU and FP waste from the fuel fabrication process. (4) Coolant and structural material activation products. We have not, as yet, characterized and quantified this waste.

5. PROLIFERATION RESISTANCE

The ENHS offers a unique combination of technological barriers and material barriers that, along with adequate institutional barriers can make the nuclear energy system extremely proliferation resistant. The technical barriers are of three categories: limited access to fuel (See Sec. 3.3 to 3.4 as well as Sec. 1 and 2), limited access to neutrons (See Sec. 3.4 and 3.5) and reduction of the host-country incentives for construction of sensitive fuel cycle facilities. Constructing and operating ENHS power plants will not give the country sensitive technologies or infrastructure that can be used for clandestine production of strategic nuclear materials. Specifically, the client country does not need any fuel fabrication or handling facilities.

The material barriers include the following: (1) The even isotopes always constitute more than 26 weight % of the Pu. (2) The concentration of ^{238}Pu keeps building up. (3) The Pu is mixed with many MA; it is never separated from the uranium and MA. (4) The overall inventory of Pu is smaller than that to be accumulated in an equal capacity nuclear energy system based on once-through LWRs. (5) The fuel always contains FP and thus has a radiation barrier. In fact, as the fuel is to be shipped to the site imbedded in Pb or Pb-Bi, it is possible to seed it with additional radioactive isotopes so as to increase the intensity of the radiation barrier without interfering with the shipment and installation of the ENHS module.

For the unique technological and material barriers offered by the ENHS to be most effective, it is necessary to come up with certain international agreements. For example: (1) The establishment of well safeguarded nuclear waste centers as well as of a limited number of co-located fuel fabrication and ENHS module fabrication facilities. These are to be located either in internationally recognized economically and politically stable weapons states, or in economically and politically stable non-weapon states with well developed nuclear power programs. (2) A ban on the development of enrichment, fuel reprocessing and fuel handling capabilities in countries that do not belong to the two categories defined above. Ideally, the fuel reprocessing and fabrication facilities are to be co-located with the ENHS module fabrication plant. The fuel feed to such a site will be spent fuel from LWRs and, when needed, natural or depleted uranium. The only fuel sent out from these sites will be embedded in ENHS modules. The radioactive waste from the fuel processing and from the used ENHS modules will be shipped for disposal in either the country that used the ENHS or to international or regional waste centers. Alternatively, fuel reprocessing could be done in the international waste centers from which the fuel will be sent to the ENHS fabrication facility.

Table 4 is a summary comparison of the proliferation resistant characteristics of ENHS-based energy system to current LWRs. The comparison has been made for a U-Pu fuel cycle but a fuel cycle. The ENHS can use even more proliferation resistant fuel cycles such as denatured Th-^{233}U.

Table 4 Proliferation resistant attributes of ENHS-based energy system relative to once-through LWR

Key: 0 same, ++ much more proliferation resistant than LWR, + more -+ some aspects less others more, - less, -- much less, N/A not applicable.

Stage of the fuel cycle	Material Barriers					Technical Barriers						Institutional Barriers		
	Isotopic	Radiological	Chemical	Mass and Bulk	Detectability	Facility Unattractiveness	Facility Access	Available Mass	Diversion Detectability	Skills, Knowledge, Experience	Time	Safeguards	Access control and Security	Location
Beginning of the cycle[(c)]	N/A													
Transport of LWR Spent Fuel	0	0	0	0	0	0	0	0	0	0	0	0	0	0
Fuel Fabrication	-+	--	--	+	--	--	--	+	--	--	-	-	-	-
Storage of Fuel	-+	--	--	+	--	--	--	+	--	--	-	-	-	-
Transport	-+	--	--	+	--	--	--	+	--	--	+	-	-	-
Reactor operations														
Fresh fuel storage	-	--	-	+	--	--	--	+	--	--	--	--	--	--
Fuel handling	No fuel handling on site; fuel is brought in and taken out in sealed ENHS module.													
Reactor irradiation	-	--	-	+	--	--	--	+	--	--	--	--	--	--
Spent fuel handling[(a)]	-	--	-	+	--	--	--	+	--	--	--	--	--	--
Pool storage of spent fuel	No fuel is stored in pools. Spent fuel is left inside ENHS sealed module embedded in Pb-Bi.													
On-site spent-fuel dry storage[(b)]	-	--	-	+	--	--	--	+	--	--	--	--	--	--
Back-end of the cycle														
Transport (of spent fuel)[(b)]	-	--	-	+	--	--	--	+	--	--	--	--	--	--
Storage (of spent fuel) **Once-through:**	It is assumed that the spent fuel will be transported in the spent ENHS module to a regional or international spent fuel center. After removal of the spent fuel from the ENHS module, the fuel is recycled without separation of Pu, MA and non-volatile FP using a dry process such as AIROX.													

(a) Spent fuel is removed from the Pb-Bi filled reactor pool inside the Pb-Bi filled ENHS module.
(b) Spent fuel is transported inside the ENHS module, imbedded in solidified Pb-Bi.
(c) Mining, transport, milling, transport, conversion, storage, transport, uranium enrichment, storage.

6. DISCUSSION

There are many open questions at this stage of ENHS development that need be answered before the practicality and viability of the ENHS-based energy system described above can be established. Among these questions are the following: (1) The feasibility of recycling the fuel limitless number of times. If not feasible, Accelerator-Driven Transmuting (ATW) reactors should be added to the system. (2) The quantity and characteristics of the waste from the system. (3) How does the ENHS system waste compares with that from a system of once-through LWR supplemented by ATW reactors. (4) The economic viability of the ENHS.

7. CONCLUSION

The novel reactor concept ENHS offers a unique combination of useful features that may enable the development of a highly safe nuclear energy system that is characterized by low waste, high proliferation resistance and high uranium utilization. These potential attributes justify pursuing, if not widening, the present feasibility assessment.

ACKNOWLEDGMENT

This work was supported by the DOE NERI program under contract No. DE-PS03-99SF21764. Contributed to the study of the ENHS are D. Wade and J. Sienicki of ANL, N. Brown and Q. Hossain of LLNL, M. Carelli, L. Conway and M. Dzodzo of Westinghouse, and D. Saphier and H. Shimada of UC Berkeley.

REFERENCES

Adamov E.O. et al. (1997), The Next Generation of Fast Reactors. *Nuclear Engineering and Design 173 143.*

Aoki K. et al. (1995), Design of Small and Simple LMR Cores for Power Generation in Remote Communities. *Proc. 3rd ASME-JSME Int. Conf. On Nuclear Eng., ICONE-3*, Tokyo, Japan, October.

Feinroth H. et al. (1999), Dry Recycle of Light Water Reactor Spent Fuel. *Proc. of 8th International Conference on Nuclear Engineering: ICONE-7*, paper 7485, Tokyo, Japan, 19-23 April.

Greenspan E. et al. (1998), Multi-Recycling of Spent Fuel with Low Proliferation Risk. *Proc. Int. Conf. on Emerging Nuclear Energy Systems, ICENES-98*, Herzelia, Israel, 28 June – 2 July.

Greenspan E. and Wade D.C., (1999), Encapsulated Nuclear Reactor Heat Source Module. *Tran. ANS, 80, 197.*

Greenspan E. et al. (2000a), The Encapsulated Nuclear Reactor Heat Source Reactor Concept. *Proc. of 8th International Conference on Nuclear Engineering: ICONE-8*, paper 8750, Baltimore, MD, 2-6 April.

Greenspan E. et al. (2000b), Long-Life Cores with Small Burnup Reactivity Swing. *Proc. Int. Topical Mtg. On Advances in Reactor Physics and Mathematics, PHYSOR 2000*, Pittsburgh, PA, 7-11 May.

Greenspan E. et al. (2001), Promising Design Options For The Encapsulated Nuclear Heat Source Reactor. *9th International Conference on Nuclear Engineering*: ICONE-9, paper 417, Nice, France, 8-12 April.

Hattori S. and Handa N. (1989), Use of Super-Safe, Small and Simple LMRs to Create Green Belts in Desertification Area. *Transaction Amer. Nucl. Soc., 60, 437.*

Hattori S. and Minato A. (1995), A Large Modular LMR Power Station Which Meets Current Requirements. 3rd *International Conference on Nuclear Engineering: ICONE-3,* Tokyo, Oct.

Hofman G. L, Walters L. C. and Bauer T. H. (1997), Metallic Fast Reactor Fuel. *Prog. Nucl. Energy 31, 83.*

Orlov V.V. et al. (1994), Lead-Cooled Reactor Core, its Characteristics and Features. *Int. Top. Mtg. On Advanced Reactor Safety, ARS 94*, Pittsburgh, PA, 17-21 April.

Saphier D. et al. (2001), Some Safety Aspects of the Encapsulated Nuclear Heat Source LMR Plant. *9th International Conference on Nuclear Engineering: ICONE-9*, paper 22, Nice, France, 8-12 April.

Sekimoto H. and Su'ud Z. (1995), Design Study of Lead and Lead-Bismuth-Cooled Small Long-Life Nuclear Power Reactors Using Metallic and Nitride Fuels. *Nuclear Technology, 109, 307.*

Sienicki J.J. and Wade D.C. (2001), Thermal Hydraulic Analysis of the Encapsulated Nuclear Heat Source. *9th International Conference on Nuclear Engineering: ICONE-9*, paper 771, Nice, France, 8-12 April.

Pergamon
www.elsevier.com/locate/pnucene

Progress in Nuclear Energy, Vol. 40, No. 3–4, pp. 441–448, 2002

Printed in Great Britain
0149-1970/02/$ - see front matter

PII: S0149-1970(02)00036-7

SUSTAINABILITY OF FISSION ENERGY IN FUTURE EQUILIBRIUM SOCIETY

HIROSHI SEKIMOTO

Research Laboratory for Nuclear Reactors

Tokyo Institute of Technology

O-okayama, Meguro-ku, Tokyo, Japan

hsekimot@nr.titech.ac.jp

The system of a fast reactor fed with natural uranium is discussed as a future sustainable energy producing system, concentrating especially on the waste problem. The following conclusions have been derived:

1. When the reprocessing with very small leakage rate can be employed, the conventional fast reactor has a potential to confine all actinides in the core.
2. The total toxicity of LLFPs is smaller but comparable to the toxicity of burned natural uranium with its daughters.
3. When the isotope separation is employed for each LLFP, whole LLFPs can be incinerated using the conventional fast reactor system with natural uranium.
4. When only the element separation is employed, the incineration of ^{107}Pd should be given up even by employing several design changes, but its toxicity is only 0.0002 of the incinerated uranium toxicity.

1. FUTURE NUCLEAR EQUILIBRIUM SOCIETY

If we use fast reactors with reprocessing facility, we can use the nuclear energy for almost a million years (Sekimoto, 1999). The environmental impact of nuclear energy is considered less than fossil fuel. However, the fission energy has three peculiar problems; safety, nuclear proliferation and radioactive waste. In the present paper sustainability of fission energy for such a long period is discussed.

From the finiteness of the earth our society ultimately should become a steady state society where the energy consumption rate is a constant (Sekimoto and Takagi, 1991, Sekimoto, 1999). If this energy supply is satisfied by the fission energy, the fresh fuel charge rate and waste removal rate are also constants. Furthermore, the reaction rate of each nuclear reaction and abundance ratio of each nuclide in the reactor are also constants. This kind of state is called as the nuclear equilibrium state.

The author calls the society in this state the nuclear equilibrium society and continues to work under this research title for long period. In this society the energy production system is considered to affect the social system strongly. It is interested to know what kind of society will be realized as such a future society. However, we have many scientific and engineering problems to be investigated before studying the society itself.

Among the three peculiar problems mentioned above, the safety is partly accepted even in the present time. The amount of total energy production in the future nuclear equilibrium society may be larger than the present time, but the safety technology will be also advanced. Even at the present time we have many reactor concepts much safer than the present operated reactors. We can expect that in the future sufficiently safe reactors will be available.

For the nuclear proliferation, this risk will become more independent from the choice whether we will continue the nuclear reactor utilization or not, since the bomb making technology will be advanced and necessity of plutonium from the reactors will decrease. The important item will become rather the condition of society. An important part of the condition is stability of the society, which must be insured with economics. The enough inexpensive energy is necessary to keep this society in the condition that the nuclear bombs will be never used.

The waste problem, the last of the above three problems, will become severer in the future than the present time, since the long-life radioactive waste will accumulate as time passing. In the present paper, the sustainability of fission energy is discussed from this point.

2. MODEL SYSTEM

We consider a nuclear energy center, in which several nuclear reactors and handling and storage facilities of nuclear fuels and wastes are operated. In the present scenario any radioactive materials will not be taken out of the nuclear energy center.

We can consider many kinds of the nuclear equilibrium systems, but start the study from simple systems. Though the system with several kinds of reactors are possible and may work as an efficient synergetic system, the number of reactor types is fixed to be one in order to know the fundamental characteristics of the effect of nuclear reactor to the system. The number of fuel types is also chosen to be one, natural uranium or thorium.

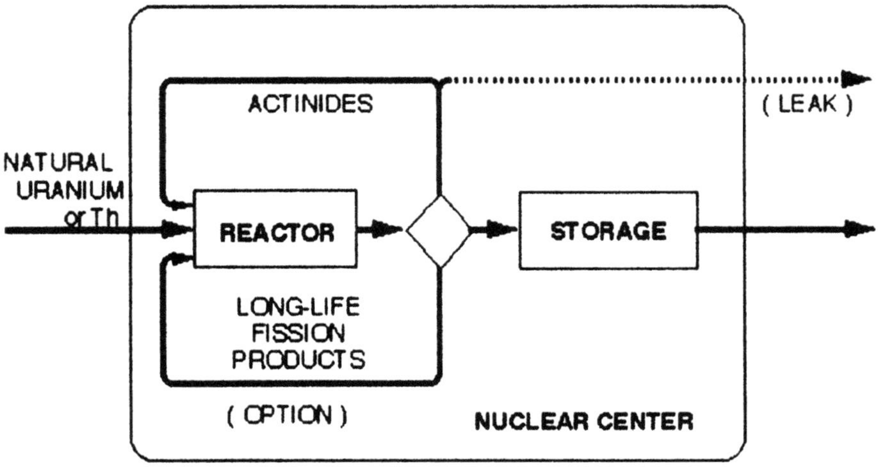

Fig. 1 A simple model of nuclear equilibrium system

In the present model, the natural uranium or thorium (and the daughter nuclides in the secular equilibrium with this nuclide) are supplied to the reactor in the center as shown in Fig. 1. All the actinides are confined in the reactor. In this system the final nuclides, which are withdrawn from the reactor, are fission products (FPs) and the last nuclides of actinide decay chain, lead and bismuth. The medium-life FPs (^{137}Cs, ^{90}Sr and ^{151}Sm) and the short-life FPs whose half-life is shorter than medium-life FPs are stored in the center until they will decay enough to be discharged from the center as common wastes. Almost 1000 years are required for enough decay of the medium-life FPs, but the total amount of stored medium-life FPs is only about 1t/GWt for each nuclide and acceptable, since they are produced continuously but decayed at the same time. On the other hand, the long-life FPs (7 LLFPs), ^{129}I, ^{135}Cs, ^{107}Pd, ^{93}Zr, ^{99}Tc, ^{126}Sn and ^{79}Se, can not be expected to decay out naturally. However, the radioactive toxicity of these nuclides is smaller by a small amount than one of the incinerated natural uranium. The choice of simple discharge from the center or incineration is discussed later.

3. CALCULATION CONDITIONS

Two types of fast reactors and two types of thermal reactors are investigated. For the fast reactors a sodium-cooled oxide-fuel fast reactor (Soft Fast) and a sodium-cooled metallic fuel fast reactor (Hard Fast) are treated, and for the thermal reactors a PWR (Hard Thermal) and a high moderation-ratio graphite-moderated gas-cooled reactor (Soft Thermal) are treated. Fundamental design parameters of the reactors are shown in Table 1. The removal half life is the mean time for which each nuclide is discharged continuously from the core. These parameters are enough for the present stage of the study, since only general aspects of each nuclear equilibrium system are interested in. The neutron spectra of these reactors are shown in Fig. 2.

Table 1. Fundamental reactor design parameters

	Thermal Reactor	Fast Reactor
Total Power Output	3 GWt	3 GWt
Power Density	50 W/cc	300 W/cc
Removal Half Life	1 year	1 year

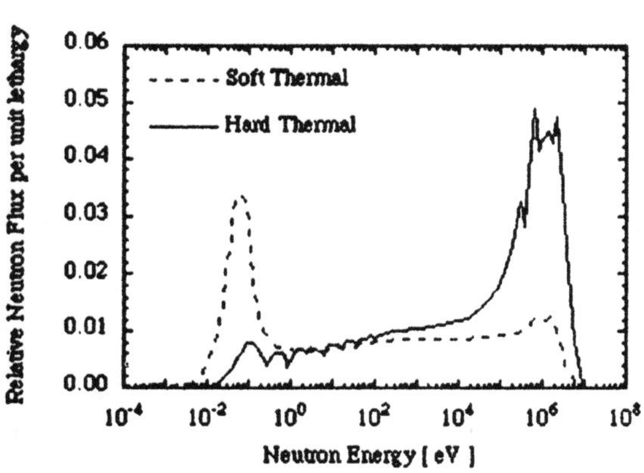

(a) Thermal reactors

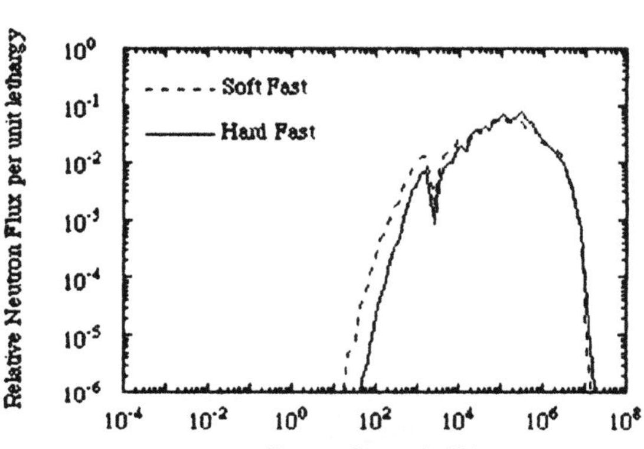

(b) Fast reactors

Fig. 2. Typical neutron spectra

4. NUCLIDE DENSITIES AND CRITICALITY

The atomic density of each nuclide can be easily obtained from so-called nuclear equilibrium equation (Sekimoto 1999). The obtained nuclide densities of all actinides more than $10^{18}/cm^3$ are shown in Table 2. The densities less than this value are also shown by the small figures in the same figure. These values do not depend on the FP discharge constant.

Table 2. Equilibrium nuclide densities for different nuclear system

Fuel	Uranium feed				Thorium feed			
Reactor			Thermal reactor				Thermal reactor	
Spectrum	Hard	Soft	Hard	Soft	Hard	Soft	Hard	Soft
Th 232	3.12E+13 *	1.72E+13	2.16E+12	7.40E+11	1.47E+22	1.01E+22	2.35E+22	1.56E+22
Pa 231	2.11E+14	1.36E+14	2.53E+12	8.28E+10	1.42E+19	7.51E+18	2.61E+18	8.08E+16
233	6.31E+11	4.09E+11	5.97E+10	8.63E+09	3.04E+19	3.06E+19	4.64E+18	5.06E+18
U 232	1.36E+14	1.09E+14	5.25E+12	1.55E+11	8.59E+18	5.72E+18	5.37E+18	1.51E+17
233	1.61E+15	1.13E+15	1.15E+13	1.13E+11	1.61E+21	1.37E+21	1.19E+20	4.34E+19
234	8.40E+18	6.94E+18	2.90E+17	1.90E+16	5.52E+20	4.72E+20	3.97E+19	2.01E+19
235	1.75E+19	1.39E+19	1.46E+18	3.60E+17	1.07E+20	1.06E+20	1.86E+19	5.14E+18
236	2.21E+19	1.74E+19	1.45E+18	7.99E+17	1.13E+20	1.13E+20	1.80E+19	1.13E+19
238	1.49E+22	1.02E+22	8.61E+21	7.37E+21	8.64E+16	1.51E+17	1.23E+17	2.87E+17
Np 237	1.65E+19	1.04E+19	1.63E+18	1.84E+17	2.49E+19	2.72E+19	5.06E+18	1.95E+18
239	2.42E+18	2.50E+18	3.95E+17	4.44E+17	1.49E+13	3.96E+13	6.35E+13	1.95E+14
Pu 238	4.85E+19	5.08E+19	5.01E+18	3.39E+17	1.98E+19	2.62E+19	5.52E+18	1.06E+18
239	1.91E+21	1.61E+21	4.95E+19	1.95E+19	5.30E+18	8.50E+18	1.09E+18	3.47E+17
240	9.77E+20	1.03E+21	1.71E+19	7.57E+18	2.69E+18	5.42E+18	3.77E+17	1.34E+17
241	1.39E+20	1.84E+20	1.87E+19	6.02E+18	3.82E+17	9.68E+17	4.12E+17	1.07E+17
242	1.10E+20	1.49E+20	2.15E+19	1.59E+19	3.02E+17	7.84E+17	4.73E+17	2.81E+17
Am 241	5.85E+19	5.72E+19	1.50E+18	1.91E+17	1.61E+17	3.00E+17	3.30E+16	3.39E+15
242m	2.83E+18	3.26E+18	2.37E+16	2.17E+15	7.78E+15	1.71E+16	5.22E+14	3.83E+13
243	2.99E+19	4.29E+19	1.23E+19	8.11E+18	8.24E+16	2.25E+17	2.70E+17	1.44E+17
Cm 242	2.56E+18	3.52E+18	4.09E+17	1.27E+17	7.06E+15	1.85E+16	8.99E+15	2.25E+15
244	2.55E+19	4.41E+19	2.49E+19	1.99E+19	7.02E+16	2.32E+17	5.48E+17	3.52E+17
245	5.44E+18	1.13E+19	3.65E+18	9.62E+17	1.50E+16	5.91E+16	8.03E+16	1.70E+16
246	3.18E+18	6.53E+18	1.45E+19	1.59E+19	8.75E+15	3.43E+16	3.19E+17	2.82E+17
247	3.44E+17	9.23E+17	1.37E+18	9.40E+17	9.48E+14	4.85E+15	3.01E+16	1.67E+16
248	3.05E+17	7.83E+17	3.10E+18	3.53E+18	8.39E+14	4.11E+15	6.83E+16	6.24E+16

* 3.12E+13 is read as 3.12×10^{13}

For the discussion on the criticality, the following performance index, which is called h-value, is employed:

$$h = \frac{\nu\Sigma_f}{\Sigma_a} = \frac{(\nu\sigma_f, \mathbf{n})}{(\sigma_a, \mathbf{n})},$$

where **n** is the vector, whose element is the atomic density of each nuclide, and (,) is an inner product.

The obtained h-values are shown in Fig. 3. The horizontal axis denotes the neutron flux level. When the flux level changes, the ratio of the neutron induced reaction to radioactive decay and the nuclide densities change inevitably, which lead to a different h-value.

When we consider the neutron economy, roughly 10% of produced neutrons are usually absorbed by construction materials, coolant, moderator, control materials, etc., and leakage from the system, though this amount depends strongly on the reactor design. Therefore the h-value should take a value more than 1.1 to perform the criticality. Judging from these calculation results, the fast reactor system fed with natural uranium can realize the criticality with some surplus. The system with thorium may perform the criticality for both fast and thermal neutron system only with small surplus.

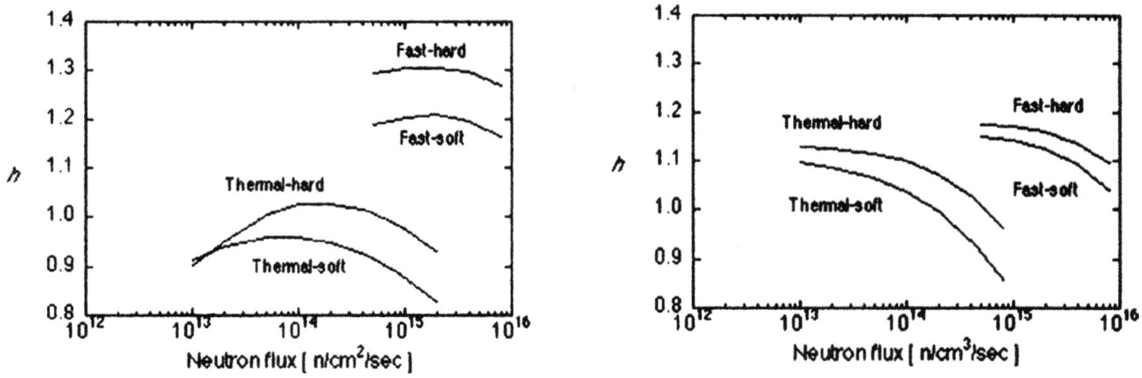

(a) Natural uranium feed case (b) Thorium feed case

Fig. 3. h-values for different nuclear system

5. RADIOACTIVE TOXICITY

For evaluating radioactive toxicity several methods can be considered. Their toxicity is measured with the annual limit on intake (ALI) employed in the Japanese regulation, which is based on ICRP recommendation (Sekimoto et al., 1996).

Even natural uranium and thorium introduced into the system have radioactive toxicity. It means that the reactor incinerates the toxicity of these nuclides and reduces the toxicity of the environmental toxicity if it does not produce any toxic materials. The toxicity of natural uranium is only 0.026ALI/g but becomes 1.16ALI/g when contributions from all daughter nuclides are taken into account. For the thorium these values are 0.14ALI/g and 0.24ALI/g, respectively. When the reactor is operated with 3GWt, the total amount of natural uranium charged to the reactor is about 1t per year. Then the total amount of incinerated toxicity is about 10^6ALI/year for natural uranium feed case, if all daughter nuclides are also charged.

Table 3. Toxicity of each nuclide from the nuclear energy center for different nuclear system

Fuel	Uranium				Thorium			
Reactor	Thermal		Fast		Thermal		Fast	
Spectrum	Soft	Hard	Soft	Hard	Soft	Hard	Soft	Hard
Se-79	4.20E+04	5.11E+04	5.46E+04	5.69E+04	1.12E+05	1.15E+05	1.14E+05	1.17E+05
Zr-93	3.31E+04	3.24E+04	3.36E+04	3.45E+04	5.40E+04	5.33E+04	5.39E+04	5.43E+04
Tc-99	8.79E+04	9.31E+04	9.98E+04	1.02E+05	6.19E+04	6.23E+04	6.84E+04	6.90E+04
Pd-107	2.43E+02	2.30E+02	1.88E+02	1.80E+02	1.11E+01	1.45E+01	1.33E+01	1.27E+01
Sn-126	2.02E+05	2.23E+05	2.32E+05	2.33E+05	2.31E+05	2.10E+05	2.22E+05	2.21E+05
I-129	1.98E+05	2.28E+05	2.35E+05	2.43E+05	2.22E+05	2.39E+05	2.52E+05	2.56E+05
Cs-135	1.31E+04	2.15E+04	5.58E+04	5.68E+04	1.34E+04	2.31E+04	5.90E+04	5.95E+04
Total	5.76E+05	6.49E+05	7.11E+05	7.26E+05	6.94E+05	7.03E+05	7.69E+05	7.77E+05

When all the actinides are confined in the reactor and their leakage from the nuclear energy center can be ignored, only the contribution of LLFP is necessary to be considered for the toxicity production in the present system. The toxicity removed yearly from the reactor is shown in Table 3. It can be seen that the produced toxicity is less than the incinerated toxicity, if all the daughter nuclides of natural uranium are inserted into the reactor. If the subject is only not to increase the total toxicity outside the nuclear energy center on the earth, the scenario in which LLFPs are not incinerated can be acceptable. For thorium cycle the produced toxicity is larger than incinerated one, even if all daughter nuclides are inserted into the reactor, and some LLFP should be incinerated to reduce the toxicity outside the center.

In this scenario all radioactive materials are assumed not to leak from the nuclear energy center. However, it is very severe problem. Especially the leakage of the actinides is important in the reprocessing process for their separation from the nontoxic wastes taken out of the center. The condition that the amount of toxicity is the same as the incinerated toxicity at the time 1000 years later is that the rate of actinide leakage is 1.1×10^{-6} for the uranium feed case, and 1.2×10^{-5} even for the thorium feed case where the amount of higher actinides is small (Sekimoto et al., 1996).

6. TRADE-OFF BETWEEN TOXICITY AND CRITICALITY

In this section we discuss several options for each LLFP in the nuclear equilibrium system; incineration or discharge, and isotope separation or element separation if incineration scenario is adopted. A sodium cooled oxide fuel fast reactor is considered in this section. The fuel pellet diameter, fuel pin diameter and pin pitch are 6.00, 7.64 and 9.40 cm, respectively. The pellet power density is set to be 810 W/cc and the cell power density is 300 W/cc. As mentioned in the previous section the fast reactor requires $h > 1.1$ for attaining criticality.

The absolute value of ALI may contain considerable uncertainty, then in this section we will evaluate the toxicity of LLFP normalized by the toxicity of incinerated natural uranium with its daughters in the equilibrium state.

At the beginning h-values and total toxicities of the discharged 7 LLFPs are calculated, where all fission products are discharged from the core with the same discharge constant. The calculated results are shown in Table 4 for different discharge constants. Taken out of the LLFP from the nuclear center may be one option, but the possibility of incineration is also studied in the present paper. Both toxicities and h-values decrease gradually with decreasing the discharge constant. The amount of decrease of toxicity may be ignorable. However, the decrease of h-value is considerable. From the criticality constraint the discharge constant of 1/4 $year^{-1}$ gives almost lowest limit condition for realization of a reactor.

Table 4. Toxicities and h-values for different discharge constant

r (year-1)	toxicity*	h-value
1	0.48850	1.23147
1/2	0.47390	1.19342
1/4	0.44792	1.12848
1/8	0.40932	1.02449

*Total toxicity of 7 LLFPs taken out from the nuclear center normalized by the toxicity of incinerated natural uranium with its daughters in the equilibrium state

The ratio of toxicity decrement to h-value decrement is calculated for each LLFP for both isotope confinement and element confinement cases. The isotope with the higher ratio is more effective to be incinerated in the reactor. A rule of thumb may be incinerating the LLFPs in the order of decreasing the ratio. When this rule of thumb is employed, for each of isotope and element confinement cases the relation between discharged normalized toxicity and h-value is shown in Fig. 4 for the case of the discharge constant of 1/2 year^{-1}. The incineration of ^{129}I and ^{126}Sn is efficient to reduce more toxicity with less h-value decrease even for element confinement case. The incineration of ^{107}Pd and ^{93}Zr seems inefficient especially for element confinement case.

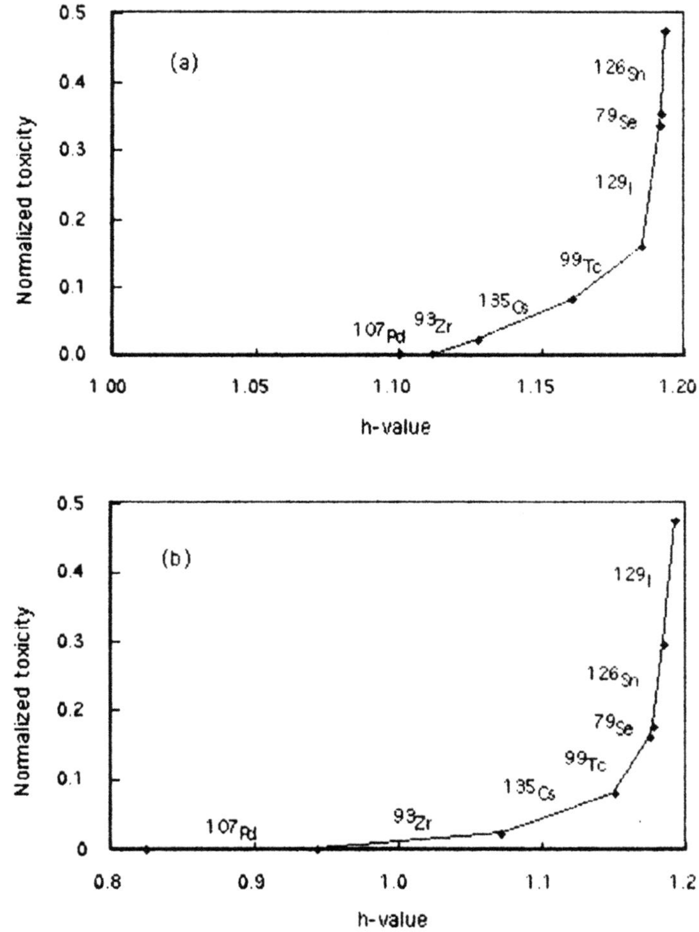

Fig. 4. Toxicity and h-vaue reduction for successive LLFP incineration for the case of the discharge constant of 1/2 year^{-1} for (a) isotope confinement and (b) element confinement.

7. FAST REACTOR OPTIONS

In the previous sections, it is shown that the fast reactor can confine all actinides but difficult to incenerate all LLFPs from the limitation in neutron economy. This reactor employs oxide fuel and sodium coolant. We can propose several alternatives. In this section neutron economy of some other fast reactors is shown. The calculated h-value and infinite neutron multiplication factor are shown in Fig. 5.

The metallic fuel shows an excellent neutron economy, and about 0.1 of neutron multiplication can be expected to be gained by changing the fuel from oxide to metallic. It is not clear yet whether this value is sufficient or not, but it is considerable from the discussion in the previous section. The change of coolant is not so sensitive compared to the change of fuel. Note that the neutron multiplication factor shown is the infinite neutron multiplication factor. For small reactor lead and lead-bismuth are known to show excellent neutron economy compared to sodium.

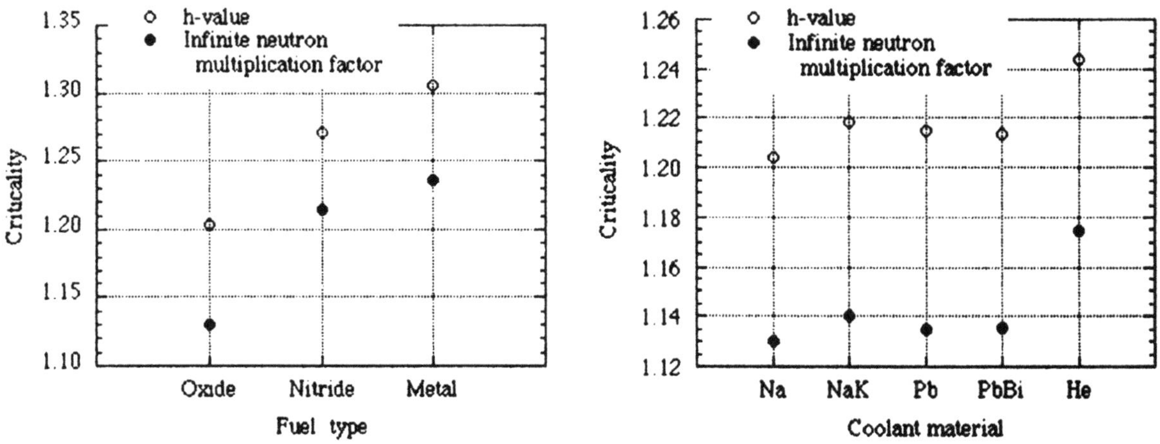

(a) for different fuels (b) for different coolants

Fig. 5 calculated h-value and infinite neutron multiplication factor for different fast reactors

8. CONCLUSIONS

When we use fission energy for extremely long period such as a million years, the radioactive waste problem becomes much more severe than the present time. The toxicity balance between charged and discharged nuclides in a nuclear energy center is studied for the nuclear equilibrium state. The following conclusions have been derived:

1. When the reprocessing with very small leakage rate can be employed, the conventional fast reactor has a potential to confine all actinides in the core.
2. The total toxicity of LLFPs is smaller but comparable to the toxicity of burned natural uranium with its daughters.
3. When the isotope separation is employed for each LLFP, whole LLFPs can be incinerated using the conventional fast reactor system with natural uranium.
4. When only the element separation is employed, the incineration of ^{107}Pd should be given up even by employing several design changes, but its toxicity is only 0.0002 of the incinerated uranium toxicity.

REFERENCES

Mizutani A. and Sekimoto H. (1998), Core Performance of Equilibrium Fast Reactors for Different Coolant Materials and Fuel Types, *Ann. Nucl. Energy*, ***25***, *1011*.

Sekimoto H. and Takagi N. (1991), Preliminary Study on Future Society in Nuclear Quasi-Equilibrium, *J. Nucl. Sci. Technol.*, ***28***, 941.

Sekimoto H. *et al.* (1996), Toxicity of Radioactive Waste Discharged from Nuclear Energy Center in the Future Equilibrium State, *Ann. Nucl. Energy*, ***23***, 663-668.

Sekimoto H. (1999), Equilibrium Models for the Fuel Cycle and Sustainability. *The 1999 Frederic Joliot / Otto Hahn Summer School In Reactor Physics*, p. 191. Karlsruhe, Germany, 18-27 August.

Takagi N. and Sekimoto H. (1992) *J. Nucl. Sci. Technol.*, ***29***, 276.

Pergamon
www.elsevier.com/locate/pnucene

Progress in Nuclear Energy, Vol. 40, No. 3–4, pp. 449–456, 2002

Printed in Great Britain
0149-1970/02/$ - see front matter

PII: S0149-1970(02)00037-9

INTERPRETATION OF ACTINIDE TRANSMUTATION IN THERMAL AND FAST REACTORS

TOSHIKAZU TAKEDA, TOSHIHISA YAMAMOTO and MAIKO MIYAUCHI

Department of Nuclear Engineering, Graduate School of Engineering,
Osaka University, Yamada-oka 2-1, Suita, Osaka, 565, Japan

Abstract

The commonly used transmutation rate of minor actinides in nuclear reactors is decomposed into four components, overall fission rate, Pu production rate, MA production rate, and element production rate. The physical meanings of these factors are described. The transmutation rates of minor actinides in two types of highly-moderated PWRs, a MOX fueled Na cooled fast reactor, and a metal fueled Pb cooled fast reactor are interpreted using the four components. The metal fueled Pb cooled fast reactor can incinerate minor actinides most (79kg/GWth/year), and this amount is about 4 times larger than the thermal reactors. The thermal reactors have large relative overall fission rates for ^{241}Am and have a potential for the incineration of ^{241}Am.

1. INTRODUCTION

A lot of studies have been carried out to investigate the effective way of transmutation of minor actinides (MAs) in thermal and fast reactors.

The utilization of soft neutron spectrum to transmute MAs was investigated by Salvatores *et al.*, Iwasaki and Hirakawa (Salvatores *et al.*, 1994; Iwasaki and Hirakawa, 1995) and also by the authors (Takeda and Yokoyama, 1996). On the contrary, the utilization of hard neutron spectrum was investigated by Hayase *et al* (Hayase *et al.*, 1990), Masumi and Aoyama (Masumi and Aoyama, 1995), and Mukaiyama (Mukaiyama *et al.*, 1990).

As general opinion, the transmutation in LWR and FR are discussed as different processes and comparison using common 'indexes' has not been done. Actually, the process of MA transmutation is not simple as described using only one 'index', such as commonly used 'transmutation rate'. For example, the incineration of ^{241}Am through the fission via ^{242m}Am which is not so expected for the hard spectrum reactors because ^{242m}Am produced by the capture reaction has large fission cross section in the lower energy range. To compare the potential of transmutation of MAs in thermal and fast reactors we have to calculate incineration rates of MA by fission, Pu buildup rates, higher MA buildup rates, MA production rates from U, Pu and other MAs. It is difficult to estimate these reaction rates by using only the commonly used transmutation rate, and it is required to introduce new 'indexes' in addition to the 'transmutation rate'.

For this purpose we decompose the commonly used transmutation rate which is the relative difference of individual MAs or total amount MAs included in reloaded fuel and discharged fuel into four components; overall fission rate, Pu production rate, MA production rate, element production rate. The overall fission rate represents the incineration rate of a relevant MA through direct fission or through indirect fission of other MAs transmuted from the relevant MA. The Pu production rate and the MA production rate represent the rates of transmutation of the relevant MA to Pu and to MA, respectively, through several capture reactions. The element production rate represents the production rate of the relevant MA from Pu, U and other MAs.

Using these four components the transmutation rates of MAs in thermal and fast reactors are compared. The advantage and disadvantage of the thermal and fast reactors are discussed from the view point of the overall fission rate, the Pu and MA production rates and the element production rate.

2. CALCULATIONAL METHOD AND MODEL

2.1 Definition of the four components of transmutation

The commonly used transmutation rate of the nuclide l is defined by

$$TR^l = \frac{N_l(0) - N_l(T)}{N_l(0)}, \tag{1}$$

where

$N_l(0)$: number density of nuclide l of loaded fuel

$N_l(T)$: number density of nuclide l of discharged fuel at $t=T$.

In order to decompose the transmutation rate we introduce the production rate $P_{lk}(t)$ which is defined by the production of a nuclide k at time T from an initial nuclide l. The production rate is calculated by performing an alternative burnup calculation under the condition that only the relevant nuclide l is present at time t=0 and the fluxes and the microscopic cross sections are the same as the conventional burnup calculation'; the fluxes and the cross sections obtained by the conventional burnup calculation are stored for each burnup step for this purpose.

In the alternative burnup calculation, the initial number density is denoted by $\tilde{N}(0)$ and the number density $\tilde{N}_k(T)$ at time T is calculated to evaluate the production rate

$$P_{lk} = \frac{\tilde{N}_k(T)}{\tilde{N}_l(0)}. \quad (2)$$

Using the relation

$$N_l(T) = \tilde{N}_l(0)P_{ll}(T) + \sum_{k \neq l} \tilde{N}_k(0)P_{kl}(T), \quad (3)$$

we can rewrite Eq. (1) as

$$TR^l = \{1 - P_{ll}(T)\} - \frac{1}{N_l(0)} \sum_{k \neq l} \tilde{N}_k(0)P_{kl}(T). \quad (4)$$

Here we assumed that $\tilde{N}_k(0) = N_k(0)$ for any k. It is noted, however, that $N_k(T) \neq \tilde{N}_k(T)$. The first term of the above equation corresponds to the incineration rate of the nuclide l by fission plus the transmutation rate to other nuclides:

$$1 - P_{ll}(T) = OF^l + \sum_{k \neq l} P_{lk}(T) \quad (5)$$

where $OF^l(T)$ is the overall fission rate of nuclide l, corresponding to the fission rates of the nuclide l.

Introducing Eq.(5) to Eq.(4) leads to

$$TR^l = OF^l + PuR^l + MAR^l - EPR^l \quad (6)$$

Where

PUR^l:Pu production rate from nuclide l.

MAR^l: MA+U production rate from nuclide l.

EPR^l: Element production rate of nuclide l from the other nuclides.

They are calculated by

$$PuR^l = \sum_{k \in Pu} P_{lk} \quad (7)$$

$$MAR^l = \sum_{\substack{k \in MA,U \\ l \neq l}} P_{lk} \quad (8)$$

$$EPR^l = \frac{1}{\tilde{N}_l(0)} \sum_{k \neq l} \tilde{N}_k(0) P_{kl} \quad (9)$$

2.2 Calculational Model

The main design parameters of PWRs and FRs used in the calculations are given in the previous study done by the authors (Takeda, Nakabayashi and Hirokawa, 1998). Here, only the outline is shown. For thermal reactors, PWR fuel cells with the moderator to fuel volume ratio Vm/Vf = 1.7 are used as the reference cell. Then the Vm/Vf ratio was increased in order to soften the neutron spectrum and enhance the transmutation of MA. Two optimized parameters were selected from the preceding studies by the authors: Case-A with the cell pitch of 1.45cm, which was found best in transmutation rate under the condition of Vm/Vf = 3.0 while changing the cell pitch (Takeda and Yokoyama,1996), and Case-B with the cell pitch of 1.21cm, which was best under the condition Vm/Vf = 4.0 while changing the fuel pellet diameter (Takeda and Nakabayashi, 1997). The reactor power is 1000MWe and the reactor operation time is 13 months with three batch refueling for the both cases. The composition of the MA loaded to the fuel corresponds to discharged fuel after 35 GWd/t burnup and 5 years cooling of currently used uranium fuel in PWRs: 49.2%^{237}Np, 30.0%^{241}Am, 15.5%^{243}Am, 5.0%^{244}Cm and negligible amount of ^{243}Cm and ^{245}Cm. The Pu vector for the fuel was taken as 56.2% ^{239}Pu, 24.3%^{240}Pu, 12.2%^{241}Pu and 7.3%^{242}Pu. We considered an equilibrium core with three batch fuel loading, and the Pu enrichment was determined so as to make average K_{eff} unity at the end of an equilibrium core. The burnup calculation was performed by the neutron transport calculations based on the continuous energy Monte Carlo code MVP (Mori and Nakagawa, 1994) with the ORIGEN2 (Croff, 1980) burnup chain.

As fast reactors (FRs) we considered MOX and/or metal fuel reactors with Na or Pb coolant. The cores are 2-region homogeneous cores. MA additions were set to 7.5 wt% in the inner core and 15.0 wt% in the outer core. The Pu enrichment was so determined that the eigenvalue at the end of equilibrium cycle becomes unity under the condition that the power distribution becomes as flat as possible. The Pu enrichment came to be 16.0wt% (inner core)/18.4wt%(outer core) for the MOX-Na-FR, and 9.7wt%/11.1wt% for the Metal-Pb-FR. The compositions of the MA and Pu are the same as those loaded to the thermal reactors.

The burnup of discharged fuel of FRs and PWRs was fixed to 95.2 GWd/t and 44.4 GWd/t, respectively. The electric power of the thermal and the MOX fueled Na cooled fast reactor was set to 1000 MWe. However that of the metal fueled Pb cooled FR was raised to 1200 MWe so as to make the burnup of discharged fuel 95.2 GWd/t.

The burnup characteristics of FRs were analyzed by using an RZ model 70-group diffusion calculation with the ORIGEN2 burnup chain. Both the thermal and fast reactor burnup calculations were carried out based on the JENDL-3.2 (Shibata *et al.*, 1990) library.

3. NUMERICAL RESULTS AND DISCUSSIONS

3.1 Comparison among the four reactors

Figure 1 shows the absolute values of the transmutation rates of MA, and the four components per GWth · year for the four reactors. The transmutation rates are denoted by black dots with solid lines. In the transmutation rates the element production contributes in negative as shown in equation (7), so the transmutation rates are less than the sum of the overall fission and the Pu production. For the fast reactors the sum is nearly equal to the total MA transmutation rate because of the small production of MA from Pu. However the production rate of MA from loaded Pu is not negligible for the thermal reactors.

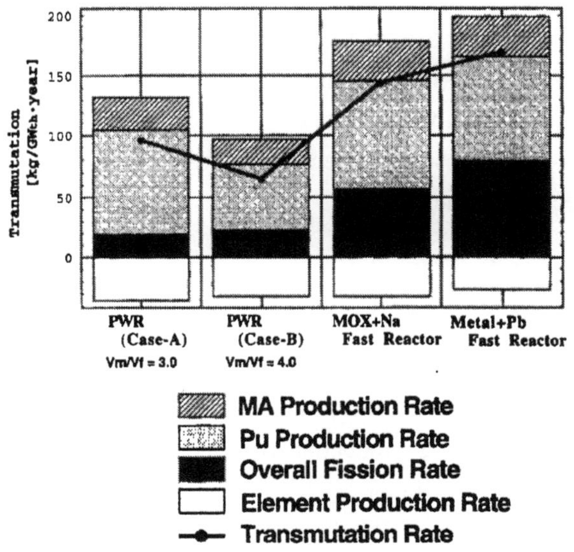

Fig. 1. MA absolute transmutation rate and four components

The metal fueled Pb cooled fast reactor can transmute MA by 169kg/GWth · year , and can incinerate MA through the overall fission by 79kg/GWth · year. The overall fission rate is larger than that of the MOX fueled Na cooled fast reactor, and is about 3.5~4.0 times larger than the thermal reactors. For the thermal reactors the Pu production rates and the element production rates are almost equal to those of the fast reactors. However, the overall fission rates are small by about 60~75 % than those of the fast reactors. The PWR of Case-A can transmute MAs by 95kg/GWth · year and this amount is larger than the PWR of Case-B which transmutes MAs by 65kg/GWth · year. This is because 101kg MA is loaded every cycle for Case-A, and 70kg for Case-B, and the absolute transmutation rate is larger for Case-A. It is noted, however, that Case-B has the same overall fission rate as Case-A.

Figure 2 shows the absolute transmutation rates and the four components of ^{237}Np and ^{241}Am. ^{237}Np is included in MA by 49.18%, and its transmutation rate is large particularly for the fast reactors. The incineration of ^{237}Np by the overall fission is small for the thermal reactors, but is rather large for the fast reactors because of the threshold fission of ^{237}Np. In the thermal reactors ^{237}Np is transmuted mainly to Pu isotopes, and the Pu production rate is large. Furthermore the MA production is small for both thermal and

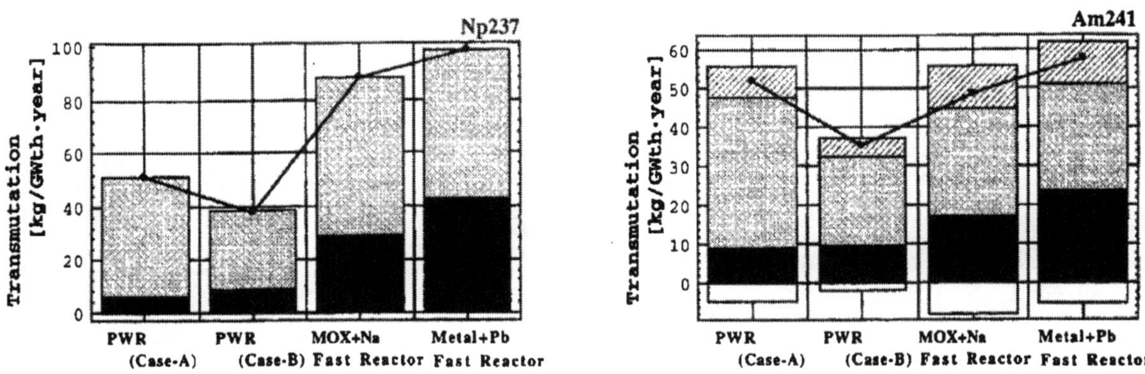

Fig. 2. Absolute transmutation rate and four components of individual MA nuclides

fast reactors.

The overall fission rates of ^{241}Am for the thermal reactors are larger than those of ^{237}Np, though those for the fast reactors are smaller than those of ^{237}Np. The rather large overall fission of ^{241}Am for the thermal reactors is caused by the large contribution of the fission of ^{242m}Am produced by the $^{241}Am(n,\gamma)$ reaction whose cross section is large in thermal energy range. The Pu production is also large for ^{241}Am because of the reaction

$$^{241}Am \xrightarrow{(n,\gamma)} {}^{242}Am \xrightarrow{\beta^-} {}^{242}Cm \xrightarrow{\alpha} {}^{239}Pu$$

The MA production is not negligible unlike ^{237}Np where the ^{241}Pu production is small and there is a little production of ^{241}Am through the decay of ^{241}Pu.

In the following let us compare the relative MA transmutation rate. Figure 3 shows the relative transmutation rate and the four components. Unlike the absolute transmutation rate Case-B has the largest transmutation rate. The overall fission rate for Case-B is almost equal to that of the fast reactors, and larger than that of Case-A. The large overall fission is caused by the ^{237}Np and ^{241}Am overall fissions. It is of interest that the relative overall fission for Case-B is the largest among the four reactors. The flux for Case-B is larger than that of Case-A by 50% because the linear power is the same between Case-A and B, and the fissile number density of Case-B is less than Case-A. Therefore the Pu production is larger for Case-B, and there are much fissions of ^{239}Pu. This is verified from Fig. 4 where the element-wise contributions to the overall fission rates of ^{237}Np are shown. For ^{237}Np, the metal fueled fast reactor has the largest overall fission rate, and the main contributor is ^{237}Np itself. Case-B has the almost equal overall fission rate as the MOX fueled fast reactor, and the main contributor is ^{239}Pu. Case-A has a small overall fission rate. Though the relative transmutation rate is largest for Case-B, the MA inventory is small because of the small fuel radius of 0.305cm and the absolute transmutation rate is small as shown in Fig. 1.

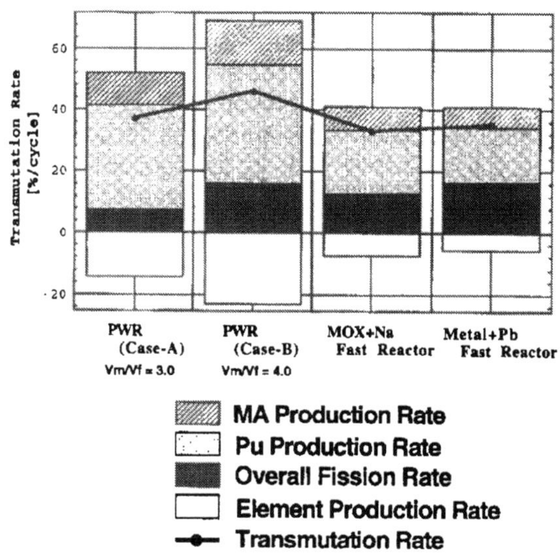

Fig. 3. Relative MA transmutation rate and four components

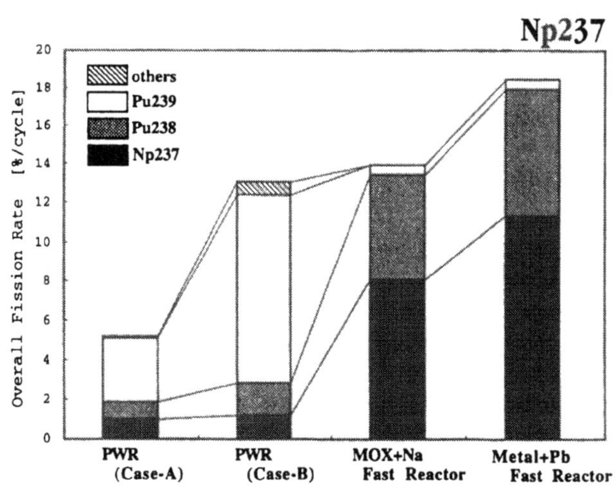

Fig. 4. Contribution of individual nuclides to ^{237}Np overall fission rate

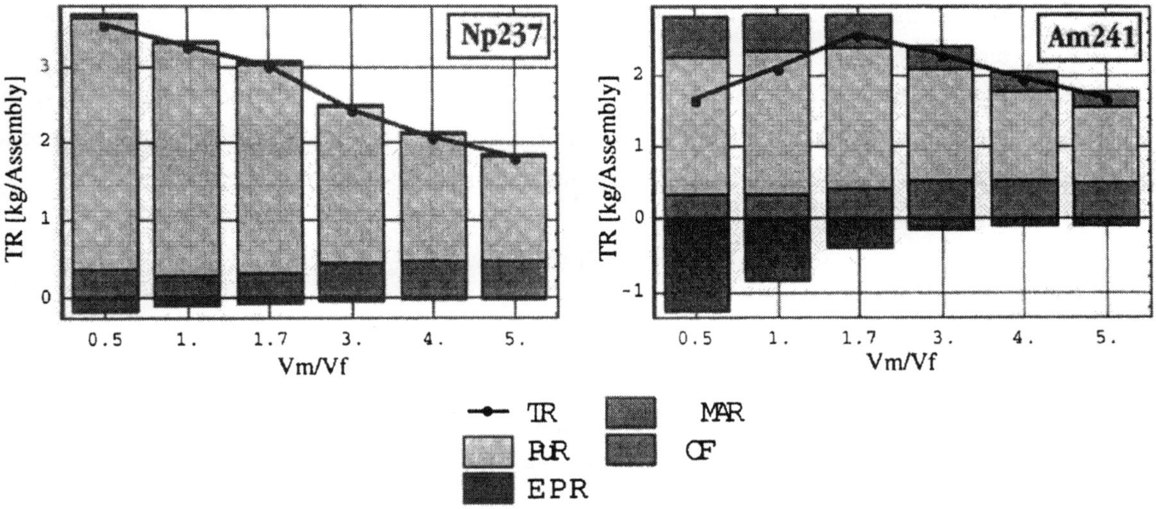

Fig. 5. Absolute transmutation rate and four components as a function of V_m/V_f

3.2 Effect of Vm/Vf ratio in the thermal reactors

Because the over-moderated LWRs are considered as optimal design for MA transmutation, it is of interest to look into the effect of Vm/Vf ratio using the four components. Figure 5 shows the four components for ^{237}Np and ^{241}Am as a function of Vm/Vf. Here, let us concentrate to OF, which is subordinated by multiple reactions and is not simple as the PuR. Figure 6 shows the fraction of each nuclide in the OF. As the Vm/Vf becomes large and the neutron spectrum becomes soft, the direct fission and the fission via ^{238}Pu decreases. On the contrary, the fission via ^{239}Pu increases because the ^{239}Pu production through double neutron capture increases. The two effects are compensated and the overall fission is almost constant as seen in Fig. 5. For ^{241}Am, as the Vm/Vf

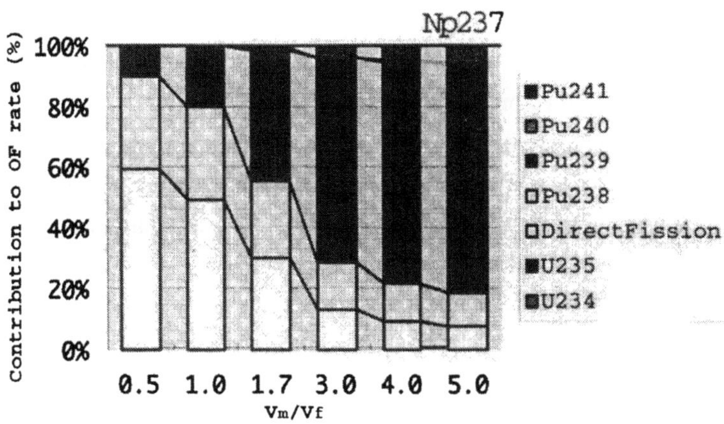

Fig. 6(a). Contribution of individual nuclides to OF rate of ^{237}Np

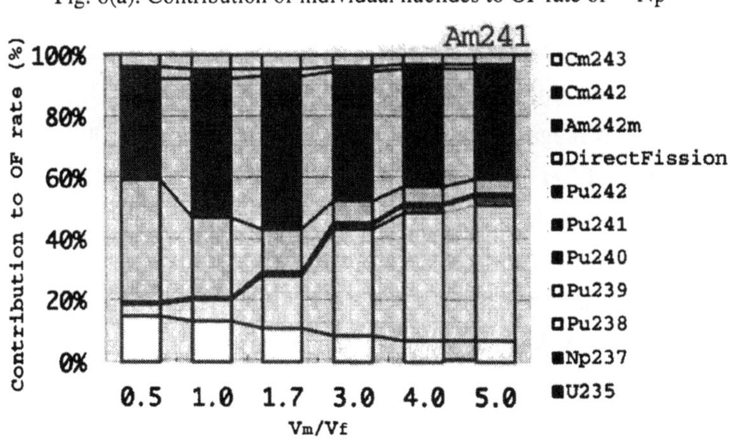

Fig. 6(b). Contribution of individual nuclides to OF rate of ^{241}Am

becomes large, the direct fission and the fission via Pu-238 decrease. On the contrary, the fission via ^{239}Pu increases rapidly. The fission via ^{242m}Am is almost flat. Then the overall OF shows only slight increase as the Vm/Vf becomes large.

From Fig. 5, it became clear that the reason of good MA transmutation characteristics is due to the enhancement of the capture reactions, and not the fission reactions. This reasoning has been discussed as general opinions, but it has becomes clear by the interpretation using the four components.

4. CONCLUSIONS

In order to evaluate the incineration of MA by fission, the Pu buildup, the higher MA buildup, and the MA production, we have decomposed the commonly used MA transmutation rate into the four components; the overall fission rate, the Pu production rate, the MA production rate and the element production rate. The physical meanings of these factors were discussed. The potential of MA transmutation in two PWRs with MOX fuel and in two fast reactors was compared using the four components.

It was found that the metal fueled Pb cooled fast reactor was the best to transmute MA among the four reactors. It can incinerate MA through the overall fission by $79\text{kg/GWth}\cdot\text{year}$. The overall fission rate is 41% larger than that of the MOX fueled Na cooled fast reactor, and is about 3.5~4.0 times larger than the thermal reactors. The PWR of Case-A can transmute more MAs than the PWR of Case-B, but the two PWRs have the same overall fission rate of $20\text{kg/GWth}\cdot\text{year}$.

The incineration of ^{237}Np by the overall fission is small for the thermal reactors, but that of ^{241}Am for the thermal reactors is larger than that of ^{237}Np particularly for the PWR of Case-B. This indicates the potential of the thermal reactors to incinerate ^{241}Am.

The effect of Vm/Vf ratio on transmutation rate was investigated. It was found that the overall fission reaction is almost unchanged and the overall tendency is subordinated by capture reactions.

REFERENCES

Croff, A. G. (1980), ORNL-TM-7175.

Hayase, T. *et al.* (1990), *Int. Conf. on Physics of Reactors*, Vol. 4, PIII-1.

Iwasaki, T. and Hirakawa, H. (1995), *J.Nucl. Sci. Technol.*,. 32[1], 8.

Masumi, R. and Aoyoma, M. (1995), *J. Nucl. Sci. Technol.*, 32[10], 965.

Mori, T. and Nakagawa, M. (1994), JAERI-DATA/Code94-007.

Mukaiyama, T. *et al.* (1990), *Proc. Int. Conf. on the Physics of Reactors, Operation, Design and Computation,* Vol. I, p.97.

Salvatores, M. *et al.* (1994), *Nucl. Sci. Eng.*, 115, 1.

Shibata, K. *et al.* (1990), JAERI-1319.

Takeda, T. and Nakabayashi, H. (1997), Optimum Transmutation of Minor Actinides in Thermal Reactors with Varios Neutron Spectra. *Proceedings of International Conference on Future Nuclear Systems, Gloabl '97*, Oct. 5-10, 1997, Yokohama, Japan.

Takeda, T. and Yokoyama, K. (1996), *Ann. Nucl. Energy*, Vol. 24, No. 9, pp. 705-719.

Pergamon
www.elsevier.com/locate/pnucene

Progress in Nuclear Energy, Vol. 40, No. 3–4, pp. 457–463, 2002

Printed in Great Britain
0149-1970/02/$ - see front matter

PII: S0149-1970(02)00038-0

Transmutation Characteristics of MA and LLFP in a Fast Reactor

TOSHIO WAKABAYASHI

Japan Nuclear Cycle Development Institute
4-49, Muramatsu, Tokai-mura, Naka-gun, Ibaraki, 319-1184, JAPAN
Tel: +81-29-282-1122/FAX: +81-29-282-4917/E-mail: waka@hq.jnc.go.jp

Abstract

Systematic studies were implemented to investigate the flexibility and attractive core concepts of MA and LLFP transmutation in fast reactors. The MA transmutation in the fast reactor core has no serious drawbacks in terms of core performance, provided that the homogeneous loading method can be employed with a small fraction of MA fuel (2~5wt%). The recycling of MA in the fast reactor is feasible from neutronic and thermal-hydraulic points of view. For FP transmutation, the introduction of target subassemblies using duplex pellets - a moderator annulus surrounding a Tc-99 core - gives the maximum transmutation rate of Tc-99 in the radial shield region of the fast reactor. The fast reactor has an excellent potential for transmuting MA and LLFP effectively. The fast reactor will be able to play an important role for reduction of environmental burden in future energy system.

1. INTRODUCTION

Partitioning and transmutation of minor actinides (MA) and long-lived fission products (LLFP) are attracting attention at present as an option to reduce the long-term radiological hazard of the high-level nuclear waste. There is general agreement that the implementation of partitioning and transmutation in radiotoxicity reduction is technically feasible.

Means of reducing the radiotoxicity of the MA and LLFP are presently under investigation. The MA nuclides could produce useful energy if converted into short-lived fission products by neutron bombardment. From this standpoint, a nuclear reactor provides the obvious means for transmutation of MA nuclides. Among the various nuclear reactors, a fast reactor is considered to have the greatest potential to transmute MA effectively, because of its hard neutron spectrum. For LLFP transmutation, due to the small neutron cross sections, the transmutation efficiency in LWRs is low. Moderated subassemblies loaded at the periphery of core and/or the shield region in a fast reactor are more appropriate devices for the transmutation of LLFP.

Systematic studies were implemented to investigate the flexibility and attractive core concepts of MA and LLFP transmutation in fast reactors. The following items were selected in the studies;

(i) loading patterns of MA subassemblies in the core (homogeneous, heterogeneous, hybrid ,etc.) or blanket/ shield region,
(ii) arrangement of MA pins in a subassembly,
(iii) effect of rare earth content in MA,
(iv) effect of MA multi-recycling,
(v) effect of fuel material (MOX, inert matrix, etc.),
(vi) influence of MA containing fuel on reactor plant and fuel cycle,
(vii) arrangement of LLFP pins and moderator pins in a subassembly.

The neutronic feature of MA nuclides enhances the potential of a fast reactor to extend operation cycle length between refueling. Taking into account the potential, a super long life core (SLLC) concept loaded with MA fuel have been studied aiming at continuous operation without refueling during plant lifetime and efficient reduction of MA and LLFP nuclides. It can confine MA and LLFP nuclides with high radiotoxicity in the reactor during plant lifetime as well as transmuting them and so there is no need of managing them during plant lifetime. The SLLC has also a large possibility to achieve a higher plant availability factor and to eliminate or reduce the ex-vessel fuel storage and the reactor refueling system.

2. MA TRANSMUTATION[(1)-(4)]

2.1 MA Loading methods

Since MA loading considerably affects not only core characteristics but also fuel material properties, it is necessary to investigate MA loading methods taking into account this influence upon core characteristics and fuel material properties. Possible MA loading methods (homogeneous, heterogeneous, hybrid, blanket, etc.) were investigated for fast reactor cores with no special design adaptation for MA loading. The MA fuel is dispersed uniformly throughout the core in the homogeneous method. In the heterogeneous method, a small number of subassemblies with concentrated MA fuel (target S/As) are loaded into the core. The hybrid MA loading method is a combination of the homogeneous and heterogeneous methods: the Np nuclide is loaded in the core region uniformly and a small number of subassemblies containing Am, Cm and RE nuclides are loaded into the blanket region.

The comparison of core performance for various MA loading methods is shown in Table 1. The MA transmutation in a fast reactor core has no serious drawbacks in terms of core performance, provided that the homogeneous loading method can be employed with a small ratio of MA to fuel (~5wt%). Since a 1000MWe-class LWR produces about 26 kg of MA per year, a fast reactor with 5wt% MA loading can transmute the MA mass from six LWRs.

The heterogeneous MA loading method can be made feasible by optimizing the fuel design, loading pattern and the coolant flow of the MA-loaded fuel subassemblies. The reduction of the fuel pin diameter and the Pu enrichment is essential to reduce the power of MA-loaded fuel in the heterogeneous MA loading method.

The hybrid MA loading method can transmute a large amount of MA without serious drawbacks in terms of core performance. The transmuted mass of MA is about 530kg/cycle, which is almost 16 times the mass produced by an LWR of the same power output.

The MA loading in the blanket region causes no problems from the viewpoint of core performance. Minor actinides are transmuted at a rate of 6% per cycle in the axial and radial blanket regions.

It was found that the hybrid MA loading method, where the Np nuclide is dispersed uniformly in the core and target subassemblies containing Am, Cm and rare earth nuclides are loaded into the blanket region, has the potential to achieve the maximum transmutation of MA with no special design considerations.

Table 1 Comparison of core performance for various MA loading methods

Item	Reference (No MA)	Homogeneous loading	Heterogeneous loading	Hybrid loading	Super long life core
MA and RE loaded in the core region	-	Np,Am,Cm:5% RE:10 %	Np,Am,Cm :49 % RE:0% (Number of target S/As:39)	Np:9.8% RE:0%	Inner core :15%MA Outer core: 30%MA RE:0%
Matrix of target	-	-	UO_2	Al_2O_3	-
MA and RE loaded in the blanket region	-	-	-	Am,Cm:46% RE:46% (Number of target S/As:72)	-
Pu enrichment (inner core /outer core) (wt%)	15.4/18.6	20.0/24.2	15.4/18.6	19.0/23.4	16.0/19.8
B.U. reactivity loss (%δk/kk')	3.31	3.71	1.83	0.90	2.5
Void reactivity	1.0	1.4(1)	1.3(1)	1.5(1)	1.0(1)
MA transmutation Amount (kg/cycle)	-	164	186	529	5340 (34 years)
Rate(%/cycle)	-	10.3	11.3	14.0	60

2.2 <u>Permissible RE level in homogeneously loaded MA</u>

Systematic parameter survey calculations were performed to investigate the basic characteristics of a fast reactor core loaded homogeneously with MA which contains RE, and also to establish a MA and RE loading method which has no serious influence on the core design. The homogeneous loading of MA and RE has no serious effects on the reactor core performance, provided that the amounts of MA and RE in the fuel are less than 5 and 10wt%

respectively. In the case of adding Am, Cm and RE in the radial blanket region, it is possible, from the viewpoint of core performance, to insert ~50wt% of Am and Cm, and ~50wt% of RE in the target assemblies.

2.3 Effect of MA recycling on core characteristics and fuel cycle system

The effects of MA recycling on the core characteristics in the homogeneous loading method and the fuel cycle system were evaluated. The absolute value of the Doppler coefficient is increased by MA recycling, the value at the 8th recycle is ~14% larger in comparison with that in the initial core. This is caused by the reduction in Pu enrichment with MA recycling; this increases the resonance absorption of U-238. Sodium void reactivity decreases with MA recycling, and the value at the 8th recycle is ~7% smaller than that in the initial core. The recycling of MA in a fast reactor is feasible from neutronic and thermal-hydraulic points of view.

However, the multi-recycled Np is significantly depleted compared to the unirradiated feed, and the fraction of Cm is greatly increased because of neutron capture in Am. The accumulation of Cm as a result of the MA recycling will bring about some problems concerning fuel handling and reprocessing, because of an increase in both the decay heat and the neutron emission rate from Cm-244.

2.4 Influence of MA containing fuel on reactor plant and fuel cycle

Both the decay heat and neutron emission rate of the MA (Np, Am and Cm) loaded fuel are very large in comparison with MOX fuel without MA. The dominant element of these fuel properties is Cm. If it is possible to remove Cm from the MA-loaded fuel, the decay heat value will decrease by one order and the neutron emission rate by three orders. Since the dominant isotope, Cm-244, has a relatively short half-life of 18 years, there might be another possibility of the fuel cycle, i.e., partitioning of Cm and Am from MA in the reprocessing and storage of Cm and Am for a period.

2.5 Selection of fuel material for MA transmutation

Different types of inert matrices, instead of uranium, for the heterogeneous MA-loading method have been investigated; they avoid the buildup of higher actinides via U-238 and achieve a high MA transmutation rate. Inert matrices of Al_2O_3 and CeO_2 were examined in this study. The MA transmutation rate of the target subassembly using inert matrices is larger than that of the target subassembly using UO_2. The inert matrices in the target subassembly effectively increase the MA transmutation rate.

3. FP TRANSMUTATION[5],[6]

A moderated target subassembly was used for FP transmutation. The subassembly consists of moderator pins and FP target pins distributed between the moderator pins. The moderated target subassemblies were loaded in the radial blanket and/or the shield region of the fast reactor as shown in Fig.1.

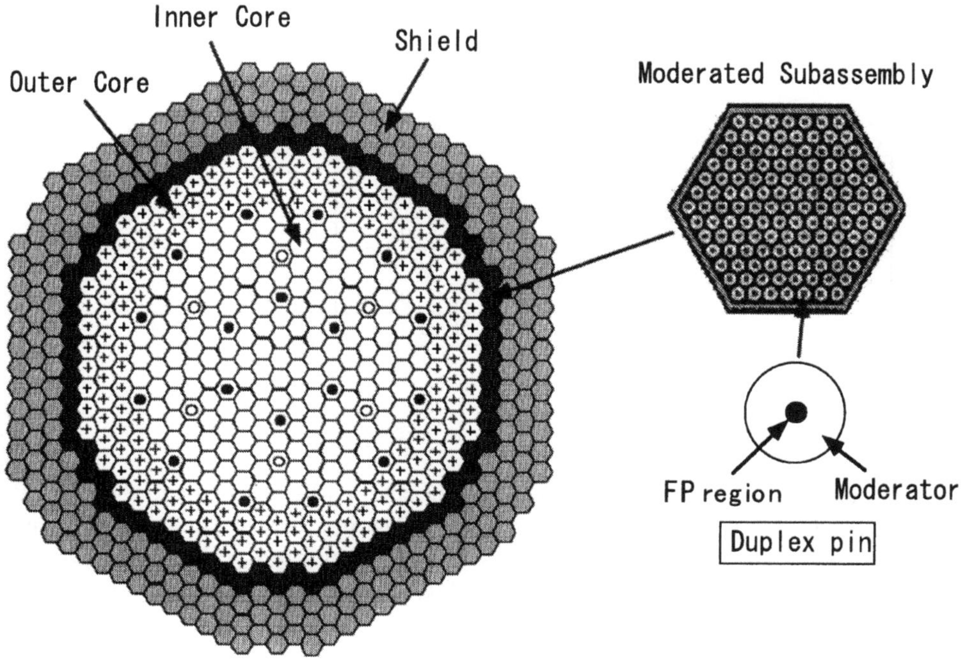

Fig.1 Moderated target subassemblies loaded in the radial blanket region of the fast reactor for LLFP transmutation

3.1 Transmutation of Tc-99

Systematic parameter survey calculations were performed to investigate the basic characteristics of FP transmutation in the blanket region of a fast reactor. The arrangement of the moderator and the target pins in the subassembly, the moderator material and the volume ratio of target to moderator were selected as parameters. The results of the calculations are shown in Table 2. A new concept of Duplex pellets was also examined: a moderator annulus surrounding a Tc-99 adopted to get a better Tc-99 transmutation performance as shown in Fig.1. The transmutation rate of Tc-99 in the new target subassembly is higher than that in the subassembly consisting of separate $ZrH_{1.7}$ moderator pins and Tc-99 target pins. A maximum Tc-99 transmutation rate of about 10%/year was obtained by using the new target subassembly loaded in the blanket region of the fast reactor. The new target subassembly can achieve an optimum transmutation performance by adjusting the volume ratio of $ZrH_{1.7}$ to Tc-99 in the duplex pellet.

The effects on main core characteristics of loading target subassemblies were also analyzed. It was found that the power density of the core fuel adjacent to the target is rather high and is about the same as the maximum in the core. However, the power spike is much mitigated compared to the case of loading target subassemblies in the core region.

Table 2 Results of Tc-99 transmutation performance

Loading method of FP pins	Number of pins in subassembly	Number of FP pins	Radius of FP pin	Transmuted amount (kg/y)	Transmutation ratio (%)
Heterogeneous	127	37	0.5	41.1	1.8
Heterogeneous	127	22	0.5	27.2	2.5
Duplex	127	127	0.2	38.1	3.5
Duplex	127	127	0.063	10.8	9.8
Duplex	217	217	0.2	46.7	2.5
Duplex	217	217	0.063	17.1	9.1

3.2 Transmutation of I-129

Several calculations were performed to determine the I-129 transmutation performance. Iodine-129 was loaded as NaI in the blanket region of another core. The transmutation rate of I-129 was 5.2% and the transmuted amount was 18kg in a year. The amount of I-129 produced by a 1000MWe -class PWR is about 5.0kg, so the transmuted amount of I-129 was equal to the output from 3 PWRs.

4. POTENTIAL OF SLLC FOR MA AND LLFP TRANSMUTATION

MA nuclides have much different characteristics compared with isotopes of Pu and U from neutronic viewpoint. The capture cross sections of Np-237 and Am-241 are very large compared with that of U-238 and many of the daughters have large fission cross sections. Therefore, MA nuclides are fertile that burn much faster than U-238.

Oxide-fueled super long life cores (SLLC) loaded with MA fuel were studied aiming at continuous operation without refueling during plant lifetime and efficient reduction of MA nuclides. The feasibility of a 300MWe-class SLLC was studied from nuclear and thermal characteristics assuming that high burnup fuel loaded with MA nuclides can be employed(7).

It was found that 300MWe-class SLLCs are feasible from neutronic view point by multiplying core volume by three compared with the conventional core. Reactivity change and power swing due to burnup were remarkably decreased by optimizing the loading fractions in the inner and outer regions of the core. Uniform loading of MA nuclides significantly increases power swing, whereas non-uniform loading where the fraction of MA nuclides in the outer region is twice more than in the inner region yielded much smaller power swing. The 300MWe-class SLLC can burn 5.3tons of MA nuclides during 34-years operation as shown in Table 1.

The influences on core safety and control is one of the issues to be further studied. The linear heat rate is about half of that in the conventional core, which is advantageous from the safety viewpoint. The 34-years core life leads to the average fuel burnup of 200GWd/t and the maximum fast neutron fluence of over $6x10^{23}$nvt. The ferritic steel under development for high burnup shows promising irradiation properties on swelling and sodium corrosion. Oxide-fueled SLLC loaded with MA nuclides is an attractive fast reactor concept as both no-refueling core and MA burner, although further research and development are required.

Parameter survey calculations were also performed to investigate the potential of FP (Tc-99) transmutation in the 300MWe SLLC. The moderated target subassemblies were loaded in the radial blanket and/or the shield region of the SLLC. It was found that the SLLC can transmute a large amount of LLFP without serious penalties in terms of core performance.

5. CONCLUSION

Systematic studies were implemented to investigate the flexibility and attractive core concepts of MA and LLFP transmutation in fast reactors. The MA transmutation in the fast reactor core has no serious drawbacks in terms of core performance, provided that the homogeneous loading method can be employed with a small fraction of MA fuel (2~5wt%). The hybrid MA-loading method, where Np nuclide is dispersed uniformly in the core and target subassemblies containing Am, Cm and rare earth nuclides are loaded into the radial blanket region, has a great potential to achieve effective transmutation of MA. Super long life fast reactor cores (SLLC) loaded with MA fuel were studied aiming at continuous operation without refueling during plant lifetime and efficient reduction of MA. 300MWe SLLCs with small reactivity change and power swing during plant lifetime of about 30 years were found feasible. A 300 MWe SLLC can transmute minor actinides of 5.3 tons. For LLFP transmutation, it was found that moderated target subassemblies loaded in the radial blanket and/or the shield region of fast reactor cores are effective for both a conventional fast reactor core and an SLLC.

REFERENCES

(1) T. Wakabayashi et al , "Characteristics of an LMFBR Core Loaded with Minor Actinide and Rare Earth Containing Fuels", GLOBAL'93 (Seattle,USA) 1993.9.12.
(2) T.Wakabayashi, S.Ohki and T.Ikegami ,"Feasibility studies of an optimized fast reactor core for MA and FP transmtation", Proc. Int. Conf. on Evaluation of Emerging Nuclear Fuel Cycle Systems (GLOBAL'95),Vol. I, Sep. 11-14 , Versailles, France (1995.)
(3) T. Wakabayashi , K.Takahashi and T. Yanagisawa, "Feasibility Studies on Plutonium and Minor Actinide Burning in Fast Reactors", Nuclear Technology Vol.118, p14-25, Apr.1997.
(4) M.Ozawa and T.Wakabayashi, "Status on Nuclear Waste Separation and Transmutation Technologies in JNC", GLOBAL'99,1999.
(5) N. Higano and T.Wakabayashi ,"Feasibility Study on the Transmutation of Long Lived Fission Products in a Fast Reactor", Proc. Int. Conf. on Future Nuclear Systems (GLOBAL'97),Vol. 2, p1322-1326, Oct. 5-10 , Pacifico Yokohama, Yokohama, Japan (1997).
(6) T. Wakabayashi and N. Higano , "Study on MA and FP Transmutation in Fast Reactors", Progress in Nuclear Energy,Vol.32, No.3/4, p259-264 , 1998.
(7) M. Yamaoka and T. Wakabayashi , "Study on Super Long Core Loaded with Minor Actinide Fuel " , Nuclear Engineering and Design 154(1995) , 239-250.

Pergamon

www.elsevier.com/locate/pnucene

Progress in Nuclear Energy, Vol. 40, No. 3–4, pp. 465–472, 2002

Printed in Great Britain
0149-1970/02/$ - see front matter

PII: S0149-1970(02)00039-2

RECYCLE TRANSMUTATION OF MA AND LLFP USING BWR FOR SUSTAINING GEOLOGIC DISPOSAL

ASASHI KITAMOTO, M. BUDI SETIAWAN

Research Laboratory for Nuclear Reactors
Tokyo Institute of Technology
2-12-1 O-okayama, Meguro-ku, Tokyo 15-8550 Japan
Tel. +81-3-5734-3073, Fax. +81-3-5734-2959
e-mail: akitamot@nr.titech.ac.jp

ABSTRACT

The actinides and fission products produced in nuclear fuels constitute an important part of the HLW. Therefore, methods for reducing the radiotoxicity of the MA and LLFP in HLW are presently under investigation. The purposes of this study are to evaluate the effectiveness of MA transmutation by taking advantage of neutron spectrum hardening due to void fraction along BWR axial direction; to understand the effectiveness of LLFP transmutation in BWR considering the large capture cross section of FP in thermal region; and to evaluate the macroscopic characteristics of longer residential period of LLFP target in the high burnup BWR core. Conceptual B/T BWR supposed in this study was reactor which the performance comparable to the current BWR. In MA transmutation case, the calculation was focused on varying the void fraction of 0 to 40% along the axial direction, which were directly associated to the lower and upper region of the BWR core. The performance of B/T BWR was evaluated in which four components of MA (^{237}Np, ^{241}Am, ^{243}Am, and ^{244}Cm) with fixed fraction were blended with UO_2 in B/T fuel. While, for LLFP transmutation, the B/T BWR was assumed to have two homogeneous regions: {1} the region for UO_2 driver fuel (99% of fuel weight), and {2} the region for LLFP (^{99}Tc and ^{129}I) target capsules (1% of fuel weight), in which metallic Tc rods and iodine in the form of CeI_3 was contained in cylindrical target capsules. The evaluation functions are {1} *fission-to-transmutation ratio,* [F/T ratio]$_{MA}$, and {2} *transmutation fraction*, TF_{LLFP}. Results show that the hardening neutron spectrum due to increase of void fraction in B/T BWR would result a higher [F/T ratio] of MA transmutation performance. Np and Am would be effectively loaded in the upper region of the core, while Cm could be loaded in any region of the core. At the EOC of equal or more than 50 GWd/Mg(HM), technetium has a higher *transmutation fraction* compared to iodine. To obtain higher LLFP transmutation fraction, the residential time in the LLFP targets in the core, should be kept for long time, for instance about 10 to 30 years. For that purpose, it was proposed that the number of B/T BWR system for LLFP treatment corresponds to the residential time of the LLFP target, i.e. 10 to 30 units.

1. INTRODUCTION

1.1 MA Transmutation in BWR

The actinides produced in nuclear fuel constitute an important part of the long-lived HLW. The radiotoxicity of MA may not lead to select the direct storage of the spent fuels for very long time due to LLMA. Therefore, the conceptual studies to reduce the radiotoxicity of those actinides are underway.

Various types of reactors and accelerator driven transmutation (ADT) facilities are among the possible choices for carrying out this extensive endeavor. However, a fully new technology, e.g. fusion reactor, fast reactor or ADT can only now be assessed in a generic way since their development is still in their preliminary stage. Adaptation of existing technologies will be more simply approached, such as reprocessing of high burnup fuel from LWR-MOX, fabrication of actinide targets, minor modification of current thermal reactors to suit the use of burning and transmutation (B/T) fuels, etc. A promising scenario studied herein is by transmuting the long-lived isotopes of MA in BWR core by taking advantage of neutron spectrum hardening due to void fraction along the axis.

1.2 LLFP Transmutation in BWR

Fission products (FP) are the most limiting factor in designing the repository facilities due to the decay heat emission that increases proportionally to the burn up. After some 500 years the major part of the FP have decayed, except for some long lived nuclides, such as Cs-135, Tc-99, I-129, Zr-93 etc, which half-lives are in the order of 10^5 to 10^6 years. Some of them are relatively mobile in the geosphere and contribute to the dose to man in the surroundings and even beyond the repository facilities.

Transmutation of long-lived fission products (LLFP) is a very difficult task, since the capture cross-section to transmute the radioactive nuclide into a short-lived fission products (SLFP) or stable nuclides requires large reactor capabilities. Thermal reactors and accelerator driven transmutation (ADT) facilities are the possible choices for carrying out this very expensive endeavor (Baetslé, 1996). A completely new technology, e.g. accelerator driven transmutation, however, needs a longer technological development since its low transmutation rate performance due to the characteristics of intermittent pulsation of ADT. Thus at the present time evaluations have to be done on the existing technologies which are more easily approached, such as minor modification of current LWRs to suit the use of LLFP targets, reprocessing of high burnup fuel from LWR-MOX and fabrication of actinide or LLFP targets.

1.3 Objective of the study

The objective of this study is (1) to find the optimized MA loading in the axial direction, taking into the account of the void change, and (2) to obtain the optimal discharged period of MA in the burning and transmutation (B/T) BWR. (3) to understand the effectiveness of LLFP transmutation in LWR considering the large capture cross section of FPs in thermal region; and (4) to understand the macroscopic characteristics of longer residential period of LLFP target in the high burnup BWR core. The summary of MA and LLFP transmutation studies is given in Table 1.

Table 1. Summary of MA and LLFP transmutation studies

	MA	LLFP
nuclide	^{237}Np, ^{241}Am, ^{243}Am, ^{244}Cm	^{99}Tc and ^{129}I
in-core	Loaded in B/T fuel blended homogeneously with UO_2	Loaded in capsule targets: metallic Tc and CeI_3
evaluation function	*fission-to-transmutation ratio*, F/T-ratio	*transmutation fraction*, TF

2. EVALUATION FUNCTION

The function of (F/T) ratio (introduced by Kitamoto et. al., 1999) was used to evaluate performance of the reactor for MA transmutation by taking the ratio of *total fission rate* to *total transmutation rate*. If the reactor criticality is kept under the averaged neutron flux, the reactor can be optimized by this evaluation function. The function, (F/T) ratio, is defined as $\Sigma_i \Sigma_j \{N_i \sigma_{f,i}^{\,j} \phi^j\} \;/\; \Sigma_i \Sigma_j \{N_i (\sigma_{c,i}^{\,j} + \sigma_{f,i}^{\,j} + \sigma_{2n,i}^{\,j})\, \phi^j\}$ where N = nuclide number density, σ_f = fission cross section, σ_c = capture cross section, σ_{2n} = n,2n cross section, i = MA nuclide and j = neutron energy group. The evaluation function of *transmutation fraction*, TF, is defined by the ratio of the *transmuted LLFP* to the *initial LLFP loading*, calculated at the various burn-up times. $[TF] = \{\Sigma(N_{LLFP})_{trans} \;/\; \Sigma(N_{LLFP})_{init}\} \times 100\%$ where N_{LLFP} = LLFP number density which is obtained from the calculation using the following equation: $N_i = F_i \rho\, N_A / M_F$ where N = atomic number density of nuclide-i [atom/cm^3], ρ_F = fuel density [g/cm^3], F_i = weight fraction of nuclide-i in fuel [-], M_F = average molecular weight of fuel [g/gmol], and N_A = Avogadro's number [6.024×10^{23}mol^{-1}].

3. MA AND LLFP SELECTION

In this study, MA selected was four components: Np-237, Am-241, Am-243 and Cm-244. U&Pu, separated with MA (U&Pu unrecovered in a reprocessing plant) and UO_2 blended with MA, are considered as fuel, though they are reacted with neutron in B/T fuel, and UO_2 and its transmuted component to TRU in B/T fuel, is considered as fuel. In case of short-lived MA nuclide, there is no influence on the inventory reduction caused by the short-lived MA nuclide, because R&P process was assumed not to having ability to separate the isotopic component. In case of the unrecovered plutonium from the reprocessing process were not included in the B/T fuel, since their influences in the fission rate were quite dominant, so that the B/T reactor performance could not be clearly shown.

LLFPs to be treated in the BWR are selected using *two-level selection*, as following.

Selection-1 comprises three criteria, which are (1) the natural decay of the radionuclides should be more than 100 years, in the view that the deep-geological waste repository system would prevent the release of the radionuclides for 100 years, (2) the fission yield of RI (radio isotope) should be more than 10^{-3}, and (3) RI are selected according to their hazard index.

Based in selection-1 criteria, six radioisotopes were selected. They are ordered according to the magnitude of Hazard Index as following: I-129, Tc-99, Zr-93, Sn-126, Cs-135, Pd-107. Up to this level, the following FPs is to be considered: iodine (I-129), technetium (Tc-99) and cesium (Cs-135).

Selection-2 comprises four criteria as following: (1) the decay acceleration, which is defined as the ratio of *total decay rate under neutron flux* ϕ to *natural decay rate* should be greater than 10, (2) the nuclides should have a high transmutation fraction in one fuel life, (3) the partitioning of the radionuclides from spent-fuel can be done by chemical separation; not isotopic separation, and (4) for the technological and economical reason for no further separation process, their isotopic composition have to be simple.

From these selections, two nuclides were chosen as candidates to be transmuted, which are: *iodine* and *technetium.* Technetium and iodine are among the most important nuclides that dominate the β activity of spent fuel after a hundred thousand years. Since their solubility in ground water is high, Tc-99 and I-129 are easily mitigated to the environment once they are released from the geological waste repository. To reduce the dose risks, Tc-99 and I-129 should be partitioned from the spent fuel and treated separately.

4. CALCULATION METHOD

The performance of B/T (burning and/or transmutation) BWR was estimated by the conceptual BWR with the specifications comparable to the current 1,000 MWe-BWR, including the size of the core and fuel cell geometry. The core was assumed to be built with two homogeneous regions: (1) the region for standard UO_2 fuel (90% of core volume) with standard enriched U-235, and (2) the region for B/T fuel (10% of core volume) in which four components of MA (^{237}Np, ^{241}Am, ^{243}Am, and ^{244}Cm) with fixed fraction were blended with UO_2 in the oxide form. Calculations were done for several void fraction conditions of 0, 10, 20, 30 and 40%. The void fractions of 0 to 40% were directly associated to the lower and upper regions of the BWR core.

Conceptual BWR-type transmutation reactor only for LLFP transmutation proposed in this study was supposed as reactor which the performance comparable to the current 1000 MWe BWR. The sizes of the core and the fuel cell geometry of conceptual transmutation reactor was kept to the same as current ones. The core of transmutation reactor was assumed to have two homogeneous regions: (1) the region for UO_2 driver fuel (99% of fuel weight) with 4.5% of enriched U-235, and (2) the region for LLFP target capsules (1% of fuel weight), in which metallic Tc rods and iodine in the form of CeI_3 was contained in cylindrical target capsules (Babelot, et al., 1995).

The numerical calculations were done by using JAERI-SRAC95 code (Okumura K., 1996). The mass balance of nuclides in the core was calculated by the cell depletion calculation for the homogeneous core model. The time needed to attain the equilibrium state of the core can be estimated by making a series of cell depletion calculations starting with a typical cell composition.

The reactivity swing for different MA or LLFP loading in the B/T fuel or target capsules was examined in order to obtain a maximum loading of MA fraction which yields a safe and maintainable reactivity swing over the fuel life time. The criterion for determining a maximum acceptable content of MA or LLFP loading was settled from the aspect of reactor operability.

5. RESULTS AND DISCUSSION

5.1 MA Burning

The hardening neutron energy due to the void fraction in BWR would result in a higher transmutation fraction of MA transmutation performance. Calculation results show that (F/T) ratio of total MA increased for the higher void fraction (Fig.1). It is also observed for individual specified nuclide, i.e. ^{237}Np, ^{241}Am, ^{243}Am, and ^{244}Cm, that in case of ^{237}Np, the higher void fraction of 40% would effectively increase its (F/T) ratio. Am increased its (F/T) ratio with hardening spectrum which correspond to 20% void fraction. In case (F/T) ratio of Cm, however, there was no observable difference in all void fractions, as concluded in Table 2.

MA nuclei are not directly fissioned instead they change to fissionable nuclide by neutron capture before fissioned. Figure 2a shows the main reactions of Np-237. As shown in this figure, Np-237 changes to fissionable Pu-239 by neutron capture and then fission. Figure 2b shows the main B/T reaction of Am-241 and Am-243 in the core. As can be observed in Fig. 2b, Am-241 has three cases of termination. Firstly, Am-241 changes to fissionable Am-242 by neutron capture; the main part proceed with α-decay to fissionable Pu-239, and some parts go to Cm-243 by neutron capture and fission due to very small capture cross section of Cm-242. Main B/T reaction of Am-243 is terminated by fission of Cm-245.

5.2 LLFP Transmutation

Calculation results of first loading and remained of Tc-99 and I-129 in the target of both BWR and PWR are shown in Fig. 3. At the BOC of both reactors the initial mass of Tc-99 and I-129 were 1200 kg respectively, which is equivalent to 1% of the 120 Mg fuel inventory in the core (Babelot, et al., 1995).

5.2.1 Technetium. At the EOC which is equal to 50 GWd/Mg(HM), technetium mass decreased quite rapidly compared to iodine. It is also shown that Tc-99 remained in BWR was lower than those in PWR. In suggests that more technetium is transmuted in BWR core. The production of technetium in the driver fuel of both reactors was comparably similar, in which production in BWR is slightly higher. The amount of technetium remained in target capsules and technetium produced in driver fuel was differed in the order of 10^2 kg. These conditions were thought to be suitable to the purpose of transmutation reactor for technetium transmutation, in which the productions of Tc-99 by fission in the

UO_2 driver fuel decreases the net transmutation rate and reduces the necessities of large technetium loadings.

5.2.2 Iodine. At the EOC, the remained of I-129 in target capsules of both BWR and PWR were comparably the same. As in the case of technetium, the production of iodine in driver fuel of both reactors was comparably similar. Production of iodine in the driver fuel is much lower than those of technetium.

As a results the overall Tc-99 and I-129 remained in the target of BWR core was much lower compared those in PWR. In the other words, more iodine is transmuted in the BWR core.

5.2.3 TF evaluation function. The evaluation of *transmutation fraction*, TF, was calculated at the EOC of 50 GWd/Mg, as provided in Fig. 4. The [TF] value for BWR (with 40% void fraction) is higher than that of PWR, which are 13% and 10%, respectively.

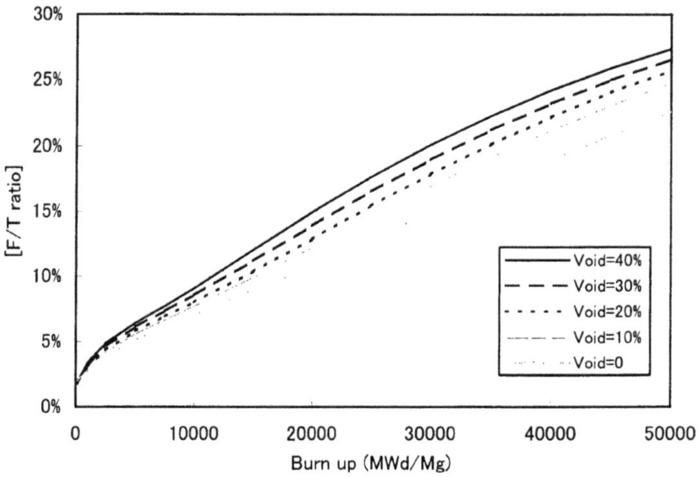

Fig. 1 F/T ratio of MA in B/T BWR (MA=5%, void 0%-40%)

Table 2. Effective loading region for B/T fuel

Nuclide \ Void(%)	0	20	40
	Lower	⟵ ⟶	Upper
^{237}Np	○	○	◎
^{241}Am	○	◎	◎
^{243}Am	○	◎	◎
^{244}Cm	○	○	○

Remarks:
◎: the effective region for B/T treatment
(activated by the void fraction)
○: the normal region for B/T treatment

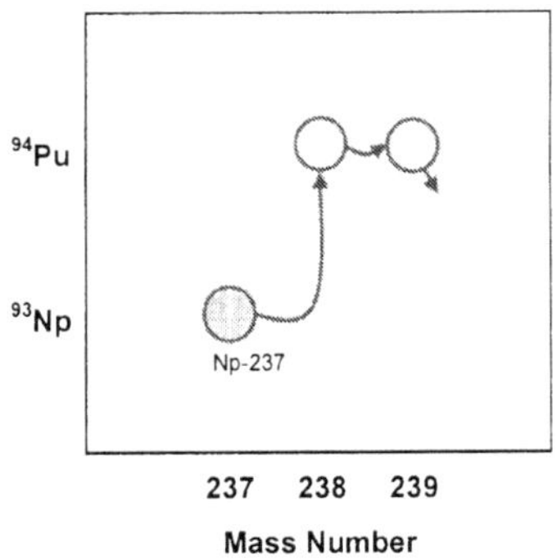

Fig 2a. Main B/T reactions of Am-241, terminated by fissions of Am-242, Pu-239 and Cm-243. And those of Am-243 terminated by fission of Cm-245

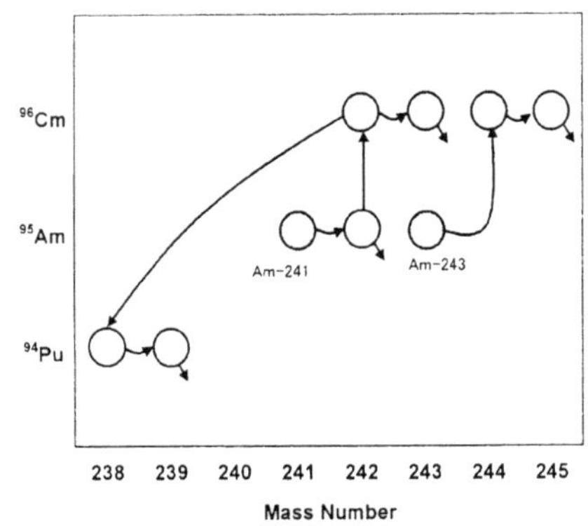

Fig 2b. Main B/T reactions of Np-237, terminated by fission of Pu-239

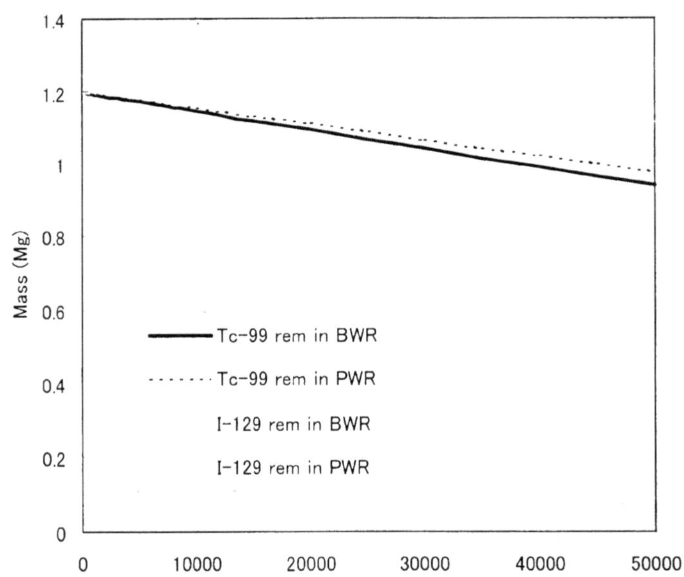

Fig. 3 Mass of Tc-99 or I-129 in the BWR or PWR with respect to burn up. The initial mass of Tc and I are equal to 45 GWd/MgHM discharged fuel from each 24 units of 1000 MWe LWR.

6. CONCEPT OF TRANSMUTATION SYSTEM FOR LLFP IN LWR

From the calculation and evaluation results of technetium and iodine in light water reactors, a concept of transmutation system for LLFP in LWR is proposed. In the LLFP transmutation treatment, the residential time should be kept for long time, for instance about 10 to 30 years, for obtaining higher transmutation fraction. Consequently, for maintaining the constant treatment for such a long time, it is suggested that the number of reactor units should be proportional to the time of transmutation treatment. For straightforwardness purpose, it is proposed that the number of multi-recycle transmutation reactor system for LLFP treatment simply corresponds to the irradiation time of the LLFP target, i.e. 10~30 units.

The upper limit of technetium and iodine loading in the core, with satisfy the critical condition was 1 w% of normal fuel, i.e. 1.2 Mg/unit transmutation BWR. As mentioned before this amount is equivalent to mass of Tc-99 produced by 57 unit of 45 GWd/MgHM LWR or to mass of I-129 produced by 237 unit of 45 GWd/MgHM LWR.

In case of 50 units of LWR, a brief concept for LLFP transmutation could be drawn as follows (Table 2). For maintaining a higher transmutation fraction, the irradiation time of technetium and iodine in the reactor is settled on 10 years and 30 years, respectively. The total inventory for one irradiation time would be similar to 50 LWR×10 = 500 LWR equivalent and 50 LWR×30 = 1500 LWR equivalent. The number of transmutation BWR system needed for transmutation is about (500 LWR equiv./57 LWR) = 9 units of transmutation BWR for technetium and about (1500 LWR equiv./237 LWR) = 6 units of transmutation BWR for iodine. As a whole, approximately 15 units of transmutation BWR are required for transmuting technetium and iodine discharged from 50 units LWR. In conclusion, it is not so easy to transmute technetium and iodine produced in 50 units LWR, since it would require large number of transmutation BWR.

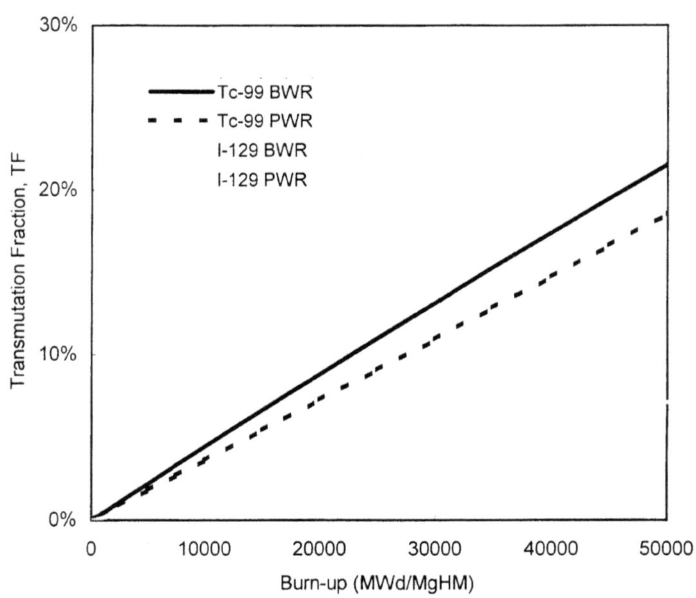

Fig. 4 *Transmutation Fraction* (TF) of Tc-99 and I-129 in BWR and PWR (1%w loading)

Table 2. The number of transmutation BWR for ^{99}Tc and ^{129}I, produced by 50 units 1GWe LWR, based on once through and multi-recycle.

TF(-)	Nuclide	Irradiation Time (y)	Total mass produced by 50 unit LWR	NR_1	Sum of NR_1 for ^{99}Tc & ^{129}I	NR_C	Sum of NR_C for ^{99}Tc & ^{129}I
0.4	^{99}Tc	10	10,525	8.8	15.1	21.9	37.7
	^{129}I	30	7,590	6.3		15.8	
0.5	^{99}Tc	12.5	13,156	11.0	18.9	21.9	37.7
	^{129}I	37.5	9,487	7.9		15.8	

Notes:
NR_1 = Number of BWR in once through
NR_C = Number of BWR in multi-recycle
TF = Transmutation Fraction

7. CONCLUSIONS

(1) The hardening neutron energy due to the void fraction in B/T BWR would result in a higher (F/T) ratio of MA transmutation performance. Np and Am would be effectively loaded into the top region of the core, while Cm could be loaded in any region of the core. The top region of the core is optimized for the position of loading in B/T fuel.

(2) In case of technetium, BWR is able to transmute more effectively than those of PWR. However, in case of iodine, the significant difference of its remained in target capsule of both reactors was not observed.

(3) BWR has a comparably higher *transmutation fraction* [TF] value for LLFP transmutation compared to PWR.

(4) To obtain higher transmutation fraction, the residential time in the LLFP transmutation treatment, should be kept for long time, for instance some 10 ~ 30 years. The number of transmutation reactor for LLFP treatment simply corresponds to the residential time of the LLFP target capsules, i.e. 10 ~ 30 units.

References

Babelot J.F (1995) Development of Fuels for Transmutation in the frame of EFFTRA European Collaboration. *International Conference on Evaluation of Emerging Nuclear Fuel Cycle Systems (GLOBAL 1995),* pp 524-529. Versailles, 11-14 September

Baetslé L.H. (1996), A Comparative Systems-Analysis Approach on Fuel Cycles with Partitioning and Transmutation. *The Fourth International Information Exchange Meeting on Actinide and Fission Product Partitioning and Transmutation.* p. 45. Mito City, Japan, 11-13 September

Kitamoto A., Setiawan M.B., Taniguchi A. (1999), *Proposal of New Evaluation Function to Indicate the Performance of B/T Reactor.* 1999 Annual Meeting of the Atomic Energy Society of Japan., F1, p. 299

Okumura K., et al. (1996), *SRAC95 General Purpose Neutronic Code System*, Japan Atomic Energy Research Institute JAERI Data/Code 96-015

Pergamon
www.elsevier.com/locate/pnucene

Progress in Nuclear Energy, Vol. 40, No. 3–4, pp. 473–480, 2002

Printed in Great Britain
0149-1970/02/$ - see front matter

PII: S0149-1970(02)00040-9

SYMBIOSIS SYSTEM FOR TRANSMUTATION ON NITRIDE FUEL FBR AND ADS

Hideki TAKANO and Kenji NISHIHARA

TOKAI RESEARCH ESTABLISHMENT
JAPAN ATOMIC ENERGY RESEARCH INSTITUTE

Tokai-mura, Naka-gun, Ibaraki-ken, 319-1195, JAPAN

ABSTRACT

A conceptual scheme for mass flow of transmuting Plutonium (Pu), minor actinides (MA) and long-lived fission products (LLFP) is studied. In this feature, the existing light-water reactors (LWRs) cycle will be main stream for nuclear electric generation during a long-term period more than 50 years, and Pu will be utilized in mixed oxide fuel (MOX)-LWRs. In future, when Pu recycling system will be achived by introducing high-conversion LWRs (HCLWRs) and/or fast breeder reactors (FBRs), the accelerator driven transmutation system (ADS) transmutes Pu, MA and Iodine from Purex or Dry reprocessing. This is due to reduce burden for transmuting the excess or remained Pu, MA and LLFP by commercial reactor plants in Pu-recycling system. For this purpose, we introduce a concept of symbiosis system for transmutation based on nitride fuel FBR and ADS. The core design for lead-bismuth (Pb-Bi) cooled FBRs and ADS, Pb-Bi technologies, 15N enrichment and 14C toxicity are studied. And the mass flows for MA and Iodine are discussed based on an estimated scenario for nuclear electric plants introduction in future.

1. INTRODUCTION

The management of the long-lived radioactive nuclides becomes a key issue of the nuclear society in present day. Various concepts for transmuting them to shorter-lived or stable nuclides have been proposed to reduce the risk from the long-term toxicity. Due to hard neutron spectrum, a fast breeder reactor (FBR) and accelerator driven system (ADS) have capability to burn minor actinides (MA) such as neptunium, americium and curium which dominate the long-term toxicity of spent fuels.

Recycling the all actinides and some long-lived FPs into FBR and ADS to close the fuel cycle from a view point of actinide confinement is, therefore, one of the promising options to be considered in solving the problem. A concept of nitride fuel cycle system based on pyrochemical reprocessing has been proposed, and excellent core performance of the lead-bismuth cooled nitride FBR could provide design flexibility of reactor systems for energy production and/or reactors for burning or transmuting long-lived radioactive nuclides (Takano and et al, 1993, Osugi and et al., 1995). For utilizing the nitride fuel, the effect of ^{15}N enrichment on nuclear characteristics and the evaluation of toxicity of ^{14}C generated from ^{14}N was appeared, and excellent performance for the minor actinide (MA) transmutation was shown (Takano and Osugi, 1995). Furthermore, the current status for the liquid heavy metal Pb-Bi technologies were investigated (Takano and et al., 2000).

In the symbiosis concept, the base load reactors of FBRs produce electric power and extra plutonium fuel if needed, and the MA (about 1 wt%) generated by themselves are recycled (Osugi and et al., 1995). The MA transmuters have a role of incineration for MA and Iodine of ^{129}I and ^{127}I generated from UO_2 and MOX fueled LWRs, and FBRs systems. A part of the recovered Pu and minor actinides by pyrochemical reprocessing are recycled into the ADS transmuters to transmute the Iodine.

Furthermore, in future, accumulation and transmutation of MA and Iodine based on the symbiosis recycle system are been investigated for introducing the accelerator driven system (ADS) with 800 MWt. It is shown that the introduction of ADS can play a significant role as "Transmuter" in the back-end of fuel cycle in the scenario of nuclear plant capacities for maximum 140 GWe, which consists of LWRs, MOX-LWRs and FBRs.

2. CONCEPT OF SYMBIOSIS SYSTEM FOR TRANSMUTATION

In this study, we introduce a concept of symbiosis system for transmutation on nitride fuel FBRs and ADS as shown in Fig. 1. The objectives are to transmute Pu, MA and LLFP produced by Pu recycling and utilization. Pu will be utilized in mixed oxide fuel (MOX)-LWRs. In future, when Pu recycling system will be achived by introducing high-conversion LWRs (HCLWRs) and/or fast breeder reactors (FBRs), the accelerator driven transmutation system (ADS) transmutes Pu, MA and Iodine from Purex or Dry reprocessing.

We select MA (Am, Cm and ^{237}Np) and Iodine of ^{129}I and ^{127}I as the most important target nuclides for transmutation. Am and Cm are the most dominant contributors for long-term potential hazard in spent fuel. On the other hand, Iodine is soluble in water and one of the most troublesome nuclides on the geological disposal technology, though its potential hazard is smaller than those of MA.

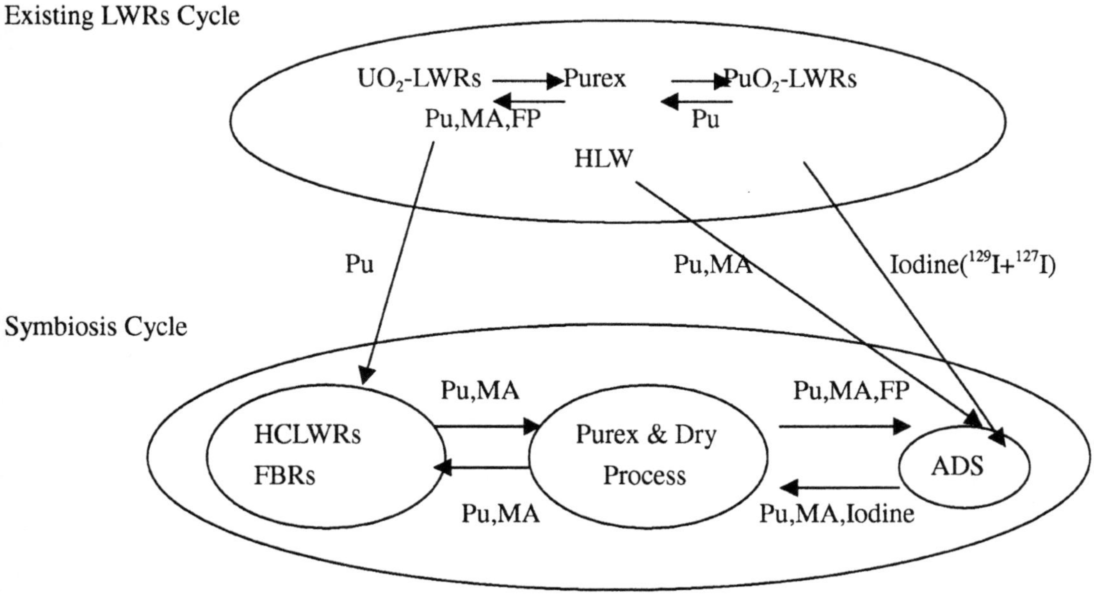

Fig. 1 Conceptual scheme for symbiosis system of transmutation

3. CONCEPT OF LEAD-BISMUTH COOLED NITRIDE FBR

The design study of Pb-Bi cooled nitride FBR was performed. The characteristics of Pb-Bi coolant are no active with water, high boiling temperature of 1700℃ and high gamma and neutron shielding effect. And the core characteristics are a harder neutron spectrum, a moderated coolant void reactivity and a small burnup reactivity swing. Table 1 shows the core performance and the core has very small burnup reactivity swing of 0.3 $ and negative coolant void reactivity.

These characteristics for this FBR give a safety aspect. In FBR design, the most troublesome issues of safety are the core distructive accident (CDA) and the vapor explosion. In an event sequence of an anticipated transient without scram (ATWS), the most probable sequences are shown with thick line in Fig. 2. This core will be very low probability for the CDA event.

A seismic design study shows that the reactor structure has the integrity against 2G seismic acceleration condition.

Table 1 The core performance of the lead-bismuth cooled FBR

Reactor power	1500 MWt
Core height	100 cm
Diameter	355 cm
Blanket thickness	30 cm
Pin diameter	10.2 mm
Pitch/diameter	1.51
Fuel	2w%Pu (U,PU)15N
Loaded Pu	4.0 ton
Burnup	92 GWd/t
EFPD	2000 days
Burnup swing	0.14 (%dk/k)
Fissile Pu gain	1.26
Void reactivity	-1.5 (%dk/k)
Doppler coe.	-3.7 x 10-3 Tdk/dT
Pb-Bi density coe.	3.55x10-4 (%dk/k)
Fuel temperature coe.	-7.88 x 10-4 (%dk/k)
Beta-eff	0.44 (%dk/k)

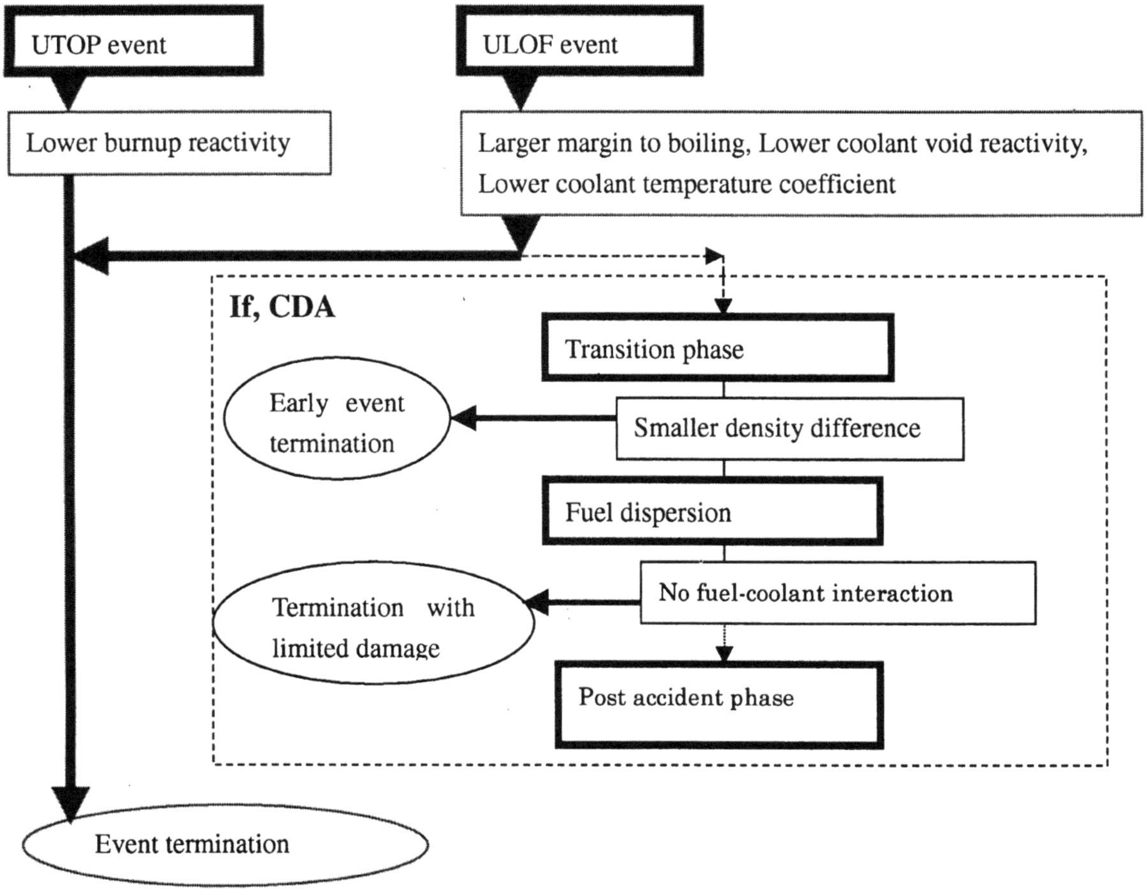

Fig. 2 The most probable sequence in ATWS events

Investigation of Pb-Bi technologies (Takano and et al., 2000):

(1) In Japan, the electric generation system in a refinery factory has being operated by using liquid Pb-Bi in 17 years. We can find no corrosion for the thermal conductive materials (1Cr-0.5Mo steel) used under the condition of natural convection with temperature around 430-530 ℃.

(2) In Russia, extensive experience in Russian submarines and in R & D during about 50 years are available: zero corrosion for Cr-Si materials by adjusting oxygen film with oxygen concentration control between 10^{-7} to 10^{-5} % mass.

(3) Polonium forms PbPo in Pb-Bi, and the evaporation rate become less three factor than that of Po, and furthermore, the rate decreases in the atmosphere. The effects of Po on employee and environment will not be dominant in comparison with those of fission products.

(4) AS for the Bi-resource, a confirmed amount will be 260,000 ton and an estimated amount will become ten times of the confirmed ones by including resources in Russia. This shows there are enough amounts for ADS developments

N-15 enrichment and C-14 toxicity:
The toxicity of ^{14}C becomes almost equivalent to that of Am + Cm, and smaller by a factor than Pu, after 5000 years. If 90% enriched ^{15}N is used in the fuel, the ^{14}C toxicity becomes smaller and similar to that of FPs after 1000 years. This means the problem of ^{14}C is not radiotoxicity, but capability of recovering ^{14}C in reprocessing system. The pyrochemical process has a possibility of 100% recovering ^{14}C , because the carbon is kept in salt with a chemical form of ZrC with Zr in fission product.

4. A CONCEPT OF LEAD-BISMUTH COOLED ADS

In Table 2, the excess and leakage neutrons are compared for a typical FBR and ADS. These values of ADS are considerably larger than them of FBR, then ADS will be advantage for transmutation of Iodine.

Table 2 Comparison of excess neutron in FBR and ADS

	Spallation Source	Neutron yield /fission	Fission	Capture in core	Excess neutron	Leakage
FBR		2.92	1.00	1.54	0.38	0.06
ADS	0.21	2.98	1.00	1.59	0.61	0.29

Transmutation of MA and Iodine was studied by using a lead-bismuth cooled ADS with 800 MWt. MA(Am, Cm and ^{237}Np) are most dominant contributors for long-term potential hazard in spent fuel. On the other hand, long-lived fission product nuclide Iodine-129 will be recovered as AgI formation in reprocessing system for spent fuels. And the Iodine is soluble in water and one of the most trouble some nuclide on the geological disposal technology, though its potential hazard is smaller than those of MA Thus, transmutation for this nuclide is strongly expected as one of troublesome isotopes from the view-point of waste disposal. The Iodine are loaded axially and radially with the form of NaI around the MA-fule core in ADS, and it was shown that an ADS can transmute the MA and Iodine generated from 9 or 10 units of LWR with 33 GWd/t per year as shown in Table 3. The conceptual design of the ADS is shown in Fig.3. The detailed reactor structure will be discussed at this seminar by Takizuka and et al., 2000.

Table 3 Core performance of ADS for MA and I transmutation ADS with proton beam power of 30 MW (1.5 GeV, 20 mA)

Core		
Thermal power	800 MW	
Keff	0.95	
core height / diameter (cm)	100 / 115	
Fuel compositions	60%MA+40%Pu	
Initial heavy metal loading (kg)	4000	
Transmutation of MA (kg/300 days)	250	
Blanket	Axial	Radial
Thickness (cm)	25	16
NaI/(NaI+ZrH) (%)	50	40
Initial Iodine loading (kg)	765	839
Transmutation of I (kg/300 days)	35	22

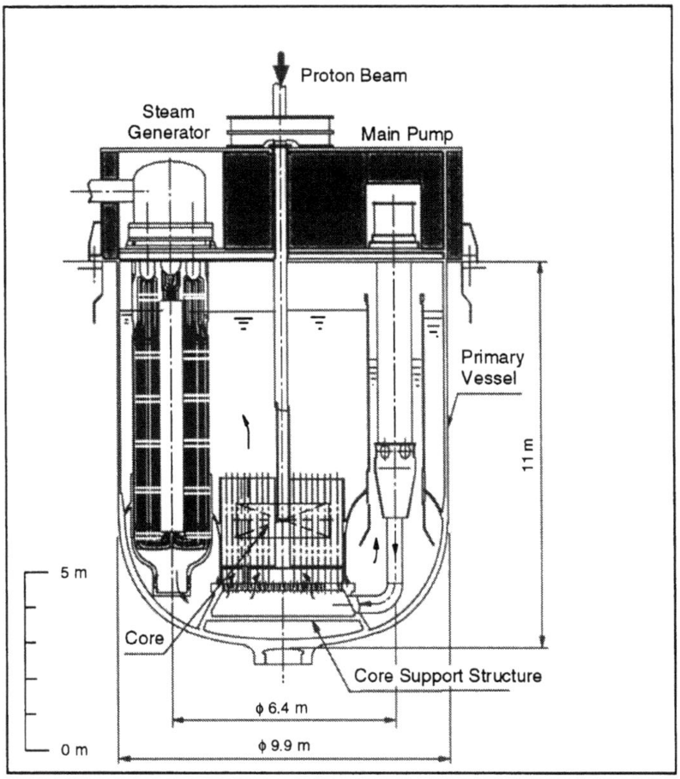

Fig. 3 Concept of Pb-Bi cooled ADS for transmuting MA and Iodine

5. EFFECT OF ADS INTRODUCTION ON TRANSMUTATION OF MA AND IODINE

Dependence of MA and Iodine buildup on fuels, burnup and cooling times: The amount of the MAs produced in LWRs with UO_2 / MOX fuels and an FBR with MOX fuel have been investigated for various conditions such as fuel compositions, burnup and cooling times as shown in Table 4. For the UO_2-LWRs, MA buildup increases

Table 4. Dependence of buildup for MA and Iodine burnup, cooling times and fuels in various reactors

Reactor	PWR	PWR	PWR	PWR	FBR
Fuel	UO2	UO2	MOX	MOX	MOX
Burn-up(GWd/t)	33	60	33	60	140
Cooling(y)	5	5	5	5	3
MA-buildup(kg/y/plant)	22.2	26.3	91.8	88.1	39.1
Iodine-buildup(kg/y/plant)	5.7	5.5	7.5	7.3	8.6

slightly with increasing burnup. The MA buildup quantities for full MOX fueled LWR are more than three

.imes for the UO_2 fueled LWRs. In addition, MOX-FBR produces about two times MA comparative with JO$_2$-LWR.

The effect of ADS introduction: An estimation of the nuclear plant capacities introduced in future will be lepend on a requirement from CO_2 reduction scenario decided in the COP3 conference at Kyoto and imitation of uranium resources. Here, based on these conditions in Japan, the nuclear plant capacities are lecided as maximum 140 GWe as shown in Fig.4. The mass flows of MA and Iodine are investigated in a future symbiosis system for transmutation consisting of UO_2/MOX-LWRs, FBRs and ADSs. The accumulations of MA and Iodine become about 150 and 20 tons in 2040 and 900 and 200 tons in 2200, respectively, as shown in Fig. 5. When the ADS will be introduced from 2040, the system inventories of MA with Pu and Iodine in the ADS cycle will become about 250 and 50 tons during a time period of 2200. The mass for Pu, MA and Iodine are balanced by introducing 22 units of ADS until 2160.

CONCLUSIONS

The concept of symbiosis system combined with FBR and ADS was studied based on the nitride fuel cycle in order to reduce inventories and long-term hazard of HLW by transmuting long-lived nuclides.

The liquid heavy metal (Pb-Bi) cooled / nitride fuel FBR had some superior characteristics for safety, Pu breeding, MA transmutation, etc.. The R & D of Pb-Bi technologies will be necessary, though the employment experiences in a refinery, submarines etc. were useful.

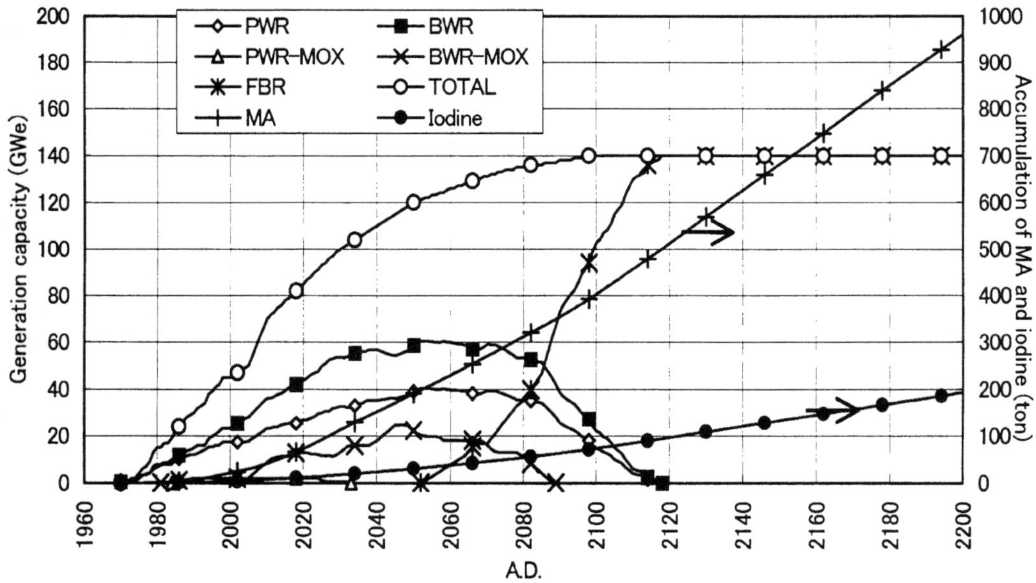

Fig. 4 Nuclear electricity capacity composed of UO_2/MOX-LWRs and FBRs introduced based on CO_2 reduction scenario in Japan

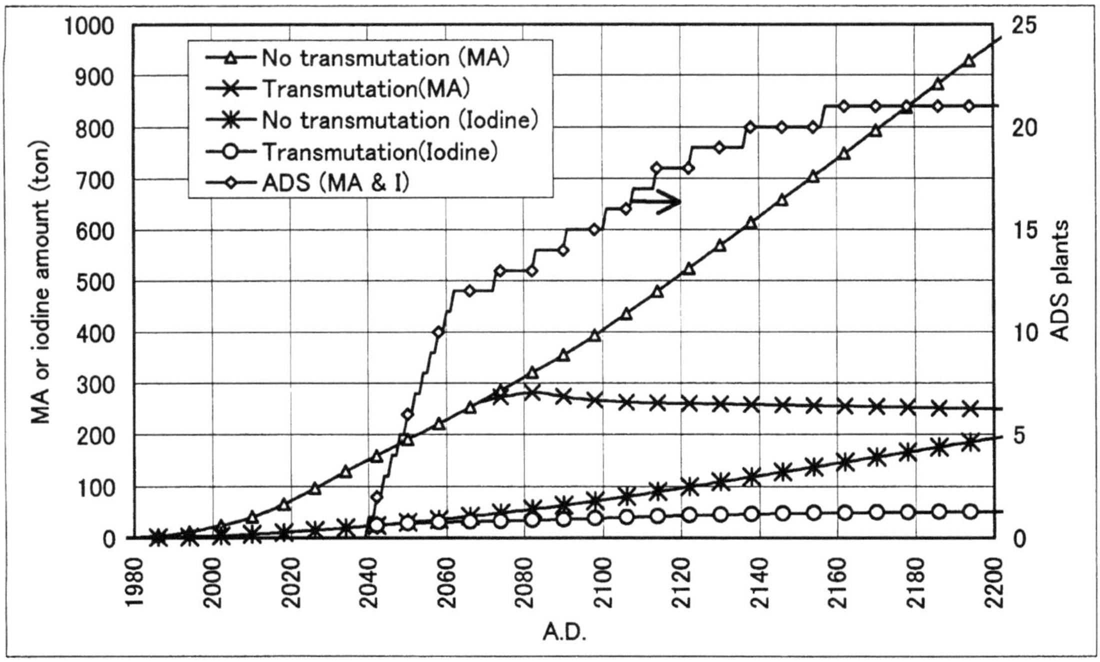

Fig. 5 Comparison of MA and Iodine accumulations and transmutation by introducing ADS from 2040.

It was shown that in the future scenario of nuclear plant capacities for maximum 140 GWe, which consists of LWRs and FBRs, the introduction of ADS can play a significant role as "Transmutor" in the back-end of fuel cycle.

REFERENCES

Takano, H., Akie, H., Handa, M. et al.(1993), A Concept of Self-Completed Fuel Cycle Based on Nitride Fuel Lead-Cooled Fast Reactor, Proc. Seventh International Conf. on Emerging Nuclear Energy Systems, ICENES'93, p.308, World Scientific Press.

Osugi, T., Takano, H., Ogawa, T., et al.(1995), A Conceptual Design Study of Self-Completed Fuel Cycle System, Proc. International Conf. on Evaluation of Emerging Nuclear Cycle Systems, Global'95, Sept. 11-14, Versailles, France, 1995, Vol.1, p.181.

Takano, H. and Osugi, T.(1995), A Concept of Nitride Fuel Actinide Recycle System Based on Pyrochemical Reprocessing, presented to Japan-Russian Seminar on Fast Breeder Reactor, 11-15 Dec. 1995, Oarai, Japan .

Takano. H. Takizuka, T. and Kitano,T.(in Japanese) (2000), Investigation of Corrosion, Water Reaction, Polonium Evapolation and Bismuth Resource in Liquid Metal Lead-Bismuth Technology, JAERI-Review 2000-014.

Takizuka T., et al.(2000), Design Study of Lead-Bismuth Cooled ADS Dedicated to Nuclear Waste Transmutation, at this Susono Seminar.

Mukaiyama, T., et al.(2001):" Review of Research and Development of Accelerator-Driven System in Japan for Transmutation of Long-Lived Nuclides," to be published in Progress of Nuclear Analyses.

Pergamon
www.elsevier.com/locate/pnucene

Progress in Nuclear Energy, Vol. 40, No. 3–4, pp. 481–488, 2002

0149-1970/02/$ - see front matter

PII: S0149-1970(02)00041-0

A STUDY OF TRANSMUTATION OF MINOR-ACTINIDE IN A THERMAL NEUTRON FIELD OF THE ADVANCED NEUTRON SOURCE

TOMOHIKO IWASAKI

Department of Quantum Science and Energy Engineering,
Tohoku University
Aoba 01, Aramaki, Aoba, Sendai, 980-8579, Japan
TEL: +81-22-217-7910, FAX: +81-22-217-7908
E-MAIL: tomohiko.iwasaki@qse.tohoku.ac.jp

ABSTRACT

Since Minor Actinides (MAs) have a large cross section in a thermal energy region, a thermal neutron field has possibility to provide high relative transmutation rate. The transmutation of MA in a thermal neutron field was investigated in this study, focusing on relative transmutation rate, transmutation rate of weight and neutron economy in a thermal neutron field in comparison with those in a fast neutron field. The followings are the major results: (1) a thermal neutron field, especially a well-thermalized neutron field, provides high relative transmutation rate of MA. (2) However, the transmutation rate of weight of MA is limited since the region with high flux is restricted even in the ANS. (3) The accumulation of ^{246}Cm slows the relative transmutation rate of MA in a thermal neutron field. However, it can be compensated by providing a neutron field with high flux in keV region. (4) The neutron economy of the transmutation of MA is not so bad in a thermal neutron field, and is rather good when the neutron flux is high.

1. INTRODUCTION

The problem of the management of long-lived radioactive waste is one of the difficult issues associated with fission energy. The most likely concept for the waste is to solidify it and to confine it in deep geological formations over a ten or hundred of thousands years. It is very difficult to verify the assurance of confinement for the long period. Therefore, the transmutation concept has been proposed as an alternative concept for the problem.

A fast neutron field has been mainly studied for transmuting Minor Actinides (MAs). However, a thermal neutron field has possibility to provide high relative transmutation rate (that is, high reduction rate of weight and toxicity of MAs by transmutation) since MAs have a large cross section for thermal neutrons. Therefore, it is important to clarify basic features of the transmutation of MAs in a thermal neutron field.

We have studied the transmutation of MAs in a thermal neutron field[1,2,3,4]. This paper describes some

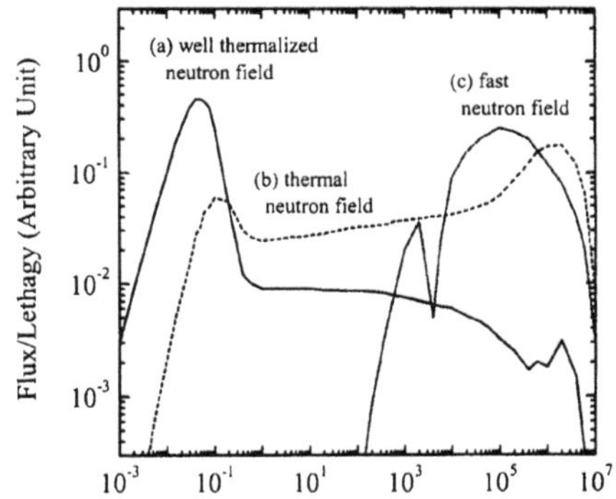

Fig.1 Neutron spectra of well thermalized, thermal and fast neutron fields

of the features of the transmutation of MAs in a thermal neutron field such as the relative transmutation rate (ex. %/year), the transmutation rate of weight (ex. kg/year) and the neutron economy in comparison with those in a fast neutron field.

2. NUCLIDE CHARACTERISTICS OF MA IN THERMAL NEUTRON FIELD

To clarify the nuclide characteristics of MAs related to transmutation, the following neutron fields were considered: (a) a well-thermalized neutron field in a large D_2O tank, which is installed in the ANS explained in Sec.3, (b) a thermal neutron field in a conventional light water reactor and (c) a fast neutron field in a typical Na-cooled and MOX-fueled fast reactor. The neutron energy spectra for the three fields are shown in Fig 1.

2.1 Larger cross section

Table 1 shows the effective one-group cross sections (fission and capture) of MAs for the three neutron fields. Those cross sections were calculated under the assumption of infinite dilution using the spectra in Fig.1. Cross section data were taken from JENDL-3.2[5] and ENDF/B-VI[6]. It is found from the table that the cross sections of the thermal neutron fields (the thermal and the well-thermalized neutron fields) are significantly larger that those of the fast neutron field. The cross sections of the well-thermalized neutron field are very large. Those of the fast neutron field are very small for all the MAs.

Table 1 Effecticve on -group cross section for three neutron fields (JENDL-3.2)

	Fission cross section			Capture cross section		
	Well-thermalized	Thermal	Fast	Well-thermalized	Thermal	Fast
Np237	0.02	0.5	0.3	117.2	35.1	1.8
Np238	1243.0	160.0	0.7	124.6	16.3	0.1
Am241	2.2	1.1	0.3	412.3	115.0	2.0
Am242m	4899.0	665.3	3.3	966.5	130.1	0.6
Am242	1322.0	187.5	3.4	3400.0	372.8	0.7
Am243	0.1	0.4	0.2	71.8	51.3	1.8
Am244m	1220.0	191.3	3.4	306.2	48.0	0.8
Am244	1528.0	214.7	3.4	394.1	54.7	0.9
Cm243	405.3	86.4	3.2	79.4	13.7	0.4
Cm244	0.7	0.8	0.4	17.3	16.0	0.8
Cm245	1076.0	126.9	2.8	181.6	20.1	0.4
Cm246	0.1	0.5	0.3	2.1	3.1	0.4

2.2 Large fission cross section

Many MAs have a very large thermal fission cross section. Table 2 shows the thermal fission cross sections of MAs in order of the value of cross section. The cross sections in the table are the values at the energy of 0.025eV and were taken from IENDL-3. Thirteen MAs have the fission cross section over 1000 barn. In the table, the fission probabilities, i.e. the ratio of fission cross section to total cross section (=fission+capture) are also shown. The MAs with the large fission cross section have the large fission probability at the same time.

2.3 Large ν value

Generally, the ν values, i.e. the number of neutrons emitted per fission, becomes larger for higher MAs, i.e. MAs with a larger mass number. Table 3 shows the ν values of MAs for thermal (0.025 eV) and fast (average between 1 and 2 MeV) neutrons. The differences of ν value between Np and Cm isotopes are almost 1.0.

2.4 Large spontaneous fission probability

Some MAs have a large spontaneous fission probability. Table 4 shows the spontaneous fission probabilities, i.e. the decay fraction of spontaneous fission to total decay mode, and the half-lives of MAs. These values were taken from Ref.4. It should be noted that there are the nuclides with the spontaneous fission probability of 100% since "perfect (100%)" transmutation of MA by fission may be possible owing to the nuclides with the spontaneous fission probability of 100% such as ^{254}Cf and ^{250}Cm (^{254}Cf's daughter).

Table 2
Fission cross section and ratio of MA

No.	Fission Cross Section (barn)*	Fission Cross Section Ratio
Am242m	6409	0.83
Cf251	4935	0.63
Fm255	3360	0.99
Am244	2300	0.79
Pu237	2100	0.79
Am242g	2100	0.28
Np238	2026	0.9
Cm245	2001	0.85
Es254	1966	0.98
Cf249	1666	0.77
Am244m	1600	0.8
Cf253	1136	0.74
Pu241	1015	0.73
Bk250	959	0.73
Pu239	746	0.73
Cm241	700	0.82
Cm243	617	0.82
U235	584	0.84
Pu243	181	0.63

* JENDL3.2
** = Fis / (Fis+Cap)

Table 3
ν values for thermal and fast neutrons

No.	ν Thermal*	ν Fast**
Np-237	2.54	2.75
Np-238	2.79	2.99
Am-241	3.22	3.49
Am-242g,m	3.27	3.53
Am-243	3.20	3.45
Am-244g,m	3.14	3.42
Cm-243	3.43	3.70
Cm-244	3.24	3.52
Cm-245	3.60	3.72
Cm-246	3.19	3.49
Bk-249	3.41	3.74
Cf-251	4.10	4.44
Cf-252	4.00	4.30

* at 2200 m/sec
** average between 1 and 2 MeV

Table 4
Spontaneous fission probability and half life

No.	Spontaneous Fission Probability	Half-life
Cf-256	1.000	12 min
Cm-250	1.000	1.1E+4 year
CF-254	0.997	60 day
Fm-256	0.919	2.6 hour
Cm-248	0.083	3.4E+5 year
Cf-252	0.031	2.6 year
Pu-244	0.012	8.3E+7 year
Cm-246	0.003	4.7E+3 year

* Table of isotope 8th Ed.

3. TRANSMUTATION OF MA IN THERMAL NEUTRON FIELD

3.1 Burnup Calculation

Using the effective cross sections in Table 1, burn-up of MAs was calculated using ORIGEN-2[7] under the assumption of constant flux. By the calculation, the weight and the toxicity variations for all actinides were obtained. However, this paper only concerns the total weight, i.e. the sum of the weight of all actinides.

Figure 2 (a) and (b) show the total weight after burning ^{237}Np(100%) and a typical MA (^{237}Np 49.35 %, ^{241}Am 37.58 %, ^{243}Am 10.59 %, ^{244}Cm 2.25 %, etc). The composition of the typical MA is that for a discharged fuel from a conventional PWR (33GWd/t and 5 years cooling). The initial weight of ^{237}Np and the typical MA were set to 1 kg in the calculation. The values in the figures are the total weight after burring up to the fluence of 8.64×10^{22} n/cm^2, which is the fluence of 365 days irradiation at the flux of 2.7×10^{15} n/cm^2/sec. Each figure includes three curves for the three neutron fields. The flux level was changed between 10^{14} and 10^{17} n/cm^2/sec.

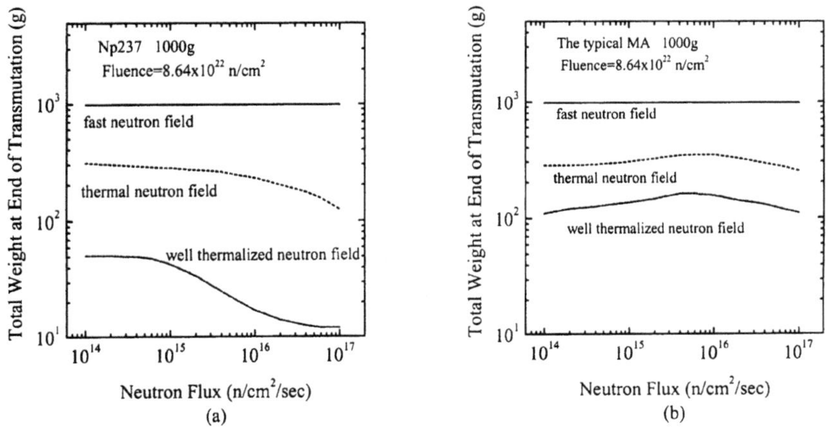

Fig.2 Total weight after transmutation of (a) Np237 and (b) typical MA

3.2 ^{237}Np

For ^{237}Np (Fig.2 (a)), in the well-thermalized neutron field, the total weight after burring are below 50 g More than 95% of Np^{237} was burned by fission at every flux level. In the thermal neutron field, the remaining after burning is 100 - 200 g. However, 970g remain after burning in the fast neutron field. A thermal neutron field is efficient to the transmutation of ^{237}Np.

It should be remarked that the relative transmutation rate of ^{237}Np largely depends on flux level. This dependence is owing to ^{238}Np, which has a large fission cross section and a short half-life of 2.1 day. This behavior has been pointed out by LANL [8] and E. D. Arthur [9].

3.3 Typical MA

As shown in Fig.2 (b), the similar conclusion is remarked for the typical MA. A thermal neutron field is useful for burning the typical MA, but only a little amount is burned in a fast neutron field. In comparison with the result for ^{237}Np, however, larger amount remains after burning the typical MA. This is because of the accumulation of ^{246}Cm as described in Sec.5 of the present paper.

4. TRANSMUTATION BY ANS

4.1 Advanced Neutron Source (ANS)

Aiming the thermal flux around 10^{16} n/cm^2/sec, the Advanced Neutron Source (ANS) [10,11] had been designed at Oak Ridge National Laboratory in 1980's. The ANS consists of a small hollow core (1 m high and 0.5 m in diameter) with 93%-enriched ^{235}U and a large D_2O reflector (4 m in height and in diameter) as shown in Fig.3. The core is operated at the power of 350 MWth and provides the thermal flux around 10^{16} n/cm^2/sec in the reflector near the core. The neutron flux distribution is shown in Fig.4. The neutron energy spectrum is already shown in Fig.1 as the well-thermalized neutron field.

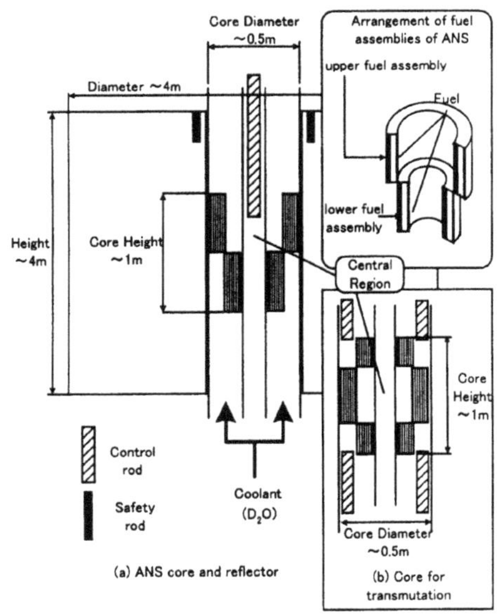

Fig.3 Schematic view of (a) ANS core and reflector and (b) core for transmutation of MA

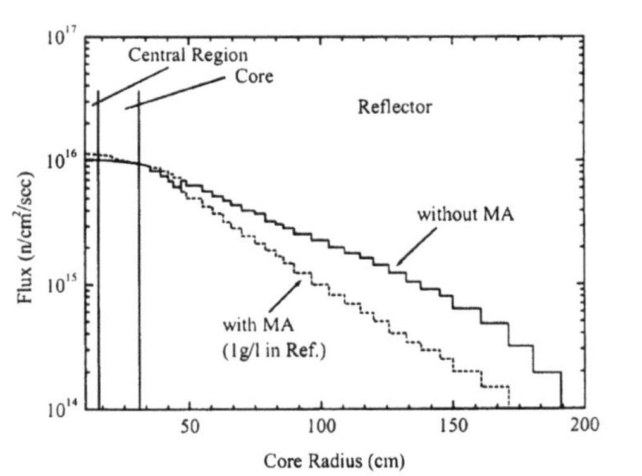

Fig.4 Radial flux distribution with and without the typical MA

4.2 Analysis of transmutation of MA

The transmutation of the typical MA using the ANS was analyzed. In the analysis, it was assumed that the typical MA are mixed to the D_2O reflector at the concentration of 1 g/liter. This concentration was determined to avoid large flux depression and to obtain maximum transmutation rate of weight. Although the ANS reflector is about 50,000 liter, the MA was added into almost half the reflector (20,000 liter) near the core. This was because transmutation in the outer region of the reflector was inefficient since the high flux region was limited and the flux level was very low. The total amount of MA mixed into the reflector was 20 kg. The flux distribution with MA is also shown in Fig.4. Slight flux depression is observed, however, the degree of the depression is not large since the concentration of MA is low.

The analysis was performed by the one dimensional cell calculation by SRAC[12] using the 107 group nuclear data set, the core calculation by CITATION[13] using 5 group neutron data set and the burn-up calculation by ORIGEN2 using one group cross section. Since the flux in the reflector largely decreases with radial distance as shown in Fig.4, the reflector was divided into 8 sub-regions in the burn-up calculation and the burn-up of the typical MA was calculated for each sub-region independently.

4.3 Transmutation weight by ANS

The transmutation rate of weight by fission was about 650 g per cycle. The weights of Pu and Cm generated were about 550 g (^{238}Pu and ^{239}Pu) and 100 g (^{244}Cm and ^{246}Cm) per cycle, respectively.

The ANS was designed to operate for 14 days. One operation cycle of the ANS is 17 days including 3 days for fuel discharge and loading. The transmutation rate of weight of MA per year was obtained supposing this operation cycle, i.e. 21 cycles per year. The transmuted weight of MA by fission was 14 kg/year. The weights of Pu and Cm generated were 12 kg/year and 2.2 kg/year.

For each cycle, about 15 kg of ^{235}U and 1.1 kg of ^{238}U were loaded to the ANS core. After operating the ANS for one year, 145 kg of ^{235}U was consumed. The weights of Np and Pu generated in the ANS core were both 2.1 kg/year. No Am was generated.

5. ACCELERATION OF TRANSMUTATION OF ^{246}CM

5.1 Accumulation of ^{246}Cm

Long-term transmutation of MA in the three neutron fields was investigated assuming a flux level of 1.0x10^{16} n/cm^2/sec. Figure 5 shows the burn-up variations of the total weight. In the well-thermalized and thermal neutron fields, the total weight rapidly decreases during the first year. The relative transmutation rate is high in the first year, but it becomes low after the first year. In the same figure, the variation of the weight of ^{246}Cm in the well-thermalized neutron field is also shown. It is found from the curve of ^{246}Cm that the slow-down of the relative transmutation rate is due to the accumulation of ^{246}Cm. The accumulation of ^{246}Cm is a very serious problem for the transmutation of MAs in a thermal neutron field.

5.2 Improvement of neutron field for ^{246}Cm

^{246}Cm has the small cross section in all neutron fields as shown in Table 1. The accumulation C cannot be avoided for the transmutation in a thermal neutron field. However, it is observed from Table 1 that the cross section in the thermal neutron field is largest among the three neutron fields. This is because several resonances exist between 1 and 100 eV. It may be possible to make the cross section (one group) ^{246}Cm larger by utilizing those resonances.

Neutron fields with a different neutron spectrum were studied to obtain larger cross section of ^{246}Cm. In Fig.6, the spectrum and the effective cross section of ^{246}Cm are shown. Maximum cross section of ^{246}Cm was obtained by a neutron field with high flux between 1 and 100 keV. The maximum cross section was 5.5 barn, which is 2.5 times larger than that in the well-thermalized neutron filed (see Table1). It should be noted that the one-group cross section more than 5.5 barn is impossible for ^{246}Cm when a realistic neutron field is supposed.

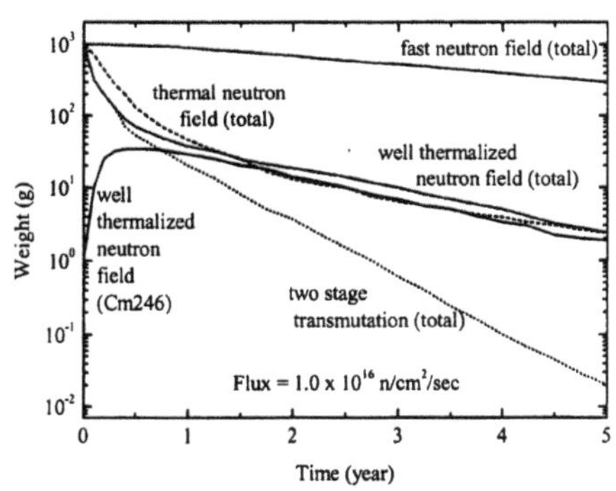

Fig.5 Variation of total weight of all radioactive nuclides above Tl and weight of ^{246}Cm during transmutation in well thermalized, thermal and fast neutron fields

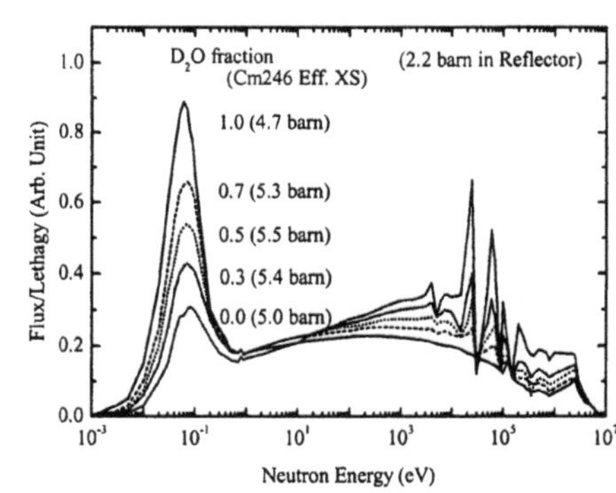

Fig.6 Neutron energy spectra and effective cross section of ^{246}Cm

5.3 Acceleration of burnup of ^{246}Cm

Using the improved neutron field for ^{246}Cm, the burnup of the typical MA was analyzed assuming a "two stage transmutation" method where MA is burned first in the well-thermalized field, then in the improved neutron field for ^{246}Cm.

One result of the two stage transmutation is shown in Fig.5. This result is for the typical MA irradiated for 100 days in the well-thermalized neutron field, then in the improved neutron field for ^{246}Cm. The burnup was calculated up to 5 years. The weight by the two stage transmutation after the 5 years transmutation was about 1/100 of that only in the well-thermalized neutron field. The total weight at the end of the 5 years transmutation was about 10^{-5} of the initial MA weight. The acceleration by the improved neutron field for ^{246}Cm (i.e. the two stage transmutation) is clearly shown from Fig.5.

6. NEUTRON ECONOMY OF TRANSMUTATION OF MA

6.1 Neutron economy of transmutation

For a thermal neutron field, it is anticipated that many neutrons are consumed to transmute MAs. To investigate this problem, in this study, the following numbers are calculated: (1) the number of neutrons absorbed while a MA is converted to higher nuclide and is transmuted by fission and (2) the number of neutrons produced by fission reaction when a MA is transmuted by fission, then (3) the net number of neutron produced, i.e. the difference of those numbers of neurons. The similar study has been performed by M. Salvatores[14] and Ref.4.

6.2 Analysis of neutron economy

In this study, first, "fission probability" P_f, "capture probability" P_c, "decay probability" P_d of a nuclide were defined as follows:

$$P_f=\frac{\sigma_f\phi}{\sigma_f\phi+\sigma_c\phi+\lambda_d} \qquad P_c=\frac{\sigma_c\phi}{\sigma_f\phi+\sigma_c\phi+\lambda_d} \qquad P_d=\frac{\lambda_d}{\sigma_f\phi+\sigma_c\phi+\lambda_d}$$

where σ_f is effective fission cross section (barn), σ_c is effective capture cross section (barn), λ_d is decay constant (1/sec) and ϕ is neutron flux (n/cm^2/sec). Second, we introduce "existence probability" N. Existence probability is the probability that a nuclide is produced in a burn-up chain of the nuclide. Using an example burnup chain in Fig.7, we explain the method to obtain the numbers of neutrons produced and absorbed. The existence probability is assumed to be 1 for a first nuclide "A", i.e. $N^A = 1$.

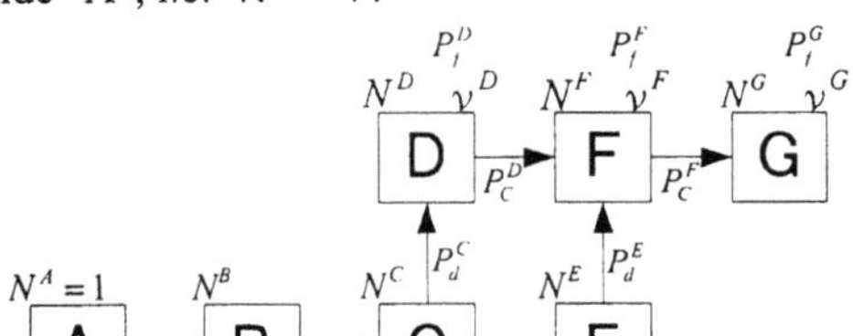

Fig.7 Burn-up chain for explanation of calculation of neutron economy

Let us consider the existence probability of the nuclide "C". Since "C" is produced after the capture reactions of "A" and "B", the existence probability of "C" is given by the products of the capture probabilities of the two nuclides of "A" and "B": $N^C = P_c^A \cdot P_c^B$. The nuclide "C" has two paths of capture reaction or decay. If "C" decays, "D" is produced, then "D" is converted to nuclide "F" by capture reaction. The nuclide "E" is generated by the capture reaction of "C", then "F" is produced by decay of "E". The existence probability of "G" N^G, which is the daughter of "F", is given by the products of the capture probabilities or the decay probabilities along the chain from "A" to "G": $N^G = P_c^A \cdot P_c^B \cdot \left(P_c^C \cdot P_d^E + P_d^C \cdot P_c^D\right) \cdot P_c^F$.

Each nuclide from "A" to "G" certainly absorbs one neutron except when it decays. Therefore, the number of neutrons absorbed in the chain n_{abs} is obtained by the summation of the existence probability times the probability that it does not decay.

$$n_{abs} = N^A + N^B + N^C \cdot \left(1-P_d^C\right) + N^D + N^E \cdot \left(1-P_d^E\right) + N^F.$$

On the other hand, each nuclide emits neutrons when fission reaction is occurred. The number of neutrons produced in the chain n_{prd} is obtained by the following equation.

$$n_{prd} = \sum_{i=A}^{G} N^i \cdot P_f^i \cdot \nu^i .$$

The net number of neutrons produced n_{net} is obtained by

$$n_{net} = n_{prd} - n_{abs} .$$

6.3 Neutron economy of transmutation of MA

Figure.8 (a) and (b) shows the net neutron number n_{net} for ^{237}Np and the typical MA in the three neutron fields changing the neutron flux levels form 10^{14} to 10^{17} n/cm^2/sec. The n_{net} values of the fast neutron field were positive (~1.0). This is the merit of the transmutation of MAs in the fast neutron field.

In the well-thermalized and thermal neutron fields, the n_{net} value was negative (~-1.0) at the flux level of 10^{14} n/cm^2/sec. Nearly one neutron is consumed for transmuting one MA in the thermal neutron field. It is remarked that this situation is the same as the transmutation of fission products where more than one neutron is consumed and no neutron is produced.

In the high flux region, the n_{net} values were larger owing to the nuclides with a short half-life such as ^{238}Np. At 10^{16} n/cm^2/sec, the n_{net} value was almost zero (~0). Zero of the n_{net} value means that no additional neutron is required to transmute MA by fission. The transmutation of a thermal neutron field with high flux around 10^{16} n/cm^2/sec has a good neutron economy.

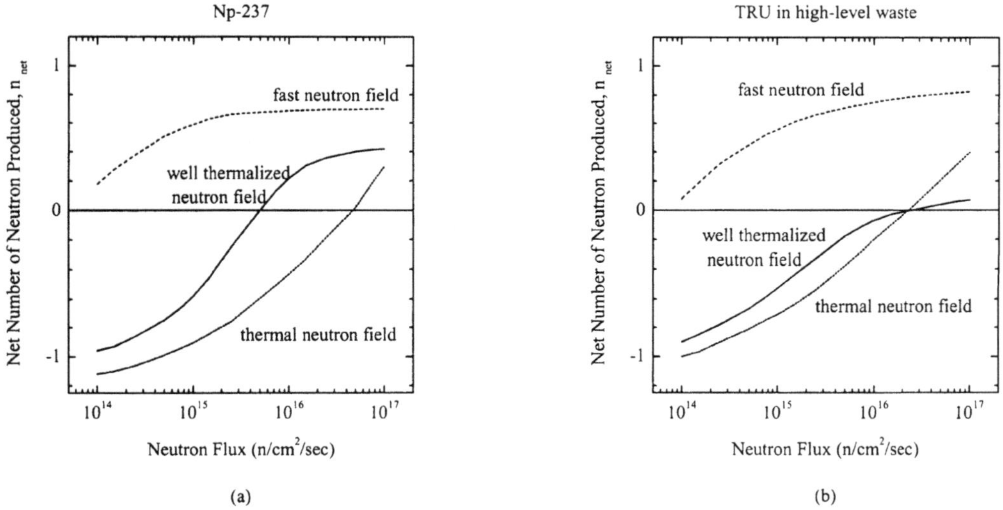

Fig.8 Net number of neutrons produced in transmutation of TRU in high-level waste

7. CONCLUSION

The transmutation of MA in a thermal neutron field was investigated in this study. The followings are concluded: (1) Transmutation rate of MA is very high in a thermal neutron field. (2) Transmutation weight of MA in a thermal neutron field is limited since the region with high flux is restricted in the current reactor technology as in the design of the ANS. (3) Accumulation of Cm246 slows the relative transmutation rate of MA in a thermal neutron field. However, it can be compensated by a neutron field with high flux in keV region. (4) Neutron economy of transmutation of MA is not bad in a thermal neutron field, and is rather good when neutron flux is high.

REFERENCE

[1] T. Iwasaki: Journal of Nuclear Science and Technology, 32,8-17 (1995)
[2] T. Iwasaki: Journal of Nuclear Science and Technology, 31,1255-1264 (1994)
[3] T. Iwasaki: Proc. Annual Meeting of Atomic Energy Society of Japan, Tokyo April 2-4 (1990) Vol.1 78
[4] R B. Firestone and V. S. Shirely (ed.): "Table of Isotopes, 8th Ed.", Wiley-Interscience (1996)
[5] K. Schibata et al.: JAERI-1319, NEANDC(J)-150/U, INDC(JPN)-137/L (1990)
[6] P. Rose and C. Dunford: ENDF-102, Brookhaven National Laboratory (1991)
[7] A. G. Croff: ORNL-TM-7175 (1980)
[8] C. D. Bowman et al.: Nuclear Instruments and Methods, A320 336 (1992)
[9] E. D. Arthur et al.: Fusion Technology,20, 641 (1991)
[10] F. C. Difilippo et al.: Nuclear Instruments and Methods, A249 58 (1986)
[11] J. M. Ryskamp et al.: Nuclear Technol. 93 330 (1991)
[12] K. Tsuchihashi et al.: JAERI-1302 (1986)
[13] T. B. Fowler et al.: ORNL-TM-2496 Rev.2 (1971)
[14] M. Salvatores et al.: Nuclear Science and Engineering, 116, 1-18 (1994)

Pergamon
www.elsevier.com/locate/pnucene

Progress in Nuclear Energy, Vol. 40, No. 3–4, pp. 489–496, 2002

Printed in Great Britain
0149-1970/02/$ - see front matter

PII: S0149-1970(02)00042-2

BASIC STUDY ON NEUTRONICS OF FUTURE NEUTRON SOURCE BASED ON ACCELERATOR DRIVEN SUBCRITICAL REACTOR CONCEPT IN KYOTO UNIVERSITY RESEARCH REACTOR INSTITUTE (KURRI)

S. SHIROYA, A. YAMAMOTO*, K. SHIN[1)], T. IKEDA[2)], S. NAKANO[2)],

AND H. UNESAKI

Research Reactor Institute, Kyoto University
Kumatori-cho, Sennan-gun, Osaka 590-0494, Japan
Tel: (+81) 724-51-2353; Fax: (+81) 724-51-2603
e-mail: shiroya@kuca.rri.kyoto-u.ac.jp
[1)]Dept. of Nucl. Eng., Graduate School of Eng., Kyoto Univ.
Yoshida-Honmachi, Sakyo-ku, Kyoto 606-8501, Japan
[2)]Dept. of Fundamental Energy Sci., Graduate School of Energy Sci., Kyoto Univ.
Yoshida-Honmachi, Sakyo-ku, Kyoto 606-8501, Japan

ABSTRACT

A basic study on the nuclear characteristics in the accelerator driven sub-critical reactor (ADSR) was performed through a series of neutronics design calculations and reactor physics experiments. Calculations were executed mainly by the MCNPX code, and experiments were performed at the Kyoto University Critical Assembly (KUCA). Some nuclear features of the research reactor type ADSR were revealed through the present study. The following facts were found: 1) Further studies are necessary concerning the nuclear data in the high energy region and the generated neutrons through the spallation reactions especially by the light nuclei and the lower energy protons. 2) The adjustment of subcriticality by the control rod significantly affects the reactor power of ADSR because of the distortion in the neutron flux distribution caused by the control rod insertion. 3) An accurate calculation is essential to evaluate the neutron multiplication in the ADSR. 4) The neutronics behavior after a pulse injection can be approximately simulated by the calculation. © 2002 Published by Elsevier Science Ltd.

* Guest Associate Professor of Research Reactor Institute, Kyoto University for a period of October 1999 through September 2000. Present affiliation: Nuclear Fuel Industries, Ltd., 950 Ohaza-Noda, Kumatori-cho, Sennan-gun, Osaka 590-0481, Japan

1. INTRODUCTION

In Kyoto University Research Reactor Institute (KURRI), an accelerator driven subcritical reactor (ADSR) is considered to be a final goal of a future plan named as the "Neutron Factory" project (Kawase *et al.*, 1997, 1998). It is anticipated that this ADSR will become a substitute for the current Kyoto University Reactor (KUR) established in 1964 for promoting the joint use program among university researchers in Japan. "Neutron Factory" will be a synergetic system of a high energy proton accelerator and a spectrum controlled type research reactor (SCRR), which allows operation in either independent or combined mode to generate so-called "controlled neutrons" in the energy and time spaces. A conceptual neutronics design study was initiated in 1997 by a working group of a future plan in KURRI (Shiroya *et al.*, 1998, 1999, 2000). In the early stage of this study, preliminary calculations were carried out for the injection of monoenergetic neutrons below 20 MeV into the ADSR. Last year, the MCNPX code (Hughes *et al.*, 1999) developed in Los Alamos National Laboratory was introduced to promote the neutronics design study.

The ADSR has attractive features as a future system for generating nuclear energy (Rubia *et al.*, 1995). Therefore, a preliminary study on the thorium fuel cycle and the ADSR was performed for two years from 1998 as a framework of the Research for the Future Program in the Japan Society for the Promotion of Science (JSPS). As a result, a study on the development of a high-energy particle accelerator with a high efficiency based on the concept of the Fixed-Field Alternating Gradient (FFAG) Synchrotron, which Dr. Chihiro Ohkawa proposed in 1953, was approved as a JSPS project for three years from 2000.

Moreover, a series of basic reactor physics experiments on the ADSR was launched in 1999 by using the Kyoto University Critical Assembly (KUCA) as a joint use project among university researches in Japan.

The present paper provides some results of neutronics design calculations by MCNPX and those of the KUCA experiments.

2. NEUTRONICS DESIGN CALCULATION BY MCNPX (1)
- Neutron Generation by Spallation Reaction -

The MCNPX code was utilized in this study, since this code is merger of a continuous energy Monte Carlo code MCNP4B (Breismeister *et al.*, 1997) and a high-energy particle transport code LAHET (Prael, 1989). This code was applied to analyze the angular dependent neutron spectra from the thick target bombarded with high energy protons to check the accuracy of neutron generation calculation.

Table 1 Comparison between the calculated and measured numbers of neutrons

Target material	P energy [MeV]	Calculated result	Experimental result[1)]	C/E value
Pb	540	8.36	7.90	1.058
	720	12.81	12.05	1.063
	960	18.16	16.82	1.080
	1,470	27.74	26.82	1.034
Depleted U	540	13.12	18.64	0.704
	720	20.11	29.09	0.691
	960	29.08	40.46	0.719
	1,470	45.77	57.73	0.793

[1)]: Landeyro P. A. *et al.* (1998) Simulation of Thick Target Experiments for the Validation of Spallation Codes, *Int. Conf. on Physics of Nuclear Science and Technology*, Long Island, NY, USA, Oct. 5-8, Vol.2, 1327.

Table 1 shows the comparison between the calculated and measured numbers of neutrons leaked from the Pb and depleted U targets of 10.1 cm in radius and 61 cm in thickness for the 540, 720, 960 and 1,470 MeV proton bombardment, respectively. It was found that the C/E value is close to unity for the Pb target, whereas the calculated result underestimates the number of neutrons generated as large as 30%. This fact indicates that the further improvement would be necessary to simulate the generation process by MCNPX.

Figure 1 shows the comparison between the calculated and measured results of angular dependent neutron spectra from the Pb target bombarded with 500 MeV protons which were measured in KEK (Meigo *et al.*, 1995) and those from Al target bombarded with 68 MeV protons at TIARA in JAERI. The size of Pb target was 15 cm in diameter and 20 cm in thickness, and that of Al target was 3 cm in diameter and 3.7 cm in thickness. The neutron spectra were measured by the TOF method. This figure shows that the agreement is better for the Pb target than for the Al target. It was pointed out that the agreement between the calculated and measured result becomes worse with decreasing energy of protons bombarded the target and with decreasing mass number of the target material. This fact indicates that the further study on the generation process of neutrons by the spallation reaction as well as that on the nuclear data in the high energy region above 20 MeV to improve the accuracy of the MCNPX calculation.

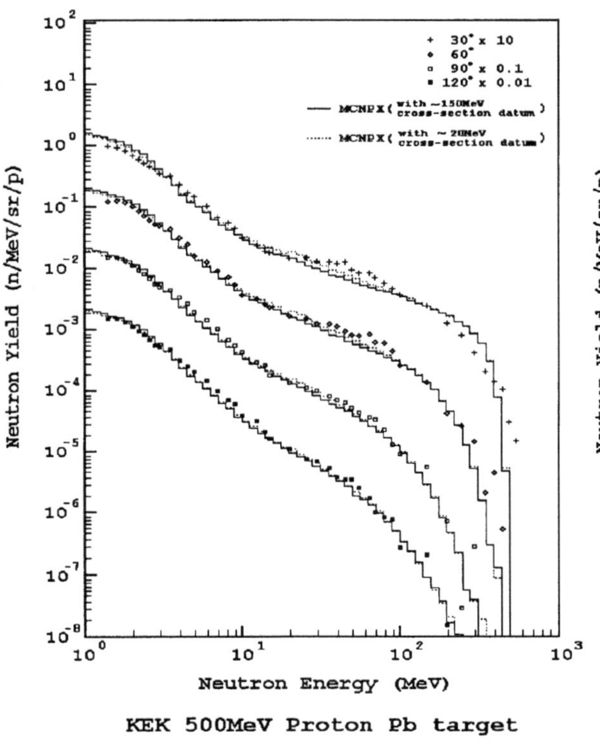

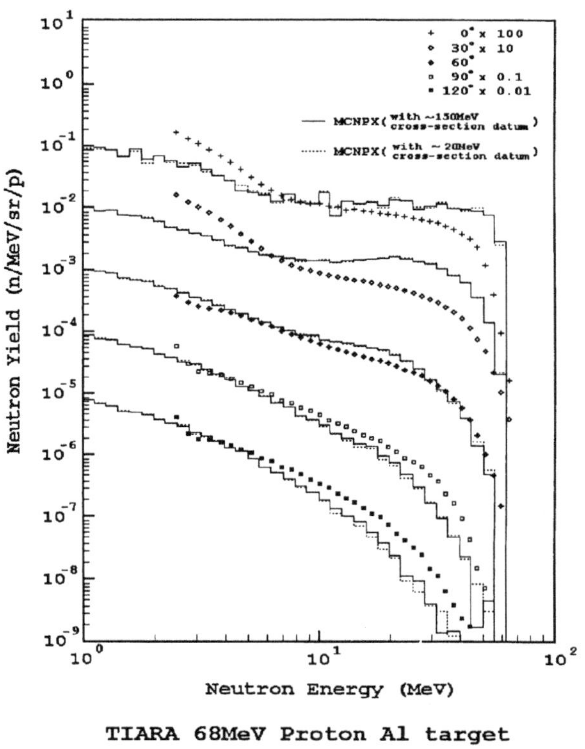

Fig. 1 Comparison between the calculated and measured results of angular dependent neutron spectra

3. NEUTRONICS DESIGN CALCULATION BY MCNPX (2)
- Effect of Control Rod Insertion on the ADSR Power -

Some results of the conceptual neutronics design calculation on the basis of the current KUR design was previously presented in the international conference (Yamamoto A. *et al.*, 2000). Through this previous study, the following precious knowledge on the ADSR was obtained:

1) The neutron spectrum obtained by the fixed source calculation including the spallation process was similar to that obtained by the eigenvalue calculation.
2) The higher neutron flux level than the current KUR under 5 MW operation can be realized in the ADSR, and this flux level can be varied with adjusting subcriticality of the ADSR.
3) The neutron flux and thermal power depend considerably on the selection of target material.
4) The power distribution in the core does not change significantly in the ADSR of research reactor type regardless of the subcriticality or the incident proton energy.

In this previous study, the subcriticality of ADSR was adjusted by changing the fuel enrichment. However, in the actual ADSR, it was impossible to adjust the subcriticality by this method. It is considered to be realistic that the subcriticality will be adjusted by the stroke of control rod insertion. Therefore, the effect of control rod insertion on the ADSR power was examined by assuming the core configuration shown in Fig. 2.

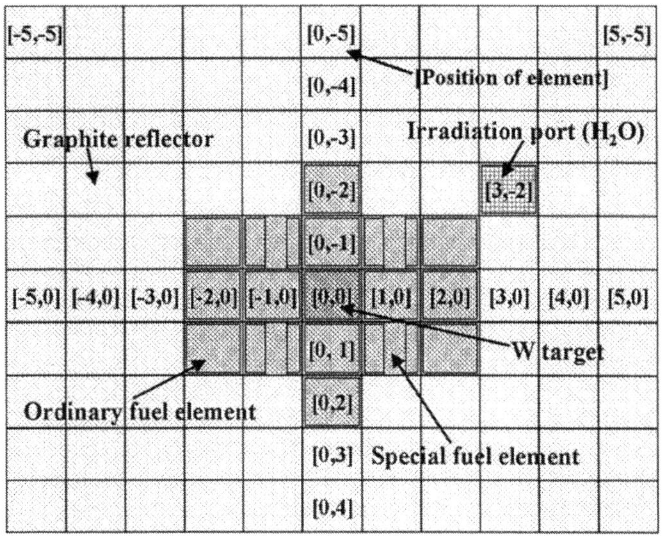

Fig. 2 Core configuration of ADSR employed in the present study

The core is composed of 12 ordinary fuel, 4 special fuel and 1 target elements. Ordinary fuel elements are located at positions [0,-2], [-2,-1], [0,-1], [2,-1], [-2,0], [-1,0], [1,0], [2,0], [-2,1], [0,1], [2,1] and [0,2], where [x, y] corresponds to the core position as shown in Fig.2. Special fuel elements for control rod insertion are located at positions [-1,-1], [1,1], [-1,1] and [1,1]. The target element is located at the position [0,0]. The size of each element is 7.7×8.1 cm^2. The fuel is MTR-type of ~20 wt% ^{235}U enrichment (LEU). The fuel meat is made from siliside (U_3Si_2) and covered with aluminum (Al) cladding. The thickness of the fuel meat, Al cladding and fuel plate are 0.076, 0.038 and 0.152 cm, respectively. The active height of fuel meat is 59.4 cm. The pitch between fuel plates is 0.433 cm. Light water acts as both the moderator and coolant. It should be noted here that the core size of the present ADSR is much smaller than that considered in a common energy amplifier concept where the typical size is a few meters in diameter.

Table 2 Calculated effective multiplication factor and control rod worth

Status of control rod	k_{eff}	ρ [%$\Delta k/k$]
Full withdrawal	0.995	0.0
Half insertion	0.952	4.3
Full insertion	0.894	10.1

Table 2 shows the calculated effective multiplication factor k_{eff} and the calculated control rod worth ρ for the three conditions of control rod insertion. It was found that the control rod worth in the ADSR core is very similar to that in the current KUR although the ADSR core has the W target element at the center of the core.

Table 3 shows the comparison between impacts on the ADSR power for two methods of subcriticality adjustment; one is the adjustment by the fuel enrichment and the other by the stroke of the control rod insertion. It was found that the difference in ADSR power between the two methods becomes as large as

24% when the status of control rod insertion is close to the half insertion. This fact indicates that the distortion in neutron flux distribution significantly affects the ADSR power. Therefore, one should be mind that it is necessary to optimize the location of control rods with taking into account the control rod worth for the neutronics design of the ADSR. It should be noted here that there was no remarkable difference in the shape of neutron spectrum between two methods of subcriticality adjustment.

Table 3 Difference in ADSR power [MW] in accordance with the method of subcriticality adjustment

k_{eff}	Method of subcriticality adjustment		Difference in %
	Fuel enrichment (1)	Control rod insertion (2)	(1) – (2)
0.98	14.37	12.38	14
0.95	7.56	5.77	24
0.90	3.75	3.55	5

4. NEUTRONICS DESIGN CALCULATION BY MCNPX (2)
- Neutronics behavior after the pulse injection of neutrons and protons into the ADSR -

The neutronics behavior after the pulse injection of neutrons and protons into the ADSR was examined by MCNPX, since the pulse operation of ADSR would be desirable in consideration of usage as a neutron source for the joint use program among universities in Japan.

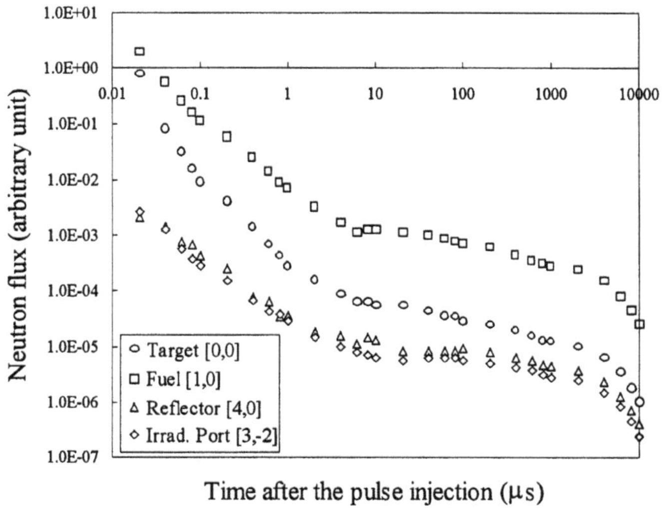

Fig. 3 Behavior of neutron decay in the energy range of 20 MeV – 1 eV after the injection of 500 MeV protons into the ADSR at various positions in the core (refer to Fig. 2)

Figure 3 shows the behavior of neutron decay in the energy range of 20 MeV – 1 eV after the injection of 500 MeV protons into the ADSR. It was found that the neutrons with energy above 20 MeV decay out in the period of less than 0.02 μs. Delayed neutrons were not taken into account in the calculation by MCNPX. Note that the behavior of neutron decay in the energy range of 20 MeV – 1 eV after the injection of 14 MeV neutrons showed the similar tendency as that for the injection of 500 MeV protons shown in Fig. 3. Note also that the isotropic generation of 14 MeV neutrons was assumed at the center of W target, whereas protons bombarded the W target from the top.

Figure 4 shows the behavior of neutron decay in the energy range below 1 eV after the injection of 500 MeV protons into the ADSR. The behavior of neutron decay in the energy range below 1 eV after the injection of 14 MeV neutrons also showed the similar tendency as that for the injection of 500 MeV protons shown in Fig. 4.

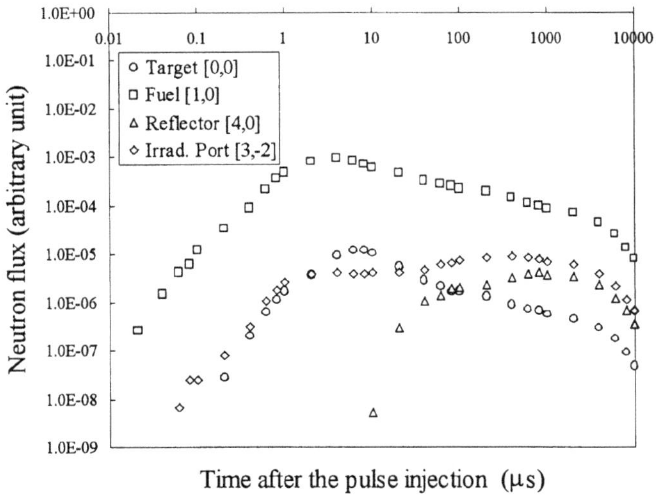

Fig. 4 Behavior of neutron decay in the energy range of below 1 eV after the injection of 500 MeV protons into the ADSR at various positions in the core (refer to Fig. 2)

From Figs. 3 and 4, the following phenomena are indicated:

1) Source neutrons are dominant in the time period of less than several μs after the pulse injection and the fission neutrons becomes dominant after that time in the ADSR core.
2) In other words, the diffusion process of source neutrons is terminated in around 10 μs and the neutron multiplication process by the fission reaction becomes dominant from several μs after the pulse injection.
3) The shape of the neutron spectrum in the pulse operation becomes similar to that in the constant operation after around 10 μs.

5. BASIC EXPERIMNTS ON THE ADSR IN THE KUCA

The KUCA experiments were carried out with the combination of a pulsed neutron generator of the Cockcroft-Walton type and a solid moderator core A-core among the three cores of the KUCA [Shiroya S. et al., 1988]. In the KUCA experiments, 14 MeV neutrons were injected into the polyethylene moderated and reflected core loaded with a highly enriched U-Al alloy fuel. The unit cell of the fuel rod employed in the present study was consisted of a 9.386 mm thick and 5.08×5.08 cm^2 polyethylene plate and a 1.5875 mm thick and 5.08×5.08 cm^2 U-Al alloy fuel plate. The repetition time of unit cells was selected to be 36 in the fuel rod to form the fuel region which was sandwiched by the upper and lower polyethylene reflector layer of around 50 cm in thickness. The core configuration employed in the present study is shown in Fig. 5.

The deuteron beam accelerated up to around 200 keV was led to the tritium target placed outside the polyethylene reflector to generate 14 MeV neutrons by the fusion reaction. An optical fiber detector system [Mori C. *et al.*, 1994] was utilized to measure the neutron decay constant by the pulse operation of the KUCA accelerator. A mixture of ^{6}Li enriched LiF and ZnS(Ag) scintillator was pasted on the end of an optical fiber of 1 mm in diameter with the instant adhesive. The arc pulse was utilized to trigger the multi-channel scaler (MCS) unit where the neutron signals from the fiber detector were accumulated. The measurement of neutron decay constant was carried out several times by changing the subcriticality of the core with adjusting stroke of the control rod insertion. The subcriticality was measured by the area ratio method [Sjöstrand N. G., 1956] which is one of the pulsed neutron methods. The measurement of neutron flux distribution in the core was attempted by the fiber detector to observe the neutron multiplication in the subcritical system by varying the subcriticality.

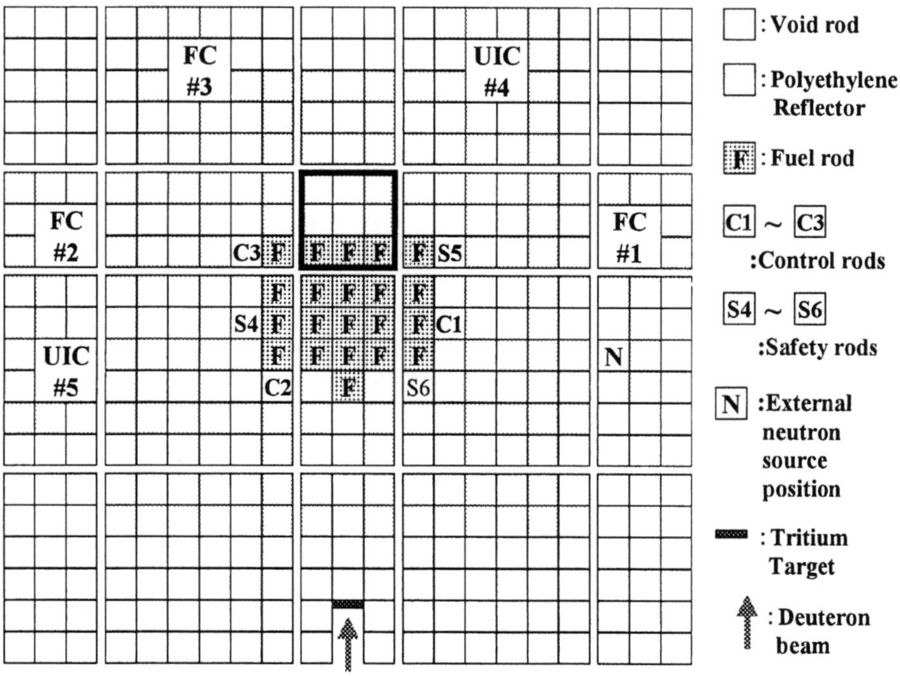

Fig. 5 Core configuration employed in the basic experiments for the ADSR in the KUCA

The analyses of these experiments were performed by using a continuous Monte Carlo code MVP [Mori T. et al., 1993] based on the JENDL-3.2 library [Shibata K. *et al.*, 1990]. Through these analyses, it was recognized that the accuracy of k_{eff} calculation is essential to examine the neutron multiplication in the ADSR system as well as the accuracy of subcriticality measurement. The calculated neutron decay constant agreed well with the measured one, although the correction for the effect of delayed neutrons is necessary since the MVP code can not be taken in to account that of delayed neutrons.

6. CONCLUSION

Although the further study would be necessary both in the calculation and experiment to promote the neutronics design study of the ADSR, some features of the ADSR were revealed through the present study. The results obtained in the present study can be summarized as follows:

1) The MCNPX code is powerful for the neutronics design study of the ADSR, however, the improvement of neutron generation calculation and the nuclear data in the energy range above 20 MeV would be necessary to promote the study into the further step.
2) There is a significant effect of the control rod insertion for the adjustment of subcriticality on the ADSR power because of the distortion in neutron flux distribution. The optimization of control rod worth and location would be necessary to realize a research reactor type ADSR.
3) The rough image of neutron behavior in the pulse operation was grasped through the calculation by MCNPX and the KUCA experiments, although the calculation taking into account delayed neutrons would be necessary to promote the study into the further step.

4) The accuracy of k_{eff} calculation is essential to examine the neutron multiplication in the ADSR system in comparison with the experimental result.

ACKNOWLEDGMENT

The authors are grateful to the members of the working group on the future plan in KURRI, including Former Director Y. Maeda, Present Director M. Inoue, Prof. Y. Kawase, Prof. H. Moriyama and Prof. K. Mishima for invaluable discussion and advice. They thank to the Nagoya University group including Prof. Y. Yamane and Mr. H. Kataoka, the Osaka University group including Prof. T. Takeda, Mr. T. Kitada, and the KUCA staff including Assoc. Prof. T. Misawa, Mr. C. Ichihara, Mr. K. Kobayashi, Mr. H. Nakamura, Mr. R. Ikegawa, Mr. H. Yamamoto and Dr. Y. Kitamura, for the cooperation in the KUCA experiments. A part of this study was supported by Grant-in-Aid for Scientific Research from the Ministry of Education, Science, Sports and Culture in Japan, by Research for the Future program from JSPS, and by the JAERI's Nuclear Research Promotion Program (JANP).

REFERENCES

Briesmeister J. F. *ed.* (1997), *LA-12625-M.*

Hughes H. G. *et al.* (1999), *Int. Conf. on Mathematics and Computation for Reactor Physics and Environmental Analysis in Nuclear Application*, 963.

Kawase Y. and Inoue M. (1997), Neutron Factory Project at KURRI, *11th Symp. on Accelerator Science and Technology*. Harima, Hyogo, Japan, 21-23 October.

Kawase Y., Shiroya S. and Inoue M. (1998), An Accelerator-Assisted Nuclear Fuel Assembly for a Future Project at KURRI, *XIX Int. LINAÇ Conf. LINAC98*, Chicago, USA, 23-28 August.

Meigo S. *et al.*, (1995) Measurements of Spallation Neutrons from a Thick Target Bombarded with 0.5 and 1.5 GeV Protons, *ICANS-XII*, 11-14 Oct., PSI, Switzerland, 442.

Mori C. *et al.* (1994) Simple and Quick Measurement of Neutron Flux Distribution by Using an Optical Fiber with Scintillator, *J. Nucl. Sci. Technol. 31, 248.*

Mori T., Nakagawa M. and Sasaki, M. (1993), Vectorization of Continuous Energy Monte Carlo Method for Neutron Transport Calculation. *J. Nucl. Sci. Technol. 29, 325.*

Prael R. E., Lichitenstein H. (1989), *LA-UR-89-3014.*

Rubia C. *et al.* (1995), *Conceptual Design of a Fast Neutron Operated High Energy Amplifier*, CERN/AT/95-44(ET).

Shibata K. *et al.* (1990), *Japanese Evaluated Nuclear Data Library, Version-3*, JAERI 1319.

Shiroya S., Kanda K. and Tsuchihashi K. (1988), Analyses of Reactor Physics Experiments in the Kyoto University Critical Assembly. *Nucl. Sci. Eng. 100, 525.*

Shiroya S. and Unesaki H. (1998), Study on Conceptual Neutronics Design of "Neutron Factory" as a Future Project at KURRI, *9th Int. Conf. on Emerging Nuclear Energy Systems ICENES'98* Vol.1, p.289. Tel-Aviv, Israel, 28 June - 2 July.

Shiroya S., Unesaki H. *et al.* (1999), Accelerator Driven Subcritical System as a Future Neutron Source in Kyoto University Research Reactor Institute (KURRI) – Basic Study on Neutron Multiplication in the Accelerator Driven Subcritical Reactor -, *3rd Int. Symp. on Global Environment and Nuclear Energy Systems GENES-3*, Tokyo, Japan, 14-17 Dec.

Shiroya, S., Unesaki, H., *et al.* (2000), Conceptual Neutronics Design of "Neutron Factory" as a Future Plan on Neutron Source in Kyoto University Research Reactor Institute (KURRI), *Reactor Dosimetry, ASTM 1398*, John G. Williams, David W. Vehar, Frank H. Ruddy, and David M. Gillian, Eds., American Society for Testing and Materials, West Conshohocken, PA, 2000.

Sjöstrand N. G. (1956), *Arkiv Fysik 11, 233.*

Yamamoto A. and Shiroya S. (2000), Study on Conceptual Neutronics Design of "Neutron Factory" as a Future Project at KURRI, *10th Int. Conf. on Emerging Nuclear Energy Systems ICENES 2000*, p.66. Petten, The Netherlands, 24 - 28 Sept.

Pergamon

www.elsevier.com/locate/pnucene

Progress in Nuclear Energy, Vol. 40, No. 3–4, pp. 497–504, 2002

Printed in Great Britain
0149-1970/02/$ - see front matter

PII: S0149-1970(02)00043-4

ATW NEUTRONICS DESIGN STUDIES

D. C. WADE, W. S. YANG, H. KHALIL

Argonne National Laboratory

9700 S. Cass Avenue, Building 208

Argonne, IL 60439

ABSTRACT

The Accelerator Transmutation of Waste (ATW) concept has been proposed as a transuranics (TRU) (and long-lived fission product) incinerator for processing the 87,000 metric tonnes of Light Water Reactor (LWR) used fuel which will have been generated by the time the currently deployed fleet of commercial reactors in the US reach the end of their licensed lifetime. The ATW is proposed to separate the uranium from the transuranics and fission products in the LWR used fuel, to fission the transuranics, to send the LWR and ATW generated fission products to the geologic repository and to send the uranium to either a low level waste disposal site or to save it for future use. The heat liberated in fissioning the transuranics would be converted to electricity and sold to partially offset the cost of ATW construction and operations. Options for incineration of long-lived fission products are under evaluation.

A six-year science-based program of ATW trade and system studies was initiated in the US FY 2000 to achieve two main purposes: (1) "to evaluate ATW within the framework of nonproliferation, waste management, and economic considerations," and (2) "to evaluate the efficacy of the numerous technical options for ATW system configuration."

This paper summarizes the results from neutronics and thermal/hydraulics trade studies which were completed at Argonne National Laboratory during the first year of the program. Core designs were developed for Pb-Bi eutectic (LBE) cooled and Na cooled 840 MW_{th} fast spectrum transmuter designs employing recycle. Additionally, neutronics analyses were performed at Argonne for a He cooled 600 MW_{th} hybrid thermal and fast core design proposed by General Atomics Co. which runs critical for ¾ and subcritical for ¼ of its four year once-thru burn cycle.

The mass flows and the ultimate loss of transuranic isotopes to the waste stream per unit of heat generated during transmutation have been calculated on a consistent basis and are compared. (Long-lived fission product incineration has not been considered in the studies reported here.)

INTRODUCTION

The function of the proposed ATW is to reduce the amount of – and the long term toxicity contained in – the waste consigned to the US proposed geologic repository.[1] ATW systems are proposed to be interposed between the commercial LWR once through fuel cycle and the repository and would be used to incinerate the TRU contained in the LWR spent fuel by fission (and optionally to transmute selected long-lived fission products). Heat released by the TRU fissioning would be used to generate electricity or other energy intensive products and sold to partially offset the costs of ATW and recycle construction and operation. Two broad strategies are being evaluated: multi recycle and once-through deep burn. For recycle, an electrometallurgical process is considered and specialized waste forms are developed for the fission products; for deep burn, part of the recycle infrastructure is not required and the spent fuel triso particle fuel is considered as a waste form.

Two aspects of the transuranic content in the waste stream sent to the repository from the ATW are relevant. First is its overall mass – the fraction of TRU from the LWR spent fuel which has escaped being transmuted to fission products. For the deep burn strategy, this aspect is controlled by the achievable discharge burnup. For the multi recycle strategy, it is controlled by the loss per recycle pass and the number of passes. The second relevant aspect is the toxicity per unit mass and its time decay behavior. This aspect depends on the isotopic spectrum of the transuranic mass sent to the repository (based on differing toxicity and half-life by isotope). For the deep burn strategy this aspect is affected primarily by the neutron spectrum and secondarily by discharge burnup. For the multi recycle strategy it is affected by neutron spectrum and by conversion ratio (i.e., blend ratio of recycle and feed).[2]

The focus of this paper is a comparison of isotopic mass flows to the repository from three ATW point designs established during the first year of the 6-year science-based ATW program.

Liquid Metal Cooled Multi-Recycle ATW Point Designs

For the liquid metal cooled ATW concepts, a fertile-free fuel was selected so as to maximize the "support ratio" defined as number of LWRs that a given ATW can service (MW_{th} from LWRs/MW_{th} from ATW). A fuel form comprised of TRU-Zr alloy particles dispersed in Zr matrix in a pin geometry and clad in ferritic stainless steel has been proposed; the average discharge burnup capability for this fuel is postulated to be 30 a/o.[3] In the case of a multi recycle strategy, the goal of minimizing loss of LWR TRU feedstock to the waste stream motivates the design for maximum achievable discharge burnup to minimize number of recycle passes (to reduce opportunity for losses during recycle and refab processing). On the other hand, for fertile free fuel, the source multiplication in the subcritical blanket deceases with increasing burnup due to the reactivity loss and in order to minimize the resulting needs for increasing accelerator power and/or introducing an excess reactivity and active reactivity control, it is desirable to minimize the burnup reactivity loss. The LBE and Na cooled design optimizations were, therefore, focused on trading off two contradictory performance objectives: achieving 30 a/o discharge burnup to minimize number of successive recycle stages while minimizing burnup reactivity loss over an operating cycle – and to do so within the constraints of heat removal under acceptable temperature and coolant velocity limits, and of discharge fluence level, reactivity coefficient values, vessel size, and etc. Given each choice of coolant, a wide range of potential transmuter designs was evaluated at a fission-power level of 840 MW_{th}, driven by a 1 GeV, 11.25 mA proton beam on a LBE spallation target and operating at a neutron multiplication level at beginning of equilibrium cycle (BOEC) of 0.97.[4,5] The resulting parameters of the optimized liquid metal cooled transmuter core designs are shown in Table I. In both cases, multi batching was used to reduce burnup reactivity loss and radial power peaking; a 7 batch core/42 month fuel residence time for LBE and 8 batch core /48 month fuel residence time for sodium was necessary to hold reactivity loss to ~5% Δk/k over a 6 month burn cycle. A reduced power density/high coolant volume fraction design was used for the LBE coolant while conventional values were used for Na. Peak discharge fluence ($40 \cdot 10^{22}$ fast nvt) controlled the fuel residence time, but in both cases discharge burnup nearly attained 30% burnup as was targeted for the inert matrix dispersion fuel. The details of the optimizations are reported elsewhere.[4,5] peaking limited the residence time to 6 batches; a similar notation applies for the sodium system. See Fig. 1 for the gas system fuel loading logistics.

Table I. Main Integral Parameters of LBE, Sodium, and Gas Cooled System Point Designs

Parameter		LBE	Sodium	Gas	
				Critical	Subcritical
Reactor power (MWt)		840.0	840.0	600.0	
Cycle length (days)		137.0	135.0	270.0	
Number of fuel assemblies	Inner	36	42	6	6
	Outer	168	90	102	102
Number of batches*		7/6	8/7	3	1
Multiplication factor	BOEC	0.9703	0.9696	1.0775	0.9634
	EOEC	0.9180	0.9202	1.0062	0.7323
Burnup reactivity loss (%Δk)		5.23	4.94	7.13	23.11
Core-average total flux (n/cm^2-s)	BOEC	4.59E+15	4.41E+15	8.02E+13	1.01E+14
	EOEC	4.99E+15	4.75E+15	8.99E+13	1.63E+14
Core power peaking factor	BOEC	1.46	1.50	1.97	4.79
	EOEC	1.51	1.51	1.64	23.79
Core-average power density (W/cc)	BOEC	156.46	241.36	6.21	6.21
	EOEC	156.26	241.98	6.21	6.21
Coolant Volume Fraction	%	68.2	36.6	18.6	
Average discharge burnup	atom %	26.79	29.51	47.54	16.44
	MWD/kg	250.8	275.2	445.1	153.9
Effective cycle burnup of charged fuel (%)		3.93	3.84	15.85	16.44
% Δk / atom % burnup		1.33	1.29	0.45	1.41
Core ht/diam**	M	1/2.5	1/2.1	7.9/4.9	

*7/6 for the LBE system indicates a 7 batch core – except for the inner most fuel zone where power
**Equivalent fuel region outer diameter of annular core around central spallation target and buffer.

Gas Cooled Hybrid Once Thru ATW Point Design

General Atomics (GA) has proposed an ATW concept[6] based on a variant of the Gas Turbine-Modular Helium Cooled Reactor (GT-MHR). Four transmuters share one accelerator, and each transmuter is comprised of an outer thermal zone and inner fast zone. Each transmuter operates in a three batch-loaded critical mode for three years and in the source driven mode for the fourth year. In the three-years-long critical operating mode, the fission process is maintained by the critical thermal region driving the subcritical fast region and discrete burnable poison limits reactivity loss. After three years, the thermal region, loaded with three year burned assemblies, itself becomes subcritical and the transmuter is driven during the fourth year by the spallation source. The overall plant is comprised of four 600 MW_{th} transmuters, sharing one 15 MW accelerator with beam shifting from core to core at one year intervals. The transmuter thermal zone is fueled in TRISO coated particles with (fertile free) TRU recovered from LWR used fuel. The fast zone is fueled with four-year-burned TRISO particles which have been discharged from the thermal zone, separated from the graphite compacts and moderator, and reconfigured into fuel rods. Figure 1 illustrates the loading sequence of fuel as it progresses through its four-year burn cycle. Burnup reactivity loss is mitigated by use of discrete burnable poison (erbium) rods distributed in the thermal zone and benefits from the in situ conversion of Pu^{240} in the feedstock to fissile Pu^{241}. The absorption resonance in erbium shields the principal Pu^{239} fission resonance and produces a negative moderator temperature coefficient.

Preliminary analyses by GA have indicated that this design will achieve deep levels of transmutation without requiring reprocessing so that it can be operated once thru, and that it can employ the spent TRISO particles as a waste form thanks to encapsulation of the transmuted materials within the ceramic coated microspheres. Argonne collaborated with General Atomics staff to perform independent core performance and mass flow calculations for the gas cooled hybrid design.[7] Table I summarizes the main parameters of all three ATW point designs studied in the first year of the program.

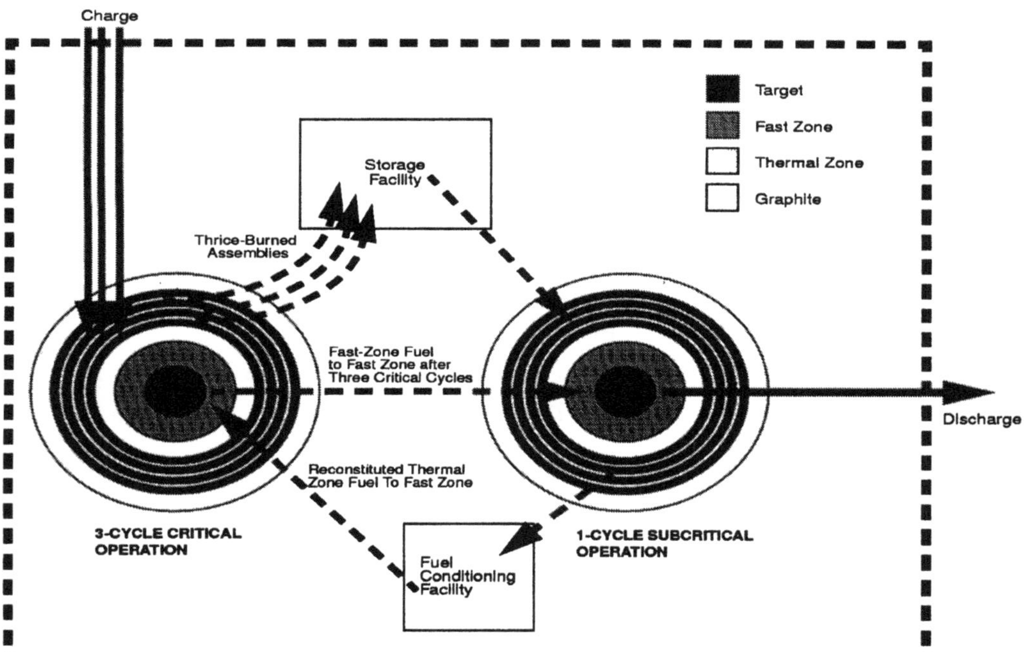

Fig. 1. 3-Batch GT-MHR Coupled to an AD-MHR

Comparison per Unit of Fission Energy of Isotopic Losses into the Waste Stream

Detailed comparisons of equilibrium-cycle mass-flows and discharges to the waste stream destined for the repository were made for the LBE, sodium, and gas cooled system point designs. Because the LBE and sodium system point designs are each 840 MW_{th} whereas the gas system point design was for a cluster of four 600 MW_{th} systems, all comparisons were made consistent by basing mass flows on a normalization to the same MW_{th} of fission power.[1]

The TRU consumption per unit of energy production is of course ~1.0 g/MW_{th} day for all systems, because the energy released per fission is approximately constant across all TRU isotopes. It is the evolution of the isotopic mix and the ultimate loss to waste which is of interest. Incore isotopic inventories and consumption rates, normalized to one MW_{th} of fission power, are compared in Table II. First comparing the two liquid metal options, the slightly lower TRU inventory of the sodium system compared to the LBE system is perhaps surprising. For a fixed blanket size, a sodium system would require a higher TRU inventory than a LBE system because of greater neutron leakage, but the sodium system design exploits a higher coolant velocity (and much lower coolant volume fraction) producing a more compact, higher power density, and less leaky core with slightly lower TRU inventory. As shown in Table I, when designed to the same limit on discharge fast fluence, the sodium system attains a slightly higher discharge burnup than the LBE system because the neutron energy spectrum is not as hard so the fast fluence to burnup ratio is slightly lower.

Table II. Isotopic Inventories at BOEC, Consumption Rates, and Annual Burnup at 75% Capacity Factor

Isotope	BOEC inventory (g/MWt)			Average consumption per day (g/MWd)			Atom % burnup per year @ 75% CF		
	LBE	Sodium	Gas	LBE	Sodium	Gas	LBE	Sodium	Gas
U-234	16.81	15.91		0.0023	0.0021		3.55	3.34	
U-235	4.07	3.97	0.03	0.0001	0.0001	0.0000	0.48	0.46	-*
U-236	5.56	5.28	0.13	0.0002	0.0002	-0.0001	0.97	0.91	-
U-238	36.69	34.64	0.00	0.0049	0.0049	0.0000	3.49	3.64	-
Np237	81.28	78.58	95.12	0.0523	0.0519	0.0437	13.74	13.60	16.60
Pu238	171.65	171.14	126.03	0.0242	0.0225	-0.0622	3.63	3.39	-80.77
Pu239	770.49	758.56	630.39	0.5486	0.5460	0.8119	14.79	14.41	24.55
Pu240	1057.65	1038.64	506.42	0.2279	0.2262	0.2903	5.44	5.45	18.95
Pu241	185.13	189.25	339.48	0.0248	0.0258	0.0614	3.50	3.54	11.98
Pu242	363.14	359.49	278.48	0.0483	0.0481	-0.0710	3.46	3.46	-22.12
Am241	191.72	185.21	85.55	0.1062	0.1047	0.0611	12.21	12.06	19.04
Am242	13.68	13.32	1.99	0.0000	0.0001	0.0012	0.09	0.13	19.05
Am243	122.33	118.70	86.81	0.0095	0.0095	-0.0502	2.07	2.11	-78.24
Cm242	12.38	12.62	18.77	-0.0136	-0.0117	-0.0046	-328.96	-296.51	-
Cm243	1.27	1.38	0.50	0.0000	0.0000	-0.0003	-0.72	-0.64	-
Cm244	91.33	91.47	47.21	-0.0050	-0.0044	-0.0343	-1.54	-1.35	-267.07
Cm245	25.10	25.40	3.51	0.0001	0.0001	-0.0095	0.10	0.10	-
Cm246	17.06	15.92		0.0000	0.0000		0.01	0.01	
Pu	2548.06	2517.07	1880.79	0.8738	0.8686	1.0305	8.21	8.11	17.95
MA	556.13	542.60	339.44	0.1495	0.1501	0.0071	6.71	6.78	1.31
TRU	3167.33	3119.46	2220.39	1.0308	1.0259	1.0375	7.86	7.80	16.22

*Isotopic burnup is not defined for these isotopes, since they are not included in the feed stream but are later produced by transmutation.

[1] Even though LWR-discharge TRU (33,000 MWd/t discharge and 25y cooling) was used for all three designs, slightly different feed compositions and depletion chain models were employed for the gas-cooled system as compared with the fast spectrum cases because of peculiarities of the differing neutronics codes used to analyze the thermal system.

The gas system BOEC inventory in Table II is the initial inventory for critical mode operation. Since the thermal region is dominant, its critical mass TRU inventory is smaller than for fast systems; the TRU inventory per unit fission power is ~70% of that of liquid metal systems. Correspondingly the atom % burnup per MW_{th} of the gas system is nearly twice that of the liquid metal systems because the TRU inventory of the gas system is significantly lower and the subcritical cycle is operated only with previously burned fuels.

The proportion of minor actinides in the gas system BOEC inventory is lower (and that of Pu-239 higher) compared to the LBE and sodium systems because the gas system is fueled with 100% LWR discharge TRU whereas the recycle fuel charged to the liquid metal systems has a significant self-recycle blending component. Furthermore, due to the large capture cross section of Pu-240 in the thermal energy range, the gas system end of equilibrium cycle (EOEC) inventory has significantly greater Pu-241 fraction than the liquid metal systems.
The gas system burns plutonium Pu-239, 40 and 41 isotopes more effectively, but minor actinides and Pu 238 and 242 less effectively than the LBE and sodium systems. In the LBE and sodium systems, all isotopes except for Cm-242 and Cm-244 are net consumed, whereas in the gas system, *net production* of all minor actinides except for Am-241, Am-242, and Np-237 occurs. It was observed that even Pu-241 and Pu-242 are net produced during the critical mode segments of operation of the gas system, so that in the accelerator-driven segment of the cycle, the minor actinide inventory further increases (negative net consumption) owing to neutron capture on Pu-241 and Pu-242.

Table III shows the annual isotopic feed of LWR-discharge isotopes compared to the annual isotopic waste stream (i.e., the "leakage loss" to the repository) for the three systems.[2] In the estimation of the isotopic losses from the ATW to the repository, all fuel discharged from the gas system was assumed to go to the waste stream. For the LBE and sodium systems, a fraction of the discharge fuel was assumed lost to the waste stream on each recycle step. The overall (per recycle step) loss factor (recycle plus refab) was arbitrarily assumed[3] here to be 0.1%. (For other recycle loss fractions, the isotopic losses are closely scaleable to this fraction because the evaluation has been done for an equilibrium cycle).

The external LWR feed per MW_{th} for the LBE and sodium system are much smaller than that of the gas system because external feed is used only to makeup the TRU consumed by fission – whereas the gas system operates once through and LWR discharge feed must therefore make up for both fissioned and discharged fuel. The LBE system releases 2.7% of the LWR TRU feed (3.3% of MA) to the waste stream, and the sodium system releases 2.4% of the TRU feed (2.8% of MA). The gas system releases only 3% of the Pu^{239} charged but releases altogether 36% of the TRU feed to the waste stream; i.e., much of its fuel has been transmuted to higher mass isotopes, incompletely burned and sent to waste. The amount of minor actinides discharged from the gas cooled system to the waste stream is comparable to the amount of LWR-discharge minor actinides initially loaded.

[2] The loss to waste reported here neglects any that would occur in processing the LWR fuel for TRU recovery and fabrication into the initial ATW transmuter loading. It also neglects losses in reconfiguring the gas system thermal zone discharge fuel into the fuel for the fast zone. Moreover, spallation and activation products in the core and accelerator are not accounted for.
[3] A design target for the loss per cycle from the ATW has been set much smaller – at 0.1% cumulative loss over multi recycle steps.[1]

Table III. External (LWR-Discharge) Isotopic Feeds and Isotopic Losses from ATW System per Year (based on 75% Capacity Factor)

Isotope	External feed per year (g/MWt)			Mass loss per year (g/MWt)		
	LBE	Sodium	Gas	LBE	Sodium	Gas
U-234	0.000	0.000		0.0046	0.0039	
U-235	0.011	0.011		0.0012	0.0010	0.0038
U-236	0.006	0.006		0.0016	0.0014	0.0300
U-238	1.349	1.343		0.0100	0.0084	0.0004
Np237	14.180	14.112	18.252	0.0162	0.0134	6.2971
Pu238	3.590	3.573	5.342	0.0469	0.0419	22.3651
Pu239	150.167	149.447	229.481	0.1472	0.1244	7.2168
Pu240	60.787	60.495	106.304	0.2731	0.2382	26.8271
Pu241	10.677	10.625	35.568	0.0500	0.0456	18.7595
Pu242	13.228	13.165	22.258	0.0987	0.0874	41.6827
Am241	25.313	25.191	22.258	0.0407	0.0341	5.5363
Am242	0.040	0.039	0.445	0.0037	0.0032	0.1107
Am243	2.615	2.602	4.451	0.0343	0.0298	18.1921
Cm242	0.000	0.000		0.0041	0.0035	1.2703
Cm243	0.006	0.006		0.0004	0.0004	0.0748
Cm244	0.293	0.292	0.890	0.0275	0.0248	10.2686
Cm245	0.026	0.026		0.0074	0.0068	2.6103
Cm246	0.003	0.003		0.0050	0.0042	
Pu	238.449	237.305	398.953	0.6159	0.5375	116.8511
MA	42.476	42.272	46.297	0.1393	0.1201	44.3602
TRU	282.291	280.937	445.250	0.7725	0.6723	161.2459

CONCLUSIONS

ATW transmuter core point designs have been developed for LBE and Na cooled concepts based on a multi-recycle strategy; a gas cooled hybrid concept based on a deep burn once thru strategy which has been proposed by General Atomics and has been independently analyzed.

Mass flows have been calculated and fractions of LWR feedstock lost to the waste stream were compared for the three concepts on a consistent per MW_{th} basis. The once thru deep burn strategy employed in the gas cooled thermal/fast spectrum hybrid concept avoids the costs of some of the recycle/refab equipment and reduces TRU in the repository as compared with an LWR once through strategy by ~60%. On the other hand, the discharged isotopic spectrum is unfavorable – with essentially unchanged MA mass flow to the repository as compared with LWR spent fuel. The presence of Pu^{241} in large amounts in the discharge is especially undesirable because its ultimate decay daughter is Np237 which, with a 2.14 million year half-life and high toxicity factor, constitutes the dominant long term (i.e., subsequent to 65,000 years) toxicity hazard in the Yucca Mountain repository source term[1]. The substantial burnup in a soft neutronic spectrum coupled with the small burnup increment attainable in the fast zone of this particular point design is ill suited to minor actinide consumption.

The fast spectrum multi recycle concepts which were examined bear the extra cost of recycle, but they achieve greater overall reduction in mass sent to the repository. Based on an assumed 1% loss fraction per recycle pass, used in these calculations, the multi recycle strategy using either of the two fast spectrum liquid metal cooled transmuter designs and a metallurgical recycle technology achieves a factor of about 400 reduction in TRU as compared with LWR once thru and achieves a reduction of about 325 in MA mass. The design goal for ATW recycle is 0.1% cumulative for multi recycle; if achieved, the reduction factors would reach a value of 1000.

The desired degree of reduction in waste mass and toxicity and the cost-to-benefit ratio of achieving any specified level of reduction are issues of public debate and are not yet resolved. The tradeoff analyses reported here are part of the technology program intended to inform that ongoing debate.[1] The point designs completed thus far focus on core neutronics, mass flows, and heat removal. Future work will address further optimization and will add dynamics, safety evaluations, and cost considerations.

ACKNOWLEDGMENTS

The detailed core design optimizations and analyses which provide the basis for this comparison were performed by Yang, Hill, Taiwo and Khalil and are reported in References 4, 5, and 7.

REFERENCES

1. US Dept. of Energy (Oct. 1999), *A Roadmap for Developing Accelerator Transmutation of Waste (ATW) Technology; A Report to Congress*, DOE/RW-0519.
2. R. N. Hill et al (September 1995), Physics Studies of Weapons Plutonium Disposition in the Integral Fast Reactor Closed Fuel Cycle. *Nucl. Sci. Eng.* 121, No. 1.
3. D. Crawford and M. Meyer (Nov. 28 - Dec. 1, 2000 to be published), Current Plans for US ATW Blanket Fuel Development, *Proceedings of the IAEA Technical Committee Meeting on Core Physics and Engineering Aspects of Emerging Nuclear Energy Systems for Energy Generation and Transmutation,* Argonne National Laboratory.
4. W. Yang, H. Khalil (to be published), *Blanket Design Studies of an LBE Cooled Accelerator Transmutation of Waste System, Nuclear Technology.*
5. R. N. Hill, et al (Nov. 28 - Dec. 1, 2000 to be published), *Physics Studies of a Sodium Cooled ATW Design," Proceedings of the IAEA Technical Committee Meeting on Core Physics and Engineering Aspects of Emerging Nuclear Energy Systems for Energy Generation and Transmutation*, Argonne National Laboratory,
6. A. Baxter and Carmelo Rodriquez March 2000, *Private Communication*, GA.
7. T. Taiwo, et al (Nov. 28 – Dec. 1, 2000 to be published), Core Physics Performance of Recycled LWR Discharge TRU Oxide Fuel in a GT/AD-MHR, *Proceedings of the IAEA Technical Committee Meeting on Core Physics and Engineering Aspects of Emerging Nuclear Energy Systems for Energy Generation and Transmutation*," Argonne National Laboratory.

Pergamon
www.elsevier.com/locate/pnucene

Progress in Nuclear Energy, Vol. 40, No. 3–4, pp. 505–512, 2002

Printed in Great Britain
0149-1970/02/$ - see front matter

PII: S0149-1970(02)00044-6

DESIGN STUDY OF LEAD-BISMUTH COOLED ADS DEDICATED TO NUCLEAR WASTE TRANSMUTATION

TAKAKAZU TAKIZUKA, KAZUFUMI TSUJIMOTO, TOSHINOBU SASA,

KENJI NISHIHARA and HIDEKI TAKANO

Japan Atomic Energy Research Institute

Tokai-mura, Naka-gun, Ibaraki-ken 319-1195, Japan

ABSTRACT

Research and development on nuclear waste transmutation are being carried out with a special emphasis placed on dedicated accelerator-driven systems at the Japan Atomic Energy Research Institute under the Japanese OMEGA Program. The reference accelerator-driven system design employs eutectic lead-bismuth as spallation target material and coolant. The fuel for the subcritical core is minor-actinide mononitride. The system consists of a 1.5GeV, 14mA proton accelerator and an 800MWt subcritical core with an effective neutron multiplication factor of 0.95. The transmutation rate of minor actinides is approximately 250 kg/y at 80% load factor. The design has salient features that the coolant inventory is large due to the tank-type configuration, the temperature rise through the core is relatively low, and the power conversion is operated on a saturated steam turbine cycle. These features make the plant response to a beam trip slow and much less demanding. © 2002 Published by Elsevier Science Ltd.

1. INTRODUCTION

Research and development on transmutation of long-lived radioactive nuclides are being carried out at the Japan Atomic Energy Research Institute (JAERI) under the Japanese long-term program for research and development on partitioning and transmutation technology (OMEGA Program). An emphasis has been placed on dedicated accelerator-driven systems (ADSs) because of its flexibility in designing a minor-actinide (MA) loaded core and large margin for criticality accidents. Earlier design approach of the accelerator-driven transmutation system (ADTS) had followed that of contemporary sodium cooled fast breeder reactors (FBRs) (Takizuka *et al.* (1989), Takizuka *et al.* (1991)). The major reasons to choose sodium as coolant were its excellent thermal performance, good compatibility with stainless steel, and technology maturity.

Recently, a preliminary design study was started for a heavy liquid-metal cooling option of ADTS at JAERI (Tsujimoto *et al.* (1998), Takizuka *et al.* (1998)). The main reasons for selecting heavy liquid-metal cooling as the primary design option are improved safety and possible reduction of construction cost of the system, together with prolonged lead time for an experimental system on tens MW scale. The experimental program for development and demonstration of lead-bismuth cooled ADTS technology is being planned under the Joint Project with KEK (High Energy Accelerator Research Organization). Conceptual design is being made for the ADS experimental facility (Joint Project Team of JAERI and KEK (1999)).

One technical issue that has recently attracted attention in the ADS area is the negative impact of very frequent beam trips as experienced in existing intense accelerator facilities (Takizuka (1998)). Frequent beam trips could cause thermal fatigue problem of ADS component materials, leading to degradation of their structural integrity and reduction of their lifetime. They could also badly erode the availability or the capacity factor of ADS, resulting in poor economics. In the development of accelerators for ADS, it is vitally important to establish technologies to achieve a very high degree of reliability. In parallel, it is also important in the development of ADS to design the plant less sensitive to beam trips and structural components to withstand possible thermal fatigues.

This paper describes a design of a lead-bismuth cooled, 800-MWt ADTS and presents an analysis of beam trip transient.

2. DEDICATED TRANSMUTATION SYSTEM IN DOUBLE-STRATA NUCLEAR FUEL CYCLE

There are two types of proposals for fuel cycle concept involving partitioning and transmutation (P&T). One is an advanced fuel cycle concept, where commercial FBRs are used for transmutation. Another is a concept of double-strata fuel cycle [Murata and Mukaiyama (1984), Mukaiyama (1994)], which combines a P&T fuel cycle (the second stratum) with the conventional fuel cycle for commercial power reactors (the first stratum). Dedicated systems specially designed for transmutation purpose could be introduced in the P&T fuel cycle.

There are several advantages of transmutation with dedicated transmutation systems over recycling MAs to commercial power reactors. Because of much lower mass flow in the P&T fuel cycle, its scale could be correspondingly smaller than the main cycle. Its separation from the main power-producing cycle, and the small number of units required, would allow the extensive innovation in fuel fabrication, core design and reprocessing technology for optimizing the P&T part of the overall fuel cycle system. It would avoid burdening the main cycle with the problems associated with higher radioactivity and decay heat, and reduced safety margins in reactor physics parameters. The P&T cycle could be made very compact by co-locating the entire facilities. This would minimize the transportation of nuclides that are troublesome with respect to waste management, and confine them effectively in the P&T fuel cycle.

A dedicated transmutation system can be either a critical system (burner reactor) or a subcritical system (ADS). A core loaded only with MAs (or with MAs and plutonium) could maximize the transmutation rate and thus the support factor. Systems with very high MA loading, however, pose crucial problems related to reactivity coefficients, and to the small delayed neutron fraction. To mitigate these problems for a critical system, a considerable amount of uranium should be added to the MA based fuel, preferably high-enriched uranium to avoid undesirable accumulation of MAs, but this results in reduced transmutation rate. For an ADS, its subcriticality helps to mitigate the problems without adding uranium and thus allows the maximum transmutation rate. For this reason, it can be concluded that an ADS is the best option for a dedicated transmutation device.

A special transmutation system operated with a hard neutron energy spectrum and high neutron flux can be efficient and effective for MA transmutation. In this context, JAERI has been pursuing the strategy of transmutation using dedicated transmutation systems.

3. DESIGN OF A LEAD-BISMUTH COOLED ADTS

Conceptual design studies have been carried out for ADTSs dedicated for transmutation of MAs and long-lived fission products. The current design is aiming at supporting about 10 units of large-scale light water reactor with an electric power of 1000 MW. This corresponds to around 800-MW thermal power of a dedicated transmutation system.

For the present, the lead-bismuth cooled option is the primary candidate for the ADTS. In ADSs, lead-bismuth eutectic can play roles of both coolant and spallation target material. It also offers the possibilities to achieve a harder neutron energy spectrum and to avoid a positive void reactivity. Its chemical inertness is particularly favorable for safety in the event of coolant leakage. Intermediate heat exchangers and secondary heat transport loops will possibly be eliminated with lead-bismuth coolant. Lead-bismuth coolant allows much lower system operating temperature than lead coolant. The lower operating temperature could help to alleviate material corrosion and erosion problems.

The design of the lead-bismuth cooled system is based on the technologies of contemporary liquid-metal cooled FBRs and ex-USSR lead-bismuth cooled submarine reactors. A conceptual drawing of a lead-bismuth cooled 800-MWt ADTS is shown in Fig. 1.

The target/core system consists of a spallation target region at the center and a surrounding annular subcritical core. The target region is a cylindrical void space through which liquid lead-bismuth flows upwards. The subcritical core is made up of pin-bundle-type fuel sub-assemblies. Actinide mononitride is adopted as the fuel material. Nitride is potentially the best fuel for dedicated transmutation systems because of its excellent thermal property and capability of pyrochemical reprocessing.

The whole target/core is contained within a primary vessel made of steel. All other components of the primary system, including steam generators, main pumps, and auxiliary heat exchangers, are accommodated within the vessel. This pool-type configuration is selected for the lead-bismuth cooled system to eliminate the need of heavy primary piping from the point of aseismatic design view.

The vertical tube for the beam path is inserted from the top into the primary vessel down to the spallation target region. The hemispherical bottom end of the tube forms the beam window. The beam window is cooled by lead-bismuth flow exited from the target region.

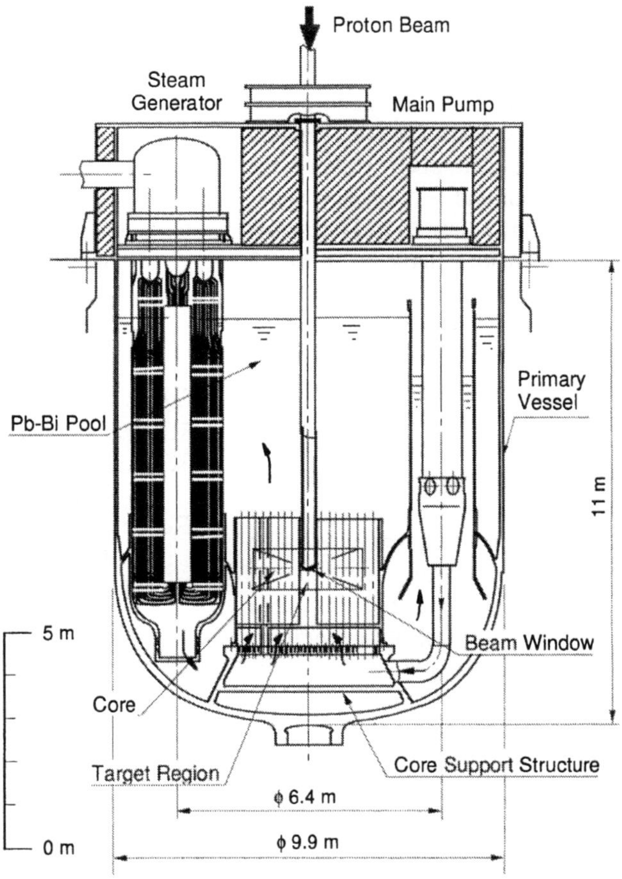

Fig.1. Conceptual drawing of lead-bismuth cooled accelerator-driven transmutation system

A high-energy beam from an intense proton accelerator is injected vertically downwards through a beam window into the target region. Spallation neutrons emitted from the target induce fission reactions in the subcritical core. Heat generated in the target/core is removed by forced upward flow of lead-bismuth coolant, and then transported to steam generators for power conversion.

The steam generators are of recirculation, separate type. Separate type steam generators allow the reduction in the depth of the beam window immersed in the lead-bismuth pool compared to integral type steam generators, and thus the reduced pressure load of the beam window. The vessel size and the lead-bismuth inventory of the system with the separate-type steam generators are much larger than those with integral-type steam generators. However, large lead-bismuth inventory slows the temperature transient caused by an accelerator beam trip. Recirculation type steam generators respond much more slowly in beam trip transient compared to one-through type steam generators because of a large heat capacity of steam drums. A disadvantage of recirculation type steam generators is a smaller temperature difference between the lead-bismuth and water/steam sides, resulting in a larger heat transfer area. Dimensions depicted in Fig. 1 are taken from the four-loop design.

The auxiliary cooling system with three lead-bismuth loops is provided as an independent means for decay heat removal in case when the water/steam system is inactive. Each loop consists of a air cooler, a lead-bismuth pump, and a heat exchanger. Heat exchangers on the auxiliary cooling system are installed in the upper plenum rather than at the top of steam generators. This allows the reduction in the steam generator height, and thus reduction in depth of the beam window. The auxiliary cooling loops are also used for preheating the lead-bismuth coolant in the standby mode.

The operating temperature range of the primary lead-bismuth is from 330 to 430 °C, which is about 100 °C lower than those for commercial LMFBR designs. Consequently, the power conversion system operates in a saturated steam turbine cycle used for conventional LWRs with efficiencies around 35% rather than a super-heated steam turbine cycle used for commercial sodium cooled FBRs with efficiencies above 43%. The lead-bismuth temperature rise through

Table 1 Major system parameters of the Pb-Bi cooled ADTS

Core Thermal Power	800 MW
Core Height/ Radius	1000/1200 mm
k_{eff} (initial/max./min.)	0.95/0.95/0.93
Linear Power Rating (max./ave.)	520/300 W/cm
Power Density (max./ave.)	310/180 MW/m^3
MA/Pu Inventory	2500/1660 kg
Coolant Temperature (in/out)	330/430 °C
Coolant Velocity (max.)	2 m/s
Coolant Flow Rate	1.98×10^8 kg/h
MA Transmutation Rate	250 kg/y

Table 2 Major design parameters of the ADTS plant

Pb-Bi Coolant Mass	
Core Section	8.34×10^4 kg
Upper Plenum Section	7.90×10^6 kg
Lower Plenum Section	3.79×10^6 kg
SG Inlet Plenum Section	7.90×10^4 kg
Core Structural Material Mass	4.65×10^4 kg
SG Heat Transfer Area	640 m^2 (per loop)
SG Heat Transfer Tube	
Cross Section/Length	0.1246 m^2/26.6 m
Recirculation Water Piping	
Cross Section/Length	0.060 m^2/38 m
Steam Piping	
Cross Section/Length	0.104 m^2/37 m
Volume of Steam Drum	35.23 m^3 (per loop)

Table 3 Nominal operating parameters of the ADTS plant

SG Outlet Steam Pressure	7.0 MPa
SG Outlet Steam Temperature	285 °C
SG Pressure Drop	1.6 MPa
Recirculation Piping Pressure Drop	0.05 MPa
Steam Piping Pressure Drop	0.05 MPa
Recirculation Water Flow Rate	7.88×10^5 kg/h (per loop)
Turbine Inlet Steam Pressure	6.0 MPa
Turbine Inlet Steam Temperature	275 °C
Turbine Inlet Steam Flow Rate	1.58×10^6 kg/h
Turbine Back Pressure	0.005 MPa
Electric Power	280 MW

the core is 100 °C, which is about 50 °C lower than those for commercial sodium cooled FBRs.

The lower operating temperature is expected to mitigate the material corrosion/erosion problem in lead-bismuth. The saturated steam turbine cycle is more tolerant in beam trip transients, since the saturated steam turbine is much more resistant to degradation of steam condition than the super-heated one. The lower temperature rise leads to lower temperature swing of the core exit lead-bismuth on beam trip and beam recovery.

Corrosion/erosion is one of the major technical issues in lead-bismuth cooling. Chromium-molybdenum steel (2.25Cr-1Mo steel, Mod. 9Cr-1Mo steel, 13Cr-1Mo steel etc.) is a candidate material for structural components in the primary system. Chromium-molybdenum steel is expected to have acceptable corrosion resistance, mechanical strength and ductility in the operating temperature range from 330 to 430 °C.

With a 1.5 GeV-14 mA proton beam, the core having an effective neutron multiplication factor of 0.95 produces about 800-MW thermal power. Assuming a load factor of 80%, the net MA transmutation rate is approximately 250 kg/y. An electric output of 280 MW is obtained at a plant thermal efficiency of 35%. The power required to operate the 1.5 GeV-14 mA accelerator is 52.5 MWe, assuming a 40% accelerator efficiency. Table 1 summarizes the major system parameters. The plant design parameters and operating parameters are listed in Tables 2 and 3.

4. TRANSIENT RESPONSE OF ADTS PLANT TO BEAM TRIP

In an ADS, a beam trip causes an abrupt drop of core thermal power very similar to the case of a scram in a critical reactor. There are two substantial differences between them from an ADS design standpoint. Beam trip analyses are reported for LANSCE by Eriksson and Piaszczyk (1998) and for SINQ by Bauer *et al.* (1998). One of the differences is the frequency of the event. Beam trips in an ADS would be much more frequent than reactor scrams. The other is the duration of down time. The main source of beam trips occurring at the existing accelerator facilities is failure of RF system. Majority of the beam trips are of short duration, say within few minutes, and the beam recovers from them automatically. For this reason, an ADS has to maintain full flow while awaiting beam recovery.

Figure 2 is a simplified flow diagram of the ADTS. Primary lead-bismuth coolant flows upward through the core and exits into the upper plenum. Then, the coolant flow divides equally among four steam generators. The flow through steam generator heat transfer sections enters a lower plenum. The coolant is then returned back to the core by primary main pumps.

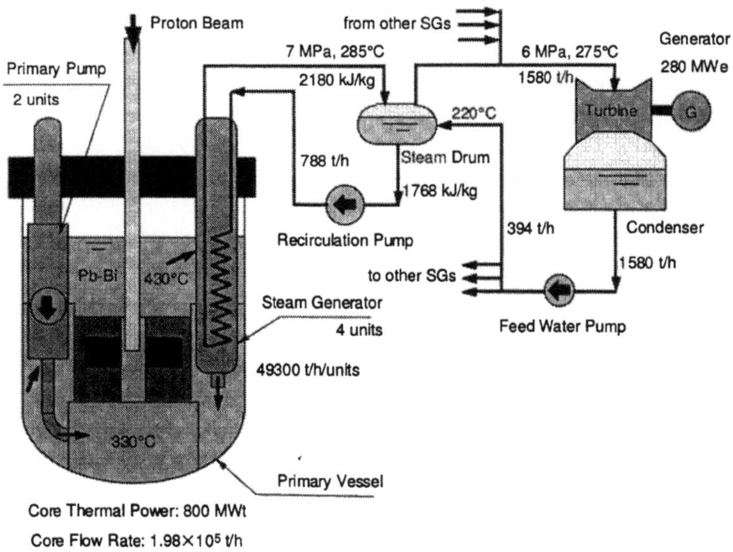

Fig.2. Simplified flow diagram of ADTS plant

The water/steam system has four steam drums, four recirculation water pumps, a turbine-generator unit, and a feed water pump. Steam raised in the steam generators is directed to the steam drums. The steam from all the four steam drums combines and drives the turbine-generator unit. The steam is cooled down into water in a condenser at the turbine exit, and the feed water pump returns the water back to the steam drums through a feed water heater not shown in the figure. The recirculation water pump delivers the water from the steam drum to the steam generator.

The analysis dealt with one single loop, on the assumption of the equal operating condition among all the four symmetric loops. The lead-bismuth primary system and the secondary water/steam system were modeled with a simple one-dimensional flow network.

The primary system was subdivided into three plenum sections (a core upper plenum, a steam generator inlet plenum, and a core lower plenum) and two components (a core and a steam generator). The primary lead-bismuth coolant was assumed to be fully mixed in the plenums, and the thermal power was given as a boundary condition in the core. In the steam generator, heat transfer was calculated using five empirical equations of heat transfer coefficient: one for the lead-bismuth side and four for the water/steam side (pre-heater region, nucleate boiling region, film boiling region and super-heater region). In the power conversion loop, the recirculation water pump head, the feed water pump flow rate and the turbine back-pressure were given as boundary condition. The steam drum was modeled as a lumped system of saturated water/steam including phase changes. The water in the drum was assumed to be controlled at a constant level. The turbine was modeled using a simple relation among the steam mass flow rate, the inlet pressure, the back-pressure, and the inlet temperature. The turbine power was assumed proportional to the inlet steam flow rate and the enthalpy difference from the turbine inlet to outlet. The conversion efficiency from thermal to electric power was assumed 35% (constant). The feed water heating was not explicitly simulated in the analysis, but the temperature of the feed water was given as a function of the turbine inlet steam flow rate, neglecting the delay due to steam-to-water heat transfer.

With this flow network model, transient pressures, flow rates, and enthalpies were solved from simultaneous equations of energy conservation, mass conservation, momentum conservation and state. The density, the thermal conductivity, the kinetic viscosity, and the Prandtl number of lead-bismuth were given as functions of temperature (JSME (1986)). Property of water/steam was based on JSME Steam Tables (1981).

Results presented here are for the case when the beam is recovered 370 s after the beam trip. A beam down time longer than 380 s requires plant and turbine trips to avoid overcooling. Overcooling could cause damages of primary system components by freezing of lead-bismuth and erosion of the turbine blades by wetting of steam.

The temperature response of lead-bismuth primary coolant is shown in Fig. 3. On the beam trip at $t = 0$, the core outlet temperature starts to drop rapidly and asymptotically approaches to the core inlet temperature within about 10 s. The coolant temperature at the steam generator inlet responds more slowly because of coolant mixing in the upper plenum. It takes about 120 s for the steam generator inlet temperature to drop down to the initial core inlet temperature. The coolant temperature at the steam generator outlet decreases almost proportionally to that at the steam generator inlet. The temperature change at the core inlet further delays because of

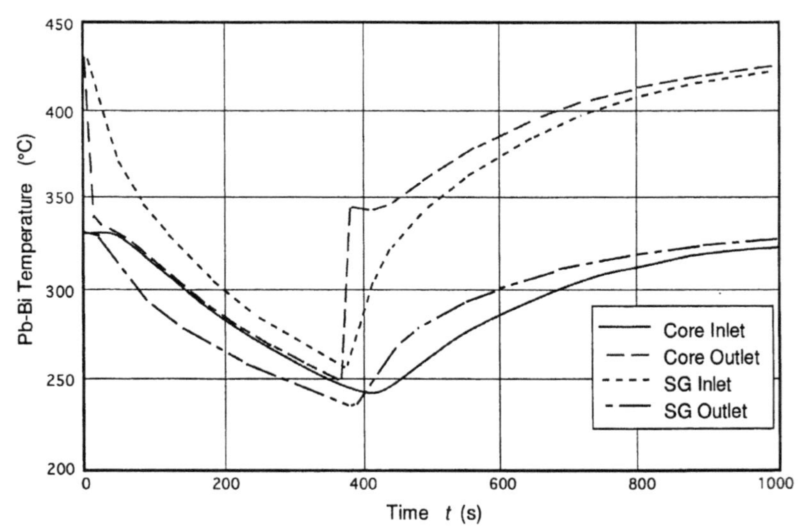

Fig. 3. Temperature change in primary lead-bismuth system during beam trip transient

coolant mixing in the lower plenum.

At t = 370 s when the beam turns on, the core outlet temperature shows a rapid rise from its minimum at 250 °C by about 100 °C, and then a slight temperature decrease due to the time lag of the core inlet temperature response. The minimums of the steam generator inlet and outlet temperatures occur at about 380 s (10 s after the beam recovery).

The core inlet temperature reaches its minimum at about 410 s. The coolant temperatures at the steam generator inlet, the steam generator outlet, and the core inlet increase monotonically from their minimums, and approach to their initial values. It takes around 1000 s after the beam-on before the initial condition is restored. Temperature swings of the primary coolant during the beam trip transient range from 88 to 185 °C.

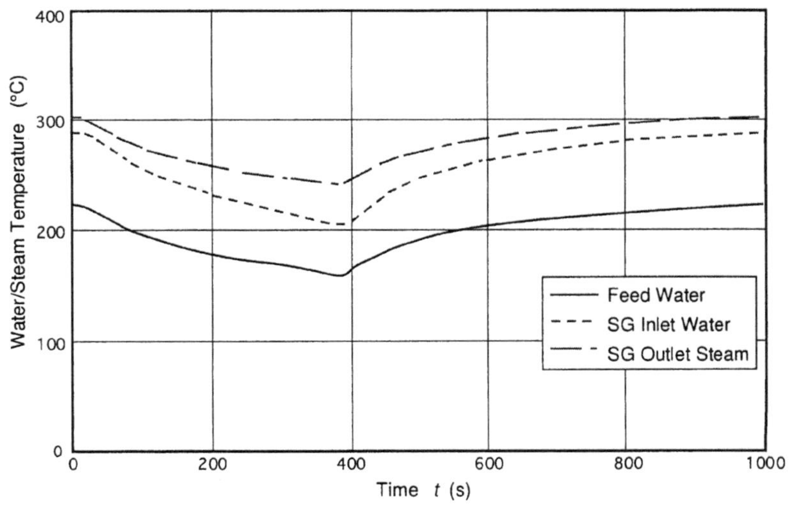

Fig. 4. Temperature change in secondary water/steam system during beam trip transient

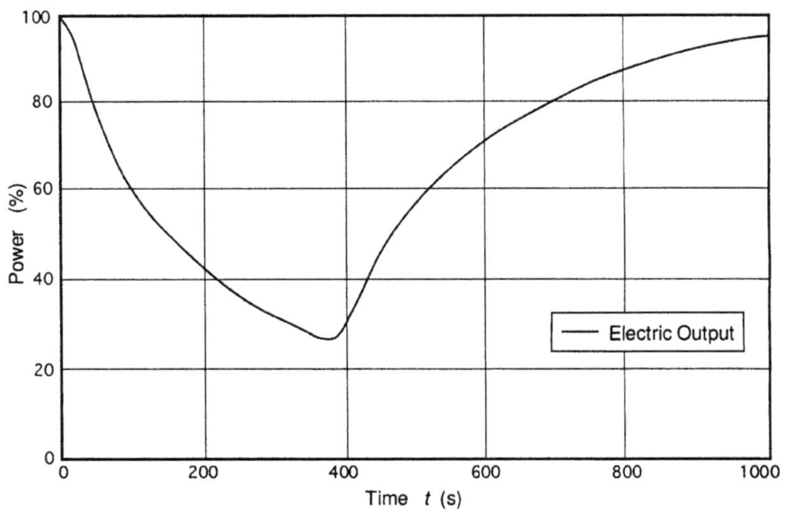

Fig. 5. Electric output during beam trip transient

The water/steam temperature in the power conversion system is shown in Fig. 4. The water/steam temperatures show similar response to the lead-bismuth temperatures at the steam generator inlet and outlet. The difference between the steam generator inlet temperature and the feed water temperature is kept almost constant at about 80 °C. The temperature rise through the steam generator increases as the temperature decreases. This indicates worsening of the steam quality at lower temperatures. Compared to the primary lead-bismuth, the water/steam shows damped temperature swings within 82 °C.

The electric power generated is shown in Fig. 5. The electric power is nearly proportional to the turbine inlet flow rate, because the enthalpy change of steam at the steam drum outlet is not so large and the enthalpy of the turbine outlet steam is kept constant. The minimum of the electric power is 27% (76 MWe) at about 380 s.

5. CONCLUDING REMARKS

A design study was made for a lead-bismuth cooled accelerator-driven system as a dedicated transmuter to be introduced in the second stratum of the double-strata nuclear fuel cycle. The system consists of a 1.5GeV, 14mA proton accelerator and an 800MWt subcritical core with an effective neutron multiplication factor of 0.95. The transmutation rate of MA is approximately 250 kg/y at 80% load factor. The system incorporates several design features to withstand the short beam interruptions that would occur much more frequently than the scrams in a critical reactor. Transient analysis showed that the plant and the turbine survive beam trip transients shorter than 370 s without trips and special controls. The beam trip statistics at existing high-power accelerator facilities indicates that most of the down time is within 5 minutes. Power ramps at a rate of about 15% per minutes are considered to be allowable because the turbine temperature remains high in a short down time after a beam trip. It is concluded that an accelerator-driven system can be designed to alleviate adverse effects of beam trips to acceptable level.

REFERENCES

Bauer G., Wydler P., Stanculescu A., Commetto M. (1998) Beam Trips and Target/Subcritical Reactor Problems in Accelerator Driven Systems, *OECD/NEA Workshop on Utilisation and Reliability of High Power Proton Accelerators,* p.355. Mito 13-15 Oct.

Eriksson M. and Piaszczyk C. (1998) Reliability Assessment of the LANSCE Accelerator System, *OECD/NEA Workshop on Utilisation and Reliability of High Power Proton Accelerators,* p.183. Mito 13-15 Oct.

JSME (1986), *JSME Data Book : Heat Transfer, 4th Edition*, JSME.

JSME (1981), *JSME Steam Tables, 4th Edition*, JSME.

Murata H. and Mukaiyama T. (1984), Fission Reactor Studies in View of Reactor Waste Programmes, *Atomkern-Energie Kerntechnik, 45, 23.*

Mukaiyama T. (1994) Importançe of the Double Strata Fuel Cycle for Minor Actinide Transmutation, *Thirrd OECD/NEA International Information Exchange Meeting on P&T*, p.30. Cadarache, 12-14 Dec.

Takizuka T., Kanno I., Takada H., Ogawa T., Nishida T. and Kaneko Y. (1989), A Study on Incineration Target System, *Fifth International Conference on Emerging Nuclear Energy Systems (ICENES '89)*, p. 70, Karlsruhe, 3-6 July.

Takizuka T., Takada H., Kanno I, Nishida T., Akabori M. and Kaneko Y. (1991), Conceptual Design of Transmutation Plant, *Specialist Meeting on Accelerator-Driven Transmutation Technology for Radwaste and Other Applications*, p. 707. Stockholm, 24-28 June.

Takizuka T., Tsujimoto K., Sasa T., Takano H., Hirota K., Kamishima Y. (1998) Heavy Liquid-Metal Cooling Option of JAERI Accelerator-Driven Transmutation Systems, *International Conference on Heavy Liquid-Metal Coolants in Nuclear Technologies (HLMC '98)*, p.143. Obninsk, 5-9 Oct.

Takizuka, T. (1998) Effects of Accelerator Beam Trips on ADS Components, *OECD/NEA Workshop on Utilization and Reliability of High Power Accelerators*, p.317. Mito, 13-15 Oct.

Tsujimoto K., Sasa T., Takano H., Takizuka T., Mukaiyama T., Hayano M., Yokobori H. Kamishima Y. (1998) Conceptual Study of the Lead-Bismuth Cooled Accelerator-Driven Transmutation System, *2nd International Topical Meeting on Nuclear Applications of Accelerator Technology (AccApp '98)*, p.137. Gatlinburg, 20-23 Sept.

Pergamon

www.elsevier.com/locate/pnucene

Progress in Nuclear Energy, Vol. 40, No. 3–4, pp. 513–521, 2002

Printed in Great Britain

0149-1970/02/$ - see front matter

PII: S0149-1970(02)00045-8

NUCLEAR FUEL CONSIDERATIONS FOR THE 21ST CENTURY

L. C. WALTERS, D. L. PORTER, and D. C. CRAWFORD

Argonne National Laboratory – West
P. O. Box 2528
Idaho Falls, Idaho 83403-2528

ABSTRACT

There are many external influences that may control the path that nuclear power deployment follows. In the next 50 years several events may unfold. Fear of the consequences of the greenhouse effect may produce a carbon tax that would make nuclear power economically superior very quickly. This, in turn, would increase the rate at which uranium reserves diminish due to the increased rate of nuclear power deployment. However, breakthroughs in the extraction of uranium from the sea or deployment of fast breeder reactors would greatly extend the uranium reserves and, as well, utilize the thorium cycle.

On the other hand, carbon sequestering technology breakthroughs could keep fossil fuels dominant for the remainder of the century. Nuclear power may only then continue, as today, in a lesser role or even diminish. Fusion power or new developments in solar power could completely displace nuclear power as we know it today.

Even more difficult to predict is when the demand for mobile fuel for transportation will develop such that hydrogen and hydrogen rich fuel cells will be in common use. When this happens, nuclear power may be the energy source of choice to produce this fuel from water or methane. In a similar vein, the demand for potable and irrigation water may be another driver for the advent of increased deployment of nuclear power.

With all these possibilities of events that could happen it appears impossible to predict with any certainty which path nuclear power deployment may take. However, it is necessary to define a strategy that is flexible enough to insure that when a technology is needed, it is ready to be deployed.

For the next few decades there will be an evolutionary improvement in the performance of uranium oxide and mixed uranium oxide-plutonium oxide (MOX) LWR fuels. These improvements will be market driven to keep the cost of fuel and the resulting cost of nuclear power electricity as competitive as possible. The development of fuels for accelerator transmutation and for reactor transmutation with inert matrix fuels is in its infancy. A great deal of research has been initiated in a number of countries, which has been summarized in recent conferences. In Europe the work on these fuels is directed at the same problem as their utilization of MOX; namely to reduce the inventory of separated plutonium, minor actinides, and Long Lived Fission Products (LLFP). In the United States there is no reprocessing and thus no inventory of separated civilian plutonium. However, in the United States there is a resistance to a permanent spent fuel repository and thus accelerator transmutation presents a possible alternative. If nuclear power does have a long-term future, then the introduction of the fast reactor is inevitable. Included in the mission of the fast reactors would be the elimination of the inventory of separated plutonium while generating useful energy. The work that is ongoing now on the development of fuel concepts for assemblies that contain actinides and LLFP would be useful for fast reactor transmutation.

There is still a great deal of work required to bring the fast breeder reactor option to maturity. Fortunately there is perhaps a fifty-year period to accomplish this work before fast breeders are necessary. With regard to fast reactor fuel development, future work should be considered in three stages. First, all the information obtained over the past forty years of fast reactor fuel development should be completely documented in a manner that future generations can readily retrieve and utilize the information. Fast reactor development came to such an abrupt halt world-wide that a great deal of information is in danger of being lost because most of the researchers and facilities are rapidly disappearing. Secondly, for all of the existing fast reactor fuels, and this includes, oxides, carbides, nitrides, and metallic fuels, the evolutionary work was far from being completed. Although mixed oxide fuels were probably the furthest advanced, there were many concepts for improved claddings and advanced fabrication methods that were never fully explored. Finally, with such an extended period before fast reactors are needed there is ample time for truly innovative fuels to be developed that are capable of performing over a wide range of conditions and coolants.

INTRODUCTION

Nuclear fuel for advanced fuel cycles is dependent on the possible paths that nuclear power development may take in the future. There are many external influences that may control the path that nuclear power deployment follows. Most authors who analyze deployment alternatives speak in 50-year intervals, which is somewhat arbitrary, but 50 years is in the range of perfecting a new technology from concept to realization.

In the next 50 years several events may unfold. Fear of the consequences of the greenhouse effect may produce a carbon tax that would make nuclear power economically superior very quickly. This, in turn, would increase the rate at which uranium reserves diminish due to the increased rate of nuclear power deployment. However, breakthroughs in the extraction of uranium from the sea or deployment of fast breeder reactors would greatly extend the uranium reserves. Development and implementation of a thorium fuel cycle would also greatly extend the nuclear power enterprise.

On the other hand, carbon sequestering technology breakthroughs could keep fossil fuels dominant for the remainder of the century. Nuclear power may only then continue, as today, in a lesser role or even diminish. Fusion power or new developments in solar power could completely displace nuclear power as we know it today.

Hydrogen and hydrogen-rich fuels have been proposed as alternatives to fossil based fuel. However, it is difficult to predict when the demand for mobile fuel for transportation will develop such that hydrogen and hydrogen rich fuel cells might be in common use. When this happens, nuclear power may be the energy source of choice to produce this fuel from water or methane, published (IAEA-TECDOC-1085, 1999). In a similar vein, the demand for potable and irrigation water may be another driver for the advent of increased deployment of nuclear power.

With all these possibilities of events that could happen it appears impossible to predict with any certainty which path nuclear power deployment may take. Adding to the uncertainty are the varying energy options, and political acceptability of some options, in different nations. In spite of these uncertainties, prudence demands that we define an energy strategy that is flexible enough to insure that when a technology is needed, it is ready to be deployed.

Future Nuclear Fuel Development

For the next few decades there will be an evolutionary improvement in the performance of uranium oxide and mixed uranium oxide-plutonium oxide (MOX) LWR fuels. These improvements will be market driven to keep the cost of fuel and the resulting cost of nuclear power electricity as competitive as possible. It is not clear at this time whether the MOX option for LWR utilization will continue. MOX utilization in Europe is a means to reduce the civilian plutonium inventory that results from the reprocessing of LWR fuels. This separated plutonium was originally destined for fast reactors, the construction of which had been anticipated. MOX utilization in the United States is simply a means to dispose of excess weapons plutonium. The economics of MOX utilization has recently been questioned for a number of reasons, published (Charpin *et al.,* 2000). In any event, the evolutionary improvements of LWR fuels will most certainly take place over the next few decades as the commercial enterprise competes for the market in both the developed and developing countries.

The development of fuels for accelerator transmutation and for reactor transmutation with inert matrix fuels is in its infancy. A broad range of smaller research effort has been initiated in a number of countries, which has been summarized in recent conferences, published (Proceedings from Global '99, 1999; Proceedings from Workshop on Advanced Reactor and Innovative Fuels, 1998; Proceedings of ICONE 8, 2000). In Europe the work on these fuels is directed at the same problem as utilization of MOX; namely, reduction of the inventory of separated plutonium, minor actinides, and Long Lived Fission Products (LLFP). The problem in Europe is envisioned to be near-term because the growing inventory of separated plutonium presents a potential proliferation and radio-toxicity problem. In the United States there is no reprocessing and thus no inventory of separated civilian plutonium. However, in the United States there remains some question regarding the licensability of a permanent spent fuel repository and thus accelerator-driven transmutation presents a means to reduce or eliminate quantities of long-lived actinides and fission products..

All of this development on fuels for actinide and waste transmutation would be completely unnecessary with the near term reintroduction of fast reactors. Maturation of these transmutation concepts, such that they could significantly begin to reduce existing inventories of actinides and LLFP, would require at least 30 years. During that period of time the long-term future of nuclear power should become evident. If a long-term future does not develop, and if reprocessing continues for the next fifty years, then a huge inventory of separated plutonium, minor actinides, and LLFP will exist that must be subjected to transmutation or immobilized for a repository. In this situation the availability of proven fuel designs for actinide and waste transmutation is crucial. Moreover, if nuclear power does have a long term future, then the introduction of the fast reactor is inevitable, published (Hori, 2000). Included in the mission of the fast reactors would be the elimination of the inventory of separated plutonium while generating useful energy. The work that is ongoing now on the development of fuel concepts for assemblies that contain actinides and LLFP would be useful for fast reactor transmutation.

A great amount of work remains to bring the fast breeder reactor option to maturity. Fortunately there is perhaps a fifty-year period to accomplish this work before fast breeders are necessary. With regard to fast reactor fuel development, future work should be considered in three stages. First, all the information obtained over the past forty years of fast reactor fuel development should be completely documented in a manner that future generations can readily retrieve and utilize the information. Fast reactor development came to such an abrupt halt world-wide that a great deal of information is in danger of being lost because most of the researchers and facilities are rapidly disappearing, published (33rd IAEA International Working Group Meeting on Fast Reactor, 2000). Secondly, for all of the existing fast reactor fuels, and this includes, oxides, carbides, nitrides, and metallic fuels, the evolutionary work which was far from being completed, should be continued. Although mixed oxide fuels were probably the furthest advanced, there were many concepts for improved claddings and advanced fabrication methods that were never fully explored, published (Nuclear Materials, 1994). Finally, development of innovative fuels should be pursued. With such an extended period before fast reactors are needed, there is ample time for truly innovative fuels to be developed that are capable of performing over a wide range of conditions and coolants. The next sections will discuss the current situation and what might be accomplished in each of the three stages.

Preservation of Fast Reactor Fuel Information

Fast reactor fuel development programs were at their peaks by 1980. Fast test reactors were operating in several countries with commercial size prototypes just under construction or coming on line. From that time onward, fast reactor development in general began to decline. By 1994 in the United States the Clinch River Breeder Reactor (CRBR) had been canceled and the two fast reactor test facilities, the Fast Flux Test Facility (FFTF) and the Experimental Breeder Reactor (EBR-II) had been shutdown; with EBR-II permanently and FFTF in a standby condition. Thus, efforts essentially disappeared for fast reactor fuel development. Similarly, programs in other nations were terminated or substantially reduced.

Not quite so obvious is the disappearance of the aging fast reactor fuel experts. With their continuing retirement from the workforce a great deal of knowledge and experience is being lost. It may be argued that all the fuel performance information gained during this period has been well documented in the open literature, conference proceedings, and company reports. In addition, there have been recent fast reactor fuel reviews. For fuel irradiation performance information this may well be the case. However, a serious effort has not been undertaken, at least in the United States, to collect and review for completeness the existing fuel performance documentation. The degree to which the information has been compiled for posterity is probably far less in the areas of fuel fabrication technology, the design and performance of in-reactor test vehicles, and the operating performance of entire cores of fuel, blanket, and reflector assemblies. Again, no assessment has been made of either the value of this information to future generations nor an assessment of the completeness of the documentation. Even now an effort to correct any deficiencies would be difficult due to the loss of the experts. Soon it will be impossible.

Based upon the author's experience with the Experimental Breeder Reactor II, EBR-II, there are several areas that need a significant effort in order to bring the more than 30 years of operating experience to a state where future generations will receive full benefit. Table 1 lists a number of areas that need to be archived in a much better manner than they are today. To a greater or lesser extent the situation is likely the same world-wide. It is simply a matter of completing the 30-year task with a small fraction of what has already been spent to allow society to realize the benefit of a generation of accomplishment.

Table 1 - Areas to Assess for Archival Completeness

Reactor Design and Construction	•	Rational for key design features.
	•	Original design drawings and justification for major changes.
	•	Specifications for materials and procurement considerations.
	•	Organization structure and evolution.
	•	Documentation of important problems and solutions.
	•	State of archival materials.
Operations	•	Operations organization.
	•	Training and procedures.
	•	Safety analysis reports and system design descriptions.
	•	Fuel handling experience.
	•	Steam system operating experience.
	•	Documentation of important problems and solutions.
Nuclear Physics and Thermal Hydraulics	•	Documentation of all computer codes along with utilization experience.
	•	Documentation of special physics and thermal hydraulic experiments.
	•	Documentation and discussion of significant safety-related events.

Fuel Fabrication	•	Fabrication organization.
	•	Training and procedures.
	•	Fuel specifications.
	•	Equipment design drawings.
	•	Materials specifications and procurement.
	•	Fabrication experience.
	•	Documentation of important problems and solutions.
Fuel Performance	•	Irradiation test results.
	•	Documentation of computer codes.
	•	Documentation of the analyses of important phenomena.
	•	Description and design drawings of in-reactor test vehicles.
	•	Documentation of important problems and solutions.

Evolutionary Improvements to Fast Reactor Fuels

As is the case for almost all products of industry, evolutionary improvements occur to make the products safer and less expensive. Fast reactor fuel development had by no means reached any sort of asymptote for ultimate performance when work in the United States was terminated. The same situation is true abroad. The problems of excessive irradiation-induced creep and swelling in fast reactor fuel cladding was well on the way to solution with the introduction of martensitic/ferritic steel cladding. However, the low swelling cladding materials had marginal high-temperature strength, so cladding improvements were desired to achieve high outlet temperatures. Although mixed oxide pellet fuel for fast reactors was the most advanced, the data base was not adequate in the United States for licensing a commercial reactor. For metallic, carbide and nitride fast reactor fuels a great deal more performance information was required before licensing could be attempted and, as well, there were many evolutionary concepts that were yet unexplored.

An international assessment should be made of all the evolutionary concepts for fast reactor fuels, and with some selectivity, the evolutionary work should move forward. However, two obstacles must be overcome. The first is a new generation of scientists and engineers must be attracted into this area to provide a continuity of expertise. Secondly, there must be fast reactor facilities to conduct the irradiation and post-irradiation examinations. Currently the United States has no fast reactor irradiation facilities, and there is a struggle to keep the hot cell capability viable. The same is true in Great Britain. France with Phenix, Japan with Joyo and Monju, and Russia with BOR-60 and BN-600 still have irradiation capability. However, the near-term future with all those reactors is uncertain.

It will be a difficult task to convince governments around the world to continue with fast reactor fuel development at some nominal level when there is no near-term foreseeable need for fast reactors while, at the same time, there is pressure from the antinuclear community to cease all fast reactor work. However, the alternative is the complete loss of facilities, expertise, and perhaps information already gained, the replacement of which will be extremely costly a few decades from now.

As pointed out earlier, a major obstacle to the development of a new fuel or the advancement of evolutionary ideas is the loss of irradiation test and examination facilities. A solution, which to some extent is already happening, is to identify the best remaining facilities, on an international basis, and devise a strategy to share in the utilization and cost of operating the facilities. There will not be a large fuel development program conducted over a relatively short period of time such as the world experienced during the 1970's for fast reactor fuel development. During that period several countries had the complete suite of facilities that included critical facilities, transient fast reactors, fast reactors for steady-state irradiations, and extensive examination facilities. Rather than the large, short-term development programs, much lower level programs will continue in each involved country in a search for the system that will replace the LWR's several decades in the future. Within this framework, a sharing of remaining facilities with minimum duplication makes the best solution. For example, fast reactor irradiation facilities such as Joyo in Japan or BOR-60 in Russia might be used in conjunction with transient test facilities such as TREAT in the U.S. to develop new fuels and to demonstrate irradiation performance under a full range of conditions. Irradiation testing programs, then, would become truly international collaborations.

An example, that falls within the experience of the authors of an evolutionary fast reactor fuel program that was not carried to completion, is that for Integral Fast Reactor (IFR) metallic fast fuel, published (Hannum, 1997). When the development program for the U-Pu-Zr metallic fuel was terminated in the United States in 1994, the fuel had been irradiated to 20 at.% burnup without failure, published (Walters, 1998). Fuel with improved cladding materials and lower smeared density was under irradiation that promised substantially greater burnup than 20 at.%. In addition, fuel with high concentrations of minor actinides was being irradiated to learn more about the fabrication and performance of fuel for minor actinide disposal.

The IFR concept was included in the United States Energy Policy Act of 1992, where a schedule was prescribed for design and construction of a prototype plant (PRISM-type). Title II design was to be initiated for the prototype ALMR in 1998. In response to the Policy Act, GE Nuclear Energy produced (early in 1993) the ALMR Fuel and Safety R&D Requirements document which outlined the issues to be addressed in order to document ALMR fuel performance sufficiently so that an ALMR could be licenced. This document was reviewed by the Nuclear Regulatory Commission against the IFR program documentation and the NRC produced NUREG-1368, published (NUREG-1368, 1994).

The issues that were raised were addressed in the Fuel Test Plan (authored by General Electric, Westinghouse, and Argonne National Laboratory) in 1994 with either technical justification or plans for additional R&D, published (Porter, 1994). Interestingly, these beginnings of the licensing process were underway just as the IFR program funding was terminated in 1994.

The Fuel Test Plan showed that U-Pu-Zr metallic driver fuel and U-Zr blanket fuel performance was known, documented, and modeled sufficiently in many areas. The fundamental areas included swelling, constituent (elemental) migration, fuel pin length changes, and fission gas release, in regards to steady-state performance. Fuel/cladding chemical compatibility and 'eutectic limits' were included in transient performance issues as well as post-failure performance (run beyond cladding breach). Proof of statistical reliability of the fuel, cladding and duct performance, and the accuracy of fuel performance modeling were also examined and the status addressed. The performance differences caused by minor actinide (Am, Np) also needed additional work based upon experiments which had been completed. The issues that remained are outlined in Table 2 below.

Table 2 demonstrates that the fuel performance was nearly complete; schedules shown with this information in the Test Plan indicated that 2-3 years were required to complete the requirements for cold-fabricated fuel, and 4-5 years for the Am, Np-bearing fuel. Note, however, that several open issues required the study of as-irradiated fuel from experiments; this fuel is no longer available. Therefore, if started now, it would require much longer times to get to the point of being ready for licensing.

Table 2

Fuel Performance Issue	Remaining Questions/Actions
Fuel Swelling	Complete exams for porosity distributions/Final Model Validation.
Constituent Migration	U-Zr complete/Complete model validation for U-Pu-Zr.
Effects of Minor Actinides	'Eutectic' temperature effects.
Fuel/Cladding Compatibility	Complete tests of high Pu (26-28 wt%) alloys with HT9 cladding at high fuel burnup/complete ex-reactor tests to be able model effects of lanthanide fission products.

Fission Gas Release	Complete
Fuel Pin Length	Complete
Transient Performance	One test at >40% overpower; steady-state irradiation after overpower tests.
Run Beyond Cladding Breach	Testing complete; complete modeling of gas release signatures, etc., for various breach locations and fuel types.
LIFE-METAL and F-PIN code development for element reliability	Complete model validation for fuels containing Am and Np; Use validated codes to run reliability analyses for prototype ALMR.
Fuel Specification	In-reactor performance-based specification complete/fuel recycle demonstration to test acceptability in remote manufacturing process.
Cladding and Duct Performance	Complete

Innovative Fast Reactor Fuels

As discussed earlier it is impossible to predict future demands on nuclear power when all the external forces are considered. However, up to this time, nuclear power has been used primarily for electricity generation. There are some desalination applications for nuclear power, such as the former operation of the BN-350 fast reactor in Kazakhstan. Breeding potential and the use of the higher temperature liquid metal coolants created additional challenges for the nuclear fuel developer. In the future, the need to produce H_2 from methane and water, compact-high energy density cores for space travel, and perhaps the wide use of nuclear power for desalinated water will further challenge the fuel designer. Concepts have been put forward for reactors that could serve these multiple uses, published (Wade, 2000). Perhaps by the end of the century nuclear parks will exist that produce electricity, breed Pu^{239} and U^{235} for further power production, H_2 by electrolysis or some other process, and heat for other uses such as water purification.

Because of the long period, 30-50 years, to bring a truly innovative concept to fruition, it is the time to begin this advanced thinking process. It would also keep a new generation of nuclear scientists challenged. However, as pointed out earlier, the larger challenge is to convince short-sighted governments that such work is needed.

Many of the fuel systems that the authors regard as "innovative" are, in fact, fuel types for which there is little or no data regarding performance, fabricability or properties. Other fuels regarded as innovative are actually evolutions of previously investigated fuel types now under consideration for new and innovative applications. Some of the requirements now emerging for new fuels are common to many applications, such as high-temperature or extended-burnup operation, while other requirements are particular to an application, such as fast transmutation rates. Some of these requirements are addressed in the following paragraphs, which also describe the fuel types being considered or proposed.

Nearly all reactor or transmutation concepts being proposed anticipate high-burnup use of fuel. For some concepts, such as those with long-life cores designed for only occasional fuel handling, extended burnup capability is essential. For other concepts, extended burnup is highly desirable to increase proliferation resistance of spent fuel with relatively high contents of higher actinides. And for many instances, extended burnup is regarded, as it has been historically, as desirable to improve economics of reactor operation by reducing the number of fuel assemblies required to produce a obtain amount of nuclear energy, with associated costs reductions for fuel handling, fabrication, recycling, and/or post-irradiation storage. Some of the design features being proposed for achieving high burnup include dispersion and vi-pac fuel forms as well as new cladding materials that are corrosion resistant or that have improved creep strength.

Some of the innovative reactor concepts being considered, or simply discussed, would employ relatively high coolant outlet temperatures to increase energy conversion efficiency or to provide heat for high-temperature processes (e.g., certain hydrogen production techniques). Accommodation of higher coolant outlet temperatures would require fuel and cladding to operate at higher temperature. For fast reactor fuel, this is a challenge to cladding performance as plenum pressures increase due to fission gas release, inducing irradiation creep deformation. Fuel materials that retain configuration under irradiation at high temperature and cladding materials with improved creep resistance will be sought. Nitride fuels are being re-considered, particularly for high-temperature applications for which metal fuels seem ill suited.

Transmutation of plutonium and higher actinides is accomplished at faster rates in systems with little or no fertile material (i.e., with no U-238). This has generated interest in non-fertile fuels, published (United States Department of Energy, 1999). There is little experience with non-fertile fuels for fast reactor application, although some efforts are underway in Europe, published (Piccard, et al., 2000), Japan, and the U.S. Effort in the U.S. is specifically directed at fuel forms for the Accelerator Transmutation of Waste concept, which will consider Zr-matrix dispersion and zirconium-actinide nitride forms. Requirements for non-fertile fuel forms are typically the same as other fast reactor fuels, but with less expectation for high-burnup or high-temperature operation.

CONCLUSIONS

There is a great deal of uncertainty regarding the future utilization of nuclear power. A popular and expanding school of thought is that the share of the world's total energy from nuclear power will increase during this century and not only for the production of electricity. If this is the case, then a strategy must be defined that includes deployment of fast reactors by mid-century.

Beginning soon there should be a concerted world-wide effort to preserve the wealth of fast reactor information while the last nucleus of researchers is still professionally active. At the same time, on a selective basis that is commensurate with available funding, a cooperative international effort should be undertaken to complete evolutionary work on existing and promising fast reactor fuel systems. This evolutionary work will require the sharing of remaining test and examination facilities. Finally, there is time to explore truly innovative fuel systems that could bring advances in safety, economics, minimization of waste, and proliferation resistance. Emerging requirements for fuel systems other than for electricity generation may lead to the innovative development of high temperature fuel and coolant systems. However, it must be appreciated that the time period from concept to a licensed fuel system is on the order of 30 years.

REFERENCES

Charpin J., B. Dessus, and R. Pellat, (2000), *Étude économique prospective de la filière électrigue nucléaire*, Rapport au Premier ministre, Juillet.

GLOBAL '99, *Proceedings of the International Conference on Future Nuclear Systems*, August 29 - September 3, 1999, Jackson Hole, Wyoming.

Hannum W. H. (guest ed.) (1997), *The Technology of the Integral Fast Reactor and Its Associated Fuel Cycle*, Special Issue, Progress in Nuclear Energy, Vol. 31, No. 1/2, Pergamon Press.

Hori M. (2000), *Role of Nuclear Energy in the Long Term Global Energy Perspective*, NEA, First Information Exchange Meeting on Nuclear Production of Hydrogen, Paris, France, October.

Hydrogen as an Energy Carrier and its Production by Nuclear Power, IAEA-TECDOC-1085, May 1999.

Nuclear Materials, *Materials Science and Technology - A Comprehensive Treatment*, Volume 10A and 10B, VCH, 1994.

NUREG-1368, *Preapplication Safety Evaluation Report for the Power Reactor Innovative Small Module (PRISM) Liquid Metal Reactor*, Final Report, January 1994.

Piccard, E., Noirot, J., Moss, R. L. , Plitz, H., Richter, K., and Rouault, J., First In-pile Experimental Results of High-Plutonium-Content Oxide Fuel for Plutonium Burning in Fast Reactors," Nuclear Technology, Volume 129, January 2000, p. 1.

Porter D. L. (1994), *Fuel Test Plan*, Contributions by Argonne National Laboratory, General Electric, and Westinghouse Hanford Co., 23 June.

Proceedings of ICON 8, 8th International Conference on Nuclear Engineer, April 2-6, 2000, Baltimore, MD, USA.

The 33rd IAEA International Working Group Meeting on Fast Reactor, Recommendations, IWG-FR, May 2000.

United States Department of Energy, "A Roadmap for Developing Accelerator Transmutation of Waste (ATW) Technology – A Report to Congress," U.S. DOE Report No. DOE/RW-0519, October, 1999.

Wade D. C (2000), *Sustainable Non-Electric Applications of Nuclear Energy*, Presented at UNESCO International School of Science for Peace Forum on, Can Advanced Energy Systems Cope with Energy Demands of Asia, OECD-Pacific and CIS?, Como, Italy, 5-8 July.

Walters L. C., G. L. Hofman, T. H. Bauer, and D. C. Wade, *Metallic Fuel for Fast Reactors*, Nuclear Materials, Materials Science and Technology - A Comprehensive Treatment, Volume 10A, p. 315.

Workshop Proceedings, *Advanced Reactors with Innovative Fuels*, Villigen, Switerzerland, October 21-23, 1998.

Pergamon
www.elsevier.com/locate/pnucene

Progress in Nuclear Energy, Vol. 40, No. 3–4, pp. 523–526, 2002

Printed in Great Britain
0149-1970/02/$ - see front matter

PII: S0149-1970(02)00046-X

RECENT ACHIEVEMENTS IN THE DEVELOPMENT OF PARTITIONING PROCESSES OF MINOR ACTINIDES FROM NUCLEAR WASTES OBTAINED IN THE FRAME OF THE NEWPART EUROPEAN PROGRAMME (1996-1999)

CHARLES MADIC[1], MICHAEL J. HUDSON[2], JAN-OLOV LILJENZIN[3], JEAN-PAUL GLATZ[4], ROBERTO NANNICINI[5], ALESSANDRO FACCHINI[6], ZDENEK KOLARIK[7] AND REINHARDT ODOJ[8]

1. DEN, Bt 523, CEA/Saclay, 91191 Gif-s-Yvette, (France), 2. University of Reading, Reading (G.B.), 3. Chalmers University of Technology, Göteborg (Sweden), 4. Institut des Transuraniens, Karlsruhe (Germany), 5. ENEA, Ispra (Italy), 6. Politecnico di Milano, Milano (Italy), 7. INE/FZK, Karlsruhe (Germany), 8. ISE/FZJ, Jülich (Germany)

ABSTRACT

Partitioning of long-lived minor actinides (americium and curium) from the nuclear wastes issuing the reprocessing of nuclear spent fuels, in order to transmute them into short-lived nuclides or to condition them into stable crystalline matrices, was the subject of intense research within the NEWPART research program of the European 4th Frame Work Program, FWP (1996-1999). The target waste considered was the acidic raffinate (HAR) issuing the reprocessing of the used nuclear fuels by the PUREX process. A two step separation process based on liquid-liquid extraction was designed. The first step consists in the co-separation of the mixture of trivalent actinides and lanthanides from the HAR by extraction with a malonamide extractant (DIAMEX process), while the second step concerns the actinides(III)/lanthanides(III) group separation (SANEX process). Several DIAMEX and SANEX processes were developed and successfully tested with cold, spiked and genuine high active effluents. The research carried out also included basic and fundamental works in order to better understand the relationships between the structures of the extractants and their affinities for the target metal ions. The lecture highlighted both the basic and applied aspects of the research. This work will be pursued (PARTNEW program) within the 5th FWP of the European Union during the period 2000-2003.

1. INTRODUCTION

One of the main concerns related to the disposal in deep geological repository of the conditioned nuclear wastes generated after reprocessing of spent fuels is related to the presence within these wastes of long lived radionuclides (LLRNs), and particularly those of the so-called minor actinides, neptunium (Np), americium (Am) and curium (Cm). Within its 4th Framework program (FWP), covering the period 1996-1999, the European Union supported several research contracts in order to define processes for the Partitioning of the LLRNs from the wastes in view of their subsequent Transmutation (P-T scenario) or Conditioning (P-C scenario). NEWPART was one of these contracts aiming to define solvent extraction processes for Am and Cm partitioning.
The main achievements obtained in the course of this research are highlighted in this article.

2. RESULTS AND DISCUSSION

2.1 Strategy of the research

The main choices of the selected strategy are the following:

(i) among the LLRNs, those of Am and Cm were selected for partitioning,

(ii) the target waste to be treated consists in the high active raffinate (HAR) issuing the reprocessing of the spent fuels by the PUREX process,

(iii) no modification of the oxidation state of the target elements to be extracted will be done; consequently the chemistry concerns only the trivalent Am and Cm (An(III)) and Ln (Ln(III)),

(iv) most of the reagents to be used should fulfill the CHON principle.

Among the difficulties to solve one can mention:

(i) how to extract the An(III) from the acidic HAR?

(ii) how to separate the An(III) from the most abundant lanthanides(III) (Ln(III)) fission products from acidic feeds?

To overcome these problems one process including two steps was studied:

(i) the first one consists in the co-extraction of {An(III)+Ln(III)} from the acidic HAR by a diamide (malonamide) extractant (Fig. 1)

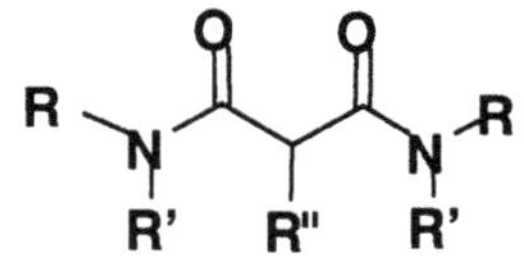

Fig.1. General formulae of a malonamide
(R, R' and R" are alkyl, aryl or oxyalkyl groups)

This corresponds to the so-called DIAMEX (DIAMide EXtraction) process,

(ii) the second process, named SANEX (Selective ActiNide EXtraction), performs the An(III)/Ln(III) group separation by the selective actinide extraction from rather acidic feeds.

Two ways were developed for the definition of SANEX processes, both based on the use of extractants bearing soft donor atoms either nitrogen of sulfur.

For both DIAMEX and SANEX research areas, fundamental studies and process developments were carried out.

2.2. DIAMEX process

Basic understanding of the extractive properties of the malonamide ligands was improved. This was done in particular by the determination of the structure of the solvates which are formed between the extractant and M(III) nitrates. In all cases, malonamides were found acting as bidendate ligands towards M(III) ions.

Process developments were carried out principally using the dimethyldibutyltetradecylmalonamide (DMDBTDMA) extractant. Several flow-sheets were developed and tested. In particular, a DIAMEX process flow-sheet (Fig. 2) was successfully tested at the ITU (Karlsruhe, Germany) in a centrifugal contactor test-loop using a genuine HAR.

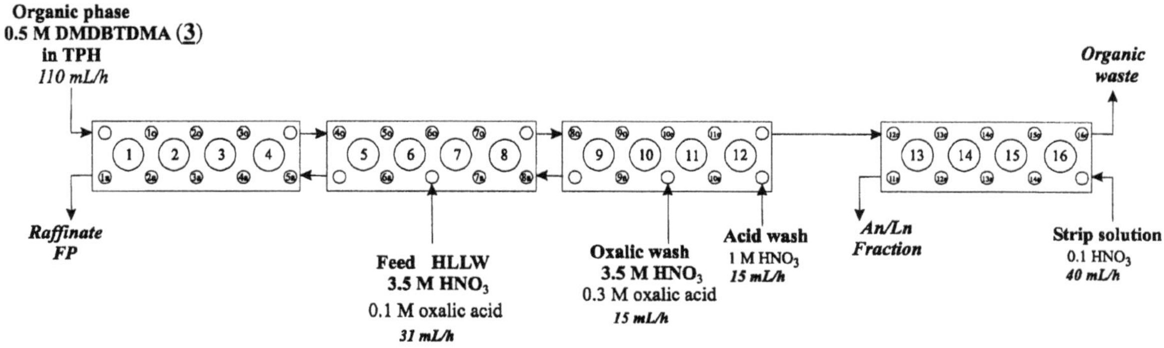

Fig. 2 Schematic drawing for the hot DIAMEX process test performed at ITU.

Later, a new diamide, the DMDOHEMA (*di*methyl*di*octylhexylethoxymalonamide), was chosen by the CEA to design an improved DIAMEX process.

2.3. SANEX processes

2.3.1 Nitrogen terdendate ligands. Several families of N-terdendate ligands were studied, but the most exciting results obtained so-far are certainly those related to a new family of ligands developed by Kolarik at FZK, the *bis*-5,6-substituted-*bis*-1,2,4-*tri*-azinyl-pyridines, the so-called BTPs, able to selectively extract the An(III) nitrates from the {An(III) + Ln(III)} mixture contained within 1 to 2 mol/L nitric acid feeds. Fig. 3 shows the formulae of these ligands.

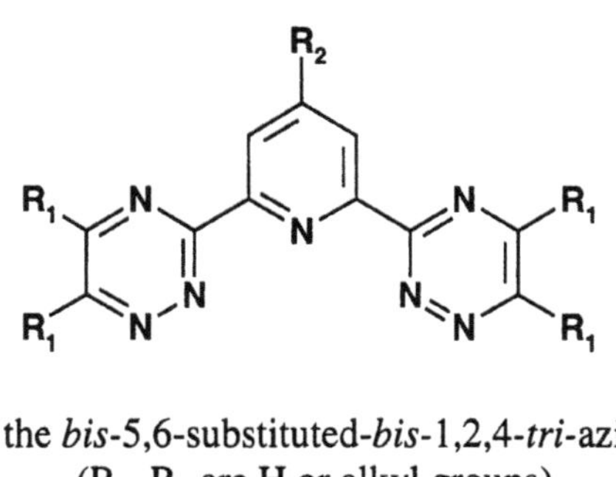

Fig. 3. Formulae of the *bis*-5,6-substituted-*bis*-1,2,4-*tri*-azinyl-pyridines, BTPs (R_1, R_2 are H or alkyl groups)

The crystal structure of a solvate formed between a Gd(III) ion and the *n*-ethyl-BTP (R_1 = ethyl, R_2 = H) ligand, with the following formulae, $Gd(BTP)_3(NO_3)_3$, has been determined at Reading. In this crystal, the Gd(III) ion is found bound to nine N atoms belonging to three BTP molecules whereas the nitrate ligands are

not bound to the Gd(III) ion. The large number of M(III)-N bonds within the extracted complexes is certainly at the origin of the high selectivity of these ligands towards the An(III) ions.
A SANEX process based on the use of the *n*-propyl-BTP was successfully designed and tested at Marcoule first, then at the ITU (Karlsruhe, Germany).

2.3.2 New acidic *di*-thiophosphinic acids. Following the discovery by Zhu in 1995 of the tremendous selectivity of the CYANEX-301, of which the "active principle" is a *di*-alkyl-*di*-thiophosphinic acid (R_2PSSH), for An(III) from mixture of {An(III)+Ln(III)}, new *bis*-substituted-*di*-thiophosphinic acids were developed by Modolo *et al.* at FZJ-Jülich (*c.f.* Fig. 4.)

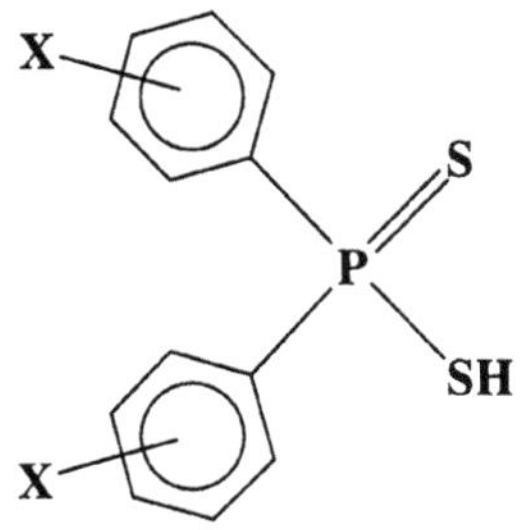

Fig. 4. Structure of the aromatic dithiophosphinic acids prepared by Modolo *et al.* at FZJ.

(X = H, methyl, Cl or F substituents)

Contrary to what was observed by Zhu with the CYANEX-301, it was found that these new acidic S-bearing ligands, in synergistic mixtures with neutral O-bearing ligands, are able to extract selectively the An(III) over the Ln(III) from highly acidic aqueous feeds. A process using a mixture of *bis*-chlorophenyl-*di*-thiophosphinic acid + *tri*-*n*-octylphosphine oxide (TOPO) was designed at FZJ and successfully tested with synthetic spiked solutions in a centrifugal contactor test loop.

3. CONCLUSIONS

The research carried out within the NEWPART program was highly successful: several DIAMEX and SANEX processes were developed and successfully tested in "hot" conditions for the partitioning of Am and Cm. About fifty papers were published in International Journals and Proceedings of International Conferences (see Madic *et al.*). The European collaboration in this field will be pursued within the 5th FWP of the European Union during the period 2000-2003.

4. REFERENCES

Madic C., Hudson M. J. , Liljenzin J-O., Glatz J-P., Nannicini R., Facchini A. , Kolarik Z., Odoj R. (2000), *New Partitioning Techniques for Minor Actinides*. EUR 19149 EN

Zhu Y., (1995), The Separation of Americium from Light Lanthanides by CYANEX-301 Extraction, *Radiochim. Act. 68, 95.*

Pergamon
www.elsevier.com/locate/pnucene

Progress in Nuclear Energy, Vol. 40, No. 3–4, pp. 527–538, 2002

Printed in Great Britain
0149-1970/02/$ - see front matter

PII: S0149-1970(02)00047-1

THE SEPARATION OF FISSION-PRODUCT RARE ELEMENTS TOWARD BRIDGING THE NUCLEAR AND SOFT ENERGY SYSTEMS

MASAKI OZAWA, YOSHIHIKO SHINODA, YUICHI SANO

Japan Nuclear Fuel Cycle Development Institute (JNC)
Ooarai-Engineering Center, Ooarai-machi, Narita 4002, Ibaraki, 311-1393, Japan
Tel:+81-29-267-4141 ext.6012, Fax:+81-29-266-3677
E-mail: ozawam@oec.jnc.go.jp

Based on the present state of the art of the separation technology, recycling of fission-product rare elements (FRE) in the FBR spent fuel is discussed. The rad.-waste fractionation is in accordance with the present society's trend toward zero-emission, and the mean of salt-free method utilizing electrochemistry agrees with the principles of the newly established *green chemistry*. A catalytic electrolytic extraction method is proposed to separate the target, radioactive but potentially strategic elements, Pd, Ru, Rh, Re (Tc), Te and Se dissolved in the HLLW. It avoids secondary waste arising. This method is particularly feasible for the separation of Pd where cyclic reaction of metal cations such as Pd(II) or Fe(II), acting as promoters or mediators and already contained in HLLW, accelerates the electrochemical deposition of Ru, Rh and Re. Elemental separation not only offers alternative material resources to meet expanding demands for catalysts in Fuel Cell / Soft Energy system but is also the first step for transmutation or other selective strategies for waste management of long-lived fission products (LLFP).

1. INTRODUCTION

An energy complex based on nuclear fission energy with some soft energy systems is a possible candidate among the various energy options in the next century, where the

Fast Breeder Reactor (FBR) serves as a powerful electricity supplier with a fuel (*e.g.*, hydrogen) generator, and the Fuel Cell (FC) as a secondary dispersed power generator. Because both are essentially sustainable and environmentally benign, the combination of FBR/FC would durably provide energy without causing carbon dioxide pollution. Thereby, allowing fossil fuels to be saved or furthermore permitting it to be used for non-combustion purposes. Recent technical advances in FC as well as FBR are dramatic and thus their feasibilities will be expected in the near future.

In the reactor spent fuel, significant amount of fission-product rare elements (FRE), such as platinum group metals (ruthenium (Ru), rhodium (Rh), palladium (Pd)), technetium (Tc), selenium (Se) and tellurium (Te), are contained, and those will likely be applied to the FC system as catalysts. If this is feasible, they will become the tertiary target substances to be recovered, subsequently to uranium (U) and transuranium elements (TRU).

As lowering the environmental burden is an indispensable requisite to realize the FBR cycle system, significant R&D efforts have been carried out to minimize rad.-waste itself as well as the release of radionuclides. The rad.-wastes minimization has to be investigated in, both, quantity and quality aspects; the quantity (bulk) of rad.-wastes would be reduced by the application of salt-free methods, and the quality can be improved by eliminating the origin of ultra long-lived radioactive-toxicity by partitioning and transmutation (P&T) methods of TRU (*i.e.*, α -free) / several long-lived fission products (LLFP) /1/Ozawa and Wakabayashi (1999).

Among the target FRE, Pd, Tc and Se are "mother" elements of long-lived ^{107}Pd, ^{99}Tc and ^{79}Se when their isotope separation is considered toward transmutation. Otherwise, if dissolved Ru and Tc can be removed from the upper PUREX stream, this will contribute to improve the decontamination performances of the PUREX process and the successive TRUEX process because ^{106}Ru and ^{99}Tc are major DF-dominant nuclides.

2. FISSION-PRODUCT RARE ELEMENTS IN THE FBR SPENT FUEL

The quantities of the target FRE are relatively rich among the ca.40 elements with ca.400 nuclides in the spent fuel. Their amounts calculated by ORIGEN II are shown in the Table 1. In the case of FBR spent fuel burned to 150,000MWd/t, the total amount of FRE becomes more than 30 kg/HMt.

Table 1 Typical Amount of Fission-product Rare Elements in Spent Fuel

FRE	Ru	Rh	Pd	Tc	Te	Se	Note
Concentration (kg/t(HM))	12.5	3.6	11.1	3.3	2.7	0.2	SF-FBR;150,000MWd/t, Cooled 4 years.

When comparing their amounts in the FBR spent fuel (α) with those in earth crust (β), concentration ratio (α /β) is particularly high for Tc to Pd and Te in the 5th series elements and Se in the 4th series, as shown in Fig. 1.

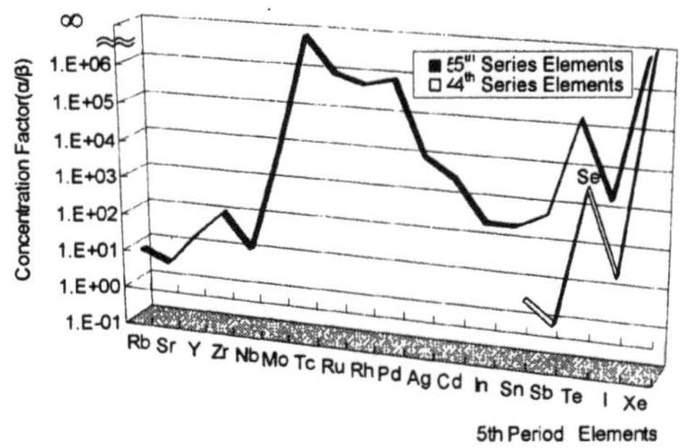

Fig. 1 **5th and 4th (Se) Period Elements in FBR Spent Fuel (α: Elements in the FBR-S.F. (g/t), β: Elements in the Earth Crust (g/t))**

Laboratory dissolution data on FRE obtained at the CPF of JNC in boiling 4M nitric acid are shown in the Table 2. It is likely that in the case of dissolving FR spent fuel by the routine dissolution mode, FRE (including Se and Te) would have a higher solubility (> *ca*.80%), whilst the dissolution ratio sometimes changes with the difference of fuel characteristics and dissolution conditions.

Table 2 Typical Solubility of Fission-product Rare Elements in Spent Fuel

FRE	Ru	Rh	Pd	Tc	Note
Dissolution ratio (%)	78-86	80-91	91-94	77-90	SF-FR ; 54,100 & 94,000MWd/t, Cooled 2.0 & 4.1 years.

As for Ru however, it must be noted that a considerable portion might be redistributed to the off-gas stream. Dissolved fractions of them were discharged to the highly active raffinate and finally gathered to the HLLW in the PUREX process.

The elemental concentration in the HLLW is shown in Fig. 2. Analyzed target concentrations of FRE were almost above

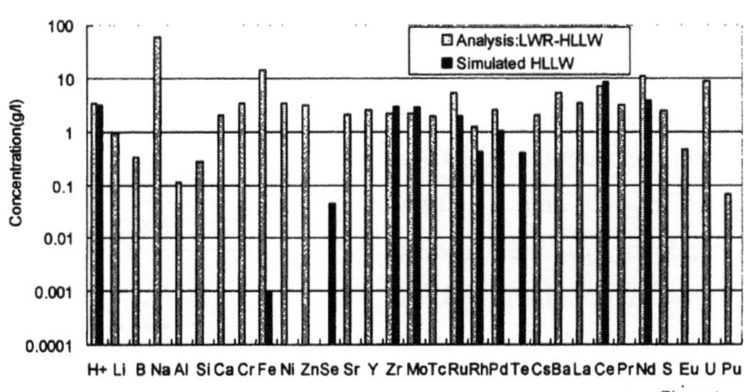

Fig. 2 Typical Component Composition of HLLW

(H+ : mol/l)

1g/l. The high concentration of sodium, due to the salt-process in the conventional PUREX steps of LWR-reprocessing, will disturb ultimate concentration of HLLW. The higher concentration of iron may, however, cause positive effect on the recovery of some of FRE during the electrolytic extraction process.

3. RADIOACTIVE-TOXICITY OF FISSION-PRODUCT RARE ELEMENTS

In the typical FBR spent fuel cooled for 4 years, the abundances of radionuclides of each FRE are rather small; *i.e.*, Ru : 0.3%, Rh : 0.004%, Pd : 17.2%, Te : 0.04% and Se : 13.9%, respectively. The time dependency of their specific radioactivity (Bq/g) is shown in Fig. 3. It shows that those of Tc, Se, Pd and Te are constant due to the ultra-long half-life isotopes. Unless isotope separation is realized, their (except for Te) specific activities are higher than any assumed "clearance" levels from 10^{-1} to 10^{3} Bq/g (*e.g.*, Japan domestic legal index, 74 Bq/g). Particularly for Tc, its specific radio activity is the highest and constant after 10 years cooling due to the β^- emitter ^{99}Tc (293kev). Nevertheless, the specific radio activities of Ru and Rh decrease sufficiently by awaiting several decades; namely, the specific activities of Ru and Rh will become far less than 10^{-1} Bq/g within 50 and 90 years, respectively.

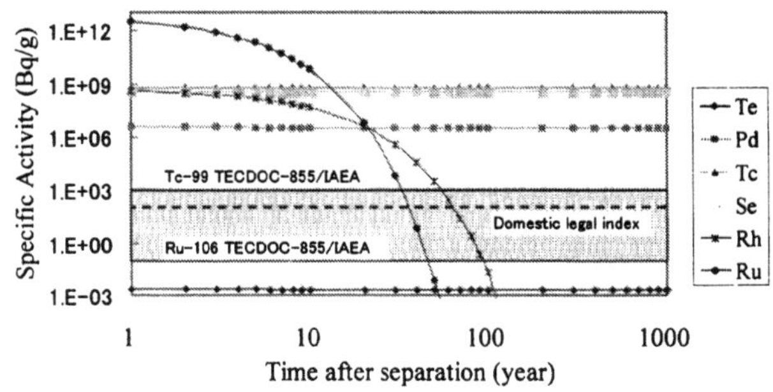

Fig. 3 Time Dependency of Specific Activities of Fission-product Rare Elements

The specific radioactive toxicity is obtained by dividing the specific activity by the isotope's annual intake (ingestion) limitation for the workers recommended by the ICRP/2/. The time dependency of the specific radioactive toxicity of FRE is shown in Fig. 4. The specific radioactive toxicity of Pd (and Te) is constant and

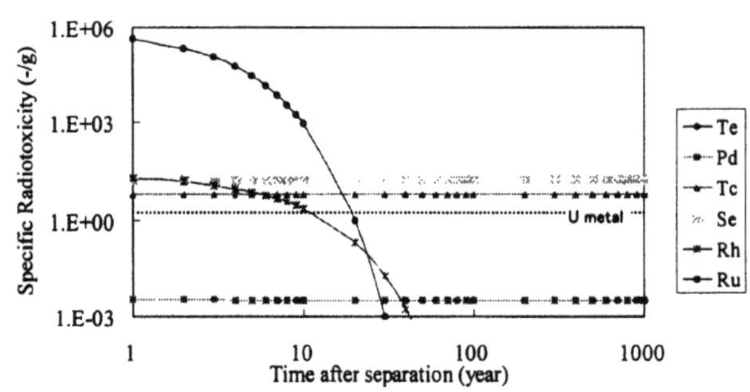

Fig. 4 Time Dependency of Specific Radioactive Toxicities of Fission-product Rare Elements

significantly low, while those of Se and Tc are also constant but close to that of natural U. After several decades, the initial high radio toxicity of Ru (and Rh) will become less than the lowest level of Pd.

Yet any index on specific radioactive toxicity is not given, fission-product platinum group elements seem to become easy to handle after several decades even when some mutual contamination among them is unavoidable in the electrochemical separation process. However, comparatively light shielding or isolation is necessary for Se and Tc against long-lived β^- emission.

4. SEPARATION OF FISSION-PRODUCT RARE ELEMENTS

4.1 Salt-free as an approach of *Nuclear Green Chemistry*

In considering the radio chemical separation process, minimization and qualification of wastes are key requisites to preserve an environmentally benign nuclear fuel cycle. In this purpose, the salt-free method, which has been applied to the core part of the FBR Purex process/3/ Ozawa and Hirano *et al* (1995), meets well with the concept of *Green chemistry* /4/Anastas and Warner (1998). The green chemistry is a new leading concept, where "molecular design" and "atom economy" will be dominant ideas replacing the classical "yield" or "cost" when reconsidering chemicals and processing procedures not to cause any negative environmental effect.

The salt-free process used to apply *in situ* electrochemistry as a reaction tool and/or waste-reagent destruction tool throughout the chemical processing. The elimination or decomposition of secondary wastes with keeping higher energy efficiency and safety was common sense in both salt-free and *green chemistry.* In the salt-free methods, plutonium or silver was used as an electrochemical catalyst; *i.e.,* redox pair of Pu(III)/(IV) as a ***mediator*** in electrochemical oxidation and decomposition of Pu(III)/HAN/Hdz in nitric acid media/3/ and Ag(I)/(II) in electrochemical decomposition of organic rad.-wastes composed of OφD[iB]CMPO/TBP/n-dodecane. Furthermore, Pd_{adatom} is expected to act as a ***promoter*** in electrolytic extraction of FRE from HLLW(*i.e.,* catalytic electrolytic extraction)/5/ Ozawa and Sano, *et.al* (2000).

4.2 Catalytic Electrolytic Extraction method

Electrolytic extraction is generally carried out under lower temperature providing a remotely controlled system by utilizing electrons to get target redox reactions. It essentially does not generate any secondary saline waste at all. In this sense, the electrolytic extraction process is a simple and peerless method for recovering FRE whose redox potentials are nobler in acidic media.

The electrochemical behavior of some FRE has been investigated in nitric acid solution and simulated high level liquid waste (sHLLW ; refer to Fig. 2) /5/,/6/ Sano and Ozawa *et al* (1998). These studies proved that the deposition ratio of FRE on the cathode generally increased with decreasing of H^+ concentration in the catholyte. And more, those of Ru, Rh and Re (Tc simulator) were significantly improved by the addition of Pd(II) or Fe(II) cations, which might be caused by co-deposition of Ru with Pd (as Pd_{adatom}) or the interaction (complexing or redox reaction) between those cations and ReO_4^- ions in the bulk of the solution/7/Funasaka and Sano, *et al* (2000). This suggested that higher recoveries of those elements became possible even from a higher acidic solution as HLLW or dissolver solution. The time dependencies of the deposition ratio for some FRE in sHLLW ($H^+_{int.}$:3.0M) during 6h electrolysis are shown in Fig. 5. Electrolysis conditions and apparatus were described elsewhere /5/. The deposition ratio of each metal was determined by the decrease in its ionic concentration in the solution with the ICP measurements. With this experimental condition, the H^+ concentration of sHLLW in the catholyte decreased along with the generation of H_2 gas on the cathode, and reached about 1M at the end of the electrolysis. The deposition behavior of Pd, Te, Se and Rh seemed to be independent of the H^+ concentration, while it greatly affected those of Mo and Zr; although the deposition yields of Mo and Zr were very low at higher H^+ concentration, their recovery yield began to increase when H^+ was less than 2M. The white precipitate was observed in this condition. This result implies that Mo and Zr do not deposit on the cathode, but precipitate as some insoluble complexes at lower H^+ concentration.

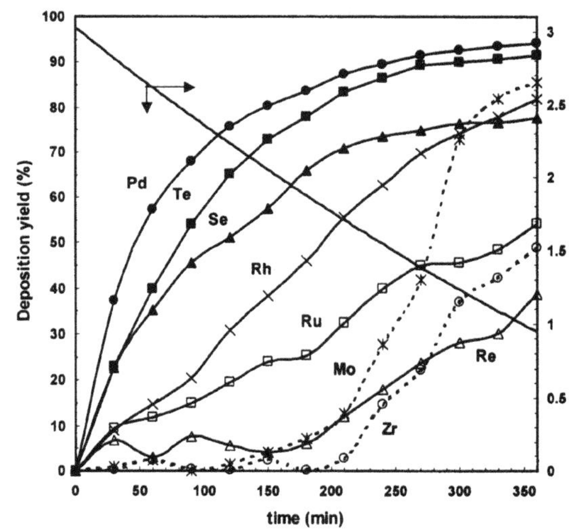

Fig. 5 Time Dependency of Deposition Yields on Fission-product Rare Elements by Electrolytic Extraction in simulated HLLW (C.D.: 500mA/cm^2, 6hr)

Figure 6 shows the deposition yields of FRE in 2.5M HNO_3 sHLLW during 6hr electrolysis under the condition where Pd(II) or Fe(II) ion was added. Adding mode of

Pd(II) ion was continuous or batch-wise, and the rate was *ca.* 0.04g/h for the continuous case. The deposition yields of Pd, Te and Se were constantly high (>*ca.*70%), which were independent of the addition of Pd(II) or Fe(II) ions. Apparent yields of Mo and Zr were *ca.*65% and 35% respectively which were due to the precipitation at decreased H^+ concentrations. On the other hand, the efficiency of the addition of Pd(II) or Fe(II) ion in the deposition of Ru, Rh and Re was obvious; the Ru deposition ratio was particularly accelerated *ca.* 1.6 times.

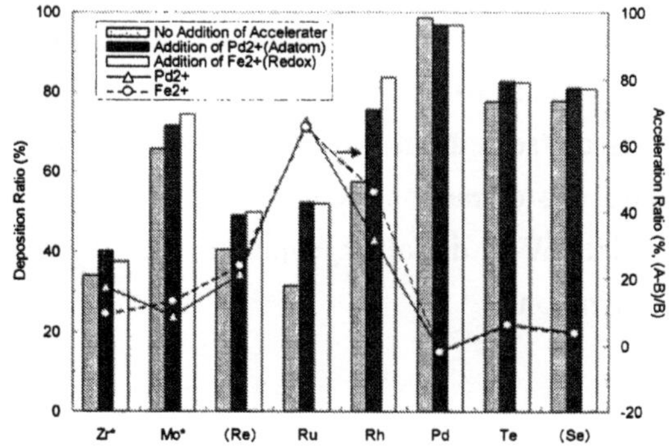

Fig.6 Recovery of Fission-product Rare Elements in simulated HLLW by Catalytic Electrolytic Extraction - Effects of the Addition of Pd^{2+} and Fe^{2+}-
A: Addition of Accelerator B: No Addition

Based on the present knowledge, the flowchart of the separation of FRE from the HLLW is designed as shown in Fig.7. The 1st electrolytic extraction step (EE(I)) is carried out for HLLW (H^+ ~ 2.5M) under the continuous addition of Pd(II). The target elements Ru, Rh, Pd, Ag, Se, Te and Tc will be obtained as a mixed deposition on the cathode. If one wants, Zr and Mo can also be separated as precipitates during lower acidity condition (H^+<1.5M). Most of FPs are removed in this step. Deposited elements on the cathode are anodically re-dissolved into HNO_3 solution for the next EE steps. In the 2nd EE(II) step, some nobler elements, such as Pd and Ag, will be readily recovered on the cathode even with low current density (CD). Selective recovery of Pd would be possible by just control of CD. The recovered Pd shall be supplied to the EE(1) step as an accelerator. Se and Te can be separated in the 3rd EE(III)

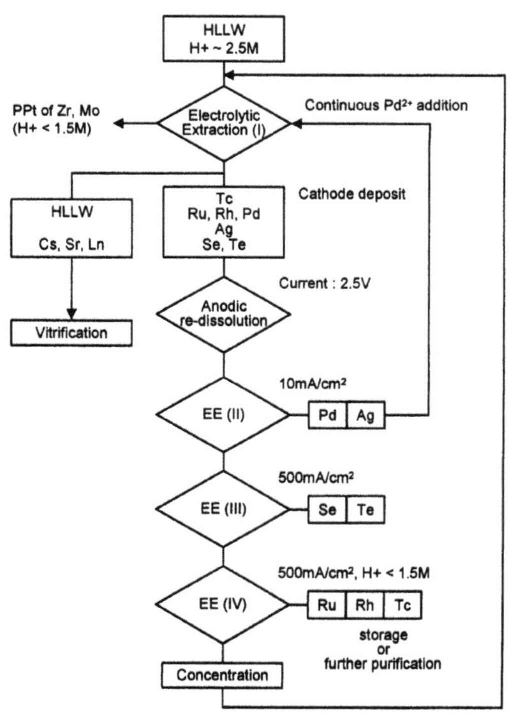

Fig. 7 Separation Flowchart for Fission-product Rare Elements by Catalytic Electrolytic Extraction from HLLW

step with higher CD. This step will be accompanied with a significant decrease of H^+ concentration by the reduction of H^+ to H_2 gas. In the last 4th EE(IV), the remainder of Ru, Rh and Tc will be recovered under high CD with low H^+ concentration.

As the separation flowchart is based on the current state of the art of the catalytic electrolytic extraction method by using simulated HLLW, extended experiments with real HLLW / dissolver solution will be required for further improvements and optimization.

5. POTENTIAL APPLICATION OF SEPARATED FISSION-PRODUCT RARE ELEMENTS TO SOFT ENERGY SYSTEMS

The platinum group metals are well known to have an extraordinary chemical property, and thus their industrial application has been versatile as such for auto catalyst (Pd,Rh), chemical (Pd,Rh), electrical (Pd,Rh,Ru), electrochemical (Ru), dental (Pd), jewelry (Pd) and glass (Rh), etc. The modern application fields for Pd, Ru, Tc, Se and Te are searched in particular with focusing on the frontier science / technology, the soft energy field (Appendix).

Tc is extremely rare in the nature and a weak β^- emitter owing to ^{99}Tc. Therefore its properties are not fully researched except for the case of ^{99m}Tc which is very short-lived γ- active and used in the medical science. However, the applicability of Tc can be foreseen by its similar property to rhenium. Little information on Te and Se is available in this field except for their promising application to the semiconductor industry.

On the contrary, very wide and promising application is foreseen for Ru and Pd as an electrode catalyst for the FC. Particularly, Ru, dispersed on a carrier such as Al_2O_3 or TiO_2, is an efficient catalyst for the hydrogen production from methane or butane by the partial combustion process with air (1) or reforming with steam (2) :

$$C_4H_{10} + 2O_2 \rightarrow 4CO + 5H_2 \quad (1)$$

$$C_4H_{10} + 4H_2O \rightarrow 4CO + 9H_2 \quad (2)$$

For the reaction (1), the Ru catalyst can suppress the subsidiary reaction (3);

$$C_4H_{10} + 6.5O_2 \rightarrow 4CO_2 + 5H_2O \quad (3)$$

Thereby, the conversion ratio to hydrogen would reach 80% at the condition of butane/air=1/10 and temp: 750~800 °C/8/Kendall and Williams(1998). The Pt-Ru catalyst is also promising as anode material by anti-CO poisoning for hydrogen fuel/9/Cameron(1997), and as a photochemical catalyst for the solar batteries in some Ru complex forms. Among the platinum group metals, the highest demand was recorded for Pd (*e.g.*, <u>*ca*</u>.254 t in 1998) with a showing sharp increase since around

1994 for the auto catalyst. Due to the sensitive property to hydrogen, the demand for Pd will further increase for the purification purpose of hydrogen.

With the expanding needs as FC catalyst, demands on Pd, Ru and Rh are estimated to be increased 5-10 fold in near future/10/Tsurumi (1999). Accordingly, the strategic elements in spent nuclear fuel should cover a worthwhile part of industrial demand.

6. CONCLUDING SUMMARY

In conclusion, a total system concept is proposed that is based on the FBR / FC energy complex bridged by the recycling of fission-product rare elements (FRE) from the spent fuel, as shown in Fig.8.

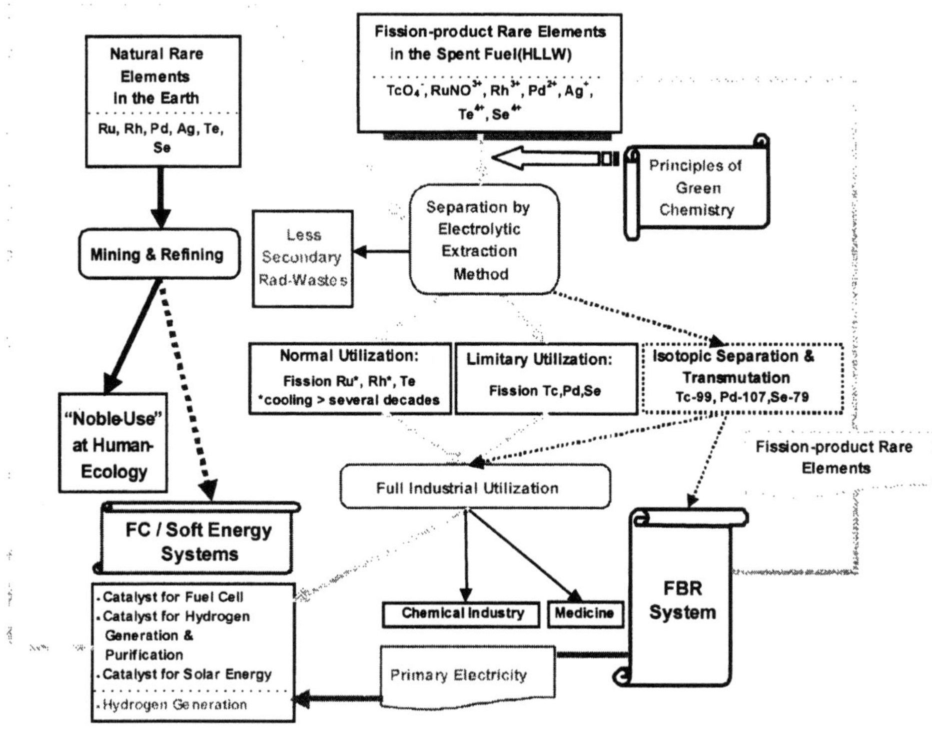

Fig. 8 Overall Concept on FBR/FC Energy Complex centered on Fission-product Rare Elements

Enhanced radioactive waste fractionation, consistent with nuclear green chemistry, is in accordance with the general trend of progress in partitioning and recycling of the wastes toward zero-emission. In this scheme, separated FRE should be utilized according to their radiochemical characteristics while allowing natural rare elements (NRE) to be preferentially conserved or used where environmentally or economically more attractive.

Elemental separation of FRE from the HLLW (Spent Fuel) not only offers alternative resources to meet widespread and increasing demands in FC/chemical industry/medical science but is also a first step for transmutation or other selective strategies for waste management of LLFP.

This paper suggests recovering and utilizing, in a novel manner, the FRP generated by nuclear fission. The *in situ* , catalytic electrolytic extraction is a feasible and promising method for this purpose.

References

/1/ Ozawa M. and Wakabayashi T.(1999), Status on Nuclear Waste Separation and Transmutation Technologies in JNC, *International Conference on Future Nuclear Systems(Global'99)*, Snow King Resort, Jackson Hole, Wyoming, Aug.29-Sept.3.

/2/ Limits for Intakes of Radio Nuclides by Workers (ICRP Publication 30 Part-1), Annals of the ICRP, Vol2, No.3-4,116(1980)

/3/ Ozawa M., Hirano H., Koma Y., Tanaka Y., Kawata.T. (1995), Enhancing Actinides Separation by Consolidated PUREX and TRUEX Processes Intensified by Salt-Free Requisite, *International Conference on Evaluation of Emerging Nuclear Fuel Cycle Systems(Global'95)*, Proc. Vol.1, pp585-594, 1995, Palais des Congres, Versailles, Sept. 11-14.

/4/ Anastas P.T. and Warner J.C.(1998), *GREEN CHEMISTRY*, Oxford University Press.

/5/ Ozawa M., Sano Y., Ohara C., Kishi T.(2000), Radioactive Wastes Minimization by Electrolytic Extraction and Destruction in a PUREX-TRUEX Actinide Separation System, *Nuclear Technology*, May 2000, Vol.130,No.2, pp196-205.

/6/ Sano Y., Ozawa M., Tanaka Y., Kato A.(1998), *The 5th International Nuclear Conference on Recycling, Conditioning and Disposal (RECOD 98)*, Nice, France, October 25-28 1998, SFEN/ENS

/7/ Funasaka H., Sano Y., Nomura K., Koma Y., Koyama T.(2000), *ATALANTE 2000*, Avignon, France, October 24-26 2000, CEA

/8/ Kendall K. and Williams D.S.(1998), Catalysts for Butane Reforming in Zirconia Fuel Cells, *Platinum Metals Review*, 42,(4),164-167.

/9/ Cameron D.C.(1997), *The Fifth Grove Fuel Cell Symposium*, *ibid.* ,41,(4),171-175.

/10/ Tsurumi K.(1999), The Circumstances of Precious Metal Catalysts Concerned with Fuel Cell System (*in Japanese*), *Catalyst*, 41,(7),547-550

Appendix ; Literature Survey on Potential Application of Separated Fission-product Rare Elements to Frontier Science & Technology

Elements/ Application	Fuel Cell (FC)				Chemical industry/ Other energy system	Medicine	Others
	Anode(Hydrogen)	Cathode (oxygen)	Fuel	Cell Structure			
Tc	N.A.(No information Available)	N.A.	N.A.	N.A.	・W/25%Tc ; Improve tungsten ductility (at Hanford /NASA) ・Improve corrosion resistance of mild steel ・Improve radiation-assisted stress corrosion cracking ・Jet engine alloys (W/Tc with detector for oxidation or wearing)	・Alloying with Ti,SS,NM ; potenial utilization of a light β − radiation field (*e.g.*, avoid plaque deposition on arterial wall)	N.A.
Ru	・Pt/Ru alloy for anti-CO poisoning for PEM (Proton Exchange Membrane) FC ・Pt/Ru-WO_3,Pt/Ru-$H_4SiW_{12}O_{40}$ for CH_3OH oxidation ・Pt(44)/Ru(41)/Os (10) / Ir(5) for DM (Direct Methanol) FC ・Pt/Ru (Pt/Ru=1/3) for PE(Polymer Electrolyte)FC	・Pt/Ru for PEFC	・Ru-Al_2O_3 ;Catalyst for Butane reforming : ①Partial Combustion Process for Zirconia FC or ②Steam Reforming Process (at LANL) ・Ru-TiO_2 (CH_4 reforming)	N.A.	・TiO_2/Ru complex/I ; Pigment-sensitized complex for solar battery ・[Ru(bipyridine)$_3$]Cl_2 ; Photocatalyst for H_2 generation ・$RuCl_3 \cdot nH_2O$; Active oxo-ruthenium for synthesis of alcohol and ketone ・Passive anode for generation of Cl_2 or O_2	・$RuCl_3$ ・Ruthenium Red ($Cl_6H_{42}N_{14}O_2$ Ru_3) ; Macro-cluster	・Sr_2RuO_4 system ; High temp. superconduction material
Pd	・Combustion catalyst for H_2 in anode off-gas	N.A.	・Pd-SiO_2/Si thin film ; Optical sensor to H_2 ・Pd-Ag thin film ; membrane material for separation of H_2 ・Purification material for high quality H_2	・Urtra thin Pd film ; prevent crossover of CH_3OH for DMFC ・Pd-SUS304 ; Corrosion-resistant separator for pure H_2 ・Pd-dispersed polymer electrolyte membrane	・Synthesis catalyst ; Carbonylation reactions for fine chemicals ・Ti/0.15%Pd alloy ; Anti-crevice corrosion material for heat exchamger for distillation of sea water ・Pd /TiO_2 macro-cluster	・Synthesis catalyst ; Carbonylation reactions for medical intermediate product (*e.g.*, substituted Succinic acid)	・Solid nuclear reaction
Te & Se	N.A.	・$(Ru_{1-x}Mo_x)_y SeO_z$ $0.02<X<0.04$, $1<y<3$, $Z \fallingdotseq 2y$; Semiconductive transient metal alloy for oxygen reduction in DMFC	N.A.	N.A.	・Cd/Te ; Semiconductive material for solar battery	N.A.	N.A.

Pergamon
www.elsevier.com/locate/pnucene

Progress in Nuclear Energy, Vol. 40, No. 3–4, pp. 539–546, 2002

Printed in Great Britain
0149-1970/02/$ - see front matter

PII: S0149-1970(02)00048-3

TRANSMUTATION OF MINOR ACTINIDES AND INNOVATIVE FUEL CYCLE CONCEPTS

T. OGAWA

Japan Atomic Energy Research Institute
Tokai-mura, Naka-gun, Ibaraki-ken, 319-1195 Japan

Both physical and chemical properties of actinides show significant variations depending on the element from uranium to curium. The materials database on minor actinides (MA:Np, Am, Cm) is very limited. Even with the scanty database, however, americium and curium did not appear to make viable fuels in the ordinary sense. The solutions may be (1) to tolerate a significantly lower fuel performance whose penalty has to be recovered by an improved overall economy of the MA-fuel cycle, and/or (2) to make a fuel concept whose performance is less restricted by physical and chemical properties of MA. In such a difficult program like actinide burning, the technology components would have to be combined by a modular approach. The system has to be integrated as a whole, but a technological module in the system has to be made as such that can be replaced by the other modules. This type of approach would be only possible if we have a technology which forms a common basis to these "modules". The advantage of taking the modular approach is to allow the system to evolve with time. The pyrochemical separation can form such a common basis.

1. INTRODUCTION

Both physical and chemical properties of actinides show significant variations from uranium to curium. In JAERI, the conceptual study of actinide burners began in 1987. The first task of a fuel engineer was to predict the behavior of minor actinides (MA: Np, Am and Cm) when incorporated into a fuel as major constituents. Even with the scanty database, americium and curium did not appear to make viable fuels in the ordinary sense. It did not seem to make sense to wish a fuel composed of these elements to have performance similar to the normal uranium-based fuels. The solutions may be (1) to tolerate a significantly lower fuel performance whose penalty has to be recovered by an improved overall economy of the MA-fuel cycle, and/or (2) to make a fuel concept whose performance is less restricted by physical and chemical properties of MA. The consideration (1) led to the selection of pyrochemical separation for the MA fuel cycle; (2), the adoption of either a soft- or a cold-fuel concept. The cold fuel reduces fission gas release and swelling; the soft fuel like metallic fuels in fast reactors allows fission gas release and alleviates the fuel-clad mechanical interaction accompanying swelling. The soft fuel may be metal and the cold fuel may be nitride, both of which appear to be processed on a common pyrochemical separation technique. Since the mutual solubility among actinides as metals is limited, we eventually adopted the nitride as the first candidate for the MA burner, while continuing the basic research on MA alloys.

2. MINOR ACTINIDES FUELS/TARGETS

Table 1 Some fundamental properties of actinides in spent fuels

Element	mp (K)	ΔH_f^o (kJ/mol)	I_1 (eV)	I_2 (eV)	I_3 (eV)	I_4 (eV)	r_m (nm)	$r_i(3+)$ (nm)	λ (W/m·K)
U	1408	536	6.194	11.9	19.7	32.6	1.542	1.025	25
Np	913	465	6.266	11.7	20.7	33.6	1.503	1.01	6~8
Pu	913	342	6.062	11.7	21.6	34.6	1.523	1.00	5~8
Am	1449	284	5.993	12.0	22.1	36.2	1.730	0.98	-
Cm	1618	387	6.021	12.4	21.0	36.8	1.743	0.97	-

mp : melting point

ΔH_f^o : enthalpy of vaporization

I_n : *n*th ionization energy

r_m : metallic radius (RT,CN=12)

$r_i(3+)$: effective radius of trivalent cation (CN=6)

λ : thermal conductivity (metal, RT)

2.1 Behavior of actinide alloys and compounds as nuclear fuels

Table 1 summarizes the fundamental properties of actinides normally found in spent fuels. Peculiarity of Am is noticed: (1) lower enthalpy of vaporization, (2) relative stability of M^{2+} state compared with M^{3+}, and M^{3+} compared with M^{4+} and (3) the jump in the metallic radius at passing from Pu to Am in contrast to the constant contraction of M^{3+} ionic radius along the actinide series. The point (1) results in the vaporization loss during fabrication; (2) the instability of cubic AmO_2 compared with less-isotropic and lower-melting Am_2O_3, and a relative stability of Am^{2+} in molten salts at lower potentials; (3) together with (1), limited mutual solubility in alloys with U and Np.

When one predicts the alloying behavior of metallic elements, one would adopt various approaches. If one has sufficient information on the thermodynamic properties of the system, one can directly calculate the phase diagram of the alloy system. An approach, which is often called the CALPHAD (CALculation of PHAse Diagrams) approach, is to find the set of free energy equations by combining experimental data on the phase diagram and the thermodynamic properties. Then one would obtain the most plausible phase diagram and thermodynamic database for the given system. This type of approach has been successfully applied to some uranium and plutonium alloy systems.

However, the experimental information on MA alloys is too limited to adopt the above approach. In this regard, the metallurgy of MA has yet to be developed. We have made a joint study with ORNL on the alloying behavior of a few Np and Am binary systems. One of the findings was that the mutual solubility between lighter actinides (U,Np) and heavier ones (Am and probably Cm) are rather limited [Gibson, Haire et al., 1994]. A Russian study on Np and Cm alloys reached the same conclusion [Shushakov, 1993]. In JAERI, the R&D was focused on the dedicated burners, where the major constituent of the cores should be transuraniums themselves with uranium content as little as possible. The limited mutual solubility among actinides as metals means that the strong peculiarities in physical and chemical properties of actinide elements persist in transuranium-rich alloys, although we wish to tame them by taking an average by alloying. Lower melting points of Np and Pu and the higher vapor pressure of Am may be still prominent in the transuranium-rich alloys. In the initial concept of MA burner core, the two parallel types of alloys, Np-Zr and (Am,Cm)-Y, had to be tentatively assumed [Ogawa et al., 1989]. Those initial survey highlighted the superiority of uranium-based alloys compared with the transuranium alloys: particularly, thermal properties of transuranium elements as metals may be far inferior to those of uranium.

From the beginning of our conceptual study, we did not consider MA oxides as the candidate fuel, due to the design targets of realizing harder neutron spectrum than the oxide-based cores. Therefore, the alternatives were narrowed to the carbides and nitrides. As far as the (U, Pu) fuels are concerned, the irradiation database was larger on the carbide than the nitride. However, the limited data showed that the thermodynamic stability of the carbides of Am and Cm is much less than the carbides of U and Pu.

Therefore, the possibility of using the nitride fuels was examined. The comparison of the data on UN, PuN and lanthanide mononitrides (LnN) suggested that, at least, we would expect that the thermodynamic stability of AmN and CmN is comparable to those of UN and PuN, although AmN and CmN like LnN would be more ionic and hygroscopic in nature. We also expected a good mutual miscibility of these mononitrides, which may give greater flexibility to the designers of dedicated burner core, where the isotope vector is strongly dependent on the past histories of the actinides.

2.2 Direction of fuel research

In JAERI the (U, Pu)N fuels program started around 1986 under the leadership of Muneo Handa [JAERI, 1988], which later grew into the irradiation program in the experimental fast reactor Joyo jointly with JNC. Thermophysical and chemical data on (U, Pu)N as well as NpN have been obtained [Suzuki and Arai, 1998]. There have been little unexpected things as far as these nitrides are concerned. The UN, PuN and NpN powders by carbothermic synthesis, their mixtures, and pellets have been successfully prepared.

With AmN and CmN, the situation is a little different. Even today, we have a very limited capability for the research of these transplutonium elements, where shielded facilities are required even when dealing a tiny amount. It has been demonstrated that AmN and CmN can be also prepared by carbothermic method. One important issue was to prove the miscibility between the mononitrides of lighter actinides (U-Pu) and heavier ones (Am, Cm). The miscibility among the actinide mononitrides has been recently demonstrated in the fabrication of (Cm, Pu)N [Takano et al., 2000]. The $(Cm_{0.4}Pu_{0.6})O_2$ was successfully converted to the mixed mononitrides by carbothermic synthesis.

Beyond the preparation of nitride powders, little has been achieved on Am and Cm. However, the experimental capability is gradually expanded. Now the light-element impurities, carbon and oxygen, which would strongly affect the property of AmN and CmN, are measured. By the end of FY 2001, the behavior of helium in these nitrides can be also studied. In NUCEF, Tokai Establishment, the high-temperature chemistry cells on transplutonium elements are to be built by the end of FY2001, for a joint program with the Japan Atomic Power Company. In the new cells, the handling and basic measurements on the transplutonium materials can be made in purified argon atmosphere.

We would have to study the feasibility of (Am, Cm) fuels containing little Np if the latter element is recycled in power reactors. The problem of higher vapor pressure of Am would become more of a problem in such fuels. High-temperature vaporization behavior of AmN is being studied in ORNL.

3. DOUBLE STRATA AND MODULAR APPROACH

Selection of fuel types is certainly an important issue in R&D of MA transmutation. However, no fuels would be ideally good. While aiming at the realization of one type of fuels and one type of transmutation devices, we would keep open the possibility of the alternative technologies. With limited funding, this requires a common base technology as well as the good international coordination.

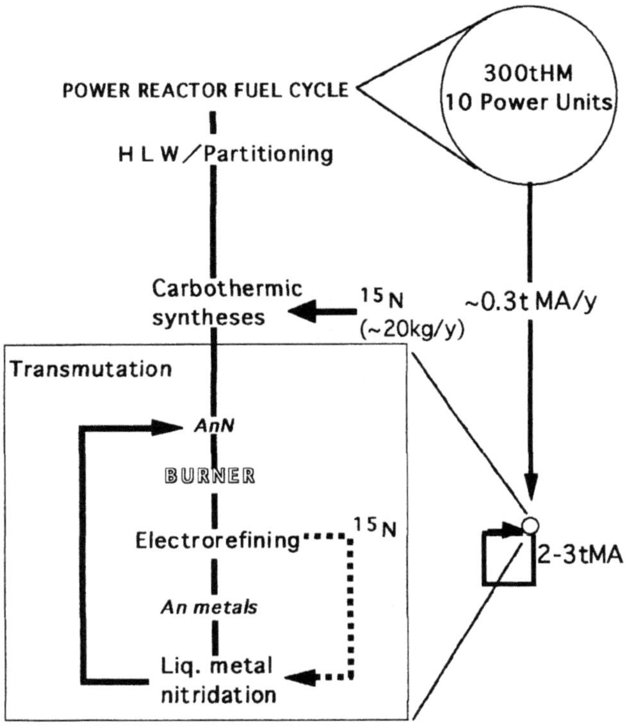

Fig. 1 Double strata fuel cycle and nitride/pyroprocess concept.

The P-T program in JAERI is based on the double-strata fuel cycle concept [Murata and Mukaiyama, 1984]. The concept gives a greater freedom in designing the whole system. We do not have to optimize a reactor system for the dual aims of realizing economically competitive power generation and the efficient MA burning, neither of which is an easy task. For instance, it would not be very easy to introduce the nitride fuels into the power reactors, considering efforts which may be required in installing the mass production of ^{15}N. With the double strata, the required plant scale for ^{15}N enrichment is much less (fig. 1), and not so widely apart from the current production capacity[+]. Idea of using ^{15}N is not so wild as using MA anyway. Particularly so, when one may recycle highly-enriched ^{15}N. There is a

[+] Shoko Co. Ltd., Tokyo, has a production capacity of 5kg/y.

reasonable possibility of recycled use of ^{15}N by applying the pyroprocess to the treatment of irradiated MA nitrides. Unlike dissolving the fuels in nitric acid in an aqueous process, ^{15}N can be recovered as nitrogen gas by treating them in molten chloride salts [Kobayashi et al., 1999].

3.1 Treatment of irrdiated MA - pyroprocess

Thus, the pyroprocess is an essential part of our transmutation study. The pyroprocesses based on the molten-salts electrolysis of irrdiated fuels have been developed on the metallic fuel in the Argonne National Laboratory [Chang and Till, 1990], and on the oxide fuel in the Research Institute of Atomic Reactors [Bychkov et al., 1993]. We are studying the possibility of treating the nitride fuels on a process similar to that developed by ANL.

An important contribution of ANL is not only their process itself but also the proposal of a modular approach to the fuel cycle in future. The same thing may be said on the concept of RIAR. Apparently the economy will be improved by enlarging the capacity of a reprocessing facility. Drawback, however, may be the total capital costs of a huge facility themselves, which make the decision of introducing new technologies extremely difficult. The modular approach makes it possible to accelerate the innovation, although the initial economy would not surpass that of the existing technology.

Anyway the transmutation fuel cycle in our double strata concept is small in nature, and we cannot rely much on the scale merit as far as the MA-fuel treatment facility is concerned. This, however, would give an opportunity to test the modular approach in the transmutation cycle.

3.2 Pyrochemistry and database

The study on pyroprocess of nuclear fuels is as old as the nuclear technology. In "Nuclear Engineering Handbook" published forty years ago [Etherington, 1958], we read: "Various high-temperature pyrometallurgical processes under development can improve the fuel-cost picture.", but, "The high-temperature processes are not developed to a point where sound economics can be applied." To develop the pyroprocess to maturity, we also have to meet the modern standard of material accountability. We would not need the extremely high technology in the pyroprocess, but the high-level science has to be involved. We have to enlarge the fundamental database on the molten salts, alloys and ceramics, particularly those containing transuraniums.

There are recent movements in this direction like OECD/NEA-NSC Pyrochemistry Working Group, CODATA Molten Salt Working Group etc. As a response, we are preparing a web-based database related to the pyroprocess as a joint program with JNC. As mentioned above, the high-temperature chemistry cells on transplutonium elements will be in operation in FY2002. We wish to place them within the internationally coordinated efforts to develop the pyroprocessing technology for the future fuel cycle systems.

4. CONCLUSION

Both physical and chemical properties of actinides show significant variations depending on the element from uranium to curium. The materials database on minor actinides (MA:Np, Am, Cm) is very limited. Even with the scanty database, however, americium and curium did not appear to make viable fuels in the ordinary sense. The solutions may be (1) to tolerate a significantly lower fuel performance whose penalty has to be recovered by an improved overall economy of the MA-fuel cycle, and/or (2) to make a fuel concept whose performance is less restricted by physical and chemical properties of MA. The combination of nitride fuel and pyroprocess is being studied as an alternative approach to transmute the minor actinides in a dedicated system. The R&D should also involve a generic approach to form a technological basis for the future nuclear systems. The pyroprocess is a technological element on which versatile nuclear systems may be developed.

REFERENCES

Gibson J.K., Haire R.G., Gensini M.M. and Ogawa T. (1994), Alloying Behavior in Selected Neptunium Binary Systems: The Role of 5f Bonding, *J. Alloys and Compounds 213/214, 106.*

Shushakov V.D. (1993), Investigation of Americium-Neptunium and Curium-Neptunium Alloys, *in summary form only, ACTINIDES-2001 ABSTRACTS, P-18.*

Ogawa T., Mukaiyama T., Takano H., Takizuka T., Suzuki Y. and Osakabe M., *Fuel Elements and Fuel Cycle Concepts of Actinide Burner Reactors*, Japan Atomic Energy Research Institute Report JAERI-M 89-123.

Japan Atomic Energy Research Institute, Oarai Research Establishment (1988), *History of Research and Development on Uranium-Plutonium Mixed Carbides and Nitrides for Fast Reactors.*

Suzuki Y. and Arai Y., Thermophysical and Thermodynamic Properties of Actinide Mononitrides and Their Solid Solutions, *J. Alloys and Compounds*, 271/273, 577.

Takano M. et al., Carbothermic Synthesis of (Cm,Pu)N, *10th Symposium on Thermodynamics of Nuclear Materials*, Halifax, 6-11August.

Murata H. and Mukaiyama T. (1984), Fission Reactor Studies in View of Reactor Waste Management, *Atomkernenergie-Kerntech.*, 45, 23.

Kobayashi F. et al.(1999), Dissolution of Metal Nitrides in LiCl-KCl Eutectic Melts, *Proceedings of the International Conference on Future Nuclear Systems*, August 29 - September 3, Jackson Hole, Wyoming.

Chang Y.I. and Till C.E. (1990), Actinide Recycle Potential in the Integrated Fast Reactor Fuel Cycle, *LMR: A Decade of LMR Progress and Promise*, American Nuclear Society, Inc., La Grange Park, IL, pp.129-137

Bychkov A.V. et al., (1993), Pyroelectrochemical Reprocessing of Irradiated Uranium-Plutonium Oxide Fuel for Fast Reactors, *Proceedings Global'93 Future Nuclear Systems*, Seattle, Washington, 12-17 September.

Etherington H. ed. (1958), Nuclear Engineering Handbook, McGraw-Hill, New York.

Pergamon

www.elsevier.com/locate/pnucene

Progress in Nuclear Energy, Vol. 40, No. 3–4, pp. 547–554, 2002

Printed in Great Britain
0149-1970/02/$ - see front matter

PII: S0149-1970(02)00049-5

ACTINIDE RECYCLING BY PYRO-PROCESS WITH METAL FUEL FBR FOR FUTURE NUCLEAR FUEL CYCLE SYSTEM

TADASHI INOUE

Central Research Institute of Electric Power Industry (CRIEPI)
2-11-1, Iwato-Kita, Komae-shi, Tokyo 201-8511, Japan
TEL: --81-3-3480-2111, FAX: --81-3-3480-7956
E-mail: **inouet@criepi.denken.or.jp**

ABSTRACT

Pyro-metallurgical technology is one of potential devices for future nuclear fuel cycle. Not only economic advantage but also environmental safety and strong resistance for proliferation are required for the fuel cycle. In order to satisfy the requirement, actinides recycling applicable to LWR and FBR cycles by pyro-process has been developed since more than ten years in CRIEPI. The main technology is electrorefining for U and Pu separation and reductive-extraction for TRU separation, which can be applied on oxide fuels through reduction process as well as metal fuels. The application of this technology on separation of TRU in HLLW through chlorination could contribute to the improvement of public acceptance on the geologic disposal.

The main achievements are summarized as follows:

- The elemental technologies, such as electrorefining, reductive extraction, injection casting and salt waste treatment and solidification, have been developed successfully with lots of experiments
- The fuel dissolution into molten salt and uranium recovery on solid cathode for electrorefining have been demonstrated by engineering scale facility in Argonne National Laboratory by using spent fuels and in CRIEPI by uranium tests.
- Single element tests, using actinides, showed the Li reduction to be technically feasible, remaining the subjects of technical feasibility on multi-elements system and on effective recycle of Li by electrolysis of Li_2O.
- Concerning on the treatment of HLLW for actinide separation, the conversion to chlorides through oxides has been also established through uranium tests.
- It is confirmed that more than 99% of TRU nuclides can be recovered from the high level liquid waste by TRU tests
- Through these studies, the process flow sheets for reprocessing of metal and oxide fuels and for partitioning of TRU separation have been established.

The subjects to be emphasized for further development are classified into three categories, that is, process development (demonstration), technology for engineering development, and supplemental technology.

The metal fuel FBR has a high potential for recycling actinides by integration with pyro-reprocessing. Alloys of U-Pu-Zr with minor actinides are investigated from points of fuel properties. The miscibility and other characteristics suggest that the maximum content up to ca. 5 wt% of minor actinides is allowable in the matrix. Nine pins of metal fuel including minor actinides are ready for irradiation at Phenix fast reactor.

1. INTRODUCTION

The future nuclear system has to be much concerned on the system safety, lower impact on environment and non-proliferation as well as economic advantage. In order to meet such requirements, CRIEPI has been proposing the "Actinide Recycling by Pyro-process with Fast Breeder Reactor", integrated metal fuel FBR cycle combined with pyrometallurgical reprocessing, reduction technology of U-, Pu- and MA- (Np, Am, Cm) oxides to metallic form for applying the pyrometallurgical technology, and TRU partitioning and transmutation technology used pyrometallurgy in order to expand the public acceptance for high level liquid waste management. In the proposed advanced nuclear fuel cycle, recovered TRU nuclides will be mixed with U-Pu-Zr to make fresh fuels for FBR and will take a role of energy production. Consequently, only the trace amount of TRU will be disposed of into geologic formation [1]. CRIEPI started the actinide recycling program 1985 including the joining in IFR (Integral Fast reactor) project of US DOE (ANL) from 1989 to 1995 [2] and the collaboration with domestic and overseas institutes.

The research and development of TRU partitioning and transmutation based on the original concept [3] in the integrated system is nominated to the OMEGA project (Long-term research / development program for the long-lived nuclides partitioning and transmutation in Japan) and the program including JAERI, JNC and CRIEPI was checked and reviewed by the Atomic Energy Commission of Japan in 1999.

2. CHARACTERISTICS OF THE SYSTEM WITH PYRO-PROCESS AND METAL FUEL

2.1 Process description

The system proposed of the actinide recycling by pyro-process with FBR is shown in figure 1.

The pyro-reprocessing consists of electrorefining with solid cathode for uranium recovery and

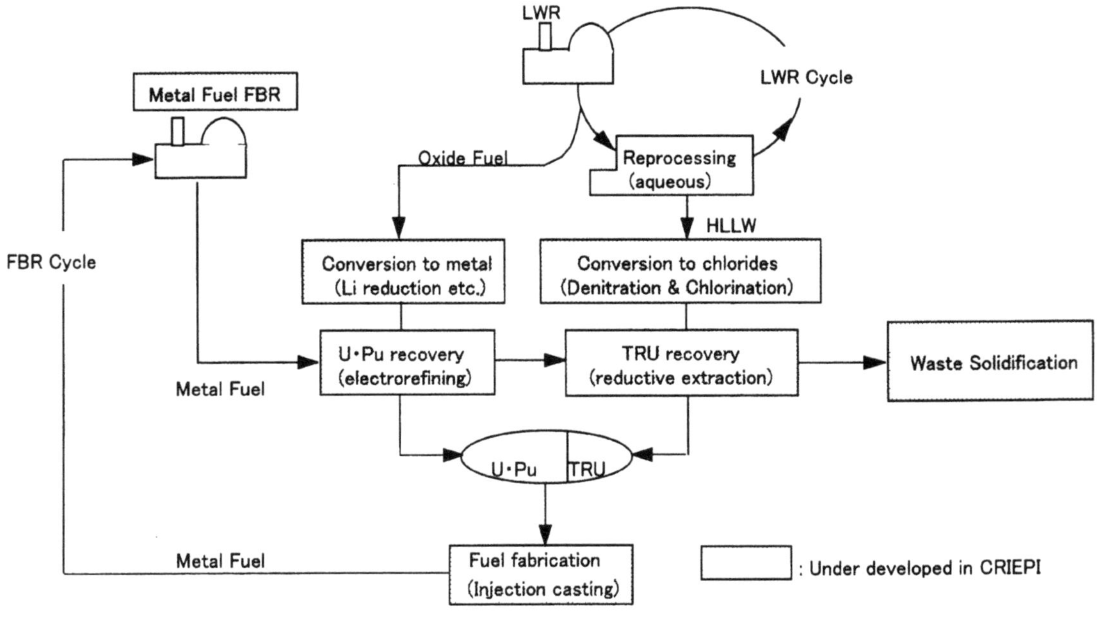

Figure 1 The system of the actinides recycling by pyro-rocess with FBR

liquid cadmium cathode for actinides collection in chloride salt of LiCl-KCl, reductive-extraction for recovering transuranic elements in LiCl-KCl/Cd and LiCl-KCL/Bi by use of lithium as reductant and fuel fabrication by casting after removing salt and cadmium by distillation for actinides products. The lithium reduction is applied to reduce actinide oxides by forming Li2O, which has to be electrolyzed for recycling to reduce the process waste production, for oxide fuel treatment

The conversion of high level liquid waste (HLLW) to chloride is required for adopting the reductive-extraction to recover transuranic elements. To facilitate chlorination, HLLW has to be denitrated to obtain the oxides. The chlorination of oxides occurs in a LiCl-KCl bath with a same kind of chloride bath operating at a lower temperature for collecting the species with high evaporation.

Metal fuels of U-Pu-Zr with minor actinides are prepared for FBR fuels, in which some amount of lanthanides contains as well, because of no large difference of free energy of formation of chlorides from actinides at the separation process of actinides.

2.2 Characteristics of nuclear fuel recycle scenario and process technology

Pyro-process technology can be applied on any type and any composition of nuclear fuels, for instance, not only on metal fuel but also on oxide fuel, like MOX, and nitride fuel, and is also applied on high burn-up fuels foreseen, and even on the short cooling fuels with high decay heat and radiation. Plutonium and neptunium cannot be separated with pure form in principle, which contributes to a high resistance for nuclear proliferation. In addition, a high potential on environmental safety is expected, because a trace amount of actinides goes into vitrification. The pyro-process facility that can be compactly designed is well fitted to the collocation with reactor(s), which fascinates to reduce remarkable number of transportation of nuclear materials. The process for reprocessing and recovering minor actinides is expected to be simple and consist of compacted facility and equipment. Small amount of waste will be generated constantly under the normal operation, because neither organic solvent nor nitric acid solution is used. A large margin for criticality will be relatively expected, because of no aqueous phase exists. The proposed TRU recovery process can be applicable not only to high level liquid waste but also to the slurry, undissolved residue and solvent washing reagent in wastes coming from aqueous reprocessing.

The metal fuel has a high potential for FBR, because of high density of fissile material and high thermal conductivity, and the metal fuel core can be expected to be inherent safety. The reprocessing of metal fuels is well fitted with pyrometallurgical method. On the other hand, we have to note that compared with Purex process and oxide fuel, less experiences has accumulated on both subjects. Large efforts shall be needed for commercial realization.

3. DESCRIPTION OF THE ACHIEVEMENT

3.1 Measurement of thermodynamic properties

In order to evaluate the separation of actinides from other elements, electrochemical potential of chloride, that is free energy of formation of chlorides, is required, especially precise measurement has to be done for actinides and lanthanides, because of the chemical similarity of element among those groups. The electrochemical potentials of those elements measured in LiCl-KCL at 450 C are given in figure 2 [4]. The distribution coefficients of actinides and lanthanides in LiCl-KCl/Cd and LiCl-KCl/Bi systems are indispensable for separation by reductive extraction. Experimental results show a higher separation factor among them in LiCl-KCl/Bi than in LiCl-KCl/Cd [5].

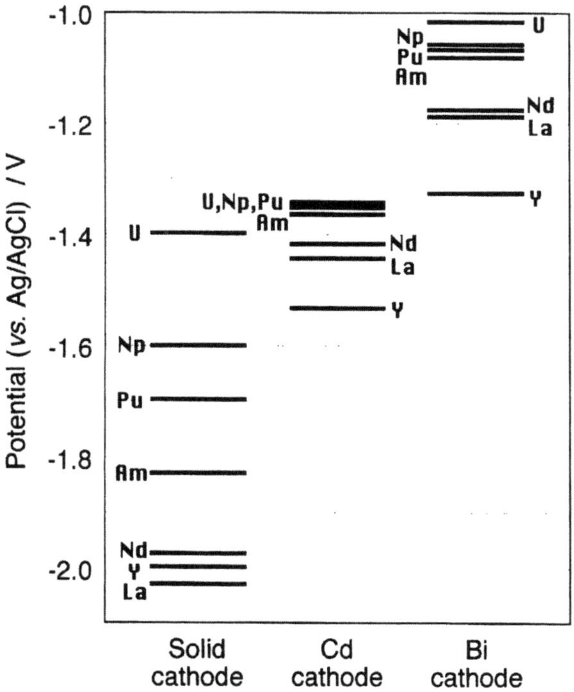

Figure 2 Reduction potentials of actinide and rare-earth elements for the solid cathode, liquid Cd cathode and liquid Bi cathode in LiCl-KCl eutectic salt

3.2 Process technology

3.2.1 U and Pu electrorefining process

Development of electrorefining technology in ANL has attained to laboratory scale test and engineering scale demonstration test [6]. Due to the policy change of no use of Pu, the demonstration of the throughout process of pyro-reprocessing has not been yet realized. The technologies, which are demonstrated in engineering scale, are the spent fuel dissolution process, U collection process by electrorefining on the solid cathode and U separation process from adhered salt.

We, CRIEPI, are aiming to optimize the operating condition on electrorefining. Uranium test with 1 kg scale has been conducted successfully [7,8]. The electrorefining with solid cathode and liquid cathode has been progressing by a joint study with JAERI using Pu and Np [9]. The figure 3 shows neptunium and plutonium recovered on solid cathode.

3.2.2 Reduction process of oxide fuel

ANL and CRIEPI, independently, confirmed that more than 99% of UO_2 was converted into metal form [10]. The experiments used single elements of PuO_2, Am_2O_3 or NpO_2 and used MOX pellet revealed the feasibility of conversion to metals, which has been conducted CRIEPI and AEAT jointly [11]. Currently, the process feasibility using simulated spent fuel is under investigation.

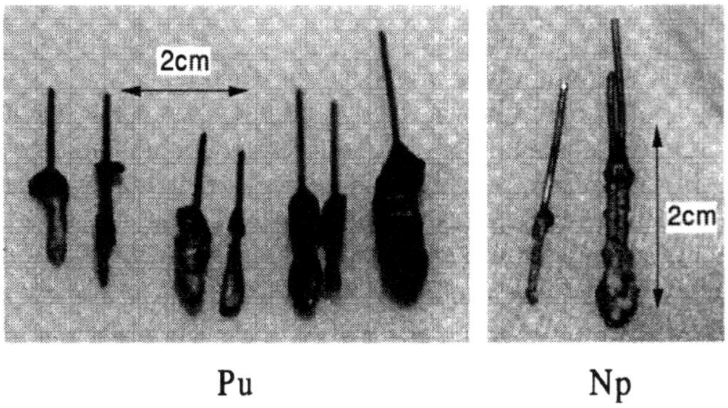

Figure 3 Plutonium and neptunium metals electrodeposited on tantalum wire cathodes

3.2.3 Chlorination process of high level liquid waste

Demonstration test using simulated high level waste indicates that almost all of lanthanides, noble metals and actinides are denitrated, remaining alkali nitrates at 500 C [12]. In this step, most of alkali elements of FP with high heat emission can be separated by water rinsing after denitration. The oxides formed are converted to chlorides by pouring chlorine gas with carbon reductant over 700 C in a LiCl-KCl bath [12]. Most of all chlorides evaporated can be captured in a LiCl-KCl bath keeping at a lower temperature.

3.2.4 Reductive extraction process for recovery of TRUs

A Multi-stage extraction, such as counter current extraction, is required to lower the lanthanide concentration in actinides recovered, when applied on waste salts concentrated fission products at electrorefining and on HLLW. Experiments for separation of TRUs used salts with simulated HLLW components were carried out and the recovery yields of all actinides are found to be over 99% with attaining the decontamination factor over 10 of lanthanides [13].

3.2.5 Waste treatment process for pyro-process

Two kinds of waste treatment process have been developed. Firstly, heated $NaAlO_2$ and SiO_2 with waste salts at high temperature, the sodalite structure is synthesized, in which waste chlorides are immobilized in an artificial mineral matrix [14]. The second process is to form borosilicate glass contained waste. Salt waste is converted to oxides through the step of electro-reduction in liquid lead and vitrified with B_2O_3 and SiO_2 at air atmosphere. Most of all solvent salt and solvent metal after electrorefining can be recycled, and Cl_2 gas generated in this process will be used at chlorination process [15].

3.2.6 Process diagram

The optimized process diagram with material balance for reprocessing of metal fuel and for TRU

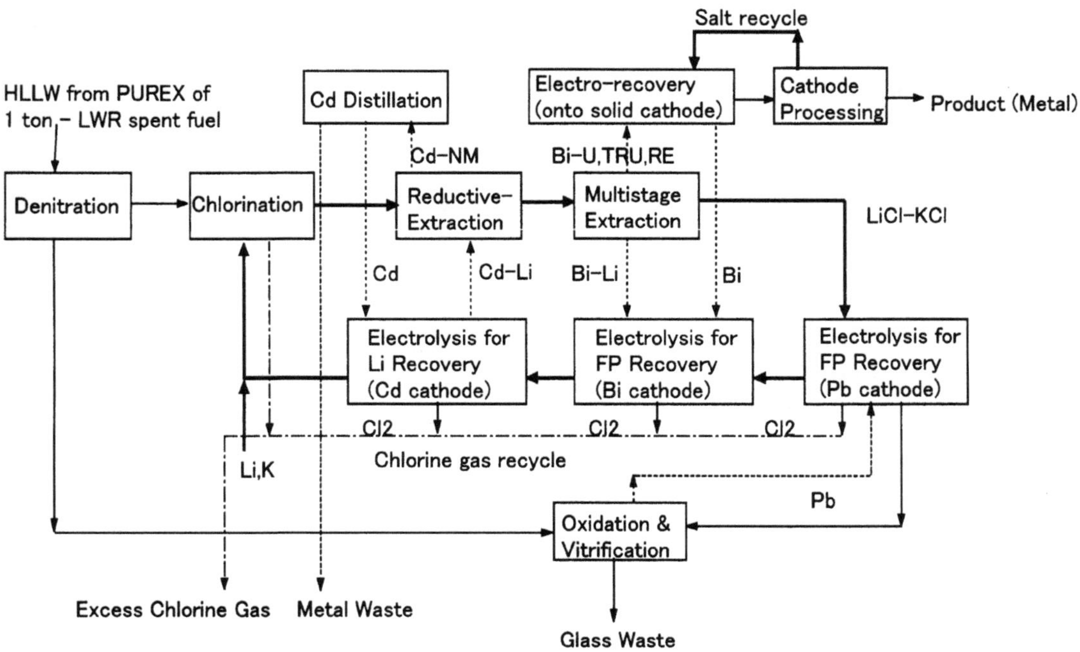

Figure 4 Process flow diagram of pyro-partitioning

separation has been established based on the experimental evidences. Figure 4 shows an optimized TRU separation process flow sheet where more than 99% TRU could be separated, maintaining a large separation from lanthanides, and solvent salt, reductant and Cl_2 gas can be recycled in the process [16].

3.3 metal fuel study

The mixing of minor actinides (MA) is unavoidable for actinide recycling in U-Pu-Zr metallic fuel. Moreover, when minor actinides removed from HLLW are transmuted in FBR, much concern has to be paid on the amount of minor actinides in fuels.

3.3.1 Eutectic formation by reacting fuel and cladding material

The liquefaction temperature by eutectic formation of U-Pu-Zr fuel and stainless steel cladding is expected to be above 600 C, which affects directly the reactor operation temperature. The liquefaction was evaluated from diffusion tests using U-Pu-Zr/Fe couple and the phase diagrams were assessed both by thermodynamic calculation and metallography [17]. The diffusion tests indicate that no melting phase appears under 650 C in a region with less than 25 wt% of plutonium in fuel U-Pu [18].

3.3.2 Allowable amount of MA and lanthanides in U-Pu-Zr

The miscibility by mixing MA and lanthanides in U-Pu-Zr has to be evaluated by observing microstructures. The morphology of U-Pu-Zr alloy contained up to ca.5wt% of MA, with the composition discharged from LWR, and of lanthanides shows the uniform distribution of rare earth-americium phase in U-Pu-Zr matrix, especially along grain boundaries [19].

3.3.3 Fuel properties of U-Pu-Zr with MA and lanthanides

The characterization study of alloys with MA was carried out in order to determine the applicability of a metal fuel for FBR. The alloys with 2wt% of MA and 2wt% of lanthanides and with 5wt% of MA and 5wt% of rare earth were prepared in order to examine the melting temperature, thermal conductivity, elongation and chemical interaction between cladding and alloys. The high evaporation rate of americium was observed above 850 C [20].

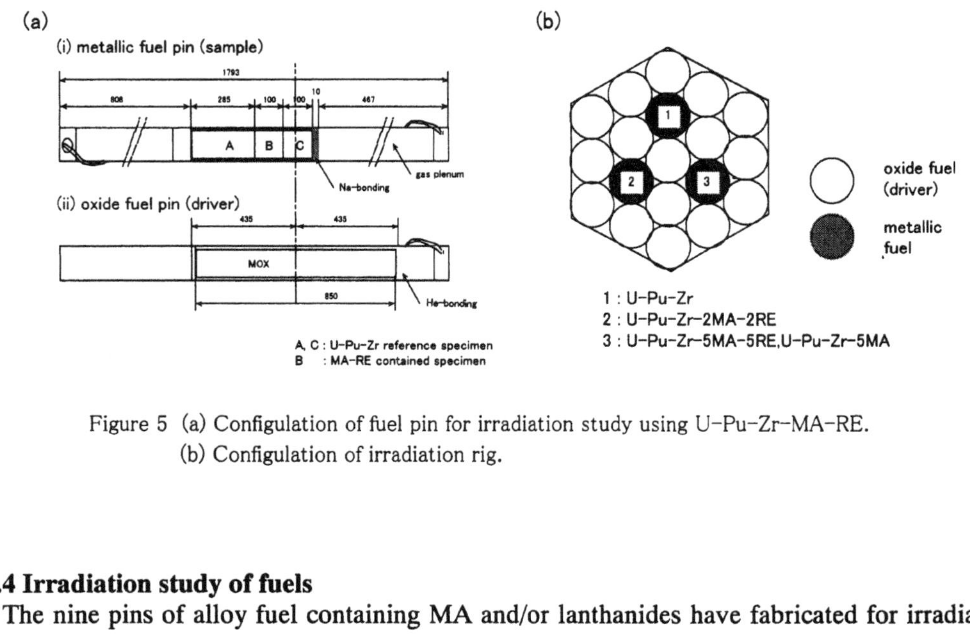

Figure 5 (a) Configulation of fuel pin for irradiation study using U-Pu-Zr-MA-RE.
(b) Configulation of irradiation rig.

3.3.4 Irradiation study of fuels

The nine pins of alloy fuel containing MA and/or lanthanides have fabricated for irradiation. The contents of MA and/or lanthanides in U-Pu-Zr are 2 or 5wt% in each. The burn-ups of test fuel alloys for irradiation is expected to be ca.1.5, 6.0 and >10 at% in a FBR. The configuration of pins is given in figure 5.

4. SUMMARY

Actinide Recycling by Pyro-process integrated with FBR has been developing since 1985 for aiming the next generation nuclear fuel cycle system with high potentials of environmental safety, non-proliferation as well as economic advantage. Through studies, elemental technologies, such as electrorefining, reductive extraction, and salt waste treatment and solidification, have been developed successfully. The fuel dissolution into molten salt and uranium recovery on solid cathode for electrorefining have been demonstrated by engineering scale facility in ANL by using spent fuels and on CRIEPI by uranium tests. Single element tests, using actinides, showed the Li reduction to be technically feasible, remaining the subjects of applicability on multi-elements system and effective recycle of Li by electrolysis of Li_2O. Concerning on the treatment of HLLW for actinide separation, the conversion to chlorides through oxides has been also established through uranium tests. It is confirmed that more than 99% of TRU nuclides can be recovered from the high level liquid waste by TRU tests. Through these studies, the process flow sheets for reprocessing of metal and oxide fuels and for partitioning of TRU separation have been established.

However, the pyro-process has much less experienced compared with aqueous process. Further efforts have to be done from engineering points of view, such as removal or dissolution of dross accumulated in equipments, handling of the cathode product and separation from salt, collection of dropped material from cathode product, transportation of the melt and remote operation technology. In addition, accounting measure of fissile material has to be developed, according to batch-wise operation. Material selection for high temperature operation is not the least issue

The examination of liquefaction temperature by U-Pu-Zr and Fe revealed no eutectic formation up to 650 C. The characterization of metal fuel with minor actinides is examined. The minor actinide content up to 5wt% in U-Pu-Zr gives no significant change of fuel properties according to the out-of core experimental results. The nine pins of alloy fuels are ready for irradiation.

References

[1] Inoue T. and Tanaka H. (1997), Recycling of actinides produced in LWR and FBR fuel cycles by applying pyrometallurgical process, *Proc. on Future Nuclear Systems (GLOBAL'97)*, pp.646-652, Yokohama, 5-10 October.

[2] Chang Y.I. (1989), The integral fast reactor, *Nucl. Technol.*, **88**, 129.

[3] Inoue T., Sakata M., Miyashiro H., Matsumura T., Sasahara A., Yoshiki N. (1991), Development of partitioning and transmutation technology for long-lived nuclides, *Nucl.Technol.* **69**, 206.

[4] Sakamura Y., Hijikata T., Kinoshita K., Inoue T., Storvick T.S., Krueger C.L., Roy J.J., Grimmett D.L., Fusselman S.P. and Gay R.L. (1998), Measurement of standard potentials of actinides (U,Np,Pu,Am) in LiCl-KCl eutectic salt and separation of actinides from lanthanides by electrorefining, *J. Alloy. Compound* **271-273**, 592.

[5]Kurata M., Sakamura Y. and Matsui T. (1996), Thermodynamic quantities of actinides and rare earth elements in liquid bismuth and cadmium, *J. Alloy. Compound* **234**, 83.

[6] Totemeier T.C. and Mariani R.D. (1997), Morphologies of uranium and uranium-zirconium electrodeposits, *J. Nucl. Mater.* **250**, 131.

[7] Koyama T., Iizuka M., Shoji Y., Fujita R., Tanaka H., Kobayashi T. and Tokiwai M. (1997), An experimental study of molten salt electrorefining of uranium using solid iron cathode and liquid cadmium cathode for development of pyrometallurgical reprocessing, *J. Nucl. Sci. Technol.* **34**, 384.

[8] Koyama T., Iizuka M., Kondo N., Fujita R. and Tanaka H. (1997), Electrodeposition of uranium in stirred liquid cadmium cathode, *J.Nucl.Mater.* **247**, 227.

[9] Sakamura Y., Shirai O., Iwai T. and Suzuki Y. (2000), Thermodynamics of Neptunium in LiCl-KCl Eutectic / Liquid Bismuth Systems, *J. Electrochem. Soc.* **147**, 642.

[10] Karell E.J., Gourishankar K.V., Chow L.S. and Everhart R.E. (1999), Electrometallurgical treatment of oxide spent fuels, *Int. Conf. On Future Nuclear Systems (GLOBAL'99)*, Aug.29 – Sep.3, 1999, Jackson Hole, Wyoming.

[11] Usami T., Kurata M., Inoue T., Jenkins J., Sims H., Beetham S. and Browm D. (2000), Pyrometallurgical reduction of unirradiated TRU oxides by lithium in a lithium chloride medium, *OECD Nuclear Energy Agency, Nuclear Science Committee, Workshop on Pyrochemical Separations,* 14-15 March, Avignon.

[12] KurataM., kato T., Kinoshita K. and Inoue T. (1999), Conversion of high level waste to chloride for pyrometallurgical partitioning of minor actinides", *Proc. of 7th Int. Conf. On Radioactive Waste Management and Environmental Remediation, ICEM'99*, 26-30 September, Nagoya.

[13] Kinoshita K., Inoue T., Fusselman S.P., Grimmett D.L., Roy J.J., Gay R.L., Krueger C.L., Nabelek C.R. and Storvick T.S. (1999), Separation of uranium and transuranic elements from rare earth elements by means of multistage extraction in LiCl-KCl/Bi system, *J. Nucl. Sci. Technol.* **36**, 189.

[14] Koyama T., Seto C., Yoshida T., Kawamura F. and Tanaka H. (1995), Immobilization of halide salt waste from pyrochemical reprocessing by forming natural occurring mineral; SODALITE, *Int. Conf. On Evaluation of Emerging Nuclear Fuel Cycle Systems (GLOBAL'95),* VOL.2, 1744, 11-14 September, Versailles.

[15] Sakamura Y., Inoue T., Shimizu T. and kobayashi K. (1997), Development of pyrometallurgical partitioning technology for TRU in high level radioactive wastes -vitrification process for salt wastes-, *Int. Conf. On Future Nuclear Fuel Systems (GLOBAL'97),* VOL.2, 1222, 5-10 October, Yokohama.

[16] Kinoshita K., Kurata M. and Inoue T. (2000), Estimation of materials balance in pyrometallurgical partitioning process of transuranic elements from high-level liquid waste, *J. Nucl. Sci. Technol,* **37**, 75.

[17] Kurata M., Nakamura K., Ogata T. (2001) , Thermodynamic evaluation of quaternary U-Pu-Zr-Fe system to design cladding temperature of metallic fuel fast reactor, *J. Nucl. Mater.* 294, 123.

[18] Ogata T., Nakamura K., Kurata M. and Mignanelli M.A. (2000), Reactions between U-Pu-Zr Alloys and Fe at 923 K, *J. Nucl. Sci. Technol.* **37**, 244.

[19] Inoue T., Kurata M., Koch L., Spirlet J-C, Walker C.T. and Sari C. (1991), Characterization of fuel alloys with minor actinides, *Trans. Am. Nucl. Soc.* **64**, 552.

[20] Kurata M., Inoue T. and Sari C. (1994), Redistribution behavior of various constituents in U-Pu-Zr alloy and U-Pu-Zr alloy containing minor actinides and rare earths in a temperature gradient, *J. Nucl. Mater.* **208**, 144.

Pergamon

www.elsevier.com/locate/pnucene

Progress in Nuclear Energy, Vol. 40, No. 3-4, pp. 555-560, 2002

Printed in Great Britain
0149-1970/02/$ - see front matter

PII: S0149-1970(02)00050-1

NEUTRON ECONOMY AND NUCLEAR DATA FOR TRANSMUTATION OF LONG-LIVED FISSION PRODUCTS

M. IGASHIRA and T. OHSAKI

Research Laboratory for Nuclear Reactors, Tokyo Institute of Technology
2-12-1 O-okayama, Meguro-ku, Tokyo 152-8550, Japan

In the study of Self-Consistent Nuclear Energy System, the following 29 long-lived fission products (LLFPs) have been selected to be transmuted into stable or short-lived nuclides: ^{106}Ru, ^{102}Rh, ^{109}Cd, ^{125}Sb, ^{134}Cs, ^{146,147}Pm, ^{154,155}Eu, ^{171}Tm, ^{85}Kr, ^{90}Sr, ^{93m}Nb, ^{113m}Cd, ^{121m}Sn, ^{137}Cs, ^{151}Sm, ^{152}Eu, ^{108m}Ag, ^{158}Tb, ^{166m}Ho, ^{79}Se, ^{93}Zr, ^{94}Nb, ^{99}Tc, ^{107}Pd, ^{126}Sn, ^{129}I, ^{135}Cs. In the present study, the number of neutrons necessary for the transmutation of the 29 LLFPs with an FBR was evaluated, and the present status of the (n, γ) and (n,2n) cross section data of the 29 LLFPs in JENDL-3.2 and ENDF/B-VI was investigated. The main results of the present study are as follows: (1)only 0.25 neutron per fission is necessary for the transmutation of the 29 LLFPs with isotopic separation, whereas 6.8 neutrons are necessary with chemical separation, (2)the accuracy of the cross sections is 30 to 100 % except for the (n, γ) cross sections of limited nuclides in limited incident neutron energy regions.

1. INTRODUCTION

A Self-Consistent Nuclear Energy System (SCNES) proposed by Fujii-e et al.(Fujii-e et al., 1992a; Fujii-e, 1992b) is an ultimate goal of nuclear energy development. The SCNES is defined as a system that realizes at least the following four objectives simultaneously:

(a) Energy generation
(b) Fuel production
(c) Burning of radionuclides
(d) System safety

If the SCNES is a fission-reactor system, the neutrons produced in the SCNES have to be used effectively to realize these four objectives. Two neutrons per fission should be used for energy generation and fuel production in an equilibrium state, and a part of excess neutrons is absorbed by materials in the reactor. Therefore, we can utilize only a small number of excess neutrons to transmute Long-Lived Fission Products (LLFPs) into stable or short-lived nuclides. The number of excess neutrons per fission is evaluated to be 0.58 for an FBR (Ohsaki and Fujii-e, 1996). This value is the maximum number of neutrons which can be utilized for the transmutation of LLFPs.

In the study of SCNES, 29 LLFPs have been selected to be transmuted on the basis of the following two criteria: $T > 1$ and $Y/T > 1.0\times10^{-10}$, where T (y) is the half life and Y (%) the cumulative fission yield (Akatsuka et al., 1994). The selected nuclides are classified with their half lives as follows:

(1) 1<T<10: ^{106}Ru, ^{102}Rh, ^{109}Cd, ^{125}Sb, ^{134}Cs, ^{146,147}Pm, ^{154,155}Eu, ^{171}Tm; [7.3]
(2) 10<T<100: ^{85}Kr, ^{90}Sr, ^{93m}Nb, ^{113m}Cd, ^{121m}Sn, ^{137}Cs, ^{151}Sm, ^{152}Eu; [13.0]
(3) 100<T<10,000: ^{108m}Ag, ^{158}Tb, ^{166m}Ho; [$<10^{-5}$]
(4) T>10,000: ^{79}Se, ^{93}Zr, ^{94}Nb, ^{99}Tc, ^{107}Pd, ^{126}Sn, ^{129}I, ^{135}Cs; [22.2]

The values in the brackets show the sums of the cumulative fission yields of the individual nuclides for the fast-neutron induced fission of ^{239}Pu (Ihara, 1989). The total of the cumulative fission yields is 42 %, which means that the number of the 29 nuclides produced per fission is roughly 0.42. This value is not so small compared with the evaluation of the number of excess neutrons per fission, 0.58. Therefore, it is very important to investigate the number of neutrons necessary for the transmutation of the 29 LLFPs.

On the other hand, many studies have been made on the transmutation of LLFPs with a variety of nuclear systems. In the studies, the accuracy of nuclear data such as neutron capture cross sections of LLFPs is very important for the evaluation of the transmutation performance of systems, because the performance directly depends on the cross section values. Therefore, it is necessary to investigate the present status of cross section data contained in the existing evaluated nuclear data libraries such as Japanese Nuclear Data Libraries (JENDL-3.2) and Evaluated Nuclear Data File (ENDF/B-VI).

In the present study, the number of neutrons necessary for the transmutation of the 29 LLFPs was evaluated through simplified-model calculations, and the present status of the (n, γ) and (n,2n) cross sections of the 29 nuclides in JENDL-3.2 and ENDF/B-VI was investigated through the check of evaluation methods and the comparison of the evaluated values with experimental ones.

2. NEUTRON ECONOMY

A simplified sodium-cooled metal-fuel FBR was adopted to evaluate neutron economy for the transmutation of the 29 LLFPs. A spherical core and uniform distribution of core materials were assumed. The volume ratio of Fuel : Na coolant : Structural materials was taken to be 35 : 40 : 25. The neutron flux and thermal output were taken to be 1.0×10^{15} /(cm^{2}s) and 2.5 GW$_{th}$, respectively.

The number of the LLFP with Z protons and N neutrons, N_{FP}, is assumed to obey the following equation:

$$\frac{dN_{FP}}{dt} = \gamma\cdot\sigma_f\cdot\phi\cdot N_{Pu} + \sigma'\cdot\phi\cdot N_{N-1} + \lambda_{\pm}\cdot N_{Z\pm1} - \sigma_c\cdot\phi\cdot N_{FP} - \lambda_\beta\cdot N_{FP} - X_{FP}\ , \qquad (1)$$

where γ is the independent fission yield of the LLFP for the fast-neutron induced fission of ^{239}Pu (Ihara, 1989), σ_f the fission cross section of ^{239}Pu, ϕ the neutron flux, N_{Pu} the number of ^{239}Pu, σ' the capture cross section of the nuclide with Z protons and (N-1) neutrons, N_{N-1} the number of the nuclide with Z protons and (N-1) neutrons, $\lambda_{\pm}$ the decay constants of the nuclides with (Z$\pm$1) protons and (N$\mp$1) neutrons, $N_{Z\pm1}$ the numbers of nuclides with (Z$\pm$1) protons and (N$\mp$1) neutrons, σ_c the capture cross section of the LLFP, λ_β the decay constant of the LLFP, and X_{FP} the extraction rate of the LLFP. One-group neutron cross sections were obtained from the cross sections evaluated by Takaki (Takaki, 1992) and the neutron energy spectrum for the model FBR. In the standard calculation with Eq.(1), 151 stable FPs and the 29 LLFPs were treated explicitly, and 1,047 radioactive FPs with half-lives shorter than 1 y were treated implicitly. In addition to the standard calculation, the calculations were performed for the transmutations of the LLFPs with half lives longer than 3, 10, 20, 30, 100, 200, 2,000 and 50,000 y.

Table 1 shows the ratio of the number of neutrons absorbed by LLFPs to that of fissions in the reactor in an equilibrium state. The second column in Table 1 shows the result for the 29 LLFPs. If the ratio is smaller than the number of excess neutrons, 0.58, the reactor can remain critical. In this case, the ratio can be interpreted as the number of neutrons per fission necessary for the transmutation of the 29 LLFPs. In the case of the transmutation with isotope separation, the ratio is 0.25: the transmutation is feasible scientifically from the viewpoint of neutron economy. On the other hand, the ratio is very large, 6.78, in the case of chemical separation, which means that the transmutation is impossible. The third column shows the corresponding ratios for the LLFPs with half lives longer than 3 y. The ratio drastically decreases in the case of chemical separation, but the value is still very large, 2.07. Even for the transmutation of the LLFPs with half lives longer than 50,000 y, the ratio is still large, 0.95. In the case of the transmutation with isotopic separation, the dependence of the ratio on the lower limit of half life is very weak.

Table 1 Ratio of the number of neutrons absorbed by LLFPs to that of fissions

Lower limit of half life (y)	1	3	10	20	30	100	200	2,000	50,000
In case of isotope separation	0.25	0.24	0.24	0.24	0.24	0.22	0.22	0.22	0.22
In case of chemical separation	6.78	2.07	1.99	1.23	1.12	1.07	0.95	0.95	0.95

3. PRESENT STATUS OF NUCLEAR DATA

The data and evaluation methods in JENDL-3.2 and ENDF/B-VI were investigated for the (n, γ) and (n,2n) reaction cross sections of the 29 LLFPs. As a result, it was found that

(1) there are no evaluated data for all meta-stable nuclides (^{93m}Nb, ^{113m}Cd, ^{121m}Sn, ^{108m}Ag, ^{166m}Ho) and the 5 nuclides with fission yields smaller than 5.0×10^{-6} (^{102}Rh, ^{109}Cd, ^{146}Pm, ^{171}Tm, ^{158}Tb),

(2) moreover, JENDL-3.2 does not contain the data for the (n,2n) cross sections of ^{125}Sb, and ENDF/B-VI does not contain the data for the (n, γ) cross sections of ^{79}Se and ^{126}Sn and the (n,2n) cross sections of ^{106}Ru, ^{125}Sb, ^{134}Cs, ^{85}Kr, ^{90}Sr, ^{137}Ce, ^{79}Se, ^{93}Zr, ^{94}Nb, ^{107}Pd, ^{126}Sn, ^{129}I, and ^{135}Cs,

(3) main evaluation methods in both JENDJ-3.2 and ENDF/B-VI are theoretical calculations.

The above (1) and (2) are the reasons why we employed the evaluated data by Takaki instead of JENDL-3.2 or ENDF/B-VI in the calculation in Sec. 2.

As an example, the capture cross sections of ^{126}Sn in JENDJ-3.2 are shown in Fig.1. There are no experimental data about ^{126}Sn. The capture cross section at thermal energy was evaluated by systematics obtained from the cross sections of neighboring nuclides, and was extrapolated up to 2 keV using the 1/v law, where v is the velocity of incident neutrons. In the region from 2 keV to several hundred keV, the statistical model calculation was adopted, and the calculated results were normalized to an evaluated value at 300 keV obtained from systematics. Above 10 MeV, the direct and semi-direct capture mechanism was taken into account using the method by Benzi and Reffo (Benzi and Reffo, 1969). This evaluation procedure is typical for nuclides with no experimental data. However, the thermal capture cross section is strongly influenced by the neighboring positive- and negative-energy resonances: the character of nuclide is strongly reflected in the thermal neutron capture cross section. Therefore, large uncertainty from several factors to one order is thought to be in the systematics of thermal neutron capture cross sections. The uncertainty of the systematics in the keV region is thought to be a factor of 2. Since the character of nuclide is not strongly reflected in the direct and semi-direct capture cross section, the uncertainty of the method by Benzi and Reffo is thought to be several 10 %.

When reliable experimental data were available, the calculated results were normalized to the experimental data in JENDL-3.2 and ENDF/B-VI. Therefore, existing experimental data were searched for the (n, γ) and

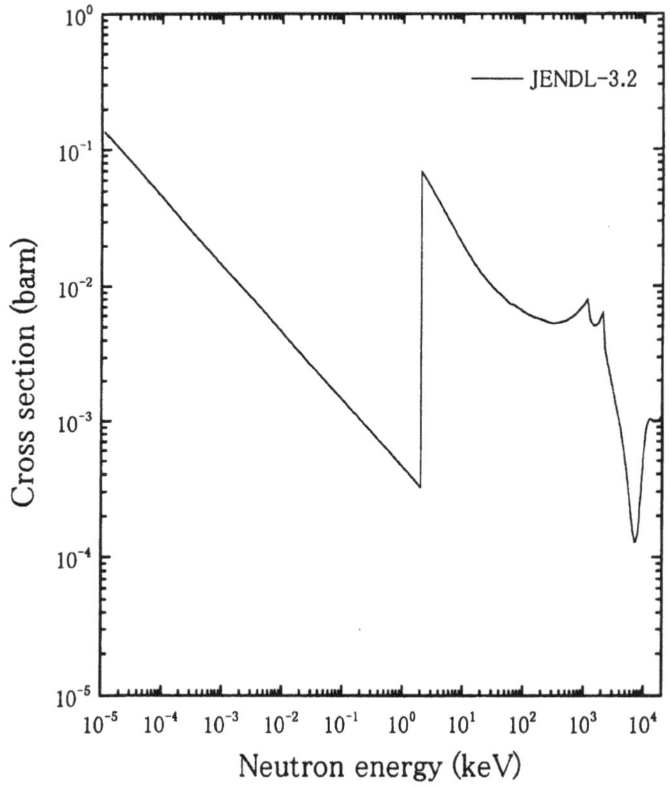

Fig.1. Capture cross sections of ^{126}Sn in JENDL-3.2

(n,2n) cross sections of the 29 LLFPs by using Compute Index of Neutron DAta (CINDA) and EXchange FORmat data library (EXFOR). As a result, it was found that

(4) the (n, γ) data are available for the 16 nuclides: ^{106}Ru, ^{134}Cs, ^{147}Pm, ^{154}Eu, ^{155}Eu, ^{85}Kr, ^{90}Sr, ^{137}Cs, ^{151}Sm, ^{152}Eu, ^{94}Nb, ^{93}Zr, ^{99}Tc, ^{107}Pd, ^{129}I, ^{135}Cs,

(5) the (n,2n) data are available for the only 2 nuclides: ^{99}Tc and ^{129}I,

(6) those experimental data exist in limited incident-neutron energy regions.

Comparisons of evaluated data in JENDL-3.2 and ENDF/B-VI with experimental ones are useful in evaluating the uncertainty of evaluated data. The evaluated capture cross sections of ^{135}Cs in JENDJ-3.2 and ENDF/B-VI are compared with experimental results in Fig.2. Both evaluations were performed before the recent measurement at 30 keV by Jaag et al. (Jaag et al., 1997). The agreement is excellent between their result and the evaluation in JENDL-3.2, although the excellent agreement is thought to be somewhat accidental. On the other hand, the evaluation in ENDF/B-VI is about one third of their result. At thermal energy, both evaluations adopted the measurement by Baerg et al. (Baerg et al., 1958), so both evaluated values are in good agreement with the recent measurement at thermal energy by Katoh et al. (Katoh et al., 1997). The capture cross sections of ^{99}Tc in JENDL-3.2 and ENDF/B-VI are compared with experimental results in Fig.3. Many measurements exist for ^{99}Tc. In particular, there are three sets of measurements in the region from 10 to 100 keV. The results from the latest measurement by Macklin (Macklin, 1982) are between the results by Chou and Werle (Chou and Werle, 1973) and by Little and Block (Little and Block, 1977), and the evaluation for JENDL-3.2 adopted the results of Macklin. However, very recent our measurements (Matsumoto et al., 2000) are larger than those of Macklin by about 20 %, and support the measurements of Little and Block.

The (n,2n) cross sections of ^{99}Tc and ^{129}I in JENDL-3.2 and ENDF/B-VI are compared with experimental results in Figs.4 and 5, respectively. In the evaluation of JENDL-3.2, the theoretical calculation with the pre-equilibrium model and the multi-step evaporation model was performed, and the calculated results were normalized to the experimental result around 14 MeV. However, the recent measurement by Nakano et al. for ^{129}I (Nakano et al., 1996) is smaller than the evaluation in JENDL-3.2 by about 30 %. Therefore, the uncertainty in the (n,2n) cross sections of other LLFPs with no experimental data is thought to be greater than 50 %, although the character of nuclide is not strongly reflected in the (n,2n) cross section.

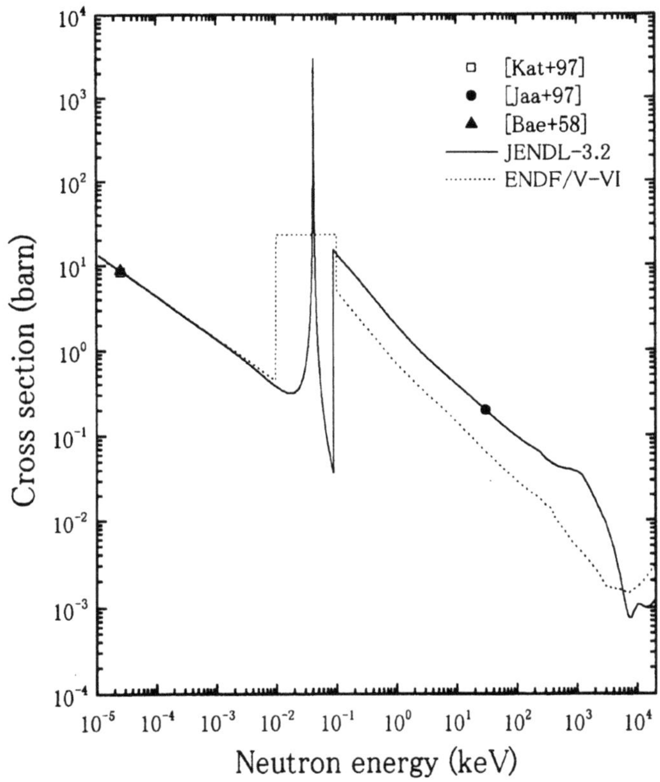

Fig.2. Comparison of evaluated capture cross sections of ^{135}Cs with experimental ones

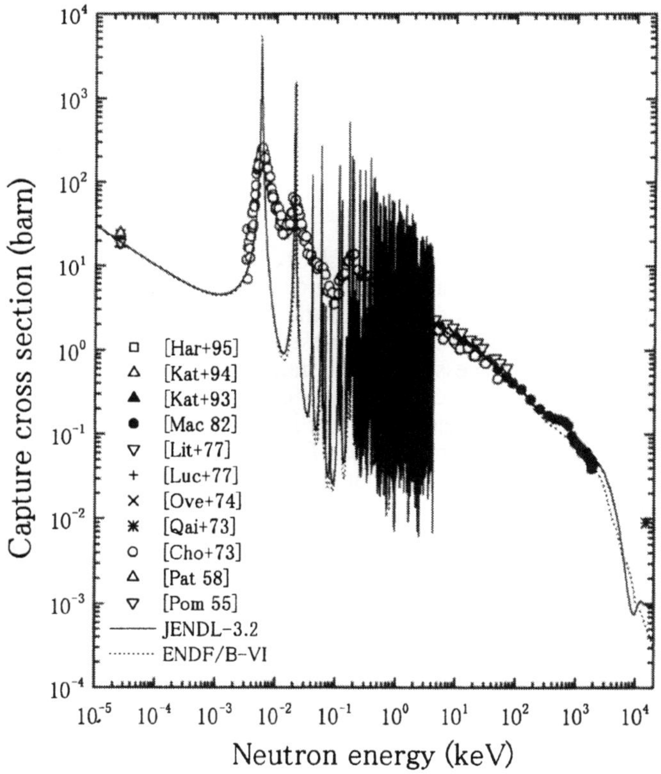

Fig.3. Comparison of evaluated capture cross sections of ^{99}Tc with experimental ones

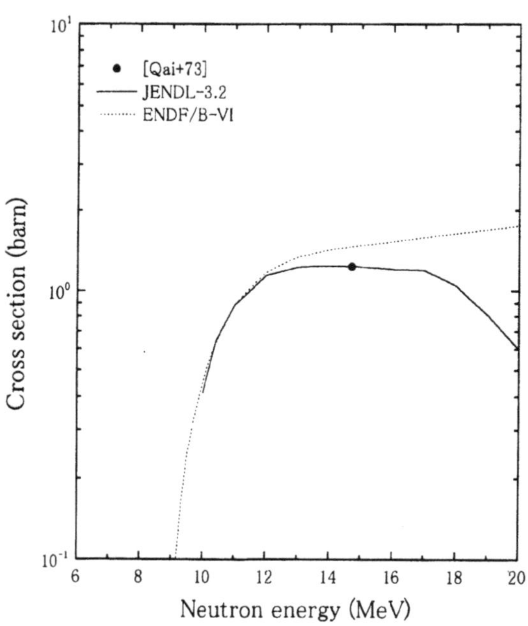

Fig. 4. Comparison of evaluated (n,2n) cross sections of ^{99}Tc with experimental one

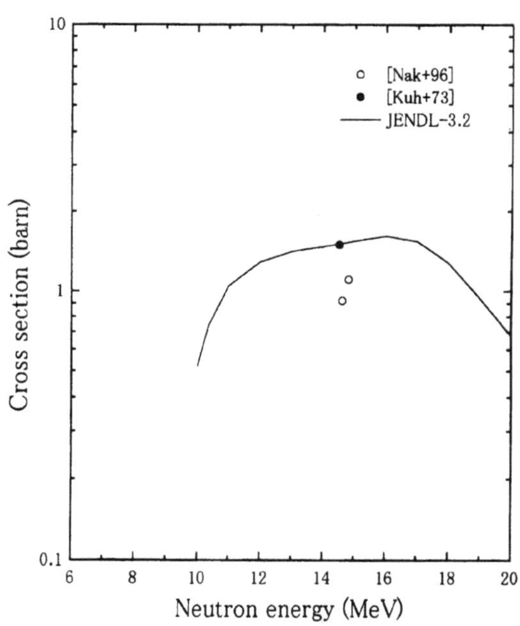

Fig. 5. Comparison of evaluated (n,2n) cross sections of ^{129}I with experimental ones

As described above, the main evaluation method in JENDL-3.2 was theoretical calculation, and the calculated results were normalized to experimental data when reliable data were available. Parameters employed in the calculation were taken from systematics obtained in the stable mass region, and there is some doubt whether they are applicable to nuclides in the unstable mass region. However, we cannot find out another good method in the present status of poor experimental data. Accurate experimental data are the basis of accurate evaluation, and the accumulation of accurate data are indispensable for the improvement of evaluated nuclear data for LLFPs.

4. CONCLUSION

The number of neutrons necessary for the transmutation of the 29 LLFPs with an FBR was evaluated, and the accuracy of the (n, γ) and (n,2n) cross sections of the 29 LLFPs in JENDL-3.2 and ENDF/B-VI was investigated. The following have been found from the present study:

(1) Only 0.25 neutron per fission is necessary for the transmutation of the 29 LLFPs with isotopic separation, although 6.8 neutrons are necessary with chemical separation.

(2) There are no evaluated data for all meta-stable nuclides (^{93m}Nb, ^{113m}Cd, ^{121m}Sn, ^{108m}Ag, ^{166m}Ho) and the 5 nuclides with fission yields smaller than 5.0×10^{-6} (^{102}Rh, ^{109}Cd, ^{146}Pm, ^{171}Tm, ^{158}Tb).

(3) Moreover, JENDL-3.2 does not contain the data for the (n,2n) cross sections of ^{125}Sb, and ENDF/B-VI does not contain the data for the (n, γ) cross sections of ^{79}Se and ^{126}Sn and the (n,2n) cross sections of ^{106}Ru, ^{125}Sb, ^{134}Cs, ^{85}Kr, ^{90}Sr, ^{137}Ce, ^{79}Se, ^{93}Zr, ^{94}Nb, ^{107}Pd, ^{126}Sn, ^{129}I, and ^{135}Cs.

(4) Main evaluation methods in both JENDL-3.2 and ENDF/B-VI are theoretical calculations.

(5) The (n, γ) data are available for the 16 nuclides: ^{106}Ru, ^{134}Cs, ^{147}Pm, ^{154}Eu, ^{155}Eu, ^{85}Kr, ^{90}Sr, ^{137}Cs, ^{151}Sm, ^{152}Eu, ^{94}Nb, ^{93}Zr, ^{99}Tc, ^{107}Pd, ^{129}I, ^{135}Cs.

(6) The (n,2n) data are available for the only 2 nuclides: ^{99}Tc and ^{129}I.

(7) Those experimental data exist in limited incident-neutron energy regions.

(8) The accuracy of the cross sections is evaluated to be 30 to 100 % except for the (n, γ) cross sections of limited nuclides in limited incident-neutron energy regions.

REFERENCES

Akatsuka, H. et al. (1994), Scientific Feasibility of Incineration in SCNES. *Int. Symposium on Global Environment and Nuclear Energy System*, p.477-484, Elsevier, Oxford.

Baerg, A. et al. (1958), *Can. J. Phys.*, 36, 863.

Benzi, V. and Reffo, G. (1969), CCDM-NW/10.

Chou, J. and Werle, H. (1973), *J. Nucl. Energy*, 27, 811.

Fujii-e, Y. et al. (1992), A Self-Consistent Nuclear Energy Supply System. *Symposium on Potential of Small Nuclear Reactors for Future Clean and Safe Energy Sources*, p.89-98, Elsevier, Amsterdam.

Fujii-e, Y. (1992), An Approach to Self-Consistent Nuclear Energy System (Potential of Fast Reactors), *Int. Conf. on Design and Safety of Advanced Nuclear Power Plants*, Vol. II, p.11.3-1-11.3-9.

Ihara, H. (1989), JAERI-M 89-204.

Jaag, S. et al. (1997), *Nucl. Phys.*, A621, 247.

Katoh, T. et al. (1997), *J. Nucl. Sci. Technol.*, 34, 431.

Little, R.C. and Block, R.C. (1977), *Trans. American Nucl. Soc.*, 26,574.

Macklin, R.L. (1982), *Nucl. Sci. Eng.*, 81, 520.

Matsumoto, T. et al. (2000), Measurement of keV-Neutron Capture Cross Sections of ^{99}Tc and ^{237}Np for the Transmutation Study. *4th Japan-Korea Seminar on Advanced Reactors*, p.208-213.

Nakano, D. et al. (1996), JAERI-97-004, 177.

Ohsaki, T. and Fujii-e, Y. (1996), *1996 fall meeting of the Atomic Energy Society of Japan*,A56.

Takaki, N. (1992), Ph.D. thesis, Tokyo Institute of Technology, in Japanese.

Pergamon

www.elsevier.com/locate/pnucene

Progress in Nuclear Energy, Vol. 40, No. 3–4, pp. 561–568, 2002

Printed in Great Britain
0149-1970/02/$ - see front matter

PII: S0149-1970(02)00051-3

ISOTOPE SEPARATION REQUIRED IN SCNES AND FUTURE SUBJECTS

Masaaki Suzuki*, Alexander Ezoubtchenko**, Hiroshi Akatsuka**
Haruaki Matsuura** and Ryuzou Takagi**

*Faculty of Engineering, Tokyo Institute of Technology,
2-12-1 O-okayama, Meguro-ku, Tokyo 152-8552, Tel/Fax +81-3-5734-2112

**Research Lab. for Nuclear Reactors, Tokyo Institute of Technology
2-12-1 O-okayama, Meguro-ku, Tokyo 152-8550, Tel/Fax +81-3-5734-3057

In the SCNES studies it has been confirmed that transmutation of radioactive nuclides and of net energy generation can be attained simultaneously. In order to realize the SCNES, material separation, isotope separation of the FP's and future subjects to be solved are discussed. In this paper we discuss about the required energy for a scheme of atomic vapor laser isotope separation (AVLIS) of FP's, the practical isotope separation system and the combination of AVLIS, plasma separation process and gas centrifugation methods for SCNES. And also we discuss about the future subject to guide to shift to SCNES smoothly in future concerning with waste management in the intermediate period to the SCNES. Consequently, the SCNES can be the ultimate goal of the nuclear energy development and it is very important to make effort to development in the intermediate period to SCNES.

1. INTRODUCTION

In the SCNES studies (Fujii-e,1992; Fujii-e et al. 1993; Akatsuka et al., 1995) it has been confirmed that realization of transmutation of radioactive nuclides and of net energy generation can be realized simultaneously. In other papers on SCNES studies, the mass of uranium and TRU's (transuranium elements) at an equilibrium state was calculated including 22 isotopes of U, Np ,Pu, Am, and Cm, in order to investigate the neutron balance in uranium and TRU's. Table 1, shows that the average number of emitted neutrons per fission is 2.95 and the average number of captured neutron by fuel, coolant and structural materials is 2.53, and that the core inventory is 72 t. It reveals that the neutron balances in uranium and TRU's are satisfied, uranium and TRU's can be contained in the core as fuel and the 0.42 excess neutrons per fission must be used for the transmutation of FP's, and the absorption by control rod.

In our previous study (Akatsuka et al., 1995), when the isotope separation is not applied to, the number of neutrons required to transmute the 29 important FP's is 1.12 per fission, much larger than the 0.42 excess neutrons per fission. Here, the 29 important FP's (fission products) are I, Pd, Cs, Zr, Tc, Sn, Se, Sm, Sr, Pm, Ru, Kr, Eu, Sb, Nb, Ho, Cd, Tb, Ag, Tm, and Rh. Namely, the neutron balance is lost. Moreover, total amount of FP's are 42.7 t/GWe, about twice larger than the amount of

Table 1. The neutron balance in the incineration of TRU's

Material	Production n/fiss	Absorbed n/fiss	Weight (t/Gwe)
Fuel TRU	2.95	2.22	22.7
Zr	-	0.03	2.5
Na Coolant	-	0.07	18.0
Struct.Mat.	-	0.21	28.8
Total	2.95	2.53	72.0

Excess Nuetron = 0.42 n/ fiss

fuel, and these FP's cannot be contained in the core. On the other hand, when the isotope separation is applied to FP's, the number of neutrons necessary to transmute the 29 important FP's is 0.25 per fission, smaller than the 0.42 excess neutrons. Here, the separated isotopes are ^{129}I, ^{107}Pd, ^{135}Cs, ^{93}Zr, ^{99}Tc, ^{126}Sn, ^{79}Se, ^{137}Cs, ^{151}Sm, ^{90}Sr, ^{147}Pm, ^{106}Ru, ^{85}Kr, ^{155}Eu, ^{125}Sb, ^{121m}Sn, ^{94}Nb, ^{166m}Ho, ^{134}Cs, ^{154}Eu, ^{113m}Cd, ^{158}Tb, ^{108m}Ag, ^{171}Tm, ^{152}Eu, ^{93m}Nb, ^{146}Pm, ^{102}Rh, and ^{109}Cd . The total amount of FP's is 6.2 t/GWe, smaller than the amount of fuel. Consequently, the neutron balance is kept and the 29 FP's can be contained in the core. It should be noted that neutron capture rate in ^{129}I, ^{107}Pd, ^{135}Cs, ^{93}Zr, ^{99}Tc, ^{126}Sn and ^{79}Se are higher than the natural decay rate of each nuclide. Therefore these nuclides are efficiently transmuted in the reactor of SCNES.

2. CHEMICAL SEPARATION AND REQUIRED ENERGY

As well known, pyrochemical treatment has a potential to offer a compact reprocessing facility, then its feasibility has been discussed (Chang *et al.*, 1990; Bychkov *et al.*, 1997). Spent metallic fuel is dissolved into a molten salt bath by anodic dissolution without any water. Transuranium(TRU) elements are eventually recovered with U. Some long-lived FP's are selected as target elements to be burnt in a reactor. Uranium and transuranium elements recovered in an electrorefining cell are transported to a fuel fabrication facility, while FP's for target and shielding assemblies are separately recovered. Stable FP's are removed from the cycle every year. In this section, how to recover uranium and TRU elements and FP's will be shown.

Spent metallic fuel pin is chopped into small pieces, and then put into a stainless steel basket. The basket is positively charged, so that uranium, TRU and some FP's are dissolved into a molten salt bath. At first, an iron rod is used for a cathode. In this stage, uranium is only deposited in the surface of the cathode, as dendrite shape. Secondly, a liquid metal cathode is used for a cathode as drawdown process. At this stage, uranium and TRU elements are deposited at the same time in a liquid metal cathode. Finally, electromigration technique is exploited in order to remove FP's from the melt bath. Multivalent or large cations are easily enriched near the anode, since they are less mobile than the solvent cations. In the final stage, Cs, Sr and rare earth's are enriched and removed from the melt bath. Fundamental studies on applicability of electromigration for enrichment have been already done (Iwasaki and Takagi, 1994; Matsuura et al., 1997).

Each process of fuel and FP's recovery processes was explained in detail (Takagi et al., 1995). The required energy for electrorefining, drawdown and electromigration process are also estimated. Especially, the required charge of electromigration is calculated by Klemm's equation approximately (Takagi et al., 1995). Although much energy consuming process is electromigration, total required energy is estimated to be about 0.4% of total electricity generated during 1year by FBR.

Table 2. The elements to which the isotope separateion must be applied

No	Element	Returned to the core	Removed from the core
1	$_{53}$I	129	125,127
2	$_{46}$Pd	107	102,104-106,108,110
3	$_{55}$Cs	134,135,137	133
4	$_{40}$Zr	93	90-92,94-96
5	$_{43}$Tc	97	97m
6	$_{50}$Sn	121m,126	114-118,119m,120,122-124
7	$_{34}$Se	79	76-78,80,82
8	$_{62}$Sm	151	146-150,152,154
9	$_{38}$Sr	90	84-89
10	$_{61}$Pm	146,147	148
11	$_{44}$Ru	106	98-104
12	$_{36}$Kr	85	80,82-84,86
13	$_{63}$Eu	152,154,155	151,153
14	$_{51}$Sb	125	121,123,124
15	$_{41}$Nb	93m,94	93
16	$_{67}$Ho	166m	165
17	$_{48}$Cd	109,113m	108,111,112,114,116
18	$_{65}$Td	158	159,160
19	$_{47}$Ag	108m	107,109,110m
20	$_{69}$Tm	171	168-170
21	$_{45}$Rh	102	102m,103

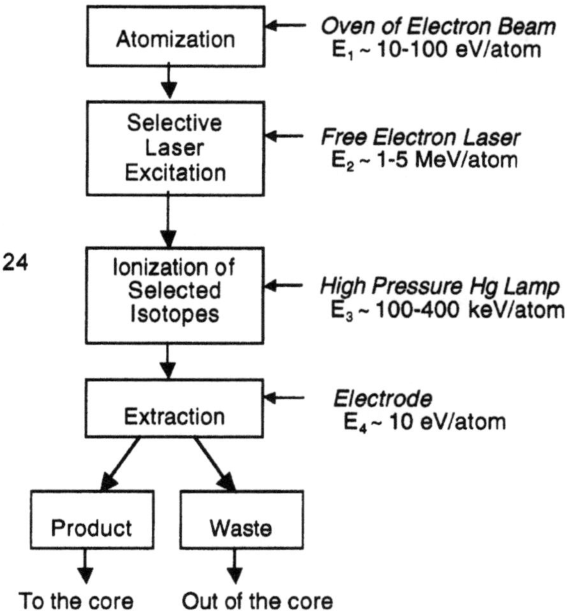

Fig.1 The schematic of AVLIS. (Akatsuka et al., 1995)

3. SCIENTIFIC FEASIBILITY OF ISOTOPE SEPARATION

In order to realize the SCNES, it is shown that material separation and isotope separation of the FP's are required. As the first stage, we investigated the required energy for a scheme of atomic vapor laser isotope separation (AVLIS) of FP's. Table 2 shows the elements to which the isotope separation must be applied. We chose AVLIS method. Here, we assume to use the two-photon excitation method, using a free electron laser (FEL) as a tunable laser and a high pressure Hg UV lamp as an ionizer for a simple analysis (Ozaki *et al.*, 1976). The scheme of the AVLIS consists of four steps, as is shown in Fig.1. Let us consider the energy Ei required for i-th stage.

(1)Atomization energy E_{atom} is about several eV, at most 10 eV for any elements. The efficiency of the heating by an oven is at least 10% (Ozaki *et al.*, 1976), then $E_1 \sim 100$eV.

(2)For the stage of the selective isotope excitation, the excitation energy E_{ex} itself is about several eV at most. However, the efficiency for this stage η_2 is very small and it can include most uncertainty. The efficiency of the selective excitation, $\eta_2 \sim 2.4 \times 10^{-6}$ was estimated and $E_2 \sim 5$ MeV was also estimated as a maximum value for the energy required for the selective excitation the energy.

(3) The energy required for the ionization of the selectively excited isotopes $E_3 \sim 400$ keV at most was obtained by using the efficiency of the Hg UV lamp (Ozaki et, al., 1976) and the absorption probability of the UV photons.

(4) As for the collection of the ionized isotopes, $E_4 \sim 10$ eV was estimated by considering the collisional loss by charge transfer collisions with other isotopes.

In consequence, the energy required for the isotope separation) or other, picking up one target isotope is considered to be $E = E_1 + E_2 + E_3 + E_4 \sim E_2 \sim 1$- 5 MeV/atom, namely, 5 MeV/atom at most. It should be noted that the calculated energy required for uranium enrichment is 0.24 MeV/atom by AVLIS, 0.30 MeV/atom by gas centrifuge process, and 3.0 MeV/atom by gaseous diffusion process (Janes et al.,

1976). It is considered that the estimated energy is comparable with these values, since the isotope shift of the FP's are about several ten times smaller than that of uranium due to the small Geld shift.

When a fissile nucleus makes one fission, the sum of the cumulative yield of the isotopes listed in Table 4 is about 0.25 for a fission of ^{239}Pu in a fast reactor (Ihara, 1989). Of course, the energy required for the isotope separation per fission should be multiplied by the fission yield, Y. On the other hand, the isotope separation should also be applied to the recycled long-lived FP's. When they are transmuted into short-lived or stable isotopes, they have to be removed from the reactor applying isotope separation. Consequently, the isotope separation should be applied, at most to the same amount of the recycled FP's as the fission originated FP,s. Therefore, the isotope separation should be applied to 0.5 FP atoms per fission.

As a result of this chapter, the energy required for the isotope separation per one fission of a fuel nucleus is estimated at (1 - 5)MeV x 0.5 ~ (0.5 - 3)MeV. The energy is 3 MeV at most which is lower than the fission energy 200 MeV, even when the thermal efficiency ~ 40% is taken into account. It means that the SCNES is feasible in terms of energy balance.

4. ISOTOPE SEPARATION SYSTEM FOR SCNES

For practical purposes, however, we should take the special feature of fission yields from SCNES core into account and suggest the candidate process for isotope separation in SCNES. We discussed about characteristics of isotope separation methods and characteristics of the radioactive isotope concentrations in the main FP's to be separated (Ezoubtchenko et al. 1998). Because the FP's have unique initial concentration and the best separation process depends on each element.

Recently plasma separation process (PSP) has been studied to obtain the good results which can lead to be one of candidate for future isotope separation. This method is based on isotopically selective ion cyclotron resonance heating of target ion component of plasma with subsequent energy and geometry discrimination of heated and cold components. The method was initially elaborated by TRW Inc.(USA) (Mussetto et al., 1983) for the candidate of uranium enrichment method operating at lower cost than gaseous diffusion.

We estimated the main energy consumption of one PSP module as follows (Ezoubtchenko et al. 1998).

1) sputtering (sputtering factor is near 2－4atoms/ion (Robinson, 1983)) = 300－400kW,
2) microwave system=200－400kW (it was needed to absorb approximately 1kW of microwave energy to obtain 1A of plasma ion flow to collector (Louvet, 1989)) ,
3) heating antenna system = 20－100kW,
4) helium liquefier = 100kW (liquid helium capacity = 40 litters/hour),
5) pumping system = 50 - 100kW.

Total energy consumption (energy cost of ion in plasma flow) is approximately 10kW/A = 10keV/ion.

In PSP method total energy consumption is weakly dependent on initial concentration of target isotope because of small relative contribution of heating antenna power. It is the difference between PSP and AVLIS, because the most part of energy in AVLIS is consumed by target isotope excitation laser systems. So AVLIS method is more preferable for low initial concentration of target isotope.

We can obtain maximum enrichment factor $\beta = 25$ (Louvet et al., 1994) in one PSP module when it operates in the enriching section and can approach depletion factor $\gamma = 8 - 10$ (Louvet et al., 1994) , (Gordon et al., 1983) when using PSP module in the stripping section (flow- ratio coefficient have to be adjusted). So it is clear that we have to use multistage (~ 10 stages) separation process with large stripping section.

Radioactive elements in FPs to be isotopically separated are approximately one half of total FPs , so energy consumption normarized to one fission (when applied only PSP method) can reach 10 keV/ion*2 ion/fiss. *0.5 * 10 stages = 100 keV/fiss, that is much less than 60 MeV electricity/fiss. in nuclear reactor. Maximum substance utilization coefficient ε in one PSP module was nearly 0.7 (Louvet, 1989). It is determined by ionization degree in ECR zone which in its turn is dependent on diameter of sputtering plate and microwave power absorbed. This coefficient does affect on above described energy estimations but represents technological benefit of PSP relatively to electromagnetic separator where ε is approximately 0.1.

The combination of PSP and AVLIS is more preferable in getting the high depletion factor in SCNES isotope separation cascade. AVLIS must be used only when the concentration of radioactive isotope in cascade became lower than 0.1%. AVLIS can be applied also to isotope separation of Rh, Ag and Ru starting from initial concentration. Furthermore, for Kr, I, Se, Cd, Sn whose gaseous compound can be obtained easily, it seems to be possible to use the gas centrifugation with having low energy consumption. The gas centrifugation can be used for these elements as a preliminary stage before AVLIS application. High performance of gas centrifugation was already demonstrated in isotope separation of sulfur, krypton, xenon and chromium(Roberts, 1989), (Szady, 1989). Energy consumption in uranium enrichment by gas centrifugation is less than 100kWh/kgSWU (Benedict et al., 1981) that implies 5 keV/atom of uranium containing 3% of ^{235}U. Of couse, separation capability of gas centrifuge in FPs isotope separation will be less than in uranium enrichment because of isotopes in FPs have mass difference $\Delta M = 1$. It must be noticed that long-term SCNES operation demands periodical removal of accumulated stable isotopes out of reactor core. Application of PSP method for this stage of SCNES operation will increase energy consumption per fission, AVLIS is to be possible method for this stage of SCNES.

5. FUTURE SUBJECTS TO SCNES

To realize the SCNES and shift to it smoothly in future (See Fig.2), we have many subjects to investigate in upstream and downstream of nuclear fuel cycle. Of course, one of the biggest subjects is the development of isotope separation method, which is the key technology for SCNES. We have to do our best to develop the free electron laser for AVLIS and the large scale of microwave generator for PSP, which are important technologies to accomplish the isotope separation. Another important issue can be the high level waste management in the intermediate period to SCNES performance. Although SCNES is our goal, it is necessary for us to offer our best technologies, which are currently available. The pyro-chemical reprocessing has a great potential and it can be the main stream of reprocessing to SCNES. We suppose that it will be introduced in near future. However, the HLW will be discharged also during intermediate period over many decades to SCNES. Problem is how to process and manage waste in this period. We have two options, (1) HLW discharged in this period is disposed underground. (2) It is stored until transmutation in SCNES.

5.1. Vitrification of Chloride Waste

In the case of option (1), the problem in the pyro-chemical reprocessing proposed here is that the wastes in the reprocessing have a form of metal chloride which is soluble generally. Many ideas have been proposed and investigated. In Japan where the vitrification is approved in the LWR waste management, the method with equivalent to or more safe than vitrification should be chosen

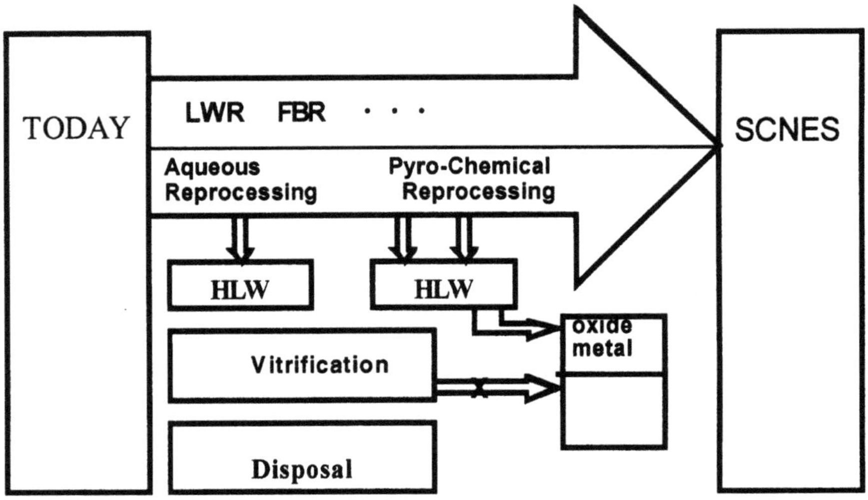

Fig.2 Concept of Downstream in Intermediate Period

also in future. This is one of the solutions that the chloride wastes are vitrified after conversion into oxide. In order to oxygenate metal chlorides or oxidize various metals, the reaction in high temperature can be effective. A method proposed (Ikeda et al. 1996) that the chloride wastes are converted into metal oxides by using the reaction with boric acid at around 1300K shows a good result.

In our previous paper (Suzuki et al. 1997) we demonstrated the conversion method of metal chloride wastes into oxides by using the microwave heated oxygen. Consequently, it was found that our oxygen plasma has a possibility to realize easily such an oxygenation reaction. Also, we proposed the direct vitrification method (Suzuki et al. 1998) which expect reacting chlorides directly with reactive oxygen atom and glass materials by plasma heating. Because plasma process requires lots of power and it is not a good way to mix oxides with glass material and vitrify them after converting metal chloride into metal oxide from the view point of energy consumption and simplicity of process. From several experimetal results we can conclude that metal chlorides can be vitrified directly by using oxygen plasma and this maehod can be useful. Therefore, in near future we can adopt the pyro-chemical reprocessing without any problem of backend.

5.2 Storage for Transmutation

In the case of option (2), to store the HLW in the form of oxide or metal is much more convenient than to extract target elements from the vitrified material. Because the vitrified material is stable and it requires lots of energy and time-consuming job. So, once the transmutation process is available, the waste should be stored in the oxide or metallic form. The conversion process of chloride-oxide is available as mentioned in above section. Also the reduction process of oxide is available. In such chemical processes energy consumption is not so large as mentioned in the pyro-chemical reprocessing. These chemical processes should be selected from a view point of the amount of secondary waste, although we have many candidates.

As mentioned above, we introduced our microwave plasma technology for oxygenation and vitrification processes. In general the plasma technology is used widely for producing many valuable materials. Especially, the plasma can supply a large amount of reactive atoms directly and stably, if molecular gas is used as a plasma gas. Generally atoms and ions in the plasma are very reactive and their generation methods have been developed and applied to various utilization. In our recent paper (Suzuki et al. 1997, Windarto et al. 2000), we reported that our microwave heated oxygen plasma fulfills its function as an oxygen atom source rather than a plasma source and its plasma flow has mild feature with not so high temperature and not so high velocity compared with the other atmospheric pressure plasma, for example, the DC arc plasma or the RF heated plasma. From a view point of safety, this mild feature has an advantage. Moreover, also it has a large potential to make processes simple and to reduce secondary waste. Thus, it is very important for us to review all of technologies available in many field and to investigate their possibilities. This kind of efforts can give us the best technology and help to shift smmothly to SCNES.

6. SUMMARY

It was shown that isotope separation was necessary in order to return the long-lived FP's into the reactor without making it subcritical. By AVLIS method, the energy required for picking the isotopes in problem out of the system was estimated at about 3MeV per fission of a fuel nuclide. And also we can confirm that when combining AVLIS, PSP and gas centrifugation methods for isotope separation, energy consumption will be less than AVLIS only. Therefore, the SCNES can be the ultimate goal of the nuclear energy development and it is very important for us to review all for technologies available in many field and to investigate their potentials. This kind of efforts in the intermediate period can help to approach the SCNES and to shift smoothly to the SCNES.

REFERRENCES

Akatsuka, H., Ohsaki,T., Obara, M., Suzuki, M. and Fujii-e, Y. (1995), Scientific Feasibility of Incineration in SCNES, *Prog. Nucl. Energy*, 29(Suppl.), 477-484.

Benedict, M., Pigford, T. and Levy, H., (1981). *Nuclear Chemical Engineering, 2nd Ed.*, McGraw Hill, New York.

Bychkov, A. V., Vavilov, S. K., Skiba, O. V., Porodnov, P. T., Pravdin, A. K., Popkov, G. P., Suzuki, K., Shoji, Y. and Kobayashi, T. (1997), *Proceedings of International Conference Future Nuclear Systems, Grobal'97*, Yokohama, Japan, 2, 912

Chang, Y. I., Walters, L. C., Battles, J. E., Wade, D. C. and Lineberry, M. J. (1990), *ANR-IFR-125*

Ezoubtchenko, A. N., Aktsuka, H. and Suzuki, M. (1998), Isotope Separation for Self-Consistent Nuclear Energy System, *Prog. Nucl. Energy*, 32[3/4], 729-735.

Fujii-e, Y. (1992). An approach to self consistent nuclear energy system – potential of fast reactors -, *Proc. Int. Conf. Design of Advanced Power Plant*, 2, 11.3, 1-9.

Fujii-e.Y., M. Suzuki, K. Arie and H. Endo (1993) An evaluation function for a SCNES. *Trans. Am. Nucl. Soc.*, 67, 353-355

Gordon, J. D., Thomson, J. J. and Suchard, S. N. (1983). Isotpically tailored HT-9 steel for radioactive waste management, IEEE Symp. on Fusion Engineering, 1260-1266.

Ihara, H. (1989). Tables and figures from JNDC nuclear data library of fission products, Ver.2. *JAERI-M*, 89-204.

Ikeda Y., Takashima Y. Kobayashi H. and Igarashi H., (1995), Conversion reaction of metal chlorides into oxides with boric acid, *J. Nucl. Sci. Technol.*, 32 [11], 1138-1145.

Iwasaki, M. and Takagi, R. (1994). Enrichment of Lanthanum in Its Dilute Molten Salt Solution,

J. Nucl. Sci. Technol., 31, 751-753.

Janes, G. S., H. K. Forsen and R. H. Levy (1976). Research and development prospect for the atomic uranium laser isotope separation process, *AIChE Symp Ser.*, 73, [169], 62-68

Louvet, P (1989). Review of Plasma Separation Process, *2nd Workshop on Separation Phenomena in Liquid and Gases, Versailles, France.*

Louvet, P (1994). Heavy stable isotope separation by ion cyclotron resonance, *4th Workshop on Separation Phenomena in Liquid and Gases, Beijin, China.*

Matsuura, H., Takagi, R., Okada, I. and Fujita, R. (1997). Enrichment of Fission Products in Ionic Salt Bath by Countercurrent Electromigration, *J. Nucl. Sci. Technol.*, 34, 304-309.

Mussetto M. et al. (1983). Isotope separation using selection ion cyclotron resonance heating, *IEEE Intl. Conf. On plasma science, San Diego.*

Ozaki, N., K. Yamada, M. Yamamoto, S. Tomiyama and K. Ueyanagi (1976). Cost analysis ofuranium enrichment by laser isotope separation process, *Appl. Energy*, 2, 279-294.

Suzuki M., Komatsubara M., Umebayashi M. and Akatsuka H.,(1997), Conversion of ChlorideWaste into Oxide by Mycrowave Heated Oxygen *Plasma, J. Nucl. Sci. Technol.*, 34 [12], 1159-1170

Suzuki M. and Akatsuka H.,(1998), Application of Microwave Heated Plasma to Vitrification and Decontamination Processes, Proc. of The 3rd Japan-Korea Seminar on Advanced Reactors, Seoul, 167-173.

Szady, A. J. (1989). Enrichment of chromium isotopes by gas centrifugation, Nuclear Instruments and Methods in Physics Research, A282, 277-280.

Takagi, R., Matsuura, H., Fujii-e, Y., Fujita, R. and Kawashima, M. (1995), A Fuel Cycle Concept for Self-Consistent Nuclear Energy System, *Prog. Nucl. Energy*, 29(Suppl.), 471-476.

Windardo H., Matsumoto T., Akatsuka H. and Suzuki M. (2000), Decontamination Process Using CF_4-O_2 Microwave Discharge Plasma Atmospheric Pressure, *J. Nucl. Sci. Technol.*, 34 [9], 787-792.

Pergamon
www.elsevier.com/locate/pnucene

Progress in Nuclear Energy, Vol. 40, No. 3–4, pp. 569–575, 2002

Printed in Great Britain
0149-1970/02/$ - see front matter

PII: S0149-1970(02)00052-5

EVALUATION OF SELF-CONTROLLABILITY OF LMFRs AGAINST ATWS

H. NINOKATA, Y. HIZUME, *H. ENDO, T. SAWADA

Research Laboratory for Nuclear Reactors
Tokyo Institute of Technology, O-okayama, Meguro-Ku, Tokyo 152-8550 Japan
Tel/Fax: 03-5734-3056; E-mail: hninokat@nr.titech.ac.jp

*Isogo Nuclear Engineering Center, Toshiba Corporation
8 Shin-Sugita-Cho, Isogo-Ku, Yokohama 235-8523 Japan
Tel: 045-770-2401; Fax: 045-770-2462

Abstract

This paper describes a simple and general method to evaluate inherent and passive safety characteristics of fast reactors that uses self-controllability limit lines. The self-controllability limit lines are derived analytically for the extreme conditions such as sodium boiling or fuel melting as linear relationships among the total reactivity, effective coolant and fuel reactivites. The analytically obtained self-controllability limit lines are compared to those obtained by numerical result from the LMFR plant system transient analysis. For a given set of various reactivity coefficients of any actual LMFRs, judgement is made on whether the reactor can maintain the core integrity under ATWS conditions.

1. INTRODUCTION

The advanced and innovative liquid metal-cooled fast reactors (LMFRs) emphasize the inherent and passive safety characteristics at least in the following way:

- Highly reliable and less complex shutdown and decay heat removal systems with the use of not only active but also inherent or passive means encouraged by including negative temperature coefficient, natural circulation, etc.;
- Designs that minimize the potential for severe accidents and their consequences by providing sufficient inherent safety, reliability, redundancy, diversity and independence in safety systems; and
- Designs that incorporate defense-in-depth philosophy by maintaining multiple barriers against radiation release, and by reducing the potential for and consequences of severe accidents.

The LMFRs under operation also provide the high level of confidence on the safety. This is achieved based on the inherent and passive safety characteristics but an emphasis is also placed on the active safety measures for shutdown and cooling functions. We consider that the active safety measures will still play an important role in realizing the defense-in-depth philosophy in the future. However, as a cost reduction requirement to LMFR plants emerges to compete with more economical LWRs as well as other fossil power plants, these active safety systems are also required to be of low cost. At the same time we need to keep the high or even higher reliability of the systems and system components. To achieve these seemingly contradictory requirements, self-actuated shutdown system (SASS), enhanced thermal expansion system (ETEM),..., etc are being considered. These additional systems are required to assure the shutdown against the possible positive reactivity insertion during anticipated transient without scram (ATWS) of low frequency occurrence. Then the subsequent core cooling is achieved by natural circulation. In a sense, these passive measures could replace some part of the active components and system, but may not lead to a drastic cost reduction. Also we note here that, unless an LMFR core have an overall negative feed back characteristics in case of any anticipated accidents, the burden onto the safety system would remain heavy against ATWS type accidents.

Thus, future generation LMFRs should provide more enhanced margin of safety not by additional safety measures but with inherent and passive safety features of the core itself. Here the "future generation" LMFRs should be interpreted as those revolutionary designs to follow the advanced and innovative LMFRs mentioned at the beginning. This would be a requirement from the viewpoint of reducing the burden onto the safety system in one way and the public concern in another. In other words, future liquid metal-cooled fast reactors would be required to have inherent and passive safety characteristics against ATWS and to render no fear against core disruptive accidents any more, even if they are extremely unlikely at the present knowledge. For instance, in case of ATWS, i.e., unprotected loss of flow (ULOF), transient over power (UTOP) and loss of heat sink (ULOHS) accidents, the reactor power during the transients should stay below the level that corresponds to the core temperatures at which sodium boiling or fuel melting is assumed to take place. We call this specific safety characteristic as "*self-controllability*" and these temperatures as "*self-controllability limits*." In order to design and evaluate the self-controllability of the future LMFRs, we have developed a general method, in which we employ self-controllability limit lines to judge whether a reactor can maintain the core integrity under the ULOF, UTOP and ULOHS conditions.

2. SELF-CONTROLLABILITY LIMIT LINES

The most straightforward approach to evaluate the self-controllability of an LMFR is to run extensively a well-validated plant system transient analysis code and find whether either coolant or fuel temperature of the LMFR exceeds the self-controllability limits or not. Every time the core neutronics or thermal hydraulics parameters are changed, this process must be repeated. Also we will not know whether the LMFR is well equipped with the self-controllability until we run the transient analysis code. Therefore it would be desired, if we have a safety map that is described by a certain set of simple parameters, preferably those which can be defined under the steady-state conditions, and that can immediately tell us the margin to the self-controllability limits.

In order to facilitate the derivation of those limit lines, we now define a coolant effective reactivity $\rho_{eff,cool}$ and a fuel effective reactivity $\rho_{eff,fuel}$. The former effective reactivity is defined as a sum of the reactivity variations due to the coolant temperature difference ΔT_C (z,0) from the core inlet temperature and the latter as the one due to the fuel temperature difference ΔT_F(z,0) from the coolant temperature at the axial elevation z and all under steady-state operating conditions:

$$\rho_{eff,cool} \equiv \int K_c(z) \cdot \Delta T_c(z,0)dz \quad \text{and} \tag{1}$$

$$\rho_{eff,fuel} \equiv \int K_F(z) \cdot \Delta T_F(z,0)dz \tag{2}$$

where K_C and K_F are defined as:

$$K_C(z) = K_{Dop}(z) + K_{F\exp}(z) + K_{Clad}(z) + K_{Na}(z) + K_{SA,\exp}(z)\text{, and}$$

$$K_F(z) = K_{Dop}(z) + K_{F\exp}(z).$$

In Eqs. (1)-(2), K's are the reactivity coefficients; and the subscripts Dop, Fexp, clad, Na and SAexp are the Doppler, fuel expansion, cladding, sodium and subassembly expansion, respectively. We later find a relationship between these two parameters $\rho_{eff.cool}$ and $\rho_{eff.fuel}$, that corresponds to such an extreme condition as sodium boiling or fuel melting initiation.

If a combination of $\rho_{eff.cool}$ and $\rho_{eff.fuel}$ obtained for a specific fast reactor is located blow the limit lines for the ATWS's, we can judge that the reactor does not suffer from coolant boiling and fuel melting. However, in the process of deriving these limit lines, we must carry out a number of time-consuming transient calculations using the transient system analysis code. One of the objectives of this work is to obtain the self-controllability limit lines with simple analytical expressions.

3. DERIVATION OF SELF-CONTROLLABILITY LIMIT LINES

In pursuit of the above objective, we introduce the following assumptions: (I) the extreme conditions are part of the quasi-steady responses under ULOF, UTOP and ULOHS conditions; (II) the total reactivity is nearly zero under the extreme conditions except for ULOF; and (III) the power transient can be approximated by a simple exponential function in case of ULOF and the reactor power is well reduced to a decay power level in case of a ULOHS extreme state. Focusing on the extreme conditions, we consider the balance of feedback reactivates and core temperature distributions with the assumptions.

3.1 General Form

The total feedback reactivity $\rho_{tot}(t_{max})$ added up to the elapsed time t_{max} from the onset of any transient at t=0 will be expressed by the sum of reactivities due to the increase of fuel, cladding and coolant temperature between t=0 and t= t_{max} as:

$$\rho_{tot}(t_{max}) = \int\left[K_{Dop}(z) + K_{F,exp}(z)\right]\cdot\left[T_F(z,t_{max}) - T_F(z,0)\right]dz + \int K_{Clad}(z)\cdot\left[T_{CL}(z,t_{max}) - T_{CL}(z,0)\right]dz$$
$$+ \int\left[K_{Na}(z) + K_{SA,exp}(z)\right]\cdot\left[T_C(z,t_{max}) - T_C(z,0)\right]dz + \rho_{ext}. \qquad (3)$$

T is the temperature, and the subscripts F, CL and C denote the fuel, cladding and coolant, respectively. t_{max} is the time when the peak temperature is attained. We can replace the peak temperature with that of coolant boiling or fuel melting. The inserted control rod reactivity ρ_{ext} should be zero in the event of ULOF and ULOHS.

Because the phenomena are considered to be slow and mild during the transients, the system response can be described by quasi-steady state calculation (Assumption I). Also we assume the power density $q'''(z, t)$ dose not change its shape very much during the transients. Therefore, we could assume that the shapes of transient temperature distributions in the axial direction are close enough to those of steady-state approximation as shown by Ninokata et al (1997). We consider the one-dimensional forced convection flow heat removal. If the coolant specific heat is assumed to be constant, the simple energy balance equation yields the following:

$$\Delta T_C(z,t) \equiv T_C(z,t) - T_{C,in} \propto \frac{1}{F(t)}\int_0^z q'''(\xi,0)d\xi \qquad (4)$$

Where $F(t)$ is the flow rate and $T_{C,in}$ is the core inlet coolant temperature. If the fuel heat conductivity $k_F(T)$ is assumed to be constant, by solving the heat conduction equation in r-z geometry, we obtain:

$$T_F(r,z,t) - T_F(R,z,t) \propto q'''(z,0). \qquad (5)$$

Assuming $T_F(R,z,t)$ is nearly equal to the coolant temperature $T_C(z,t)$, the difference temperature $\Delta T_F(z,t)$

between the fuel radial average temperature $T_F(z,t)$ and the coolant temperature $T_C(z,t)$ can be expressed as:

$$\Delta T_F(z,t) \cong T_F(z,t) - T_C(z,t) \quad \propto \quad q'''(z,0) \tag{6}$$

Introducing these difference temperatures into Eq. (3), the total feedback reactivity $\rho_{tot}(t_{max})$ is expressed by the sum of reactivities due to the power to flow ratio(ΔT_C), the power(ΔT_F) and the increase inlet coolant temperature($T_{C,in}$) as:

$$\begin{aligned}\rho_{tot}(t_{max}) = &\int\left[K_{Dop}(z) + K_{F,exp}(z)\right]\cdot\left[\Delta T_F(z,t_{max}) - \Delta T_F(z,0)\right]dz \\ &+ \int\left[K_{Dop}(z) + K_{F,exp}(z) + K_{Clad}(z) + K_{Na}(z) + K_{SA,exp}(z)\right]\cdot\left[\Delta T_C(z,t_{max}) - \Delta T_C(z,0)\right]dz \\ &+ \int\left[K_{Dop}(z) + K_{F,exp}(z) + K_{Clad}(z) + K_{Na}(z) + K_{SA,exp}(z)\right]\cdot\left[T_{C,in}(t_{max}) - T_{C,in}(0)\right]dz \\ &+ \rho_{ext}.\end{aligned} \tag{7}$$

Introducing Eqs. (1) and (2) into Eq. (7), finally we obtain the following linear relationship between $\rho_{eff,cool}$ and $\rho_{eff,fuel}$:

$$\rho_{eff,cool} = -a\rho_{eff,fuel} + b \tag{8}$$

where

$$a = \left[\frac{\int\Delta T_F(z,t_{max})dz}{\int\Delta T_F(z,0)dz} - 1\right]\Big/C, \qquad b = \frac{\rho_{ext} - \rho_{tot}(t_{max})}{C}, \tag{8a}$$

and the constant C is defined as:

$$C = \frac{\int\Delta T_C(z,t_{max})dz}{\int\Delta T_C(z,0)dz} - 1 + \frac{\int\{T_{C,in}(t_{max}) - T_{C,in}(0)\}dz}{\int\Delta T_C(z,0)dz} \tag{8b}$$

It is noted that $\rho_{eff,cool}$ is defined by reactivity due to the power to flow ratio under the steady-state operating condition and $\rho_{eff,fuel}$ is defined by the one due to the power.

Equation (8) holds for such extreme conditions as coolant boiling or fuel melting. If all material properties, i.e., the coolant specific heat and fuel heat conductivity are be assumed to be constant, with the use of Eqs. (4), (6) in Eq. (8), we obtain the following general expression for the self-controllability limit line:

$$\rho_{eff,cool} = -\frac{\left[\dfrac{P(t_{max})}{P(0)} - 1\right]\cdot\rho_{eff,fuel} - \rho_{tot}(t_{max}) + \rho_{ext}}{\dfrac{P(t_{max})/F(t_{max})}{P(0)/F(0)} - 1 + \dfrac{\int\{T_{C,in}(t_{max}) - T_{C,in}(0)\}dz}{\int\Delta T_C(z,0)dz}}. \tag{9}$$

In the following context, we assume constant material properties for simplicity, although this assumption is not appropriate in the case of metallic fuel thermal conductivity.

3.2 Simple Analytical Models for ATWS

3.2.1 Unprotected transient overpower accidents

Peak temperature will be attained within a few hundred seconds from the onset of UTOP and the flow rate

F(t) stay constant. With $T_{C,in}(t_{max})$ - $T_{C,in}(0) = 0$ and $\rho_{tot}(t_{max}) \sim 0$, then Eq. (9) is reduced to:

$$\rho_{eff,cool} = -\rho_{eff,fuel} - \frac{\rho_{ext}}{P(t_{max})/P(0)-1}. \tag{10}$$

When the coolant boiling is the limiting case,

$$\frac{P(t_{max})}{P(0)} = \frac{T_{C,BOIL} - T_{C,in}(0)}{T_C(H,0) - T_{C,in}(0)}, \quad \text{and} \tag{11}$$

when the fuel melting is the case,

$$\frac{P(t_{max})}{P(0)} = \frac{T_{F,MELT} - T_{C,in}(0)}{T_F(0, z_{max}, 0) - T_{C,in}(0)} \tag{12}$$

are used in Eq. (10). $T_F(0,z_{max},0)$ in Eq. (12) is the maximum fuel center line temperature. $T_{C,BOIL}$ and $T_{F,MELT}$ are the coolant boiling and fuel melting temperatures.

3.2.2 Unprotected loss of flow accidents

Under the ULOF condition, it is unlikely that the fuel melting takes place prior to the onset of coolant boiling. With $\rho_{ext} = 0$ and $T_{C,in}(t)$ - $T_{C,in}(0) = 0$, Eq. (9) becomes:

$$\rho_{eff,cool} = -\frac{\left[\frac{P(t_{max})}{P(0)} - 1\right] \cdot \rho_{eff,fuel} - \rho_{tot}(t_{max})}{\frac{P(t_{max})/F(t_{max})}{P(0)/F(0)} - 1}. \tag{13}$$

In Eq. (13), the following relations are used:

$$\frac{P(t_{max})/F(t_{max})}{P(0)/F(0)} = \frac{T_{C,Boil} - T_{C,in}(0)}{T_{C,out}(0) - T_{C,in}(0)} \quad \text{and} \tag{14}$$

$$\frac{P(t)}{P(0)} \approx \exp\left(\frac{\lambda \rho_{tot}(t_{max})}{\beta - \rho_{tot}(t_{max})} t\right). \tag{15}$$

Because it is impossible to obtain an analytical solution to one point kinetics equation with reactivity feedback, we used a crude approximation given by Eq. (15), where λ and β are the decay constant and delayed neutron fraction. The total reactivity ρ_{tot} can be obtained with Eqs. (14)-(15) and the following flow coast down curve equation:

$$\frac{F(t)}{F(0)} = \frac{1}{1 + t/\tau}. \tag{16}$$

Here τ is the pump coastdown half time.

3.2.3 Unprotected loss of heat sink accidents

In the event of ULOHS, the transient is slow and mild and the power will stay around decay heat level when the peak temperature is attained:

$$P(t_{max}) \approx 0 \tag{17}$$

and the inlet subcooling is already zero. This implies:

$$T_{C,in}(t_{max}) \approx T_{C,out}(t_{max}) \approx T_{C,Boil}\,. \tag{18}$$

With the total reactivity $\rho_{tot}(t_{max})$ nearly equal to zero, finally the self-controllability limit line can be derived from Eq. (9) as:

$$\rho_{eff,cool} = \frac{\rho_{eff,fuel}}{\dfrac{\int \{T_{C,boil} - T_{C,in}(0)\}dz}{\int \Delta T_C(z,0)dz} - 1}\,. \tag{19}$$

4. EVALUATION OF SELF-CONTROLLABILITY OF FBRs

Self-controllability limit lines under the UTOP, ULOF and ULOHS conditions, i.e., Eqs. (10), (13) and (19), respectively, are compared to the numerical results obtained by the ARGO code developed by Endo, et al. (1992). The UTOP lines are drawn in parallel and separated as different external reactivities are added. It was found that the analytical model Eq. (10) gives excellent agreement with the ARGO code results for the MOX fueled core. However, for the metal fueled core, the coefficient of $\rho_{eff,fuel}$ is −1 as shown in Eq. (10) and the gradient of the line is slightly steeper than that given by the ARGO results. This is considered to be due to the assumption of constant material property.

The ULOF limit lines given by of MOX- and metal-fueled cores for various pump coast-down half times are almost overlapping with each other. Agreement between analytical model and ARGO is good. A slight discepancy from the ARGO results is considered to be due to the fact that we employed Eq. (15) that really does not represent the solution of one-point kinetics equation with the feedback reactivities. This discrepancy gives the influence on the prediction of $P(t_{max})$ and the coefficient of $\rho_{eff,fuel}$ in Eq. (13). However, the value of $P(t_{max})/P(0)$ is very small compared with $\{P(t_{max})/F(t_{max})\}/\{P(0)/F(0)\}$ and, therefore, Eq. (13) gives the reasonable results.

In the same way, the ULOHS limit lines are compared between Eq. (19) and the ARGO results and in this case agreement is excellent as reported by Hizume et al. (2000). In general, the ULOHS line always stays above the ULOF line. This suggests that the ULOHS does not have to be considered in evaluating the self-controllability capabilities of fast reactors against ATWS but the ULOF and UTOP.

As a summary, the self-controllability limit lines are shown in Fig. 1 in terms of $\rho_{eff.cool}$ and $\rho_{eff.fuel}$ for ULOF with $\tau = 10$ sec, UTOP with $\rho_{ext} = 0.5$ \$ ($\dot{\rho}_{ext} = 3$ ¢/s) and $q' = 430$ W/cm and ULOHS conditions as representatives. If a combination of $\rho_{eff.cool}$ and $\rho_{eff.fuel}$ obtained by Eqs. (1) and (2) for a specific fast reactor is located blow these limit lines for the ATWSs, we can judge that the reactor is equipped with the self-controllability characteristics. Hence, this reactor does not suffer from coolant boiling and fuel melting and stay intact under any ATWS conditions.

The fuel and coolant effective reactivities $\rho_{eff.cool}$ and $\rho_{eff.fuel}$ are calculated from design specifications and are plotted in the figure for the fast breeder reactors which are already shut down, currently operating or under design stage, i.e. DFBR, FAIDUS, SPX-1, EFR, MONJU and JOYO. It is noted that the self-controllability limit lines depend on the maximum linear heat rate, the coolant inlet and outlet temperatures. Therefore comparison should be made with the self-controllability limit lines obtained in consistency with these parameters. In this paper, however, we assume that these are common for each LMFR. Also responses to UTOP, ULOF and ULOHS are different per each plant and do not always allow us to justify the assumptions (I), (II) and (III) mentioned in Section 3. Therefore assessment of the self-controllability of these LMFRs may not provide conservative results but would offer useful implications and insights into safety characteristics of each LMFR as a first step.

It is found that all LMFRs have the self-controllability for ULOHS and UTOP but for ULOF conditions.

Only JOYO has the tolerance against ULOF due to the large coolant reactivity. Thus the self-controllability limit line approach allows us immediately to assess the margin to the self-controllability limits quantitatively.

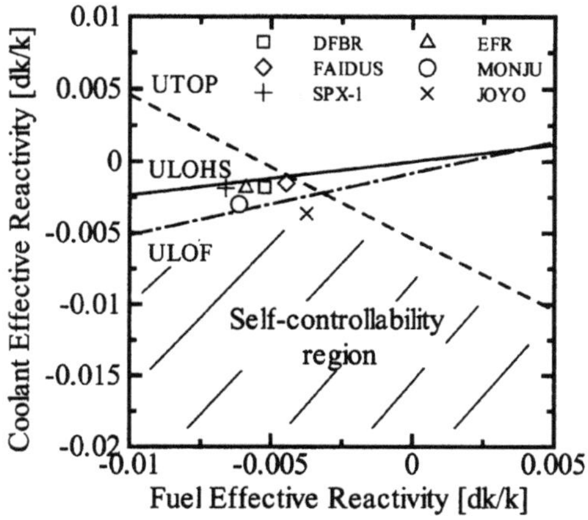

Fig. 1 Self-controllability limit lines for ULOF, UTOP and ULOHS of FBRs

5. CONCLUSIONS

In order to evaluate the inherent and passive safety characteristics of liquid metal cooled fast reactors, we have developed a simple analytical method in which we use the self-controllability limit lines. We consider that this analytical method allows us to evaluate the self-controllability of any fast reactor core without carrying out a number of time-consuming transient calculations for typical ULOF, UTOP and ULOHS accidents using plant system transient analysis codes. Also the method provides us with clear physical insights into the complicated feedback mechanisms related to the nuclear and thermohydraulics coupling phenomena.

REFERENCES

Y. Hizume, et al. (2000), Derivation of the fast reactor self-controllability limit lines by an analytical approach (3), (in Japanese), Proc. 200 Fall Mtg of AESJ, G3, September.

H.NINOKATA et al. (1997), A general method to evaluate passive safety characteristics of FBRs, Proc. Fifth International Topical Meeting on Nuclear Thermal Hydraulics, Operations and Safety (NUTHOS-5), Beijing, China, T6-1~6, April.

H. ENDO, et al. (1992), Passive safety features of bottom supported fast breeder reactor vessel, *Nucl. Tech.*, Vol. 99, 318 (1992)

Pergamon

www.elsevier.com/locate/pnucene

Progress in Nuclear Energy, Vol. 40, No. 3–4, pp. 577–586, 2002

Printed in Great Britain

0149-1970/02/$ - see front matter

PII: S0149-1970(02)00053-7

ELIMINATION OF RECRITICALITY POTENTIAL FOR THE SELF-CONSISTENT NUCLEAR ENERGY SYSTEM

H. ENDO[1*], S. KUBO[2], S. KOTAKE[2], I. SATO[3], Sa. KONDO[3], H. NIWA[3], and K. KOYAMA[4]

[1] Research Laboratory for Nuclear Reactors, Tokyo Institute of Technology
2-12-1, O-okayama, Meguro-ku, Tokyo152-8550, Japan
[2] The Japan Atomic Power Company (JAPC)
Ohtemachi Bldg., 6-1-1, Ohtemachi, Chiyoda-ku, Tokyo 100-0004, Japan
[3] O-arai Engineering Center, Japan Nuclear Cycle Development Institute (JNC)
4002, Narita, O-arai, Ibaraki 311-1393, Japan
[4] Advanced Reactor Technology Co., Ltd. (ARTECH)
15-1 Tomihisacho, Shinjuku-ku, Tokyo 162-0067, Japan

ABSTRACT

The ultimate safety goal of the Self-consistent Nuclear Energy System (SCNES) is to eliminate the recriticality-problem based on a simple safety logic. The principle of the elimination of the recriticality-problem is the Controlled Material Relocation (CMR) to establish the neutronic shutdown by removing the molten fuel to the out of core before a large scale pool formation which has potential of energetics driven by a super prompt criticality.

The CMR concept should be reliable without significant impact on the core neutronic performance. As the typical core concepts to enhance this CMR characteristic, several design options are under consideration. They are fuel assemblies with inner duct structure (FAIDUS), fuel assemblies with hollow fuel pins in the axial blanket region (ABLE) for MOX fueled cores, and fuel assemblies without fuel pin bundle structure in the lower axial blanket region (ELAB) for the metallic fueled core. Based on the core design study and accident analyses, these CMR-oriented concepts have been found feasible without significant degradation of the neutronic performance

In order to experimentally confirm the effectiveness of the CMR concept for the MOX fueled core, the EAGLE project has been started in 1998 by Japan Nuclear Cycle Development Institute (JNC) and The Japan Atomic Power Company (JAPC). The EAGLE project is the experimental program utilizing the out-of pile test facility and in-pile facility IGR of the National Nuclear Center of the Republic of Kazakhstan (NNC/RK).

I. INTRODUCTION

The Self-Consistent Nuclear Energy System (SCNES) has been proposed as an ultimate goal of the nuclear energy system/1,2/. The SCNES aims at a simultaneous realization pf four objectives, an energy production, a fuel breeding, a burning of long lived fission products (LLFPs), and a system safety. The SCNES is the system utilizing the fast neutrons, and is not in the maximum reactive configuration of the core materials. This indicates that a recriticality potential might be activated during the core damage. Therefore the safety goal of the Self-Consistent Nuclear Energy System (SCNES) is to eliminate the recriticality potential against the sever accidents such as unscrammed events.

***Present address: Toshiba Corporation,**

8, Shinsugita-cho, Isogo-ku, Yokohama 235-8523, Japan

The recriticality elimination capability of the SCNES can be achieved by the following two simple concepts, self-controllability and self-terminability, based on the inherent tolerance induced by thermal and neutronic properties of the core system:

- ***Self-controllability:*** the capability to eliminate the fuel failure propagating to the core disruptive accident (CDA) by using the negative reactivity feedback induced by the temperature increase of core materials during the unscrammed events, and
- ***Self-terminability:*** the capability to eliminate the recriticality problem, especially the energetics driven by a super prompt criticality, by using the shutdown reactivity induced by the core material relocation during abnormal events involving the fuel failure.

An aim of this paper is to describe the recent activities of the study on the elimination of the recritucality-problem (called hereinafter, recriticality elimination) in Japan, especially on the self-terminability. This paper describes the principle of the re-criticality elimination, core concepts to realize the principle, and the experimental project EAGLE which will confirm the effectiveness of core concepts.

II. PRINCIPLE OF THE RECRITICALITY ELIMINATION

Key Phenomena to be Eliminated

Once the large scale core of the fast neutron system is disrupted due to the abnormal events involving the fuel failure, the reactor power is increased by the positive reactivity feed back due to the coolant boiling, and the fuel failure takes place in about half of core region in the initial accident phase. Based on the colossal CDA study on the fast reactors, however, it is pointed out that the super prompt criticality during the initial stage of the CDA can be eliminated by the proper limitation of the design value of the positive void reactivity. The design limit for the coolant void reactivity, based on the presently available information, is evaluated about 6$ for MOX fueled cores and about 8$ for metallic fueled cores in the case of the sodium cooled fast reactors /3/.

Figure 1 shows the typical power transient of the CDA for the sodium cooled MOX fueled core. As shown in the Fig.1, the energetics driven by the super prompt criticality will be increased with increasing of the freedom of the failed fuel motion after the initial stage of the core damage. The most important energetics is induced by the self-triggered sloshing motion of the whole core pool developed from the small-scale pool formed by the combination of the failed fuel assemblies. Although the early termination of the transient without energetics is the dominant sequence of CDA due to the fuel removal to the out-of-core region (path B in the Fig.1), the possibility of the energetics due to whole core pool motion cannot be ruled out by large uncertainties involved in the failed fuel relocation phenomena (Path A in the Fig.1).

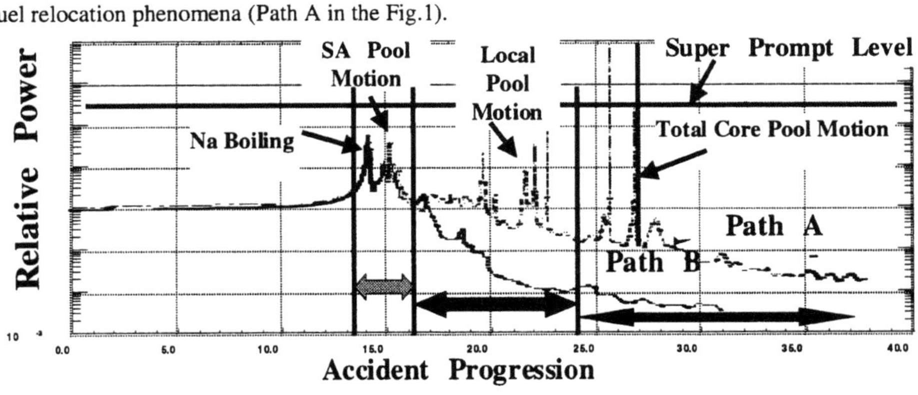

Fig.1 Typical power transient of MOX fueled core during ULOF

Therefore the key phenomena to be eliminated for the self-terminability is the molten fuel pool formation with a critical radius, which has a potential of the high energetics of the whole core pool. The critical pool is defined as the pool with minimum radius that the reactivity insertion rate exceeds the critical ramp rate to drive the super prompt criticality by the self-exiting pool motion. For the case of the sodium cooled MOX-fueled core, the critical pool radius is ranging from 30 to 50% of the total core radius as shown in the Fig.2.

Principle of the Self-Terminability and Condition to Establish the CMR

The principle of the recriticality elimination is to establish the neutronic shutdown state by the preceding removal of the molten core fuel to the out of core region before the formation of the critical pool defined above. The core material relocation intended to prevent the critical pool formation with high reliability is called as the Controlled Material Relocation (CMR) and is the basic concept to design the recriticality-free core. If the critical pool formation is prevented by the sufficient CMR, the potential of the super prompt criticality will then be eliminated with high reliability.

In order to realize the CMR concept, the following conditions must be satisfied:

i. **CMR activation condition:** CMR should be activated before the critical pool formation under the proper fuel enthalpy Ecmr, which is presently estimated to be in the range between solidus and liquids, 1.0MJ/kg< Ecmr<1.3MJ/kg,

ii. **CMR condition**: time constant of the fuel relocation path formation for CMR should be shorter than that of the critical pool formation, $\tau_{cmr} < \tau_{pool}$, and the fuel relocation velocity Vcmr should be larger than about 0.3Mo/ τ_{pool}, where the 0.3Mo is the fuel mass required to be removed from the core leading to an effective neutronic shutdown defined below,

iii. **Effective neutronic shutdown condition:** about 0.3Mo of the core fuel should be relocated by the CMR to establish the effective neutronic shutdown even in the case of the compaction of the whole of residual core as shown in the Fig. 3,

iv. **Thermal stabilization condition**: sodium mass with 3 times larger than the relocated fuel mass should be required to quench the relocated fuel to establish the thermally stabilized state.

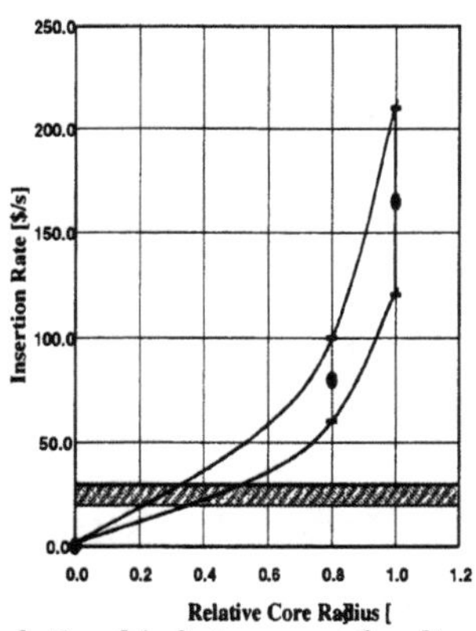

Fig.2 Relationship between pool radius and reactivity ramp rate due to self-exiting sloshing motion

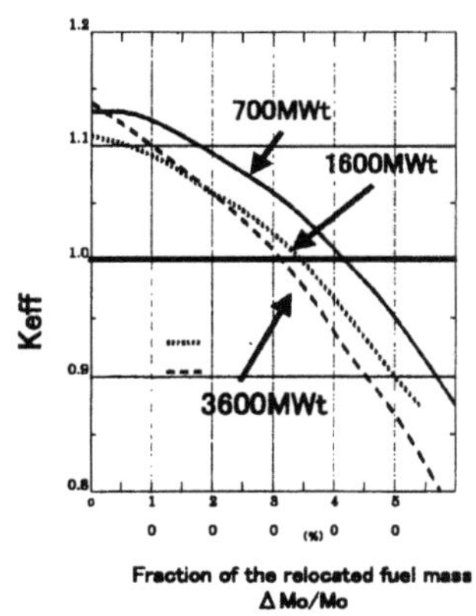

Fig.3 Fraction of fuel mass required to establish permanent shutdown

III. CONCEPT FOR THE RECRITICALITY ELIMINATION

Core Concept with CMR

From the viewpoint of the core design, the CMR should be simple and reliable with satisfying the conditions above and without degradation of the core neutronic performance. In order to establish the core concept with CRM, much effort has been concentrated on the core design of sodium-cooled FR with MOX and metallic fuel /4/-/8/. In the current proposals on the core design, the fuel relocation mode for the CMR is classified in the following categories:

i. **CMR in the fuel pin structure:** fuel relocation paths are settled in the hollow fuel pellet for the MOX fuel /4/, and in the annular fuel slug for the metallic fuel/5,6,8/. The characteristic of the CMR utilizing the in-pin fuel motion is simple one-dimensional fuel relocation in axial direction.

ii. **CMR within the fuel subassembly:** fuel relocation paths are the inner duct structure/7,9,10/ and hollow fuel pin bundle without axial blanket pellets/11/ in the fuel subassembly for the MOX fuel, and eliminated fuel pin structures in the lower axial blanket region for the metallic fuel cores/5,6/. The characteristic of the CMR within fuel assembly is also simple one-dimensional relocation.

iii. **CMR out of the fuel subassembly**: fuel relocation paths out of fuel assemblies such as inter assembly gap, control rod guide tube (CRGT), and inter assembly gap for radial blanket fuel and radial shielding/3,7/. The CMR of this type is essentially 2- or 3-dimensionalfuel relocation.

In this paper the discussion is concentrated on the CMR concept within the fuel subassembly, and three typical CMR concepts are introduced for sodium cooled MOX and metallic fueled cores.

Fuel Assemblies with Inner Duct Structure (FAIDUS) for MOX Fueled Core

The FAIDUS core/9/ is the typical CMR concept within the fuel subassembly for the sodium cooled MOX fueled core. Figure 4 shows the conceptual diagram of the FAIDUS core. The major characteristic of the FAIDUS core is the inner duct structure settled in all of fuel subassemblies in the core to provide the simple and one-dimensional fuel relocation path at the time of CDA. In order to eliminate the formation of the critical fuel pool combined with failed assemblies and to establish the independent fuel relocation path during the CDA, the difference in the duct wall thickness is designed for inner duct thickness of 3mm and for outer duct wall of 10mm.Such an independent and one-dimensional CMR concept has an advantage to realize an experimental confirmation of the phenomenological feasibility of the concept in a more effective way.

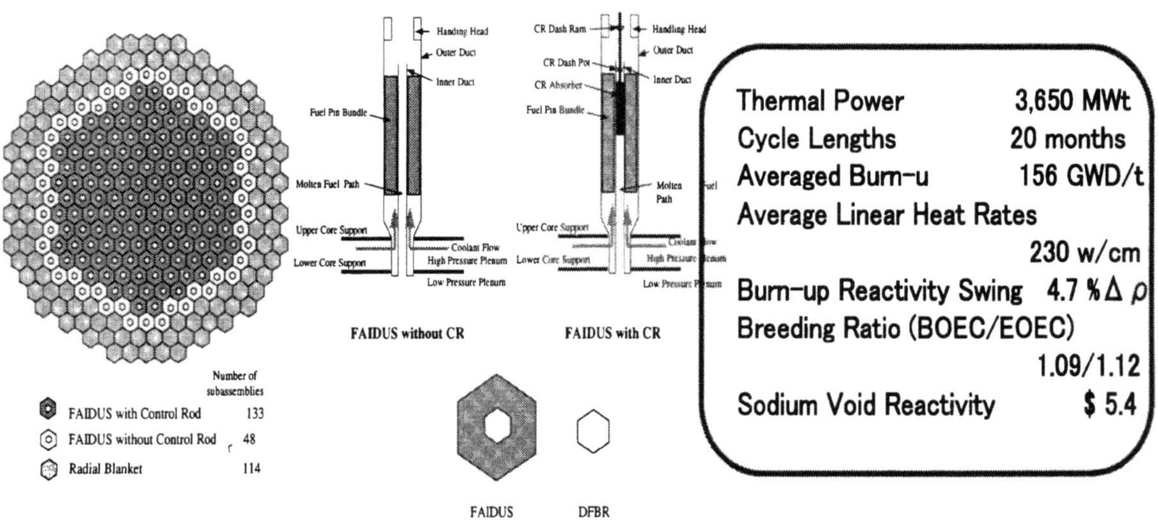

Fig.4 Concept of the FAIDUS core

As shown in the Fig.4 the FAIDUS fuel subassembly is relatively large with a diameter of ~30cm including the inner duct of ~10cm in diameter. The inner duct structure is utilized as the control rod region during the normal operation. The most suitable geometrical conditions of the FAIDUS core and fuel assembly have been systematically investigated to minimize the degradation of core neutronic performances. Figure 5 indicates the result of the optimization for the diameters of outer and inner ducts under the condition of the duct wall thickness and control rod worth required. As shown clearly in the Fig.5, combination of diameters of FAIDUS assembly with ~30cm for the outer duct and ~10cm for the inner duct indicates the maximum fuel volume fraction of 41%. Dominant core characteristics are summarized in the Fig.4. It should be emphasized that the core characteristics are not so degraded as compared with the ordinal cores especially for the breeding ratio, fuel burn-up, and void reactivity.

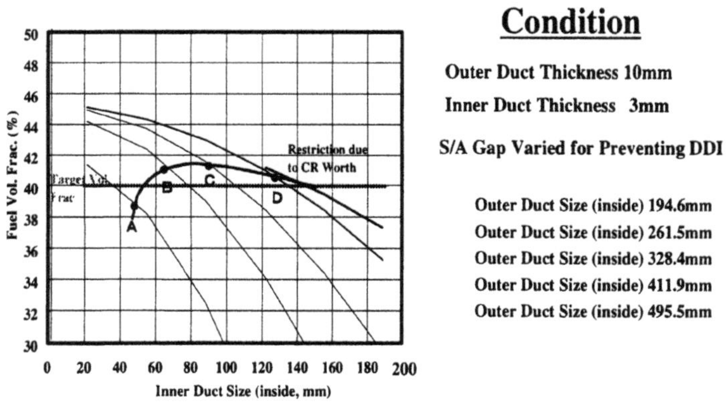

Fig. 5 Relationship between FAIDUS inner duct size and fuel volume fraction for various outer duct size

In order to evaluate the feasibility of the FAIDUS core characteristics during the typical CDA have been investigated /10/ by employing the SIMMER-II code using thermo-hydraulic models for core materials with multiple components and phases under the transport neutronics. Based on the analysis, it has been found that the FAIDUS core satisfy the CMR conditions required to establish the self-terminability against the unprotected loss-of-flow (ULOF) event as summarized below and shown in the Fig. 6:

i. The CMR is initiated at the enthalpy level of the liquidus for the average in the fuel assembly before the failure of the outer duct wall to prevent the combined pool formation of failed assemblies,

ii. The fuel relocation in the inner duct is driven by gravity and mild sodium vapor pressure generated at the time of the inner duct failure with mass velocity Vcmr of about 0.1Mo/s before the pool formation with satisfying the CMR condition of Vcmr>0.3Mo/ τ_{pool} where τ_{pool} >10s,

iii. The permanent shutdown can be established with the relocated fuel mass of 0.7Mo, and the quenching of the removed fuel to the lower core plenum can be performed by design measures to increased the sodium mass in the plenum larger than 3Mo.

Fig. 6 Power transient (A) and change of fuel distribution (B) for the FAIDUS core during ULOF event

Fuel Assemblies without Axial Blanket Fuel (ABLE) for MOX Fueled Core

Another important CMR concept is the fuel assembly without axial blanket fuel pellet in the center region of the fuel pin bundle as shown in the Fig. 7 and is called as ABLE. The structure of the ABLE core is the hollow fuel pin bundle below and above the core fuel in the center region of the fuel subassembly to provide the simple and one-dimensional fuel relocation path at the time of CDA as in the case of the FAIDUS. The most important characteristic of the ABLE concept is the limited degradation of the core performance due to less impact on the core design parameters such as fuel volume fraction and geometrical conditions.

Currently quantitative evaluations have been started to understand the safety features of the ABLE concept against ULOF/11/, and preliminary results are summarized as follows:

i. Once the fuel is disrupted, the molten fuel is axially dispersed by the fission gas pressure released from the disrupted fuel and penetrates into channels composed of the hollow fuel pin bundles,

ii. The cladding of the hollow pin bundles can immediately melt due to small thermal capacity providing effective paths for massive fuel removal into above and below the core fuel regions,

iii. Although the fuel relocation velocity is smaller than that of the FAIDUS concept, the CMR conditions can be satisfied to establish the permanent shutdown state.

Fuel Assemblies without Axial Blanket Fuel (ELAB) for Metallic Fueled Core

For the case of the sodium cooled metallic fueled core, the design limit of the sodium void reactivity is evaluated to be about 8$, and is larger than that of the MOX fueled core (~6$). This difference comes from the cancellation effect of the positive void reactivity effect by the fuel dispersion generated at nearly same temperature levels at about 1150K for the coolant boiling, fuel melting, and cladding penetration under the eutectic reaction.

The CMR of the metallic fueled core is also required to establish the neutronic shutdown before the critical pool formation. The CMR concept applicable to the metallic fueled core, however, is rather different from that of the MOX fueled core due to the thermal properties of the metallic as shown below:

i. The inner duct type CMR is not applicable to the metallic fueled core caused by the freezing of the fuel due to the lower excess enthalpy level during the CDA and due to the lower driving force by the sodium vapor pressure at the fuel temperature level of sodium boiling point. The ABLE type CMR is not also applicable to the metallic fueled core due to the high possibility of the molten fuel refreezing in the hollow fuel pin bundle region.

ii. The CMR based on the fuel assembly without fuel pin bundle structure in the lower axial blanket region (ELAB) will be applicable to the metallic fueled core due to the smaller fuel chilling effect in the fuel relocation paths. The basic concept of the ELAB is shown in the Fig. 8.

iii. The CMR concept with in-pin fuel motion is more applicable to the metallic fuel due to the minimum enthalpy loss, by employing the proper design measures such as Ta or Zr liners in the inner surface of the cladding to prevent the eutectic reaction.

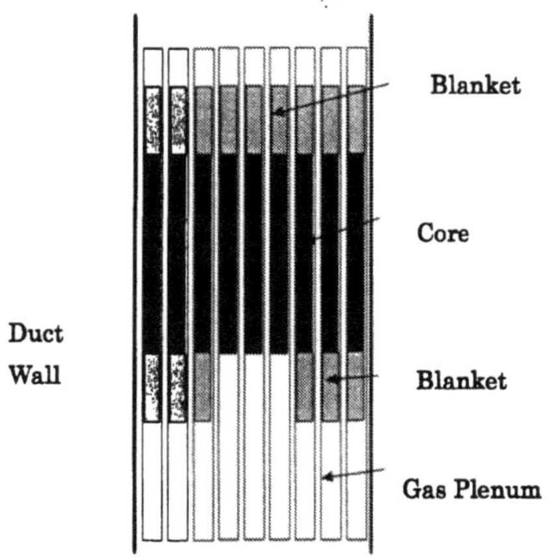

Fig. 7 ABLE concept for MOX fueled core (fuel assembly without axial blanket pellet)

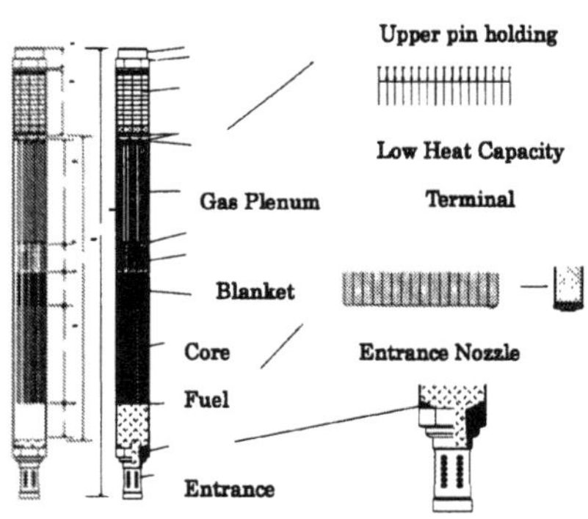

Fig. 8 ELAB concept for metallic fueled core (fuel assembly without pin bundle in lower axial blanket)

IV. EAGLE PROJECT TO CONFIRM THE RECRITICALITY ELIMINATION

Objective of the project

In order to experimentally confirm the effectiveness of the CMR concept for the MOX fueled core, the EAGLE project has been started in 1998 by Japan Nuclear Cycle Development Institute (JNC) and The Japan Atomic Power Company JAPC)/12/. The EAGLE project is the experimental program utilizing the out-of pile test facility and in-pile test facility IGR of the National Nuclear Center of the Republic of Kazakhstan (NNC/RK).

One of the major objectives of the project is to directly and experimentally confirm the CMR concept such as FAIDUS based on the in-pile integral test, supported by the in-pile and out-of pile tests culminated by a lot of fundamental experiments in step-by-step approach in a period of seven years. The other objective is to confirm the independent key phenomena controlling the CMR such as the penetration of the wall, pressures induced by fuel coolant interaction in the fuel relocation paths, fuel freezing during the relocation, relationship between fuel mobility and the enthalpy level. Such general information from the EAGLE program should be applied to the confirmation of the feasibility for wide range of CMR concepts including the ABLE.

CMR Phenomena Confirmed by the Integral Tests

The in-pile test is the final target of the EAGLE project and is aiming at the integral experiment based on the direct simulation of the CMR concept for the fuel assembly with the inner duct. Figure 9 shows the relationship between FAIDUS fuel assembly and the experimental apparatuses planned as the target tests in the in-pile experiments. The experimental apparatuses are intended to model the FAIDUS assembly, with an essentially 1/2.5 scale in the geometrical conditions. However, it is also intended to get phenomenological understanding through the test applicable to general design options for the CMR concepts.

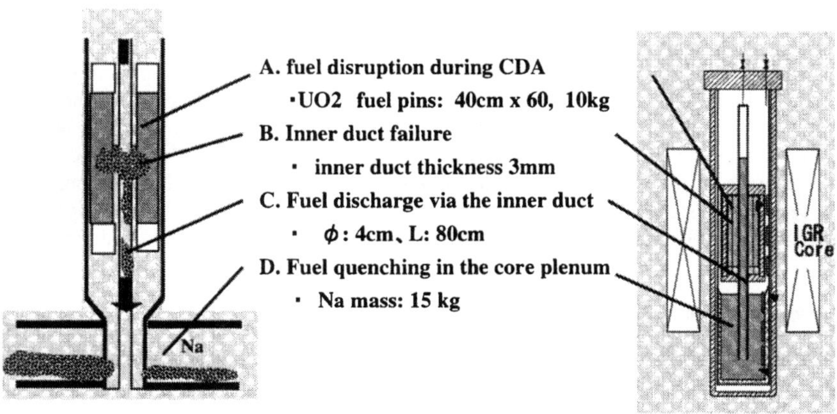

Fig. 9 Experimental concept for the in-pile integral test

In the in-pile test using the IGR reactor, the core fuel assembly is modeled by a fuel jacket with a diameter of 12cm including the inner duct with a diameter of 4.0 cm, thickness of 3mm, and length of 80cm. The inner duct is filled with the sodium at the temperature range from 700K to 900K, and is connected to the lower trap at the bottom simulating the lower core plenum and the expansion tank at the top simulating the upper sodium pool of the typical core geometry, respectively. 60 fuel pins of 40cm length are installed in the fuel jacket with UO2 pellets with the enrichment of 4.4% as the simulant materials for the MOX fuel. Total mass of the fuel is about 10.6kg.

The UO2 fuel intended to form the molten fuel pool by the IGR trapezoidal pulse before the inner duct failure at the temperature level from 3500K to 4000K as expected in the CDA of the FAIDUS core. The inner duct is evaluated to fail at about several seconds after the IGR pulse generation, and the molten fuel will be relocated to the lower trap via the inner duct involving the solidification of the liquid fuel and sodium vapor generation within the inner duct and the lower trap. Figure 10 indicates the major phenomena to be investigated in the integral test.

It should be thus emphasized that the In-pile test is just the integral test where the most of CMR phenomena can be directly and experimentally confirmed. The success criteria of the integral test are:

- Formation of the fuel pool with the average fuel enthalpy level higher than the liquidus level and inner duct failure with reasonable timing,
- Sufficient fuel relocation from the fuel jacket to the lower trap with the mass velocity close to the CMR condition discussed in the section II without noticeable fuel backward flow, and

- Fuel quenching and stabilization at the lower trap.

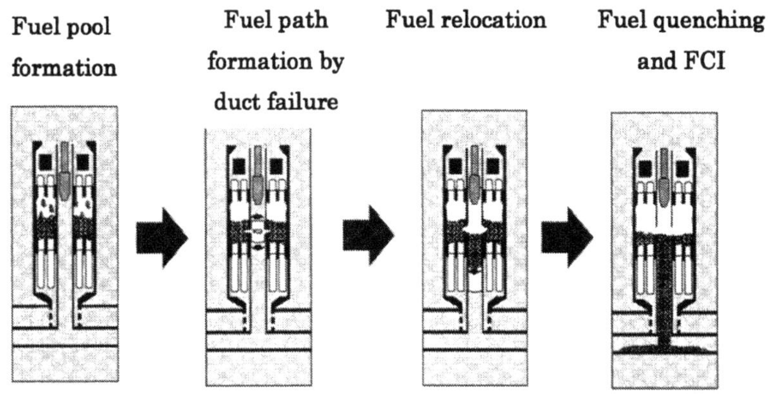

Fig. 10 CMR phenomena to be confirmed by in-pile and out-of-pile experiments

In-pile and Out-pile Tests based on Step-by-Step Approach

The out-of-pile tests are planned as the supplementary experiments to support the in-pile integral test, because the test cost is lower and a number of test shots are possible. In the out-of-pile test, the simulant material is expected to be the mixture of 78%UO2-22%ZrO2 (21kg) with steel (5.5kg). The simulant material is heated by using the electro magnetic furnace up to about 3300K with sufficient excess enthalpy above the liquidus (Tliq=2800K). Except the masses of the simulant material and the sodium, geometrical conditions of the out-of-pile test apparatus are thus designed to have the similarity with those of the in-pile test apparatus to prepare directly the technical information from the result of the out-of -pile test such as pressure and temperature distributions.

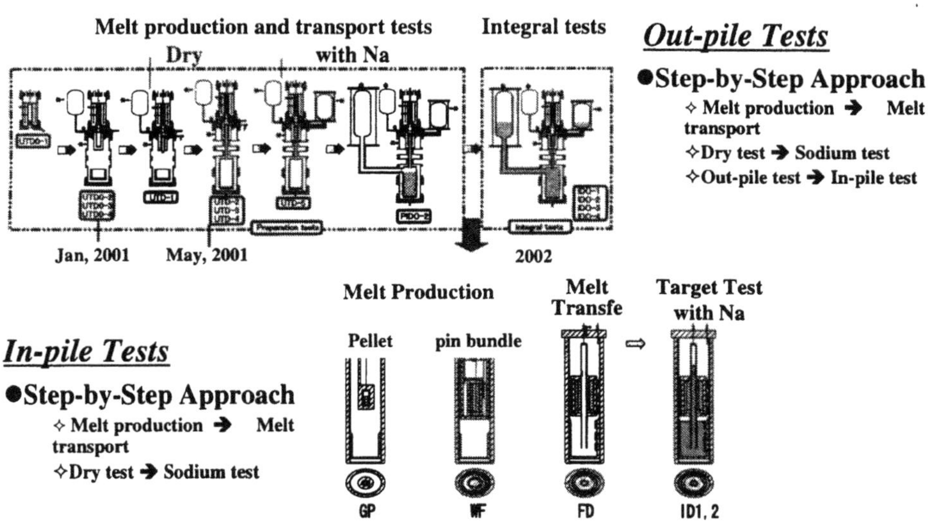

Fig. 11 Step-by-step approach in EAGLE project

In order to succeed the final integral in-pile test, the step-by-step approach is introduced aiming at the culmination of the experimental techniques by a lot of fundamental experiments as shown in Fig. 11:

i. To establish the technique for the simulant material formation and control of its relocation with the sodium handling,
ii. To establish the technique for the measurement of the high temperature materials over 3300K, and for the detection of the molten materials, and
iii. Integration of these elementary technologies for in-pile and out-of-pile experiments.

V. CONCLUSION

The ultimate safety goal of the Self-consistent Nuclear Energy System (SCNES) is to eliminate the recriticality problem under the abnormal events based on simple safety logic. The principle of the recriticality elimination is the Controlled Material Relocation (CMR) to establish the neutronic shutdown state by removing the molten fuel to the out of core region before the critical pool formation which has a potential of the severe energetics due to the super prompt criticality.

The CMR concept should be simple and reliable without too much impact on the core neutronic performance. The typical core concept with CMR is composed of fuel assemblies with inner duct structure (FAIDUS), and fuel assemblies with hollow fuel pins in the axial blanket region (ABLE) for MOX fueled cores, and fuel assembly without fuel pin bundle structure in the lower axial blanket region (ELAB) for the metallic fueled core. Based on the core design study and accident analyses, these CMR concepts have been found feasible without much impact on the neutronic performance. These CMR concepts intend to prevent the critical pool formation by the simple and independent one-dimensional axial fuel relocation, and can be experimentally confirmed by using a single test subassembly.

In order to experimentally confirm the CMR concept for the MOX fueled core, the EAGLE project has been started by utilizing the out-of pile test facility and in-pile facility IGR of the National Nuclear Center of the Republic of Kazakhstan (NNC/RK). One of the major objectives of the project is to directly and experimentally confirm the CMR concept of the fuel assembly with the inner duct such as FAIDUS based on the in-pile integral test, supported by the in-pile and out-of pile tests culminated by a lot of fundamental experiments in step-by-step approach in a period of seven years. The other objective is to confirm the independent key phenomena controlling the CMR such as the penetration of the wall, pressures induced by fuel coolant interaction in the fuel relocation paths, fuel freezing during the relocation, relationship between fuel mobility and the enthalpy level. Such general information from the EAGLE program should be applied to the confirmation of the feasibility for wide range of CMR concepts including the ABLE.

ACKNOWLEDGMENTS

We acknowledge support by O. Watanabe of ARTECH, K. Yoshimura, Y. Doi, and N. Uchiyama of Kawasaki Heavy Industries, Ltd., K. Kamiyama and K. Konishi of JNC, H. Matsumiya of Toshiba Corporation, and M. Ishida of Hitachi, Ltd.

REFERENCES

(1) Y. Fujii-e, "An Approach to Self-consistent Nuclear Energy System-Potential of Fast Reactors," Proc. Design and Safety of Advanced Nuclear Power Plant, Tokyo, Japan, Oct. (1992).
(2) M.Saito, " Multi-component Self-Consistent Nuclear Energy System for Sustainable Growth," International Seminar on "Advanced Nuclear Energy Systems Toward Zero Release of Radioactive Wastes"Susono Seminar House, Shizuoka, Japan, Nov. 6-9, 2000.
(3) H. Endo, M. Ishida, T. Hoshi and S. Kotake, "A. Study of the Initiating Phase Scenario of Unprotected Loss-of-Flow in a 600MWe MOX Homogeneous Core," IAEA IWGFR Technical Committee Meeting on Material-Coolant Interactions and Material Movement and Relocation in LMFR, PNC-OEC, Japan. June (1994).
(4) M. Ishida, H.Endo, and S.Kubo," A preliminary Study on Mitigation Effects of Annular Pellet Fuel in ATWS Events of FBR," 1999 Fall Meeting of the Atomic Energy Society of Japan, Niigata, Japan, Sept (1999).

(5) H. Endo, M.Kawashima, M.Suzuki, M.Saito, A.Shimizu, and Y. Fujii-e, "Safety Features of the Self-Consistent Nuclear Energy System," Int. Symposium on the Global Environment and Nuclear Energy System, Susono, Japan, Oct. (1994).
(6) H. Endo, M.Kawashima, M.Suzuki, H. Ninokata, T. Sawada, A.Shimizu and Y. Fujii-e, "Safety Characteristics of the SCNES Core" Int. Symposium on the Global Environment and Nuclear Energy System, Tsuruga, Fukui, Japan, Oct. (1996)
(7) H. Niwa," A Comprehensive Approach of Reactor Safety Research Aiming at Elimination of Recriticality in CDA for Commercialization of LMFBR," Int. Symposium on the Global Environment and Nuclear Energy System, Tsuruga, Fukui, Japan, Oct. (1996).
(8) T. Sawada, H. Ninokata, H. Tomozoe, and H. Endo, "A Recriticality-free Fast Reactor Core Concept," Nuclear Technology, 130(3), 242-251 (2000).
(9) K. Arie, M. Yamadate, H. Endo, M. Ishida, S. Kubo, and S. Kotake, "A Feasibility Study of MOX-Fuel FBR Concept for Enhanced Safety during Fuel Melting-/1/ Core concept," 1999 Fall Meeting of the Atomic Energy Society of Japan, Niigata, Japan, Sept (1999).
(10)S. Kubo, H. Endo, K. Yoshimura, M. Ishida, and S. Kotake, "A Feasibility Study of MOX-Fuel FBR Concept for Enhanced Safety during Fuel Melting-/2/ Evaluation of the availability for the core disruption," 1999 Fall Meeting of the Atomic Energy Society of Japan, Niigata, Japan, Sept (1999).
(11)T. Fujita and Y. Tobita, " An Evaluation Study of Measures for Prevention of Recriticality in Large Sodium Cooled FBR with MOX Fuel," 2000 Fall Meeting of the Atomic Energy Society of Japan, Aomori, Japan, Sept (2000).
(12) T. Inagaki, K. Sato, S. Kotake, Y. Sagayama, H. Endo, M. Ishida, K. Koyama, H. Niwa, Sa. Kondo, and K. Aizawa, "Role and Approach to the Recriticality Elimination with Utilizing the In-pile Test Reactor IGR," 2nd Int. Conf. on Non-Proliferation Problem, Krchatov, Republic of Kazakhstan, Sept. (1998).

Pergamon
www.elsevier.com/locate/pnucene

Progress in Nuclear Energy, Vol. 40, No. 3–4, pp. 587–596, 2002

0149-1970/02/$ - see front matter

PII: S0149-1970(02)00054-9

FEASIBILITY STUDY OF LARGE MOX FUELED FBR CORE AND APPLICABILITY OF VARIOUS COOLANTS AND FUELS AIMED AT THE SELF-CONSISTENT NUCLEAR ENERGY SYSTEM

K. FUJIMURA, T. SANDA, S. MORO[1],
M. SAITO [2] and H. SEKIMOTO [2]

Hitachi, Ltd., Power & Industry Systems, Hitachi-shi, Ibaraki, 319-1221 Japan.
Tel: (+81) 294-53-3111 Fax: (+81)294-52-9194
e-mail: kouji_fujimura@pis.hitachi.co.jp

[1] Research and Development Division, The Institute of Applied Energy,
14-2 Nishishinbashi 1-chome, Minato-ku, Tokyo 105-0003 Japan.

[2] Research Laboratory for Nuclear Reactors, Tokyo Institute of Technology,
2-12-1 O-okayama, Meguro-ku, Tokyo 152-8550 Japan.

ABSTRACT

The potential of a large MOX fueled fast breeder reactor (FBR) is evaluated with regard to its ability to transmute radioactive nuclides and its safety when incorporated in the self-consistent nuclear energy system (SCNES). The FBR's annual production amounts of selected long-lived fission products (LLFPs), Se-79, Tc-99, Pd-107, I-129, Cs-135 and Sm-151, can be transmuted by using a radial blanket region and part of a lower axial blanket region without any significant impact on the reactor's nuclear and safety characteristics. The other LLFPs are confined in the system. The hazard index level of the LLFPs per one ton of spent fuel from the system after 1000 years is as small as that of a typical uranium ore. To realize self-controllability (passive safety), the proposed FBR core concept employs gas expansion modules and a sodium plenum above the core. To realize self-terminability, even if MOX fuel melting should cause a core compaction, re-criticality of the core can be avoided by a fuel dilution and relocation module. The results show the MOX fueled FBR core has potential applicability to the SCNES. The fundamental applicability of various coolants and fuels is evaluated based on neutron balance toward the final goal of the ideal SCNES. The results show that gas coolant has a potential for increasing the transmutation efficiency of LLFPs. And an improved SCNES with several conventional FBRs and a FP transmutation reactor is also studied.

1. INTRODUCTION

The concept of the self-consistent nuclear energy system (SCNES) was proposed (Fujii-e et al., 1992) as the ultimate goal of nuclear energy development. They defined the SCNES as a system that simultaneously realizes at least four objectives: energy generation, fuel production, transmutation and/or confinement of radioactive nuclides, and system safety.

The numbers of excess neutrons required for the transmutation of radioactive nuclides were roughly evaluated for a fast breeder reactor (FBR) and light water reactor, and it was pointed out that the FBR core is more suitable for the SCNES (Shimizu and Fujii-e, 1995). The FBR core can use various types of fuels such as mixed oxide (MOX), metal and nitride fuels. Since a metallic core has a large number of excess neutrons due to its harder neutron energy spectrum, the core concept applying a metal fueled FBR to the system was proposed (Arie et al., 1995). For system safety, self-controllability was obtained by a flat core geometry, a high thermal conductivity fuel with a Pu-U-Zr alloy, and an additional neutron streaming path in a leading channel. Self-terminability could be achieved by a controlled material relocation in the leading channel (Endo et al., 1995). On the other hand, though the MOX fuel has different nuclear and thermal characteristics, such as a lower number of excess neutrons and a lower thermal conductivity, it is important to evaluate the potential for applicability of the conventional MOX fueled FBR to the SCNES.

The present study first evaluates applicability of the large MOX fueled FBR core to the SCNES in conjunction with transmutation of selected fission products (FPs) and system safety. To lessen the effects on nuclear performance parameters, the long-lived fission products (LLFPs) are loaded into target assemblies at the radial blanket (RB) region and part of the lower axial blanket (LAB) region, and the property of their transmutation by excess neutrons is discussed. Furthermore, a concept of a fuel dilution and relocation module is introduced to avoid re-criticality of the core in case of a hypothetical core disruptive accident.

Recently new FBR core concepts using various coolants (water, lead, gas, etc.) and fuels (metal, nitride) are being researched. A metal fueled core has a potential to provide more excess neutrons because the spectrum of the core is harder than that of the MOX or nitride fueled one. Nitride fuel provides good thermal characteristics and has a potential to realize a compact core, which may enhance inherent safety and/or decrease plant construction cost. On the other hand, concepts of breeder reactors using various coolants have been proposed (Adamov et al., 1992; Sekimoto, 1995; Takeda et al., 1995; Ishida et al., 1998). The present study evaluates fundamental applicability of various coolants and fuels to further the ideal SCNES based on neutron balance. And an improved SCNES with several conventional FBRs and a FP transmutation reactor is preliminarily studied.

2. CONCEPT OF THE TYPICAL SCNES

Potential applicability of the MOX fueled FBR is evaluated for four objectives in the typical SCNES concept shown in Fig. 1. The objectives are briefly described as follows.

- Energy generation: To achieve high efficiency in the utilization of fission energy.
- Fuel production: To have a breeding ratio greater than unity.
- Transmutation of radioactive nuclides: To minimize the hazard index (HI) level, which is defined in the present paper by the ratio of the FP's radioactivity to the annual limit on intake by ingestion per one ton of spent fuel, from the system. This objective is realized as follows. Uranium (U), plutonium (Pu) and minor actinides (MAs) are reutilized as fuel. Stable FPs are released from the system. Short lived FPs, whose half-lives are less than 10 years, are stored temporarily and released from the system after their hazard index level becomes negligibly small. The LLFPs having high hazard indexes are transmuted and confined in the system. The LLFPs having effective half-lives of under 10 years are transmuted in the FBR core and others are confined in the system.

- System safety: To eliminate criticality-related safety issues by natural laws.

Each item is discussed for the proposed core concept of the Na cooled MOX fueled FBR shown in Fig. 2 and Table 1 assuming that promising future technologies can be achieved.

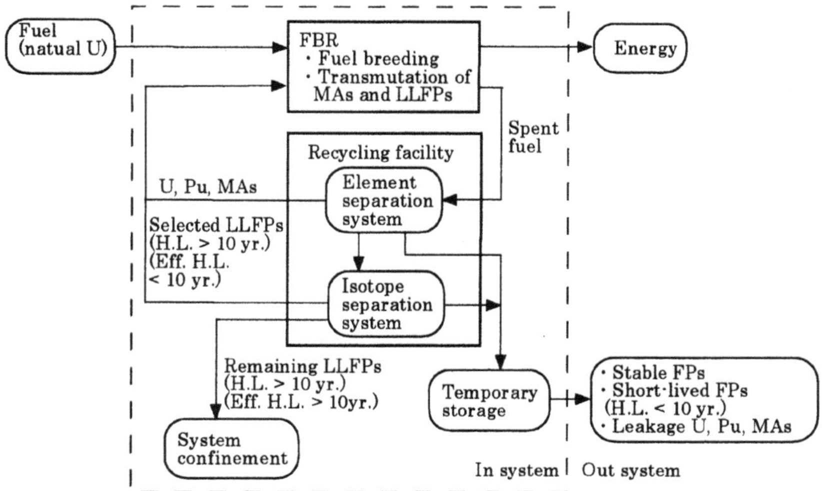

Fig. 1 Concept of the self-consistent nuclear energy system (SCNES)

Table 1 Key design parameters of t he 1300MWe Na cooled MOX fueled FBR core

Parameters	Specifications
Reactor power (MWe)	1300
No. of assemblies	
Core	438
GEM	78
Primary control rod	19
Secondary control rod	6
Fuel column height (cm)	
Core	100
Axial blanket / Inner blanket	35×2 / 20
Sodium plenum	40
Refueling interval (month)	13

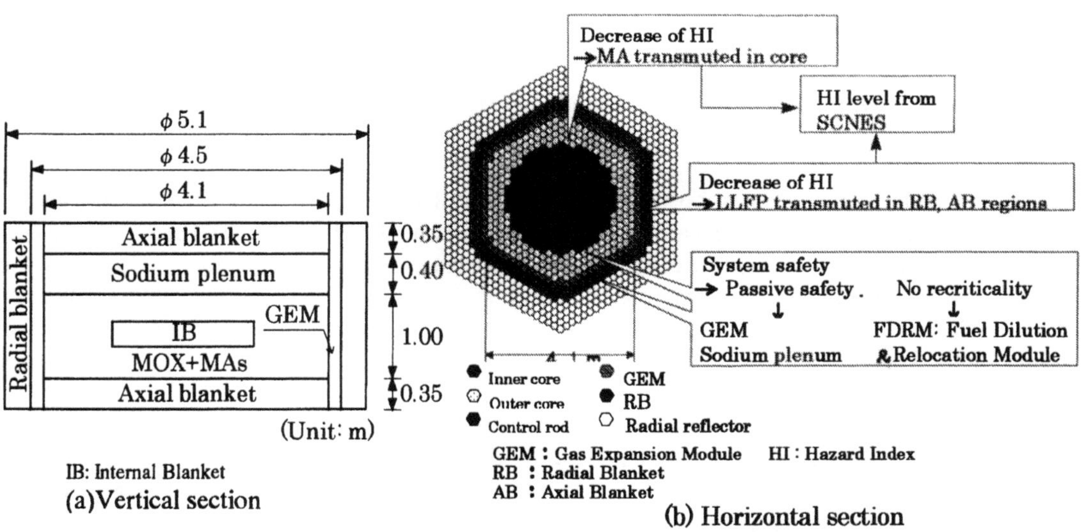

Fig.2 Core concept of 1300MWe Na cooled MOX fueled FBR

3. ENERGY GENERATION AND FUEL PRODUCTION

The goals of energy generation and fuel production are easily achieved by the FBR, since the value of the breeding ratio (BR) can be made greater than unity by using a part of the blanket regions. The fuel utilization factor FU in a FBR system is simply estimated by

$$FU = \frac{BU}{1-(1-BU)(1-L)} \tag{1}$$

where

BU: exposure (atom % $\times 10^{-2}$)
L: loss fraction for reprocessing.

The minimum BU is 0.06 and the maximum L is 0.03 for the MOX fueled FBR system. Even if FU is calculated by using these values, it is about 0.68, i. e., one ton of natural U can generate fission energy of 680 GWd/t. Net energy generation is the difference between the generation energy and the consumption energy of system loading. In order to keep the reactor critical, the SCNES requires isotope separation of the generated FPs before they are returned to the core.

According to a rough estimation (Akatsuka et al., 1995), isotope separation system could consume about 2 % of the energy produced by the fission reaction. Energy consumption in the element separation system is negligibly small due to use of chemical processes. As a result, net generated energy from the natural U is at least 660 GWd/t, and, in principle, these two objectives can be fully satisfied by using the MOX fueled FBR core.

4. TRANSMUTATION OF RADIOACTIVE NUCLIDES

The ultimate goal of transmutation of radioactive nuclides is their non-release from the system; the U and Pu separated from the spent fuel are re-utilized in the FBR fuel and the MAs are reloaded in the FBR core, which should reduce their amounts up to about 15 %/year through energy production by the fission reaction. In the present study, LLFPs to be transmuted or confined are selected from the viewpoints of their half-lifes and hazard indexes. The categories are as follows.

(1) Long-lived FPs

Sr-90, Cs-137 and Sm-151, which have a hazard index level of over 10^5 after 10^2 years, are selected for transmutation or confinement because of their high radioactivity for a long period.

(2) Ultra-long-lived FPs

Se-79, Zr-93, Tc,-99, Pd-107, Sn-126, I-129 and Cs-135, which have a hazard index level of over 10 after 10^5 years, are selected for transmutation because of their radioactivity for an ultra-long period.

Transmuted LLFPs are loaded only in the blanket region without any significant impact on nuclear and safety characteristics. As a result, not all amounts of LLFPs generated by the FBR itself are reduced, since the amounts are too large for the number of excess neutrons being used for transmutation of LLFPs in the FBR. Then some LLFPs having a large capture cross section, Se-79, Tc,-99, Pd-107, I-129, Cs-135 and Sm-151, are selected for transmutation, based on their effective half-lives of under 10 years, and the others are confined in the system.

The FBR's potential for transmutation of LLFPs is evaluated. In order to enhance transmutation efficiency, neutrons in the RB region are moderated by solid hydride $ZrH_{1.6}$. The FP target assembly configuration is determined so as to give the neutron flux and spectrum that maximize the transmutation amount. Since the selected LLFPs have relatively large capture cross sections in the low energy region, moderated neutrons are more favorable for transmutation than fast ones. Fig.3 shows the configuration of the FP target assembly and dependency of transmutation amount on the volume ratio for Tc-99.

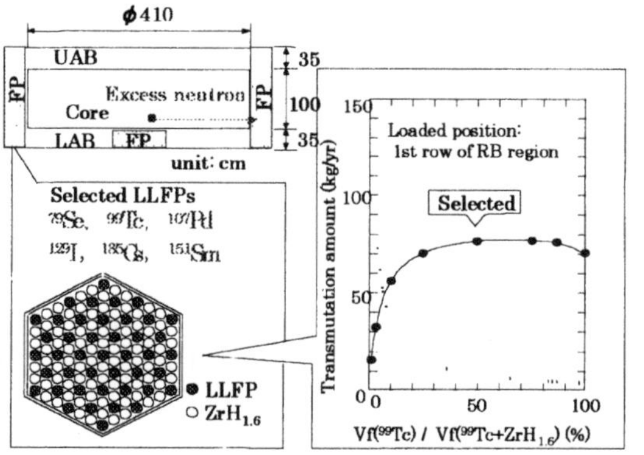

Fig.3 Optimizing the LLFP target assembly (for ^{99}Tc)

Preliminary results show that the FP target assemblies in the RB region are not sufficient to transmute the annual production amounts by the FBR itself. This is because the 1300MWe Na cooled MOX fueled core cannot provide enough radial leakage neutrons because its diameter is larger than that of the 1000MWe core (Kobayashi et al., 1997). In this study Tc-99 and Pd-107 are selected for loading in the lower axial blanket region in view of their higher melting points. $ZrH_{1.6}$ is not used because it is loaded in the same clad with fuel pellet.

The above results are used to get distributions of the FP target assemblies in the RB region for each LLFP's transmutation of the annual production amounts by the FBR itself with a view to evaluating applicability to the SCNES. Table 2 gives a distribution of the FP target assemblies for transmutation of the annual production amounts. The LLFPs annually produced by the FBR itself are transmuted by using the first row of the RB region and part of the lower AB region. Amounts of the confined LLFPs, namely Sr-90, Zr-93, Sn-126 and Cs-137, are evaluated. Assuming the life of an FBR is 50 years and 200 ages of FBRs would be constructed in the same area, about 200 tons of LLFPs are accumulated. This amount can be confined in the shielding region in the FBRs. Replacing the blanket materials with the FPs will impose some penalties on the nuclear performance parameters such as breeding ratio. The breeding ratio in the present core is about 1.02, which is smaller than the value of about 1.2 for a non LLFP transmutation core. The other nuclear performance parameters for both cores have comparable values; the self-generated LLFPs are transmuted by the FBR without significant impact on the nuclear and safety characteristics.

Table 2 Distribution of the FP target assemblies for the transmutation of annual production amounts by the FBR itself

Nuclide	Production (kg/yr)	Chem. form	No. of FP target assemblies RB region	LAB region	Loaded amount (t)	Transmutation amount (kg/yr)
Se-79	0.4	Se	1	0	0.03	0.7
Tc-99	31.2	Tc	14	33	3.5	31.4
Pd-107	16.0	Pd	0	25	1.2	16.8
I-129	7.9	NaI	20	0	0.9	8.1
Cs-135	38.9	Cs_2O	44	0	3.6	39.4
Sm-151	5.1	Sm	5	0	0.2	6.4

The hazard index of spent fuel released from the above SCNES is evaluated for the following three cases.

1) Reference case: with no recycling of MAs and LLFPs.

2) MAs transmutation case: MAs are recycled with Pu and those annually produced by the FBR itself are transmuted in the core fuel.

3) MAs+6LLFPs transmutation case: MAs are the same as in case 2) and six selected LLFPs (Se-79, Tc-99, Pd-107, I-129, Cs-135, Sm-151) annually produced by the FBR itself are transmuted in the RB and lower AB regions. Residual LLFPs, namely Sr-90, Zr-93, Sn-126 and Cs-137 are confined in the system.

As shown in Fig.4, for case 3, the hazard index level of the LLFPs per one ton of spent fuel from the system after 1000 years is as small as that of a typical uranium ore. But for c ase 2, the hazard index level is not smaller than that even after 10^6 years.

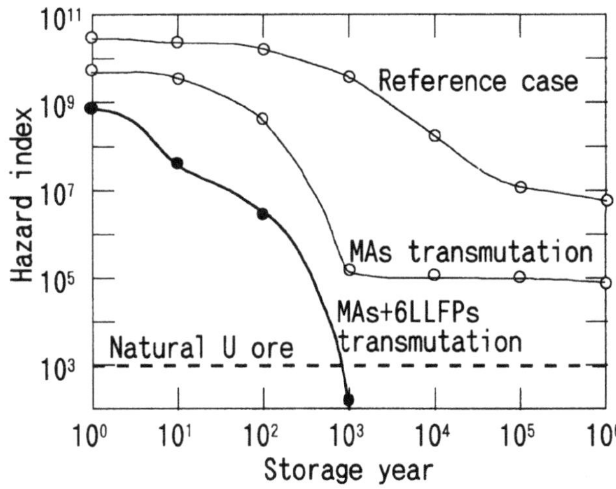

Fig. 4 Hazard index of 1 ton of spent fuel released from SCNES

5. SYSTEM SAFETY

System safety is discussed in connection with self-controllability and self-terminability. Self-controllability of the FBR core is evaluated for anticipated transient without scram (ATWS) events. A typical ATWS of the flow coast down is an unprotected loss of flow transient (ULOF). In general, the ULOF event in a large MOX fueled FBR core causes reactivity insertion due to increment of the sod ium temperature. In order to avoid sodium voiding, the core power level must be decreased to match the reduced sodium flow by inserting a negative reactivity. The negative reactivity can be inserted by the sodium plenum above the core and the gas expansion module (GEM). Another typical event of reactivity insertion is an unprotected transient overpower (UTOP). Since the UTOP is generally caused by insertion of reactivity when an inadvertent control rod withdrawal is supposed in a core of non -zero burnup reactivity, it is important to diminish the excess reactivity. In order to realize a smaller burnup reactivity, the concept of a burnup reactivity compensation module (BRCM) was proposed (Ohashi et al., 1994). In the BRCM, a neutron absorber in the module is slowly moved downwards, because the pressure of helium gas released from the B_4C in the module by the (n, α) reaction increases with burnup. As a result, the negative reactivity value of the BRCM decreases with increasing burnup. Then even if the UTOP event is postulated, fuel melting can probably be avoided inherently.

Self- terminability of the core is evaluated in relation to the re -criticality of degraded core. Since fuel in the core is not generally arranged to give the maximum reactivity value, insertion of positive or negative reactivity can be caused by a fuel relocation process such as core compaction or dispersion. In order to avoid re-criticality, fuel dilution and relocation modules (FDRMs) (Kobayashi et al., 1997) are loaded in the core. The FDRM has an inner wrapper tube containing flowing sodium and depleted uranium oxide. The depleted uranium oxide is connected to the upper part of the FDRM by a low melting point connection metal. The FDRM can be used for both fast and slow events of fuel relocation. Assuming a small part of the MOX fuel melts quickly, the FDRM can work to drop the melting fuel through the sodium in the inner wrapper tube and remove the fuel from the core. In a slow event, if high temperature sodium melts the connection metal, the FDRM works to insert the uranium into the MOX fueled core section before a compaction process; this is equivalent to diluting the MOX fuel with the depleted uranium. Hence re -criticality can probably be avoided by the FDRM, even if the core compaction process occurs in a degraded core situation. The concept of the proposed core has safety margins for self-controllability and terminability, thus sodium voiding, fuel melting and re-criticality by the compaction process can be avoided by using the BRCM and FDRM.

6. APPLICABILITY EVALUATION OF VARIOUS COOLANTS AND FUELS AIMED AT THE SCNES

As a preliminary study to pursue the ideal SCNES, fundamental applicability of various coolants and fuels is evaluated based on the Na cooled and MOX fueled FBR core proposed above with neutron balance. Fundamental nuclear characteristics (neutron balance, coolant void reactivity) for combinations of fuels (MOX, metal and nitride) and representative coolants for water, light metal, heavy metal and gas are evaluated. Reference core volume ratio of coolant to fuel (Vc/Vf) is set to that of the Na cooled and MOX fueled FBR core proposed above. For coolants other than Na, Vc/Vf values based on breeder reactor designs are considered. Nuclear characteristics are evaluated using a diffusion calculation for the one-dimensional homogenized core without blanket. The core fissile column is 100cm. Pu enrichment is adjusted to maintain core criticality at the end of equilibrium cycle for discharge burnup of 50GWd/t for water coolant s and 150GWd/t for other coolants. Conditions of the parameter survey are summarized in Table 3.

Table 3 Conditions of parameter survey

Item	Value
Volume ratio (reference)	Fuel/SS/Coolant=41/12/47%
Fuel type	MOX:PuO_2+UO_2, Metal : U-Pu-10%Zr Nitride : PuN+UN (N-15/N:100%)
Density of HM (g/cc)	MOX: 9.8、Nitride: 13.5、Metal: 14.2
Fuel smeared density(%)	MOX: 84、Nitride: 80、Metal: 75
Density of coolant (g/cc)	Na: 0.84, Pb: 10.4 (Conventional FBR conditions) Light water, Heavy water: 0.68 (PWR) Gas(He) : 6.3×10^{-3} (GCFR* conditions)
Calculation method	JENDL-3.2, SLAROM-CITATION
Dis. burnup(GWd/t)	Na, Pb, Gas : 150 、Water : 50
Pu enrichment	Maintain criticality at the end of equlibrium cycle (Core height 100cm)

* based on publication by Yoshida (1982)

The evaluated fundamental nuclear characteristics and potentials for transmuting LLFPs for combinations of MOX fuel and representative coolants (water: D_2O, light metal: Na, heavy metal: Pb and gas: He) are shown in Table 4. The results of nitride and metal fuels are compared for Na coolant. Vc/Vf for Na coolant is 1.1, for other coolants are the values are based on breeder reactor designs. Maximum BR is the BR (breeding ratio) when all leakage neutrons are captured by fertile nuclides, which is equivalent to the BR of the core surrounded by an infinite blanket. The loss of BR which arises from a core calculation with the finite blanket is assumed to be the difference between 1.42 and 1.2 of the reference Na cooled MOX fueled core without transmuting MAs or LLFPs.

The maximum BR is introduced as follows.

Leakage neutrons (excess neutrons) L can be written from the neutron balance,

$$L=\nu(\Sigma f_1+\Sigma f_2)\Phi-(\Sigma f_1+\Sigma c_1)\Phi-(\Sigma f_2+\Sigma c_2)\Phi-\Sigma cC\Phi-\Sigma cFP\Phi \quad (2)$$

divided by $\Sigma f\Phi$ $(=(\Sigma f_1+\Sigma f_2)\Phi)$

$$L/\Sigma f\Phi=\nu-1-\Sigma c_1/\Sigma f-\Sigma c_2/\Sigma f-\Sigma cC/\Sigma f-\Sigma cFP/\Sigma f. \quad (3)$$

The BR gets the maximum value when all leakage neutrons are captured by fertile nuclides. The maximum BR can be introduced as follows.

$$\text{Maximum BR}=(\Sigma c_2+L)/(\Sigma f_1+\Sigma c_1)=(\Sigma c_2/\Sigma f+L/\Sigma f)/(\Sigma f_1/\Sigma f+\Sigma c_1/\Sigma f) \quad (4)$$

where L: leakage neutrons from core (excess neutrons)

Σf、 Σf_1、 Σf_2: fission macro cross sections (1: fissile nuclides、 2: fertile nuclides)

Σc_1、 Σc_2: capture macro cross sections (1: fissile nuclide, 2: fertile nuclides)

ΣcC: capture macro cross section by stainless steel and coolant

ΣcFP: capture macro cross section of FPs, BR: breeding ratio, Φ : neutron flux

Potential for transmuting LLFPs is evaluated by comparing usable neutrons (maximum BR - neutrons necessary for breeding including loss) with neutrons necessary for transmutation of LLFPs considering isotope separation (Akatsuka et al., 1995). As shown in Table 4, the harder the core spectrum is, the larger the maximum BR would be, which makes the potential for transmutation higher. It seems that gas coolant has a large potential for increasing the transmutation efficiency of LLFPs.

Table 4 Potential for transmuting LLFP's of FBR cores using various coolants and fuels

Coolant		Light metal			Water	Heavy metal	Gas
		Na			D_2O	Pb	He
Fuel		MOX	Nitride	Metal	MOX	←	←
Volume ratio of coolant to fuel (Vc/Vf)		1.1	←	←	0.4	2.3	2.2
Discharge fuel burnup (GWd/t)		150	←	←	50	150	←
Pu enrichment(wt%)		19.4	13.2	13.9	15.5	25.5	30.1
Maximun BR		1.42	1.51	1.59	1.28	1.44	1.79
Usable neutron = Maximum BR-necessary for breeding (Including loss*1)		0.20	0.29	0.37	0.06	0.22	0.57
Necessary for transmutation of LLFPs*2 (with isotope separation)	2 isotopes*3 0.08	O	O	O	–	O	O
	6 isotopes*4 0.19	O	O	O	–	O	O
	8 isotopes*5 0.23	–	O	O	–	–	O

*1: assumed the same value for the Na-MOX core without LLFP transmutation (BR=1.2)
*2: based on publication by Akatsuka et al. (1995)
*3: Tc-99、I-129、*4 the above+Se-79、Pd-107、Cs-135、Sm-151、*5 the above+Zr-93、Sn-126

7. PRELIMINARY STUDY FOR THE IMPROVED SCNES

In the previous study, a large Na cooled MOX fuel FBR core which satisfies four objectives toward SCNES was proposed. In this section an improved SCNES with several conventional FBRs and a FP transmutation reactor such as shown in Fig.5 is preliminarily studied. In the SCNES, conventional FBRs mainly play a role of base load supplier and fuel breeder. There needs an efficient FP transmutation reactor for several conventional FBRs. Goals for energy generation and fuel production should be satisfied in the whole system. Helium is chosen as coolant of the FP transmutation reactor from the results of section 6. Recently, a fast reactor core concept for MA transmutation using hydride fuel targets was proposed (Sanda el al.,

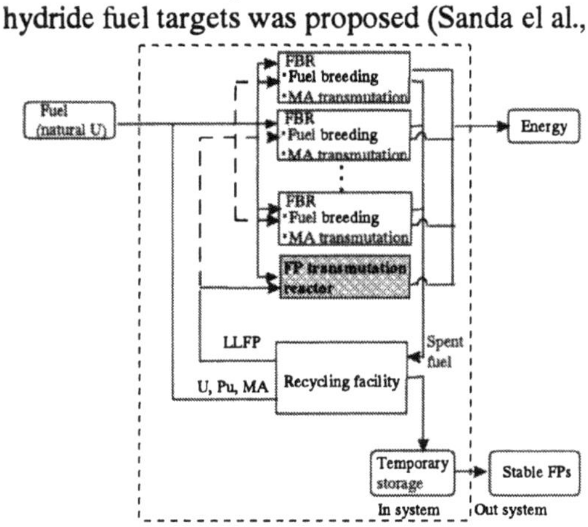

Fig 5 Concept of the improved SCNES with several conventional FBRs and a FP transmutation reactor

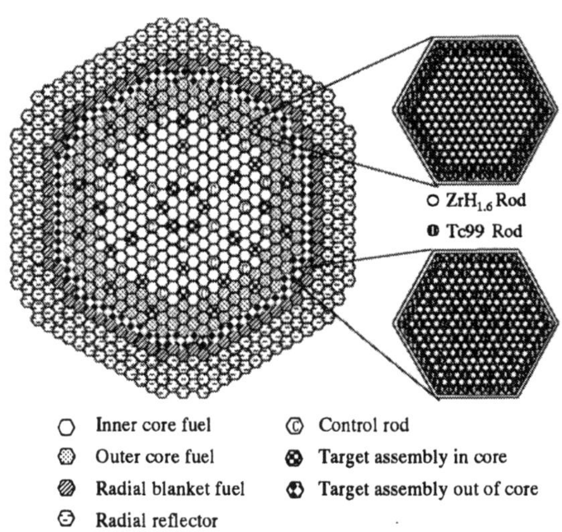

Fig. 6 Core layout of FP transmutation nitride fueled fast reactor

2000). In this concept, MAs in the form of zirconium-hydride fuel targets are loaded in the core region of fast reactor. High neutron flux and moderated spectrum are realized in those targets, which increase the MAs transmutation efficiency. In this study, a highly efficient FP transmutation core concept is studied based on the fast reactor core concept for MA transmutation using hydride fuel targets. Transmutation characteristics of LLFP is evaluated for both cores using sodium and helium as a coolant. Nitride fuel is chosen considering high thermal conductivity and large heavy metal density. Figure 6 shows the core layout. Target assemblies containing LLFP and $ZrH_{1.6}$ are loaded even in the core region to add to the out of core region. The results for Tc-99 are summarized in Table 5. The fuel volume fraction in the helium cooled core is smaller than that in the sodium cooled core (Yoshida, 1982), because thermal characteristics of helium is inferior to that of sodium. On the other hand, height of the helium cooled core can be set higher because coolant void reactivity is much smaller than that of the sodium cooled core. Transmutation rate of Tc -99 in the helium cooled core is increased by 28% in core region and 44% out of core region than those in the sodium cooled one when MLHGR are set the same value. Transformation of Tc-99 in the whole core is increased by about 40%. It is clarified that helium cooled nitride fuel core loaded target assemblies has a potential as a highly efficient FP transmutation core for the improved SCNES.

Table 5 Tc-99 transmutation characteristics for FP transmutation nitride fueled fast reactor

Case	Coolant Material	Fuel vol. fraction	Core Diameter/height	FP target Height	Pu enrich.	Tc99 inventory In core/Ex core	Tc99 trans. rate In core/Ex core	MLHGR	Average power dens.
		%	m	m	wt%	ton(initial core)	%/year	W/cm	W/cm
A	Na	44	4.1/0.6	0.6	24.0	1.8/3.7	6.0/1.6	800	380
B	He	27	3.3/1.6	1.0	24.3	1.9/3.9	7.7/2.3	750	220
C	Na	44	4.1/1.0	1.0	20.7	3.0/6.1	5.5/1.2	430	230
D	He	27	4.1/1.0	1.0	30.0	3.0/6.1	6.1/1.7	670	230
E	He	27	4.1/1.6	1.0	23.9	3.0/6.1	5.3/1.4	460	140

● Different coolant, fuel vol. fraction, core configuration(same LHGR)(Case A vs B) In core +28%, Ex core +44%
→Total transformation +40%
○Different coolant, fuel vol. fraction (same core configuration, power dens.) (Case C vs D) In core +11%, Ex core +42%
○Different coolant, fuel vol. fraction, core height (same core diameter) (Case C vs E) In core -3%, Ex core +15%

8. CONCLUSIONS

(1) Potential applicability of a MOX fueled FBR to the SCNES was evaluated assuming promising future technologies. It was found that the four objectives, i. e. energy generation, fuel production, transmutation of radioactive nuclides, and system safety could be satisfied by using the MOX fueled large FBR core. Major LLFPs (Se-79, Tc-99, Pd-107, I-129, Cs-135 and Sm-151) having a high hazard index could be transmuted in the radial blanket region and part of the lower axial blanket region without significant impact on core nuclear and safety characteristics, while the breeding ratio was kept above unity. The proposed core concept could enhance the safety margins of self-controllability and self-terminability by employing the gas expansion module, the sodium plenum above the core, and the fuel dilution and relocation module.
(2) To pursue the ideal SCNES, fundamental applicability of various coolants and fuels was evaluated based on neutron balance. Gas coolant seemed to have a potential for increasing the transmutation efficiency of LLFPs.
(3) An improved SCNES with several conventional FBRs and a FP transmutation reactor was preliminarily studied. Transmutation characteristics of LLFPs loaded in a helium cooled fast reactor was evaluated to introduce a highly efficient FP transmutation reactor toward the improved SCNES. From the results, the helium cooled nitride fueled FP transmutation reactor loaded target assemblies containing LLFP and $ZrH_{1.6}$ seemed to have a potential to increase the transmutation efficiency of LL FPs.

ACKNOWLEDGMENT

This study was sponsored by the Ministry of International Trade and Industry of Japan.

REFERENCES

ADAMOV, E. et al. (1992). Progress in Lead-Cooled Fast Reactor Design. *Proc. Int. Conf. on Design and Safety of Advanced Nuclear Power Plants,* p. 16.6-2

AKATSUKA, H. et al.(1995). Scientific Feasibility of Incineration in SCNES. *Proc. Int. Symp. on Global Environment and Nuclear Energy System,* p. 477, Elsevier, Oxford

ARIE, K. et al. (1995). A Core Concept for the Self-Consistent Nuclear Energy System Based on the Promising Future Technology. *Proc. Int. Symp. on Global Environment and Nuclear Energy System,* p. 455, Elsevier, Oxford

ENDO, H. et al. (1995). Safety Features of Self-Consistent Nuclear Energy System. *Proc. Int. Symp. on Global Environment and Nuclear Energy System,* p. 463, Elsevier, Oxford

FUJII-E, Y. et al. (1992). An Approach to Self-Consistent Nuclear Energy System (Potential of Fast Reactor). *Proc. Int. Conf. on Design and Safety of Advanced Nuclear Power Plants,* Vol. II, p. 11.3-1

ISHIDA, M. et al. (1998). Conceptual Design of recycle type PWR. *1998 Fall Meeting of the Atomic Energy Society of Japan,* E10

KOBAYASHI, K. et al. (1997). Applicability Evaluation of MOX Fueled Fast Breeder Reactor to the Self-Consistent Nuclear Energy System. *Proc. Int. Conf. on Future Nuclear Systems (GLOBAL'97)* , Vol. 2, p.1062

OHASHI, M. et al. (1994). Research of Reduced Void Reactivity Fast Reactor Core. *1994 Fall Meeting of the Atomic Energy Society of Japan,* G4

SANDA, T. (2000). Fast Reactor Core Concepts for Minor Actinide Transmutation Using Hydride Fuel Targets. *J.N.S.T,* Vol. 37, No.4, p.335

SEKIMOTO, H. (1995). Lead Cooled Fast Reactor Satisfying Both Inherent Safety and TRU+LLFP Confinement Properties in the Future Equilibrium State. *Proc. Int. Conf. on Evaluation of Emerging Nuclear Fuel Cycle Systems (GLOBAL'95)* , Vol. 1, p.923

SHIMIZU, A. and FUJII-E, Y. (1995). Self-Consistent Nuclear Energy Systems. *Proc. Int. Symp. on Global Environment and Nuclear Energy System,* p. 25, Elsevier, Oxford

TAKEDA, R. et al. (1995). General Features of Resource-Renewable BWR(RBWR) and Scenario of Long Term Energy Supply. *Proc. Int. Conf. on Evaluation of Emerging Nuclear Fuel Cycle Systems (GLOBAL'95)* , Vol. 1, p.938

YOSHIDA, H. (1982). Gas-Cooled Fast Breeder Reactor. *JAER-M 82-084* -In Japanese

Pergamon
www.elsevier.com/locate/pnucene

Progress in Nuclear Energy, Vol. 40, No. 3–4, pp. 597–606, 2002

Printed in Great Britain
0149-1970/02/$ - see front matter

PII: S0149-1970(02)00055-0

FEASIBILITY STUDY ON NITRIDE FUEL CORE AND RECYCLING SYSTEM TOWARD SELF-CONSISTENT NUCLEAR ENERGY SYSTEM

TAKASHI KAWAKITA[1], KAZUMI IKEDA[1], SATOSHI MORO[2],

MASAKI SAITO[3] and HIROSHI SEKIMOTO[3]

[1]Mitsubishi Heavy Industries, Ltd., 3-1 Minatomirai 3-chome, Nishi-ku,
Yokohama 220-84 Japan
Tel: +81-45-224-9743, Fax: +81-45-224-9962
E-mail: kawakita@atom.hq.mhi.co.jp

[2]The Institute of Applied Energy, 14-2 Nishishinbashi 1-chome, Minato-ku, Tokyo
105 Japan
Tel: +81-3-3508-8894, Fax: +81-3-3501-1735

[3]Research Laboratory for Nuclear Reactors, Tokyo Institute of Technology
Ohokayama, Meguro-ku, Tokyo 152 Japan
Tel: +81-3-5734-3074, Fax: +81-3-5734-2959

ABSTRACT

The concept of a nuclear fuel recycle system with a nitride fueled FBR core has been investigated as a part of related studies towards the Self-Consistent Nuclear Energy System (SCNES). Nitride fuel has been given attention because of its relatively high fuel density and high thermal conductivity. To materialize the SCNES concept, it is important to adequately use the excess neutrons produced in the chain reaction. The high fuel density of the nitride fuel brings out more of the excess neutrons and has a higher potential to transmute the long-lived fission products (LLFP's). The high thermal conductivity, in addition, provides margin of fuel melting, and gives negative feedback due to the Doppler reactivity in unprotected loss of flow accidents. In this paper, we discuss good use of nitride fuel in the SCNES. © 2002 Published by Elsevier Science Ltd.

1. INTRODUCTION

The Self-Consistent Nuclear Energy System (SCNES) has been proposed "published by Toukura et al. (1994), Fujii-e (1992), Shimizu and Fujii-e (1994), Shimizu and Endo (1994), Arie et al. (1994), Endo et al., (1994), Endo et el. (1990)," as an ultimate goal of the nuclear energy systems. The SCNES is defined as a system that satisfies four objectives: energy generation, fuel breeding, confinement of the minor actinides

(MAs) and the radioactive fission products (FPs), and nuclear reactor safety. The term "self-consistent" means that even if the system produces materials harmful to human beings and the environment, it can eliminate these materials by itself.

This paper discusses applicability of the nitride fueled core to the SCNES which targets on:

(1) Energy generation: the produced electric energy of the system is greater than the consumed energy.

(2) Fuel breeding: the fissile is self-supplied in the system (i.e. the breeding ratio is over 1.05 so as to recover the loss of the fissile in the recycling process).

(3) Safety of the core: potential core disruptive accident caused from an anticipated transient without scram(ATWS) is eliminated by inherent negative feedback in the intact-core state; re-criticality of the degraded core is prevented by controlled material relocation "published by Endo and et el. (1990)."

(4) Confinement of the MAs and the radioactive FPs: MAs are utilized as fuel and the radioactive FPs are burned through neutron capture reactions so as not to increase in the system.

2. FEATURES OF THE NUCLEAR SYSTEM

2.1 Reactor and core

Table 1a and 1b shows the specifications of the plant and the nitride fueled core. The thermal power of reactor is set to 3,500MW (1,500MWe), which is regarded as typical commercial plant scale. The operation period is extended to 24 months so that the duty factor exceeds 90%. The sodium primary coolant system is a four-loop type. The outlet coolant temperature of the reactor is 585℃, the main steam pressure 250kg/cm^2 and main steam temperature 540℃. These high temperature heat balance enables remarkable high thermal efficiency of about 45%.

Table 1a Major Plant Specifications

Item	Specification
Reactor type	Top entry loop-Hybrid type
Electric power	1,500MWe
Thermal power	3,500 MWth
Operation period	24 months
Number of loop	4
Coolant Temperature	453/585 ℃
Reactor shutdown system	2 independent system + SASS
Secondary cooling system	Compound type: Pump + SG
Decay heat removal system	IRACS
Reactor Containment	Concrete containment with liner

Table 1b Main Specifications of the Nitride Core

Item	Specification
Kinds of Fuel	Nitride (99.9%^{15}N enrichment)
Type of Subassembly	Ductless
Volume Fraction of Fuel	55%
Core Height/Diameter	60cm/430cm
Burnup Reactivity Swing	1.6%Δk/kk'/cycle
Average Linear Heat Rate	270W/cm

Nitride fuel is employed as core fuel because of its high nuclear and thermal performance. The nitride fuele generates remarkable amount of radioactive ^{14}C due to the ^{15}N (n,d) ^{14}C and the ^{14}N (n,p) ^{14}C reaction as described later. The 99.9% ^{15}N enriched nitride is employed to nitride fuel in order to avoid problems with ^{14}C production and adverse effect on core characteristic.

Another characteristic specification of the core is the ductless assembly which we have been studying as a key technology of future FBRs. By introducing the ductless assembly, a 55% fuel volume fraction is feasible as described in the conclusion of design study "published by Itoh and shibahara (1996)." Because of the high fuel fraction, the burnup reactivity swing diminishes to 1.6%Δk/kk'/cycle. The average linear heat rate is limited to 270 W/cm.

The core configuration is shown in Fig. 1. The core is flat consisting of two homogenous regions. In order to enhance the self-controllability "published by Shimizu and Endo (1994)" of the core, the concepts of the cold fuel (low linear heat rate), the ductless fuel subassembly and

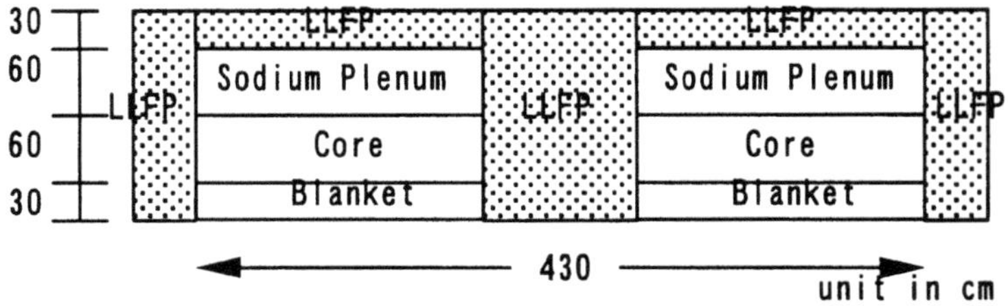

Fig.1 Core configuration of the nitride core.

the upper sodium plenum are employed. The equivalent diameter of the core is about 430cm, and the height of the core and the depth of the upper sodium plenum are 60 cm each.
The feature of this core is the smaller burn-up reactivity and negative reactivity coefficients, which are shown in Table 2.

Table 2 Reactivity Coefficients and Initial Condition

Items	characteristics
Fuel temperature coefficient	$-2.0\times10^{-6}\Delta k/k/℃$
Structure temperature coefficient	$1.5\times10^{-6}\Delta k/k/℃$
Coolant temperature coefficient(core/plenum)	$-2.1\times10^{-6}\Delta k/k/℃$ $(3.5/-2.8)\times10^{-6}\Delta k/k/℃$
Core radial expansion coefficient	$-7.4\times10^{-6}\Delta k/k/℃$
Doppler coefficient	$-4.3\times10^{-3}Tdk/dT$
Maximum coolant temperature on power	650℃
Maximum cladding temperature on power	675℃
Maximum fuel temperature on power	1160℃
Maximum linear heat rate on power	480W/cm

2.2 Material flows in the system
The material flow in this system is shown in Fig.2.

(1) Pu, U and MAs are recycled through the reactor and the reprocessing plant, and are not released outside the system.
(2) The nitride recycling system consists of a solvent extraction process, makeup of ^{15}N, and laser isotope separation. During withdrawal of ^{15}N in the reprocessing process, the concentration of ^{15}N decreases from 99.9% to 95%. The loss and depletion of ^{15}N-enriched nitrogen is recovered by the re-enrichment of ^{15}N.
(3) The efficiency of laser generation and ionization of FPs is presumed to respectively be 0.1% and 10% in the process of laser isotope separation.
(4) The FPs are loaded into the core for neutron irradiation, and reloaded after six years for separation of stable or short lived FPs.

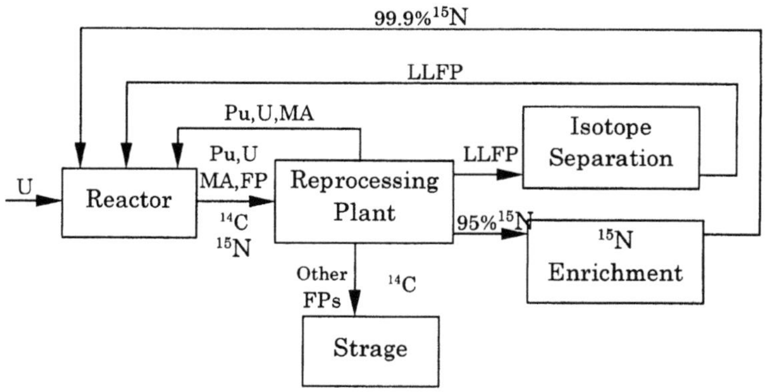

Fig.2 Material flows of the nitride fuel reactor and the recycling system.

3. SAFETY ANALYSIS OF UNPROTECTED TRANSIENTS

The effectiveness of the safety features incorporated in the conceptual core design is discussed by results of core transient analysis for ATWS.

3.1 Criteria for ATWS
The criteria for ATWS are given as follows:
(1) Sodium boiling temperature: 930℃
(2) Fuel melting point:2500℃ (helium bonded fuel) or Fuel thermal decomposition :2000℃ (sodium bonded fuel)

3.2 Coolant and fuel temperature response for the anticipated transient without scram

The transient response for the nitride-fueled core was assessed for some different input conditions. The cases in which the coolant and the fuel temperature do not exceed the coolant boiling and fuel melting points at ATWS are shown in Table 3 (ULOF; unprotected loss of flow) and Table 4 (UTOP; unprotected transient over power).

Fig. 3 shows the first 200 seconds of the temperature response during a ULOF which is characterized by the following input condition: 20 seconds for the flow halving time, 650℃ for the initial coolant temperature and -0.1(¢/℃) for the control rod drive-line expansion coefficient at the end of the cycle.

While the coolant and fuel temperature initially begin to rise dominated by the flow decrease, the negative feedback reactivity effects due to the core radial expansion and the control rod drive-line expansion gradually reduce the core power. When the flow rate falls to 15%, the pony motor starts keeping the flow rate. The coolant temperature begins to drop as the power falls. The maximum coolant temperature is 874℃ in this case. And although the maximum coolant temperature increases by 46℃ if the flow halving time is changed into 10 seconds, the coolant does not boil.

Table 3 Maximum Temperatures in a ULOF

Flow halving time (s)	Coolant temperature on power (℃)	CR driveline expansion coefficient (¢/℃)	With or without sodium plenum	Bond material in fuel gap	Maximum coolant temperature (℃)
20	650	-0.1	with	He	874
5	650	-0.1	with	He	978
10	650	-0.1	with	He	920
50	650	-0.1	with	He	837
20	620	-0.1	with	He	813
20	650	0	with	He	942
20	650	-0.2	with	He	832
20	650	-0.1	without	Na	832
20	650	-0.1	without	He	919

Table 4 Maximum Temperatures in a UTOP

Inserted reactivity (¢)	Maximum linear heat rate (W/cm)	Control rod drive-line expansion coefficient (¢/℃)	With or without sodium plenum	Bond material in gap	Maximum Fuel temperature (℃)	Maximum coolant temperature (℃)
60	480	-0.2	with	He	1760	837
100	480	-0.2	with	He	2190	967
20	480	-0.2	with	He	1360	743
60	600	-0.2	with	He	2020	899
60	480	-0.4	with	He	1620	795
60	480	0	with	He	2020	917
60	480	-0.2	without	Na	1420	928
60	480	-0.2	without	He	1830	858

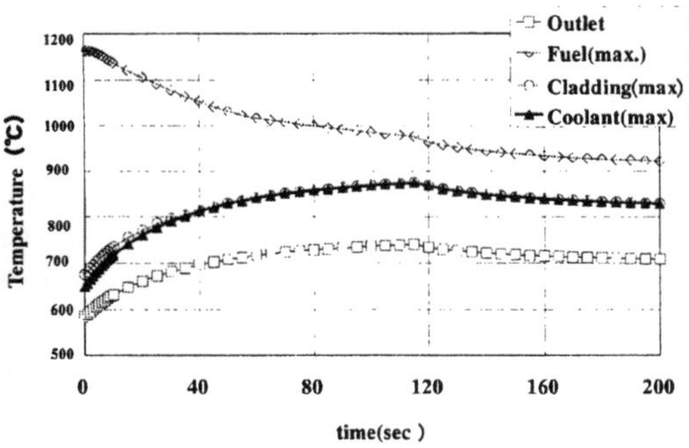

Fig.3 Response for ULOF.

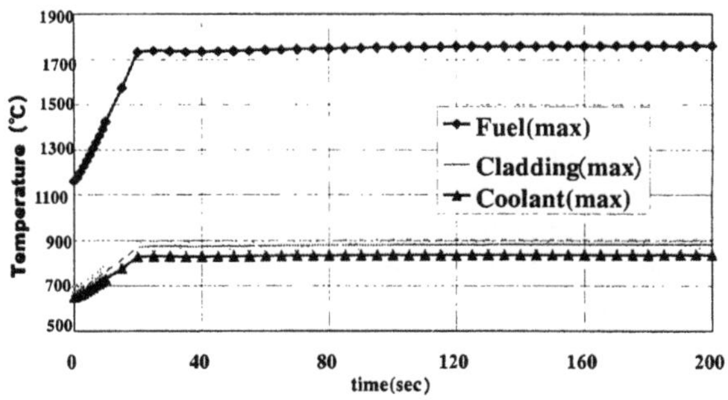

Fig.4 Response for UTOP

For the second postulated transient, Fig. 4 shows the temperature response for UTOP. In this case, the inserted reactivity is 60 ¢, the maximum linear heat rate is 480 W/cm, and the control rod drive-line expansion coefficient is -0.2 ¢ /℃ at the beginning of the cycle. The withdrawal reactivity is inserted at 3 ¢ /s in 20 seconds for a UTOP vent. While the temperature of the fuel rises in the beginning as the core power increases, the Doppler effect, the fuel axial expansion and the control rod drive-line expansion effect hold down the temperature increase with the negative reactivity. After 20 seconds from the start, the power reaches pseudo equilibrium, keeping the maximum coolant temperature; 837℃ and the maximum fuel temperature; 1,760℃. As a result of the calculation, it is shown that the coolant does not boil nor does the fuel melt in the nitride fueled core during an anticipated transient without scram.

3.3 Comparison of the helium and the sodium bonded fuel

The sodium-bonded fuel is advantageous in a ULOF as shown in Table 3, because of the small positive or even negative Doppler feedback. Where as for a UTOP, the helium-bonded fuel seems to be advantageous. The sodium-bonded fuel gives a small Doppler effect due to the low average fuel temperature as shown in Fig.5, so that increased power is shown in Fig.6. The maximum coolant temperature of the sodium bonded fuel is higher by about 70℃.

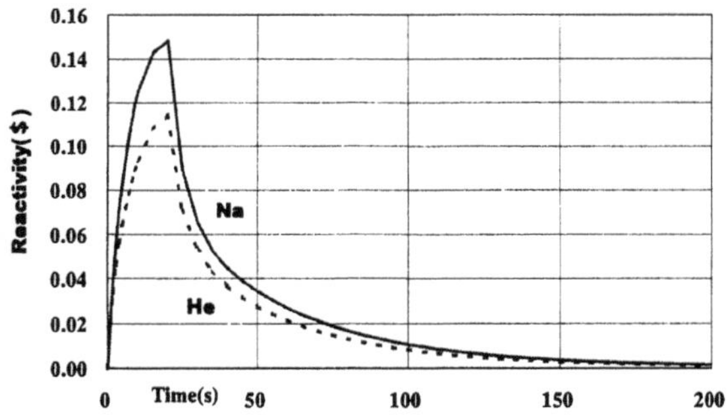

Fig. 5 Bond material and power in a ULOF.

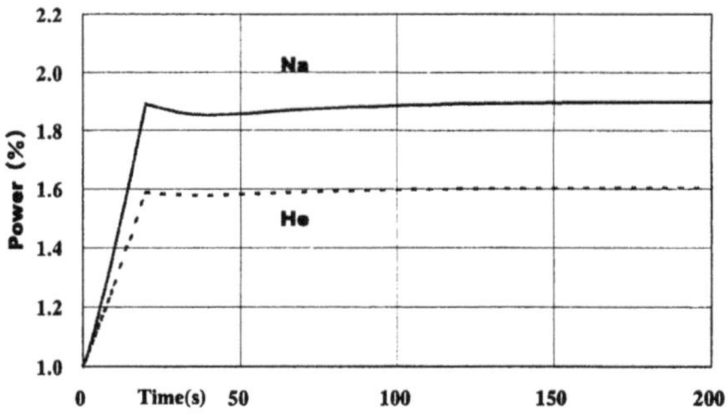

Fig. 6 Bond material and power in a UTOP.

4. SELF-TERMINABILITY OF THE DEGRADED CORE

We assumed that the re-criticality of the degraded core is determined by the self-terminability "published by Sawada et al. (1999)." Before falling to the degraded core state, material relocation occurs because of physical laws, giving negative reactivity. The mechanism of the controlled material relocation consists of a physical switch and a physical driving force.

In the cold fuel concept, the coolant temperature reaches the boiling point before fuel melting during UTOP. Some ideas about controlled material relocation have been proposed.

One idea is a self-actuating module of a neutron-absorber located in the fuel subassembly "published by Ikeda et al. (1996)." The neutron absorber is placed above the sodium plenum. If the coolant temperature rises during an anticipated transient without scram, material that has a lower melting point melts and the neutron absorber (ex. ^{10}B) falls down to just above the core, inserting negative reactivity. Another one is a self-actuating module in the fuel subassembly. If the coolant temperature rises during an anticipated transient, material of a lower melting point melts, or loses magnetic force, and then the fuel falls down to bottom of the active core, inserting negative reactivity.

Table 5 shows that a self-actuating fuel module which has 4.5% Δ k/kk' of shutdown reactivity, is sufficient for the 3.4% Δ k/kk' of excess reactivity of the degraded core. These modules will be selected and optimized so that the self-terminability of the core is enhanced and the influence of the core characteristics is small.

Table 5 Excess Reactivity of Degraded Core and Reactivity of a Self-actuating Module

Items	Reactivity
Reactivity insertion of degraded core	3.4% Δ k/kk'
Reactivity of self-actuating module of neutron absorber	2.5% Δ k/kk'
Reactivity of self-actuating module of fuel (10% of fuel core)	4.5% Δ k/kk'

5. INCINERATEATION OF LONG LIVED FP

5.1 Neutron balance

The neutron balance of the nitride core at the EOEC is shown in Fig. 7. The ^{15}N enriched fuel core is found to have approximately 0.6 excess neutrons per fission appropriate for transmutation of FPs. Whereas, the excess neutrons are estimated to be 0.4 neutrons per fission for the mixed oxide core and 0.7 for the metal-fueled core.

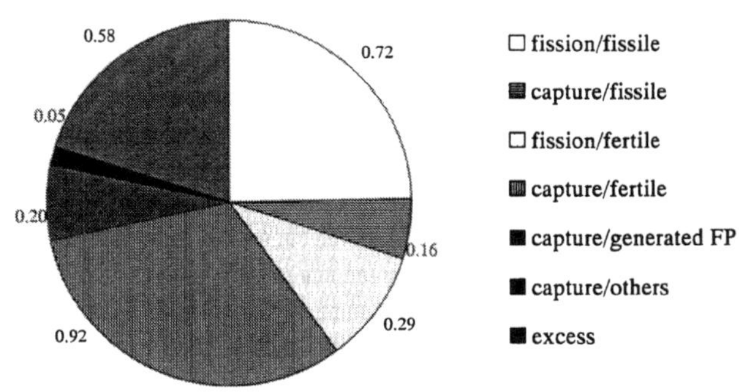

Fig.7 Neutron balance of nitride fueled core.

5.2 Selection of long-lived FPs

Selection criteria is as follows:

(1) The half-life is relatively long.
(2) The radioactive toxicity is still great after 1,000 years.
(3) The neutron capture reaction of the concerned FP decreases the radioactivity effectively in the core.
(4) Criticality and breeding (i.e. B.R. 1.05) are maintained.
(5) Element separation in the chemical process is practicable.

To select the long-lived FPs to be transmuted, the nuclear characteristics of the several radioactive fission products,

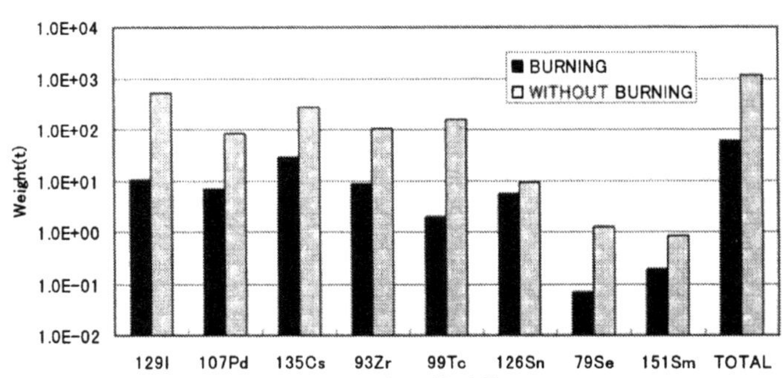

Fig.8 Comparison of FPs inventory after 5,000 years with burning in the core vs. without burning.

with half-lives over 10 years, are compared in Table 6 using the above criteria. The toxicity is defined as the radioactivity divided by the annual limited intake. The number of required neutrons is calculated from the yield of FPs per fission multiplied by the burning efficiency. Practicability of element separation is judged from the concentration in the FPs.

Table 6 Characteristics of Long Lived Radioactive Fission Products

Nuclide	Half life (y)	Yields (/fission)	Toxicity[1] (ALM/HMT)	Burning[2] efficiency	Required[3] neutron (/fission)
^{129}I	1.6×10^7	0.0142	3.5×10^4	100	0.0142
^{107}Pd	6.5×10^6	0.0274	2.5×10^3	100	0.0274
^{135}Cs	2.3×10^6	0.0716	1.0×10^4	100	0.0716
^{93}Zr	1.5×10^6	0.0396	9.7×10^4	100	0.0396
^{99}Tc	2.1×10^5	0.0591	1.5×10^4	100	0.0591
^{126}Sn	1.0×10^5	0.00261	1.0×10^5	99	0.0026
^{79}Se	6.5×10^5	0.00056	3.0×10^3	100	0.00056
^{151}Sm	90	0.00781	7.6×10^4	96	0.0075

1 Toxicity of spent fuel of 150GWd after 1,000y
2 Burning / decay + burning 4
3 Only for burning
4 Total neutron flux is $4X10^{15}n/cm^2/s$

The absorption cross-sections and half-lives of the FPs in the nitride core are shown in Fig. 9. When the total flux is 1.0×10^{15} /cm^{2}s and the absorption cross section is 1.0 barn, the half life in the core is 22 years. The minimum of the cross-section of the FPs is 0.008 barn of ^{126}Sn. Therefore, it is difficult to burn ^{126}Sn. The other nuclides, ^{129}I, ^{107}Pd, ^{135}Cs, ^{93}Zr, ^{99}Tc, ^{126}Sn, ^{79}Se and ^{151}Sm can be effectively burned by the neutron capture, and are selected as targets. The total number of neutrons required to burn these FPs is 0.22 neutrons per fission, which can be supplied by a nitride fueled core.

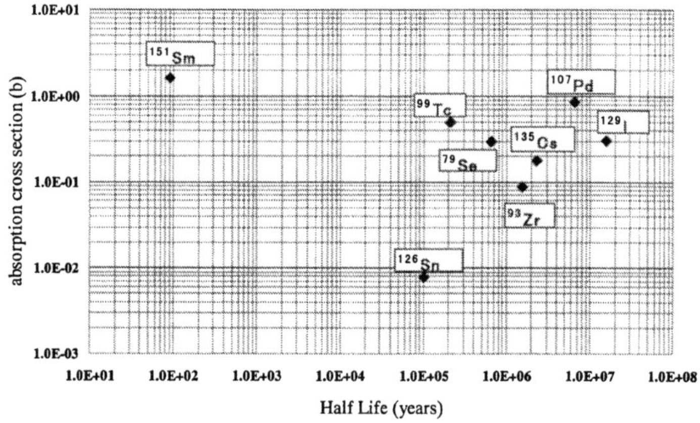

Fig.9 Half life and absorption cross section of LLFP in the nitride fueled core.

5.3 <u>Radioactive fission products</u>

Fig.10 shows the radioactivity of the long-lived FPs in the system estimated under the following conditions.

(1) Estimated uranium resources of 3,200,000t are used in this system in 5,000 years.
(2) The annual total generated power is constant.
(3) The loss of heavy material is 2% in the recycling process. The loss of FPs in isotope separation is ignored.

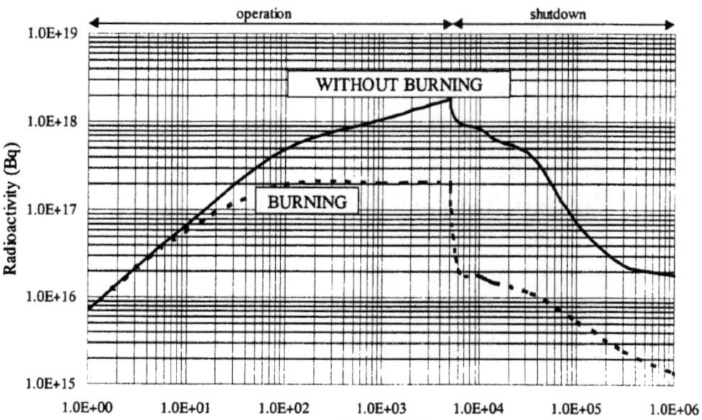

Fig.10 Effect of FP burning in the nitride fueled core on discharged radioactivity of LLFP.

(4) ^{129}I, ^{93}Zr, ^{99}Tc, and ^{126}Sn are placed inside the inner core, and ^{135}Cs, ^{107}Pd, ^{79}Se and ^{151}Sm are loaded above the sodium plenum and outside the outer core.

Even having introduced FP burning, an increase of long-lived FPs will also continue with plant operation for hundreds of years.
After the uranium resources are consumed over several thousands of years, radioactive material will be left in the future. The radioactivity of the remaining long-lived FPs is compared. With isotope separation and incineration of FPs, the amount of long-lived FPs left over is found to decrease by a factor of 50, as shown in Fig. 10.

6. ENERGY BALANCE

The enrichment of ^{15}N and laser isotope separation, which are in the R&D stage, will consume some energy. The consumed energy in the fuel recycling system has been estimated. As a result, the recycling system has been calculated to consume 2.4% of its produced electric energy, as shown in Table 7. "General" in Table 7 includes oxidation/reduction, de-cladding, dissolution, precipitation, TBP extraction, HDEHP extraction, vitrification and cooling of decay heat.

Table 7 Energy Balance

Items	Energy per Heavy metal(J/tHM)	
Production	Thermal	1.3×10^{16}
	Electric	5.6×10^{15}
Consumption	Isotope Separation	9.8×10^{13}
	Makeup of ^{15}N	3.4×10^{13}
	General	2.4×10^{12}
	(Total)	1.3×10^{14}

7. THE PRODUCTION OF ^{14}C

There are two main reactions that produce ^{14}C in the nitride fueled core. ^{14}C is produced mainly by the ^{14}N (n,p) ^{14}C reaction and a few by ^{15}N (n,d) ^{14}C reaction. While the cross section of ^{15}N (n,d) ^{14}C reaction is 5×10^{-5} barn, the ^{14}N (n,p) ^{14}C cross section is estimated to be 0.012 barn by MCNP and JENDL-3. As the concentration of ^{14}N is reduced down to 0.01% by isotopic enrichment, the production of ^{14}C decreases, as shown in Fig. 11.
The annual production of ^{14}C is 6 g (9.3×10^{11} Bq) in the nitride core with 99.9% ^{15}N enriched fuel. Fig. 12 shows the cumulative production of ^{14}C in the 1,500 MWe reactor. On the other hand, 8 kg of ^{14}C is generated annually by the cosmic neutrons and decays naturally; the global total ^{14}C was estimated to be 1.1×10^{19} Bq "published by Lal and Peters

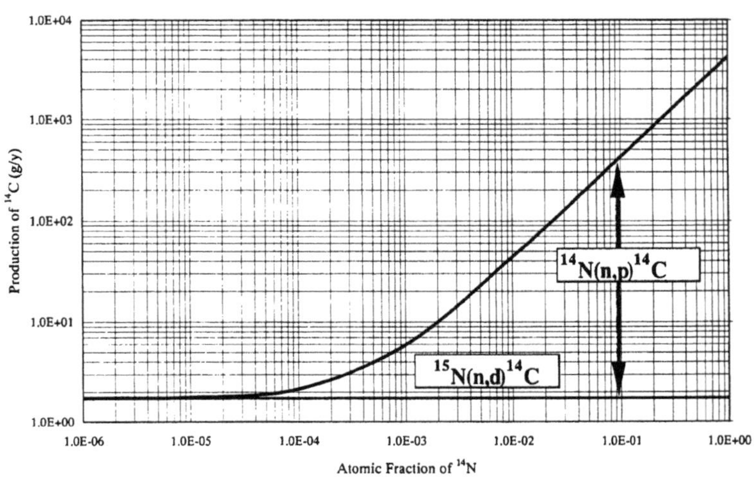

Fig.11 Production rate of ^{14}C as function of ^{14}N concentration.

(1967)." If a natural nitride fuel core were operated for 5,000 years, 2.4×10^{18} Bq of ^{14}C, whose half life is 5,730 years, would still remain. This is one of the reasons natural nitride fuel is not adopted.
When 99.9% ^{15}N enriched fuel is used and the greater part of the ^{14}C in the reprocessing plant is confined, the ^{14}C released outside the nitride fuel recycling system is much less than the ^{14}C released by natural production. The enrichment with ^{15}N improves the burnup reactivity swing and significantly reduces the global impact of ^{14}C.

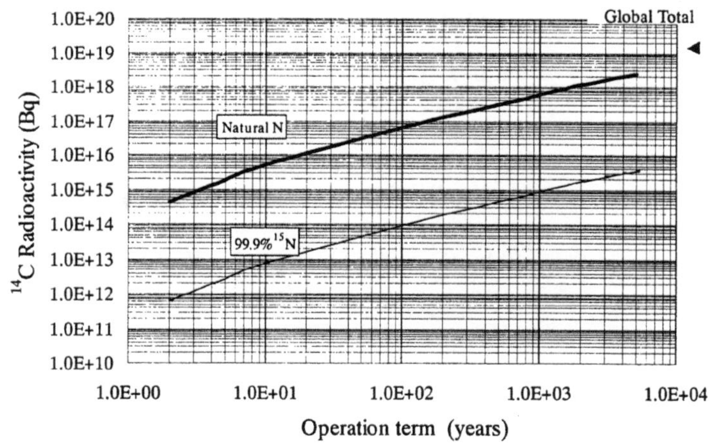

Fig.12 Accumulation of ^{14}C by one 1,500MWe plant with nitride fueled core.

8. CONCLUSIONS

It has been shown that the nitride fueled core serves the purpose of the SCNES. The defined characteristics of the SCNES that is featured by the nitride fuel are:

(1) The nitride-fueled core of 3,500MWth that consists of the ductless subassemblies with the upper sodium plenum has the feedback characteristics of low burnup reactivity swing and negative sodium temperature coefficient, and consequently have the passive shutdown capability during an ATWS. In fact, safety analyses on an ATWS show that the maximum fuel temperature of an UTOP is 1,760℃ and the maximum coolant temperature of an ULOF is 837℃.

(2) As long as the source of the MAs remaining in the fresh fuel is confined to the system itself, and the target LLFPs are located in and around the core as shown this paper, the influence of the MAs and LLFPs on the core characteristics is small. This means the core loaded with MAs and LLFPs meets the required efficiency.

(3) By the transmutation of the LLFPs in the reactor, the total radioactivity of the fission product in the system almost reaches maximum in one hundred years. At the end of the nuclear fuel cycle age, which will come several thousand years after, the radioactivity of the fission products is 1/50th that of a system without transmutation.

(4) N-15 enrichment for the nitride fuel and the laser isotope separation of the LLFP's are assumed in this system. Their energy consumption is presumed to be about 2% of the electric energy produced by the system.

This study was sponsored by the Ministry of International Trade and Industry (MITI).

REFERENCES

Books:

Lal, D. and Peters B. (1967), Cosmic ray produced radioactivity on the earth in Encyclopedia of Physics, Vol. XLVI/2 on Cosmic Rays, Spinger-Verlag, New York

Conferences:

Arie, K. et al. (1994), A Core Concept for the Self-Consistent Nuclear Energy System based on the Promising Future Technology, Int. Symp. Global Environment and Nuclear Energy Systems, Susono, Japan, 24-27 October, 455,

Endo, H. et al. (1994), Safety Features of Self-Consistent Nuclear Energy System, *ibid*.

455

Endo, H. et al. (1990), Safety Features of Fast Reactor with Metallic-Fueled Core, *Int. Fast Reactor Mtg.*, Snowbird

Fujii-e, Y. (1992), An Approach to Self-Consistent Nuclear Energy System-Potential of Fast Reactor, *Design and Safety of Advanced Nuclear Power Plants*, Tokyo, 25-29 October, 3-1, 11

Ikeda, K. et al. (1996), Feasibility Study of Nitride Fuel Core and Recycle System toward Self-Consistent Nuclear Energy System, *Global Environment & Nuclear Energy System-2*(GENES-2), Tsuruga, Japan, 29 October – 1 November, 673

Itoh, K. and Shibahara I. (1996), Feasibility Study on Application od Ductless Fuel Assembly to FBR, *The 4th Int. Conf. on Nucl. Engineering* (ICONE-4), New Orleans, U.S.A.

Sawada, T. et al. (1999), Characteristics of Fuel Melting and Relocation in Metalic Fueled Fast Reactor Core and its Feasibility for Eliminating Recriticality, *Proceedings of the Third Int. Symp.* (GENES-3), Tokyo, Japan, 14-17 December, 157

Shimizu, A. and Fujii-e Y. (1994), Self Consistent Nuclear Energy Systems, *Int. Symp. Global Environments and Nuclear Energy Systems*, Susono, Japan, 24-27 October, 25

Shimizu, A. and Endo H. (1994), Study of Self-Controllability of Fast Reactor Cores, *Int. Topical Meeting on Advanced Reactor Safety*, Pittsburgh, April,

Toukura, A. et al. (1994)., Study on Core Concept for Commercial LMFBR Plant Toward Self-Consistent Nuclear Energy System Concept, *Int. Symp. Global Enviroment and Nuclear Energy Systems,* Susono, Japan, 24-27 October, 265

Pergamon
www.elsevier.com/locate/pnucene

Progress in Nuclear Energy, Vol. 40, No. 3–4, pp. 607–613, 2002

Printed in Great Britain
0149-1970/02/$ - see front matter

PII: S0149-1970(02)00056-2

A METAL FUEL FAST REACTOR CORE

FOR THE SELF-CONSISTENT NUCLEAR ENERGY SYSTEM

KAZUO ARIE, MASAO SUZUKI, MASATOSHI KAWASHIMA

Power Systems & Services Company, Toshiba Corporation
8, Shinsugita, Isogo-ku, Yokohama 235-8523, Japan
Phone: +81-45-770-2074
FAX: +81-45-770-2462
E-mail: kazuo.arie@toshiba.co.jp

SATOSHI MORO

Institute of Applied Energy
1-14-2, Nishi-shimbashi, Minato-ku, Tokyo 105-0003, Japan

MASAKI SAITO, HIROSHI SEKIMOTO

Research Laboratory for Nuclear Reactors, Tokyo Institute of Technology
2-12-1, Ohokayama, Meguro-ku, Tokyo 152-8550, Japan

ABSTRACT

Feasibility of transmutation of the major long-lived FPs (I, Pd, Tc, Sn, Se, Zr, Cs) while maintaining fuel breeding capability for the Self-Consistent Nuclear Energy System is evaluated based on the actinide recycle metal fuel core of a fast reactor. It is shown that I, Pd, Tc, Sn, Se, and Zr can be transmuted simultaneously by an aid of the isotope separation of Pd-107, Zr-93, Sn-126 and Se-79. Cs, which is difficult to transmute with the other FPs, is planned to be utilized as an in-reactor shielding material to confine in the system The overall assessment based on those results indicates that the developed system has the great potential toward the Self-Consistent Nuclear Energy System.

1. INTRODUCTION

The Self-Consistent Nuclear Energy System (SCNES) has been proposed (Fujii-e et al, 1991, 1992ab, 1993, Shimizu and Fujii-e, 1994c) as the ultimate goal of nuclear energy systems. It is defined as the system that satisfies the four objectives, i.e., energy generation, fuel breeding, transmutation and confinement of radioactive FPs produced by itself, and system safety simultaneously. Since fission neutrons must be fully utilized to achieve the objectives, a fast reactor is the most promising concept for the SCNES because of its large number of available neutrons. TRU transmutation can be achieved by using TRU as nuclear fuel. This paper focuses on the possibility of transmutation of long-lived FPs based on actinide recycle metal fuel core. Overall assessment from a viewpoint of the SCNES is also performed.

This study was performed under the sponsorship of the Ministry of International Trade and Industry(MITI).

2. FP TRANSMUTATION

2.1 Selection of important FPs to be transmuted

The important FPs to be transmuted in a reactor core are selected by taking account of the followings.

1. Long half life of natural decay, T1/2
2. Large yields per fission, Y
3. Significant shortening of effective half life of radioactive FPs due to transmutation

Table 1 shows the major radioactive FPs in order of half life of natural decay with large yields (i.e., T1/2 x Y) (Ihara, 1989). Effective half life due to transmutation is calculated by assuming the neutron spectrum of a typical fast reactor core and a total neutron flux of 1×10^{15}n/cm^2/s. The half lives of Cs-137 and Sr-90 cannot be shortened by transmutation in the core. Therefore, it is clarified that the important FPs to be transmuted are I-129, Cs-135, Pd-107, Zr-93, Tc-99, Sn-126, and Se-79.

Table 1 Major radioactive FPs with long half life and large fission yields

Nuclides	T1/2 x yield (y/fission)	Half Life (y) Natural Decay	Half Life (y) Transmutation*1
I-129	2.5×10^5	1.6×10^7	6.2×10^1
Cs-135	2.3×10^5	3.0×10^6	3.3×10^2
Pd-107	2.1×10^5	6.5×10^6	4.2×10^1
Zr-93	5.7×10^4	1.5×10^6	2.6×10^2
Tc-99	1.3×10^4	2.1×10^5	4.9×10^1
Sn-126	2.9×10^2	1.0×10^5	3.1×10^3
Se-79	3.7×10^1	6.5×10^4	9.0×10^3
Cs-137	2.0×10^0	3.0×10^1	1.8×10^3
Sr-90	5.8×10^{-1}	2.9×10^1	1.7×10^3

*1. Total neutron flux 1×10^{15} n/cm^2/s

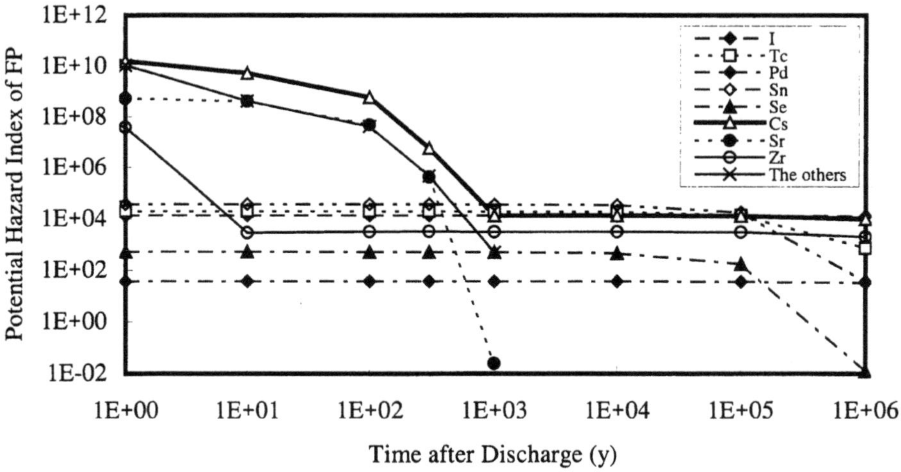

Fig. 1 The Potential Hazard Indexes of FPs as a function of the time after discharge

The Potential Hazard Index of major FPs is shown in Fig. 1. The Potential Hazard Index H(t) is defined as follows.

$$H(t) = Bq(t) / ALI \quad (1)$$

where,

H(t) : the Potential Burden Index for radioactive FP, i, at time t after reactor shutdown
Bq(t) : the radioactivity of radioactive FP, i at time t after reactor shutdown (Bq)
ALI : the Annual Limits of Intake for radioactive FP, i (Bq)

Cs (Cs-137) and Sr are dominant over the short or middle term. I, Cs (Cs-135), Tc, Sn, Zr, Se and Pd are dominant in the long term. The contribution of the other FPs are negligibly small especially after about a thousand years cooling. Therefore, the selection of the target FPs is found to be reasonable.

2.2 Effect of isotope separation of FPs

The major issues for FP transmutation are to keep neutron balance and to reduce FP inventory in a reactor at equilibrium phase. One of the method to reduce the inventory is to increase effective neutron cross sections of FPs by an aid of neutron moderator. However, neutron moderator is not necessarily a good way for the simultaneous transmutation of the seven FPs, because neutron moderator also increases parasitic neutron capture of the other materials such as structures (Arie et. al., 1996). Therefore, this study does not assume neutron moderator.

The other method against these issues is isotope separation of FPs during recovery from spent fuel. Each long-lived FP is accompanied by stable or short-lived FPs in case of no isotope separation of FPs. The accompanied stable or short-lived FPs have unfavorable influence on not only neutron balance but also equilibrium inventory. The effect of separation of the stable and short-lived FPs from the long-lived FPs, i.e., isotope separation, is evaluated from a viewpoint of inventory as shown in Table 2. The equilibrium inventories of Pd, Zr Sn, Se, and Cs can be significantly reduced by isotope separation, because they have many stable isotopes. The effect of the isotope separation on neutron balance is shown in Table 3.

The required neutrons for transmutation of Pd and Zr can be also reduced much. It should be noted that these effects are not only simply due to the reduction of the parasitic capture of the stable isotopes but also due to the reduction of the equilibrium inventory of the target FPs itself, because isotope separation eliminates neutron capture of the parent isotopes of the target FPs which results in the lower yield of the target FPs at equilibrium phase.
Therefore, isotope separation of Pd, Zr Sn, Se, and Cs has large effects on these FP transmutation.

Table 2 Effects of isotope separation (IS) on equilibrium inventory of FPs

nuclides	equilibrium inventory (t/GWe) *1		reduction rates(%)	merit
	w/o IS	with IS		
I	2.1	1.6	24	small
Cs	30	11	63	medium
Pd	35	0.92	97	large
Zr	404	8.2	98	large
Tc	3.1	3.1	0	small
Sn	52	17	67	medium
Se	3.6	0.03	99	large

*1 Total neutron flux 2.8 x 10^{14} n/cm^2/s corresponding to the first row of the core peripheral

Table 3 Effects of isotope separation on the required neutron per fission for transmutation

nuclides	required neutrons w/o isotope separation	required neutrons with isotope separation	neutron saving
I	0.025	0.015	0.010
Cs	0.216	0.075	0.141
Pd	0.433	0.031	0.402
Zr	0.454	0.037	0.417
Tc	0.060	0.060	0
Sn	0.133	0.003	0.130
Se	0.006	0.001	0.005

3. CORE CONCEPT

3.1 Target FPs Location

In developing the core layout for FP transmutation, the assumed chemical forms and the required volume for the FPs are shown in Table 4.
Among these nuclides, Cs has the largest required volume for transmutation even if the isotope separation is assumed, because it has the smallest atom density in the chemical form. Therefore, we do not assume isotope separation of Cs, and CsCl is located as in-reactor shielding material rather than transmutation target with $SrCl_2$ behind the FP transmutation region to confine them in *the system*. *The system* means the nuclear energy system including fuel cycle system. The developed core layout is shown in Fig. 2. Sn2-Zr is located above the core. CeI_3, Pd metal, Tc metal and Se4-Zr5 metal are located at the radial peripheral of the core for transmutation.

Table 4 Considarations on chemical forms, FP locations and its required volumes

nuclides	chemical form	density (g/cm3)	location	isotope separation	required volume (L)
I	CeI3	5.6	peripheral *1	no	310
Tc	metal	11.5	peripheral	no	160
Pd	metal	12.3	peripheral	yes	45
Se	Se4-Zr5	5.5	peripheral	yes	9
Zr & Sn	Sn2-Zr	7.0	above core *2	yes	580
(Cs	CsCl	4.0	peripheral	yes	2040)

*1 Total neutron flux 2.8 x 10^{14} n/cm^2/s

*2 Total neutron flux 5.0 x 10^{15} n/cm^2/s

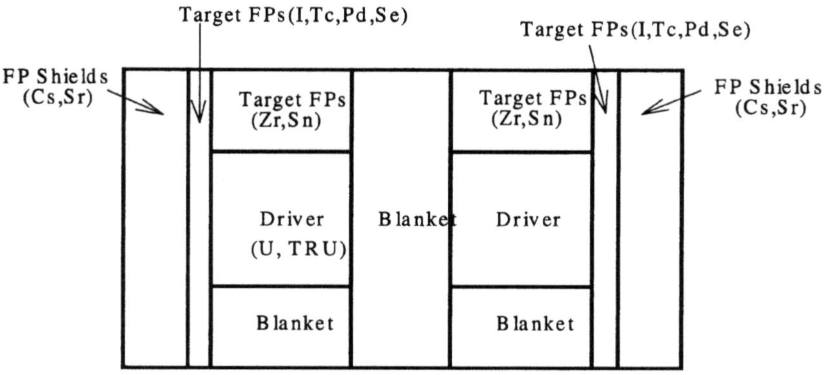

Fig. 2. Core layout for the Self-Consistent Nuclear Energy System

Table 5 Main characteristics of the core

Thermal Power	1,600 MWth
Operating Cycle	365 days
Core Diameter	4.5 m
Core Height	45 cm
Breeding Ratio	1.03
Burnup Reactivity Swing	3.5 %$\Delta\rho$
Average Burnup	100 GWd/t
Peak Linear Heat Rates	410 W/cm
Target FP Inventory	8.1 t

3.2 Core Characteristics

In the SCNES concept, TRU transmutation can be achieved by using them as nuclear fuel. The core has the annular and pancake geometry which provides the space for neutron utilization around the driver region, and is used for fuel breeding and FP transmutation

The main characteristics of the core in the equilibrium phase are shown in Table 5. A breeding ratio of 1.03 is obtained by 2 dimensional diffusion-depletion calculation. The total inventory of the target FPs in the equilibrium phase is about 8.1 t, which can be incorporated around the driver region. Thus, simultaneous fuel breeding and FP transmutation is confirmed.

4. OVERALL ASSESSMENT

The achievement of the system has been evaluated from a viewpoint of the SCNES relative to the conventional nuclear energy system (Arie et. al., 1996). Since the main feature of the SCNES is simultaneous realization of energy production, fuel breeding and radionuclide transmutation (with the system safety), the achievement V is evaluated by the ratio of the amount of energy produced from a unit amount of natural uranium to the total Burden Index left at the out-of-system due to the energy production, as follows.

$$V = E / B \tag{2}$$

where,

E : the amount of energy produced from a unit amount of natural uranium

$E \propto \eta \, \beta \, Nu$

η : energy utilization efficiency

β : natural uranium utilization factor

Nu : an amount of natural uranium used for the system

B : the Burden Index left at the out-of-system due to the energy production

B = Σ (amount of the radionuclide) * (its half life of natural decay.)

The result is shown in Table 6. The energy system developed in this study (Case 1) has approximately 9 times higher achievement than that of conventional LWR and FBR energy system mainly due to FP transmutation. The recycling loss factors for FP in Case 1 are the roughly estimated value for the FP recovery process based on current technology (Takagi et al, 1994d). Furthermore, if the loss factor can be reduced to 0.1 % in the future, the achievement of the system will increase dramatically.

Table 6 Overall assessment on the achievement of this study as SCNES

items	LWR	FBR	this study	
	(no recycle)	(Pu recycle)	(case 1)	(case 2)
energy utilization efficiency (%)	34	40	40	←
breeding ratio	0.4	1.0	1.0	←
burnup (%)	3	10	10	←
recycling loss factor for fuel (%)	-	0.2	0.1	←
natural uranium utilization factor (%)	0.43	99	99	←
produced energy, E (relative)*1	1.0	268	271	←
assumed recycling loss factor for FP (%)				
I, Tc, Pd, Sn	-	-	10	0.1
Se	-	-	1	0.1
Zr, Cs, Sr	-	-	20	0.1
Burden Index, B(relative) *1				
fuel	0.060	6.5	0.02	0.02
FP	0.940	216	29.8	0.22
total	1.0	223	29.8	0.24
Achievement, V(relative) *1	1.0	1.2	9.1	1130

*1, normalized to LWR case

5. CONCLUSION

(1) The feasibility of the long-lived FP transmutation has been clarified based on the actinide recycle metal fuel core of a fast reactor.

- Major long-lived FPs (I, Pd, Tc, Sn, Se, Zr) can be transmuted while maintaining breeding capability by isotope separation of Pd-107, Zr-93, Sn-126, and Se-79.
- The other major long-lived FP, Cs should be confined in the system by utilizing it with Sr as an in-reactor shielding material.

(2) The energy system with the developed core has the great potential to achieve the goal of the SCNES.

(3) To realize the SCNES, the following R&D are needed.

- Actinide recycling technology
- FP transmutation technology (recovery, isotope separation, fabrication, nuclear data etc.)
- System safety technology (passive shutdown, recriticality issue etc.)

REFERENCES

Arie K., Suzuki M, Kawashima M., Fujita R., Endo H., Toukura A., Saito M., and Fujii-e Y.(1996), Long-Lived FP Burning Based on the Actinide Recycle Metal Fuel Core, *Int'l Symp. Global Environment and Nuclear Energy Systems 2*, Tsuruga, Japan, 29 October - 1 November.

Fujii-e Y., Morita T., Kawakami H., Arie K., Suzuki M., Iida M., and Yamazaki H. (1991), A Self-Consistent Nuclear Energy Supply System - The Role of Excess Neutrons and the Potential of a Fast Reactor -, *Int'l. Specialists' Mtg. Potential of Small Nuclear Reactors for Future Clean and Safe Energy Sources*, Tokyo Institute of Technology, 23-25 October.

Fujii-e Y., Arie K., and Endo H. (1992a), An Approach to a Self-Consistent Nuclear Energy System, *Trans. Am. Nucl. Soc.,* Vol.66, p.342, Chicago., 15-20 November.

Fujii-e Y. (1992b), An Approach to Self-Consistent Nuclear Energy System - Potential of Fast Reactors -, *Design and Safety of Advanced Nuclear Power Plants*, Tokyo, 25-29 October.

Fujii-e Y., Suzuki M., Arie K., and Endo H. (1993), An Evaluation Function for a Self-Consistent Nuclear Energy System, *Trans. Am. Nucl. Soc.*, Vol.69, p.353, San Francisco, 14-18 November.

Ihara H., ed. (1989*), Tables and Figures from JNDC Nuclear Data Library of Fission Products, Version 2*, JAERI Report JAERI-M 89-204, Japan Atomic Energy Research Institute.

Shimizu A. and Fujii-e Y. (1994c), Self Consistent Nuclear Energy Systems, *Int'l Symp. Global Environment and Nuclear Energy Systems*, Susono, Japan, 24-27 October.

Takagi R., Matsuura H., Fujii-e Y., Fujita R., Kawashima M. (1994d), A Fuel Cycle Concept for Self-Consistent Nuclear Energy System, *Int'l Symp. Global Environment and Nuclear Energy Systems*, Susono, Japan, 24-27 October.

Pergamon
www.elsevier.com/locate/pnucene

Progress in Nuclear Energy, Vol. 40, No. 3–4, pp. 615–620, 2002

Printed in Great Britain
0149-1970/02/$ - see front matter

PII: S0149-1970(02)00057-4

A METALLIC FUEL CYCLE FOR SELF-CONSISTENT NUCLEAR ENERGY SYSTEM (SCNES)

REIKO FUJITA*, MITSUAKI YAMAOKA, MASATOSHI KAWASHIMA,

MASAKI SAITO**, HARUAKI MATSUURA, HIROSHI AKATSUKA

*Nuclear Engineering Laboratory, Toshiba Corporation
4-1, Ukishima-cho, Kawasaki-ku, Kawasaki, 210-0862, Japan
**Research Laboratory for Nuclear Reactors, Tokyo Institute of Technology2-12-1, O-okayama, Meguro-ku, Tokyo, 152-8550, Japan
Tel:+81-44-288-8153, Fax:+81-44-270-1807, E-mail:reiko2.fujita@toshiba.co.jp

ABSTRACT

A fast reactor core and fuel cycle concept has been discussed for Self-Consistent Nuclear Energy System (SCNES) concept. This paper discussed loading material candidates for long-lived fission products (LLFPs) and removal of stable nuclides from radioactive nuclides with isotope separation using tunable laser. Some of LLFPs were possible to be loaded in metal of the generated form. The potential for LLFP-confinement in the reactor system is discussed along with a metallic fuel cycle concept. The proposed fuel cycle scheme is a successful candidate for SCNES concept.

1. INTRODUCTION

Metallic fuel cycle concept for Self-Consistent Nuclear Energy System(SCNES) (Fujii-e,1992 and Shimizu and Fujii-e,1994, respectively) has been proposed in previous papers (Takagi et al., 1995 and Kawashima et al., 1998, respectively). Metallic fuel is used in the slug form as U-TRU-10%Zr alloy with a few percent minor actinides (MAs) along a homogeneous recycling scheme, in which all TRU elements are treated as homogeneously mixed fuel material without MA-separation for special target assemblies.

Se-79, Zr-93, Tc-99, Pd-107, Sn-126, I-129, and Cs-135 were selected as the target long-lived-fission-products (LLFPs) (Table 1). This study discusses on the treatment of those major LLFPs along with the metallic fuel cycle electro-refining and fuel injection casting technology application.

2. RECOVERING LLFP ELEMENTS FROM ELECTRO-REFINING

The proposed fuel reprocessing processes of metallic fuel Fast Reactor is adopted in electro-refining, retort, filtrate, electro-migration, injection casting. Metallic Tc is precipitated at the bottom of Cd pool in an electro-refining cell. Metallic Pd, Sn and Zr are dissolved in Cd pool. Iodine is recovered as NaI in a pre-treatment process before electro-refining. Se is recovered in the off-gas system during pre-treatment process. $SeCl_4$ is also recovered in the off-gas system in an electro-refining cell. Cs is remained as chlorides in the electro-migration process for salt (Fujita et al., 1997 and Matsuura et al., 1997, respectively).

3. FEASIBILITY ASSESSMENT OF LOADING MATERIALS OF TARGET LLFPS FOR BURNING

3. 1 Loading material candidates of target LLFPs

The melting points and compatibility with cladding materials were considered for selecting loading material chemical form candidates of target LLFPs recovered from the elecro-refining process.

3.1.1 Criteria of loading material candidates

The criteria of loading material candidates for SCNES is shown in Table 2. We chose four criteria ;

Table 1 Key parameters for burning' Major FPs[1)]

Nuclides	Bi (y/fission)	Half Life(y) Decay	Half Life(y) Burning*1
I-129	2.5×10^5	1.6×10^7	6.2×10^1
Cs-135	2.3×10^5	3.6×10^6	3.3×10^2
Pd-107	2.1×10^5	6.5×10^6	4.2×10^1
Zr-93	5.7×10^4	1.5×10^6	2.6×10^2
Tc-99	1.3×10^4	2.1×10^5	4.9×10^1
Sn-126	2.9×10^2	1.0×10^5	3.1×10^3
Se-79	3.7×10^1	6.5×10^4	9.0×10^3
Cs-137	2.0×10^0	3.0×10^1	1.8×10^3
Sr-90	5.8×10^{-1}	2.9×10^1	1.7×10^3

1) K.Arie,et al., Proc. of Int'l Symposium on Global Environment and Nuclear Energy System, pp455-461, Oct.24-27, 1994, Susono-city.
*1:Total neutron flux 1×10^{15} n/cm²/s.

Table 2 Criteria for selecting loading material candidates

Criteria	Loading material Candidates		
	Metal	Alloy	Compound
Minimize space-volume loading	◎	○	△
Decrease capture of parasitic neutrons	◎	○(△*1)	○
Melting point and Density	○	○	△
Compatibility with cladding	○	○	△

*1: Amount of parasitic neutron captures depends on component for alloy.

(1)minimizing space-volumes for loading, (2) decrease capture of parasitic neutrons, (3) melting point and atomic density, and (4) compatibility with cladding. Metallic form is the best choice for loading material candidates because the space-volume for the loading is minimum, and because parasitic neutron capture cross section is very small.

3.1.2 Recovering of LLFP chemical forms

The recovering of LLFP chemical forms in metallic fuel cycle is shown in Table 3. Metallic forms can be chosen for Tc, Pd and Zr. Tc is recovered by filtrating Cd-pool after major electro-refining process. Pd is recovered as metallic form by retorting Cd. Zr recovery is also as well. Those metallic targets can be irradiated for long time, because the main transmutation process for Tc-99, Pd-107 and Zr-93-incineration is neutron capture reaction which yields no gas production.

3.1.3 Loading material candidates

(a) Iodine loading candidate

Iodine is loaded as NaI, which is better than AgI and CeI_3 because of higher melting point and good compatibility with claddings. I-129 burning produces Xe-gas and after gas release sodium remains as bond material. Xe-gas releases gives strong restriction for the irradiation periods and loading quantity, avoiding pin-failure. Vibration packing technology is applicable for the target pin production with NaI particles.

(b) Se , Sn and Zr loading candidates

Se and Sn have low melting points, which are 232 ℃ and 217.4 ℃, respectively. Metallic forms can not be chosen for target material of Se and Sn. The loading material candidates for Se and Sn are examined. For Sn126 and Se79 incineration, taking into account that Zr93 is also target LLFP, Sn-Zr and Se-Zr alloys are proposed, because higher melting points and higher densities can be attained compatible to Sn and Se. Table 4 shows the result of the examination of the loading material candidates for Se and Sn.

Table 3 Recovering LLFP chemical forms in a metallic fuel cycle

Element	State	M.P(℃)	Density(g/cm^3)	Evaluation
I	NaI	651	3.67	○
Se	Se	217.4	4.8	× (low M.
	$SeCl_4$	308*1	—	× (low M
Pd	Pd	1555	12.3	○
Sn	Sn	232.0	5.80*2,7.28*3	× (low M
Tc	Tc	2250	11.5	○
Cs	CsCl	645	3.99	○
Sr	$SrCl_2$	875	3.55	△ (low de
Zr	Zr	1900	6.4	○

*1 :Closed tube, *2: α, *3: β

Table 4 Loading material candidates for Se and Sn

Element	State	M.P(℃)	Density(g/cm^3)	Evaluation
Se	Se	217.4	4.8	× (low M.P.)
	Se_4Zr_5	800	5.53	○
	$SeCl_4$	308*1	—	× (low M.P.)
	Se-Sn	SnSe 861	6.18	○
		$SnSe_2$ 651	5.0	○
Sn	Sn	232.0	5.80*2,7.28*3	× (low M.P.)
	SnO_2	1127	7.0	○
	Sn-Zr	$ZrSn_2$ 1142	6.96	○
		Zr_5Sn_3 1592	6.70	○
		Zr_4Sn 1327	6.55	○

*1 :Closed tube, *2: α, *3: β

The composition changes occur due to the subsequent beta-decays associated with Se- and Sn-isotope neutron capture. Isotopes of Se are transmuted to Br isotopes. Subsequently after neutron captures with Br-isotopes, the isotopes of Kr are yielded. The loading material candidates for Br and Kr based on Se (β decay) →Br(β decay)→Kr transmutation are shown in Table 5. There are no loading material candidates for Br and Kr in Table 5 from a viewpoint of melting temperatures. Therefore, sodium bonded targets can be appropriate to retain Br into sodium in the target pins. Reprocessing for the irradiated Se-Zr alloy can remove stable and short-lived elements after adequate intervals. As the gaseous Kr production is small, the proposed alloy can be irradiated for the appropriate terms.

Isotopes of Sn are transmuted to isotopes of Sb after neutron captures and subsequent beta-decays. Isotopes of Te are yielded by neutron captures and beta-decays of Sb isotopes. Table 6 shows melting points

Table 5 Loading material candidates for Br and Kr based on Se $\xrightarrow[-\beta]{}$ Br $\xrightarrow[-\beta]{}$ Kr reaction by transmutation

Element	State	M.P(℃)	Density(g/cm^3)	Evaluation
Br	Br_2	-7.3	—	×
	Br-Sn	$SnBr_2$ 215.5	5.12	×
		$SnBr_4$ 31.0	—	×
	$SbBr_3$	96.0	4.15	×
	$ZrBr_4$	450(15atm)	—	×
Kr	Kr	-57.4	—	×

Table 6 Loading material candidates for Sb and Te based on Sn $\xrightarrow[-\beta]{}$ Sb $\xrightarrow[-\beta]{}$ Te reaction by transmutation

Element	State	M.P(℃)	Density(g/cm^3)	Evaluation
Sb	Sb	630.5	6.68	○
	Sb_2Se_3	611	—	○
	$SbZr_2$	875	—	○
Te	Te	449.6	6.24*1, 6.0*2	× (low M.P.)
	Te-Se	221-449.6	—	× (low M.P.)
	Te-Zr	Te_5Zr 505	>6	× (low M.P.)
		Te_3Zr 630	>6	○
		Te_4Zr_5>1000	>6	○

of Sb-Se , Te-Se , and Te-Zr compounds. Sb-Zr and Te-Zr compounds can be expected to have adequate melting points in the Sn-Zr compound matrix.

Loading material candidates for Se and Sn, Se_4Zr_5 is best for Se. Sn-Zr alloy is best for Sn. Conversion process for Se from $SeCl_4$ and alloy preparation process are required. The injection casting is directly applicable to Se_4Zr_5 target preparation because there is no difference in density of Se and Zr. Sn-Zr alloy target can be also fabricated by the injection casting. For those metallic form targets, injection casting technology can be applicable for making slugs.

As for Pd-107 burning without isotope separation, Pd-108 is accompanied with Pd-107 and generated from the capture and is transmuted by the reaction ; Pd-108 (n,gamma)→Pd-109(β decay)→Ag-109. Transmutation of Pd yields isotopes of Ag. The melting point of Ag is rather low (962℃) compared with that of Pd (1552℃), but is still almost the same as sodium boiling points.

There is the lack of the data on melting points, density and compatibility with claddings for selected loading material candidates. The data for melting points, density and compatibility with claddings of loading material candidates needs to be measured by experiments including irradiation tests.

3.2 Tunable laser application

The feasibility of the stable- and short-lived-isotope a mixture group separations form the chemically separated elements from the proposed process is discussed using tunable laser application. Followings are discussed;

3.2.1 Cs-133, 134, 137 mixture separation form Cs

Residual Cs(mixture of Cs-135 and other Cs isotopes) is the target for Cs-135 incinerartion. The spectrum of Cs isotopes is shown in Fig.1(Garhardt et al., 1978). The conceptual schematic system for Cs isotope separation is shown in Fig.2.

3.2.2 Zr-90, 91, 92, 94, 96 mixture separation from Zr

Residual Zr(mixture of Zr-93 and other Zr isotopes) is the target for Zr-93 incineration. The spectrum of Zr isotopes is shown in Fig.5(Stern et al., 1986).

Similar application is also expected to be applicable from the isotope shift effects shown in Fig.3(Stern et al., 1975) for other five major LLFPs, targeting volume reduction for LLFPs to be burnt.

Preliminary results indicate that approximately 80% of Cs-133 and Cs-137 are separated from total Cs mixture using isotope shift effect. For Zr mixture, a polarized laser is effective for separation of odd-number and even-number mass isotope. Approximately, 80% Zr-91 and Zr-93 are expected to be condensed for effective burning of LLFP Zr-93.

To achieve Cs isotope separation, it is required to develop a laser with much sharper band width than that a vapor flow control system to avoid the Doppler broadening. A polarized laser is needed for Zr isotope separation with the appropriate excitation schemes.

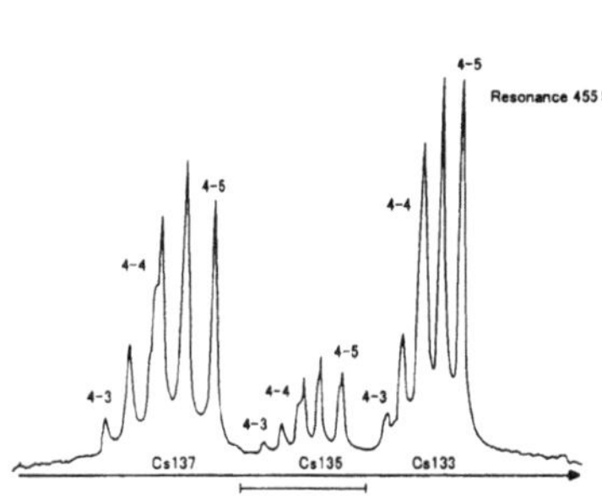

Narrow isotope shift
(Several GHz for isotope shift of ^{235}U and ^{238}U)

⇩

Requirements for separation process
•Laser with sharp band width
•Vaper flow without doppler effect

Fig. 1 Spectrum of Cs Isotope

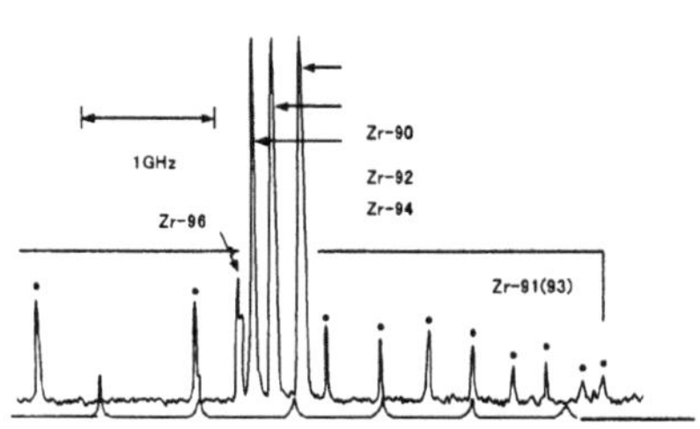

Isotope shift < Zr-91 HFS* width
*Hyper Fine Structure

⇩

Separation technology based on isotope shift is not appplicable.

⇩

Polarized laser is effective for separative of odd and even mass isotope.

Requirements for separation process
•Selection of excitation scheme
•selection of polarized laser set

Fig.3 Spectrum of Zr Isotopes

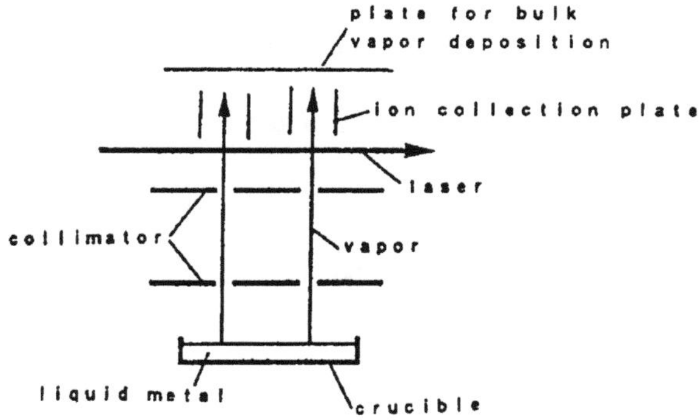

Fig. 2 **The Conceptual Schematic System**

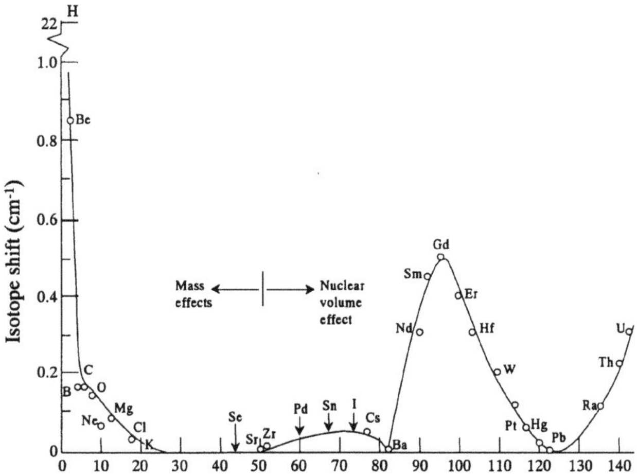

Isotope Shift vs Neutron Number

Fig.4 The possibility for I, Pd, Sn, Se of the Isotope separation

3. 2CONCLUSION

A feasibility study for SCNES concept was continued along with metallic fuel cycle technology application. For LLFP incineration, new additional processes were discussed. For burning Tc-99, Pd-107 and Zr-93, metal fuel cycle technology can be directly applicable at solid metal separation from FPs and injection casting for target preparation. For Se-79 and Sn-126 burning, Se-Zr and Sn-Zr alloys are proposed for making metal forms with appropriate melting points. The investigation suggests that it is possible for some LLFPs to be separated on the stable nuclides from radioactive ones with tunable laser. Cs and Zr can be treated in LLFP recycle scheme as well as the other major five LLFPs recycle in the proposed fuel cycle concept. This scheme results in enhancing massive FPs(Cs and Zr) volume reduction. The data for melting points, density and compatibility with claddings of loading material candidates needs to be measured by experiments. Combining the technology for partial removal of stable and short-lived FPs with isotope separation by tunable laser in the recycling schemes, the proposed fuel cycle is a candidate for realizing SCNES concept.

REFERENCES

K.Arie ,M.Suzuki, M.Kawashima, R.Fujita, and H.Endo (1998), *Progress in Nuclear Energy,* Vol.32, No.3/4,665.

H.Endo, M.Kawashima, M.Suzuki, M.Saito, A.Shimizu, and Y.Fujii-e (1995), *Progress in Nuclear Energy,* Vol.29(Supplement), 463.

H.Endo, M.Kawashima, M.Suzuki, H.Ninokata, T.Sawada, A.Shimizu and Y.Fujii-e (1998), *Progress in Nuclear Energy,* Vol.32, No.3/4, 689.

Y.Fujii-e (1992), " An Approach to Self-Consistent Nuclear Energy System –Potential of Fast Reactors", *Proc. of Int'l Conf. on Design and Safety of Advanced Nuclear Power Plants (ANP`92), Tokyo, Oct.,25-29*

R.Fujita, M.Kawashima, H.Ueda, R.Takagi, H.Matsuura and Y.Fujii-e (1997), *Proc. of Int. Conf. Future Nuclear Systems, Global'97,* Vol.1, 332. Yokohama, Japan, Oct. 5-10,

H.Garhardt, E.Mattnias, F.Schneider and A.Timmermann (1978), " Isotope Shifts and Hyperfine Structure of the 6s-7p Transitions in the Cesium Isotopes 133, 135 and 137", *Z.Physik* Vol. A288, 327

M.Kawashima, R.Fujita, H.Ueda, R.Takagi, H.Matsuura and Y.Fujii-e (1998), *Progress in Nuclear Energy*, Vol.3, No.3/4, 707.

H.Matsuura, R.Takagi ,I.Okada and R.Fujita (1997), *J. Nucl. Sci. Technol.,* Vol.34(3),304.

A.Shimizu and Y.Fujii-e (1995), *Progress in Nuclear Energy,* Vol.29(Supplement), 25.

R.C.Stern and B.B.Snavely (1975), " The LASER Isotope Separation Program at Lawrence Livermore laboratory", *UCRL-76720.*

R.Takagi, H.Matsuura, Y.Fujii-e, R.Fujita and M.Kawashima (1995), *Ibid.,* 471.

Pergamon

www.elsevier.com/locate/pnucene

Progress in Nuclear Energy, Vol. 40, No. 3–4, pp. 621–628, 2002

Printed in Great Britain
0149-1970/02/$ - see front matter

PII: S0149-1970(02)00058-6

THE IMPLICATION OF HIGH-FLUX FUSION NEUTRON SOURCE FOR THE SELF-CONSISTENCY IN MULTI-COMPONENT NUCLEAR ENERGY SYSTEM

M. SAITO*, V. ARTISYUK, A.TAKIBAEV¶, T. SAWADA, S. IIO

Tokyo Institute of Technology,
2-12-1 O-okayama, Meguro-ku,
Tokyo 152-8550, Japan
*E-mail: msaitoh@nr.titech.ac.jp

and

A. SHMELEV
Moscow Engineering and Physics Institute,
31 Kashirskoe shosse,
Moscow, Russia

ABSTRACT

Among fission products (FP) discharged from a fission reactor, long-lived fission products are considered as of primary concern. Their transmutation has been of high priority to reduce the long-term consequences of nuclear energy generation. A self-consistent nuclear energy system (SCNES) in which we center fast breeder reactor may not have enough degree of excess neutron sources to transmute the fission products that potentially would poses environmental hazards in long-term period if they are buried in geologic disposal. Here we propose a so-called multi-component SCNES in which fission reactor systems can be combined with fusion reactor systems mainly for compensating the loss of enough capability for the transmutation. Amongst long-lived fission products, major concern has been paid for iodine and technetium and little attention was given to radioactive ^{93}Zr, although its hazard appears to be rather substantial. The importance of ^{93}Zr transmutation is emphasized and the transmutation capability was examined with fusion neutron sources by incorporating adequate moderation structures. As a result, we have demonstrated that the fusion neutron sources with high-flux blanket can be applied to transmute ^{93}Zr sufficiently and resolve the problem of its accumulation within the time period of several decades. © 2002 Published by Elsevier Science Ltd.

1. INTRODUCTION

High-level radioactive wastes comprise mainly minor actinides (MAs) and fission products (FPs) among which, especially, long-lived FPs seem to be troublesome in their processing. They cannot be recycled into

¶ Currently with Institute of Nuclear Physics, Ibragimov Street 1, Almaty, Republic of Kazakhstan 480082

reactor core as nuclear fuels for energy production. Contrary to short-lived FPs, they are hard to be essentially reduced by natural decay. The long-term continuous generation of fission energy may accumulate huge amounts of long-lived radioactive nuclides. To solve this problem three concepts of future nuclear energy systems were proposed in the last decade. They are the concepts of equilibrium nuclear society (Sekimoto and Takagi, 1991) self-consistent nuclear energy system (SCNES) (Fujii-e et al., 1992) and radiologically clean nuclear power (Salvatores et al., 1994). The goal attributed to FP transmutation in these concepts is to keep their inventory at the level that is manageable within reactor fuel cycles to avoid the deep geological disposal. As a result, the focus in FP transmutation studies was shifted to providing neutrons that are enough abundant to cover required transmutation capability with the analysis of equilibrium mass balance. All the three concepts rely on the fast spectrum reactors operated either in critical or subcritical, driven by accelerators, conditions. They hinge upon the fact that fission in a hard spectrum gives the favorable degree of excess neutrons.

As for the SCNES, the four prerequisites to a self-consistent nuclear energy system are addressed, namely, (1) energy production, (2) fuel generation, (3) environment protection, and (4) safety assurance. There are two key issues to be resolved for a nuclear system to meet the four prerequisites. One is *energy balance* and the other is *neutron balance* in the system. The neutron balance means the abundance of neutrons available or *excess neutrons*. The prerequisite "environmental protection" includes the transmutation of MAs as well as FPs. It has been known that usually the degree of the excess neutrons generated by a fission reactor is not enough to transmute the MAs and FPs, irrespective of the reactor type, i.e., fast reactor or thermal reactor. Furthermore, we have a number of problems to perform the isotope separation of 29 nuclei that are identified as those to be transmuted. If we pursue a practical way to complete the fuel cycle only by elements separation without isotope separation, it is revealed that the number of excess neutrons obtained by any type of fission reactor is lacking for it. For example, if we recycle the FPs having half-lives of longer than one year without isotope separation, the absorbed neutrons to be used for their incineration are 6.7 per fission. Even if we aim at incinerating FPs with half-lives of longer than thirty years, we need 1.12 excess neutrons per fission. This means we cannot obtain necessary excess neutrons only by a fission system. Thus, in order to keep sufficient neutron balance in these facilities, transmutation should be complemented with the process of *FP isotope separation* (Akatsuka et al., 1995) that is potentially dangerous and costly.

Although ^{99}Tc and ^{129}I are considered to be comparatively transmutable among the long-lived FPs with half-life more than 100 year, which includes 7 nuclides: ^{79}Se, ^{93}Zr, ^{99}Tc, ^{107}Pd, ^{126}Sn, ^{129}I, ^{135}Cs, the recent survey concerning partitioning and transmutation (P/T) has drawn rather pessimistic conclusion against effective transmutation for the two mono-isotopes (Baetsle et al., 1999). The transmutation of ^{99}Tc or ^{129}I in the present reactors is said to be not very practical because of the long transmutation half-lives and the huge inventories required. To create favorable excess neutrons requires fast spectrum that, in general, gives an average cross-section inferior to that in thermal spectrum. In addition to that, neutron flux in fast reactors at the level of 10^{15} n/cm^{2}s has a strong limitation imposed by heat removal. As for the thermal reactors, they are not considered as possible candidates at all because of lack of excess neutrons. Namely, we have insufficient transmutation environment merely by the use of fission reactors.

This is the background motivation for us to propose a novel system constitution to improve the neutron balance in the viewpoint of the excess neutron, or neutron economy. This is a symbiotic system which can be comprised of fusion systems and accelerator-driven subcritical fission reactors outside the conventional critical fission reactors. Namely the fusion systems and subcritical reactors are used to supply external neutron source. We name such a symbiotic system as *multi-component self-consistent nuclear energy system* (multi-component SCNES) (Saito et al., 2000).

In this paper, we examine the incineration capability of the excess neutron sources by thermonuclear fusion systems as an element of the multi-component SCNES for exemplifying its usefulness. A promising choice is the use of fusion neutron source (FNS). We have emphasized that the FNS with ITER-like plasma characteristics could supply considerable excess neutrons in the nuclear energy system to avoid FP isotope separation (Artisyuk et al., 1998). Thus a multi-component SCNES can be comprised of fusion system and conventional fission systems, which may be of course symbiotically coupled with accelerator-driven subcritical systems (ADSs). In this paper we show how to create transmutation environment within the ITER fusion neutronics to make effective the transmutation of problematic FPs. Then, as an example case, we discuss the elimination capability for Zr under the transmutation environment with FNS.

2. BLANKET FOR FISSION PRODUCTS TRANSMUTATION BY FUSION NEUTRON SOURCE

Parish and Davidson (1979) proposed a concept to apply the neutron rich environment in the D-T fusion blankets to the transmutation of FPs. However, since then there are few studies on approaching the equilibrium with technologically approved plasma characteristics. The FNS considered in this paper is based on the plasma characteristics of ITER EDA (ITER NSSR-2, 1998).

In a tokamak-type fusion reactor with DT plasma, we have a rich neutron source of 14 MeV which crosses the first wall (FW) into blankets. This neutron current is often called the first wall load (FWL). It determines the neutron flux in the blanket as well as the radiation damage in structural materials. The presence of 14 MeV neutrons in the neutron population leads sometimes to confusing image of their role in FP transmutation. A meaningful example of such a conclusion is advertising the 14 MeV neutrons for FP transmutation through (n,2n). The backbone of this approach is the fact that for some isotopes (among the FP of interest they are ^{137}Cs, ^{90}Sr and ^{126}Sn) the cross-section of (n,2n) reaction significantly exceeds that of capture at the thermal point. However the contribution of 14 MeV neutrons into the total flux even in the first wall is not so large and drops to less than 10% if one takes into account backward scattering of moderated neutrons. This fact is especially stressed in Fig. 1 embracing the result of neutron transport analysis in the structure resembling the typical first wall arrangement. The first wall was modelled with Fe layer of 2 cm thickness, surrounding material is a lead representing typical neutron multiplier. All calculations were done in spherical geometry with the help of MCNP4 A code (Breimester, 1993) with nuclear data ACER files from the FSXLIB-J3R2 library (Kosako, 1994). As shown by Fig. 1, the total neutron production is stabilized at approximately 20 cm of lead thickness being around 1.6 neutrons per one fusion reaction. At this lead thickness number of 14 MeV neutrons penetrating through the lead drops to less than 0.1 per fusion (one order of magnitude compared to that of entering the first wall). Their fraction in the total flux in the first wall drops from 0.56 (no lead) to about 0.06 (with 20 cm of lead surrounding first wall).

From the consideration above it appears that the main attractive feature of the FNS is the flexibility in creating the appropriate neutron spectra. It is because original place of neutron production (plasma) and blanket (with transmutation zone) are spatially separated and, therefore, with appropriate selection of blanket material it is possible to create a very soft spectrum that is subjected to the target of transmutation.

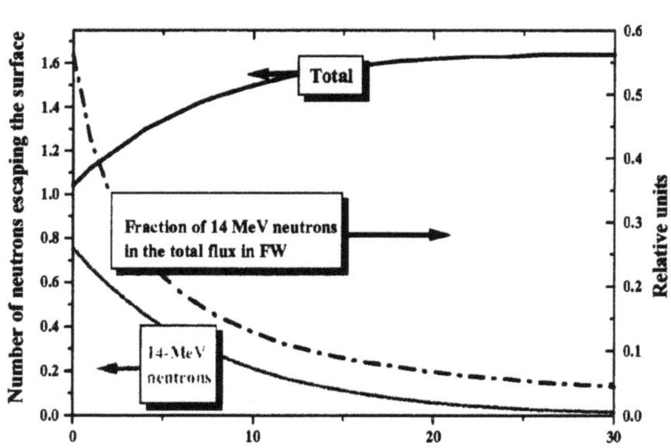

Fig. 1 Neutronics of the first wall and lead multiplier. Number of neutrons escaping the blanket surface per one source neutron and Relative neutron flux ($\varphi_{14\text{-MeV}}/\varphi_{total}$) averaged over the First Wall.

The concept of the blanket with characteristic dimensions is shown in Fig. 2. The total blanket thickness is not far from that of the ITER EDA design. The main structural material should be one with low neutron capture cross section and fine moderation properties (carbon or beryllium).

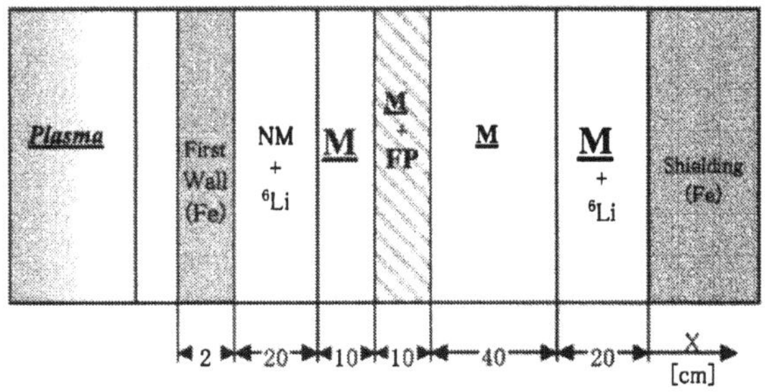

Fig. 2 The concept of high-flux blanket with characteristic dimensions (M- moderator, NM – neutron multiplier).

In general the following regions are to be distinguished:

- First Wall and shielding made of stainless steel;
- Neutron multiplier adjacent to the First Wall (lead or beryllium);
- Transmutation zone surrounded by layers of neutron moderator;
- Some fraction of ^{6}Li to breed tritium is introduced in the layer adjacent to the First Wall (to reduce parasitic capture of the softened neutrons in the First Wall) and in the layer before shielding.

Depending on the FP properties, the dimensions of the zones might be changed as well as the type of a moderator. It is quite illustrative to give an example of ^{93}Zr transmutation in such a blanket (Takibaev et al., 2000). In this case carbon moderator was considered and 0.1%Li atomic fraction in the lead and moderator layer before the shielding. Zirconium fraction of 1 % was chosen in order to avoid its degrading effect on the flux and spectrum in transmutation zone. All the dimensions are exactly the same as presented in Fig. 2. This geometrical and material arrangement results in 1.1 atoms of tritium produced per one fusion reaction that provides the self-sustaining tritium resource. Neutron moderation is quite effective as demonstrated in Fig. 3. In going to the transmutation zone the fraction of high energy neutrons (>10 MeV) drops by almost two orders of magnitude compared to the plasma region. Meanwhile the fraction of low energy neutrons (<1.13 eV) is significantly increased. As a result, the neutron spectrum in the transmutation zone manifests a large peak in the thermal region as shown in Fig. 4 in comparison to the spectra typical for fast and thermal reactors. It clearly states that the most contribution in the total neutron flux is provided by thermal neutrons with Maxwellian-like distribution.

It is obvious that neutrons in fission reactors are born and absorbed in the same core fuels so that the contribution of 2 MeV fission neutrons into the total flux is rather pronounced while in the FNS the birth place and the place of the neutron use are spatially separated thus giving a chance to obtain a very soft spectrum.

3. FISSION PRODUCTS TRANSMUTED BY FNS

From the radiological risk point of view, it has been recognized that ^{99}Tc and ^{129}I have potentially the dominant contribution to jeopardize the barriers of the deep geologic disposal to reach the biosphere (Pigford, 1990). However, recently it has been considered to be more practical to estimate the radiological toxicity in the unit of annual limit on intake (ALI). It is the threshold amount of given radionuclide taken into the body of the reference man by either inhalation or ingestion in a year that would result in a committed effective dose equivalent of 5 rem or a committed dose equivalent of 50 rem to any individual organ or tissue (ICRP, 1979). The diversity in specifying the toxicity by inhalation or ingestion as well as the different normalization, i.e., the whole body or specific organ, makes it flexible and being far from uniquely determined. To avoid ambiguity it is quite instructive to assess the maximum potential effect on human being associated with a given nuclide. The minimum intake through inhalation or ingestion for the whole body or specific organ might serve as a logically proven criterion to make such an effort. This approach reveals a remarkably high toxicity of ^{93}Zr that dominates over that of ^{99}Tc and ^{129}I up to ten million years after discharge as clearly exhibited by Fig. 5. It seems that ^{93}Zr could not be put in a shade of ^{129}I and ^{99}Tc and is worthwhile to be transmuted in the same level of priority.

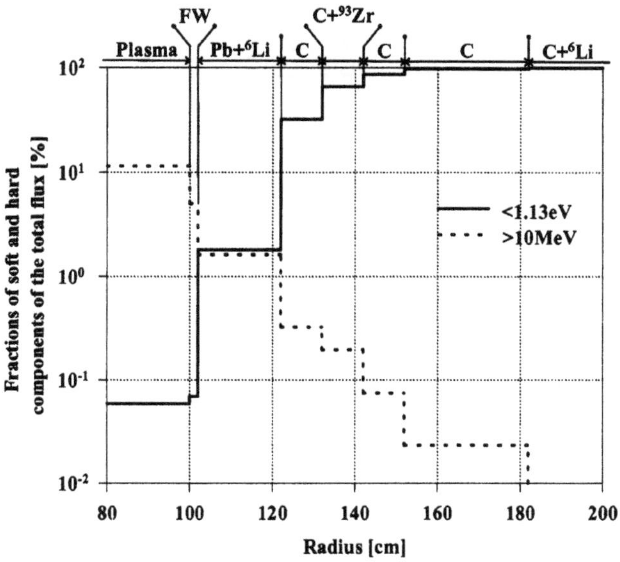

Fig. 3 Fractions of soft and hard components of neutron flux in different zones in the blanket.

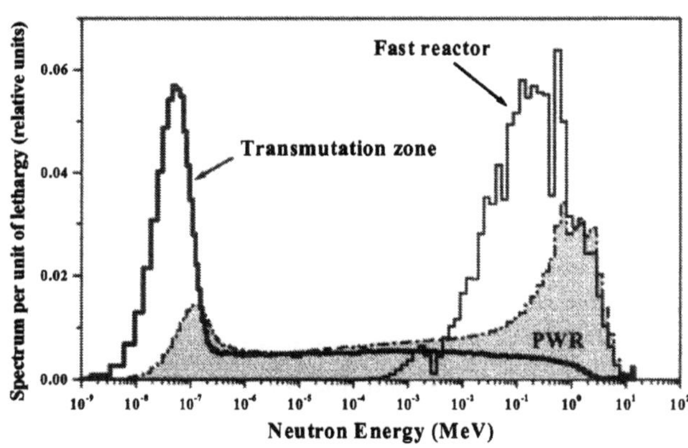

Fig. 4 Neutron spectra in the transmutation zone of FNS compared to typical spectra of the fission reactors.

4. TRANSMUTATION CHARACTERISTICS OF ZIRCONIUM-93 BY FNS

The neutronics characteristics of the long-lived FP: ^{93}Zr in the FNS are listed in Table 1, in comparison to the typical values for fission reactors.

The neutron flux achievable in the FNS corresponds to the reference FWL of the ITER EDA that is 1 MW/m^2 (or 4.6×10^{13} n/cm^2s of the current of 14 MeV neutrons). The neutron flux is of the same order of magnitude as in a thermal reactor, however, the spectrum in the transmutation zone is markedly softer as can be judged from the cross-sections averaged over the neutron spectra. The mean life-time of ^{93}Zr in transmutation zone, that singularly governs their approach to equilibrium, is, correspondingly, one and two orders of magnitude shorter than in fission reactors.

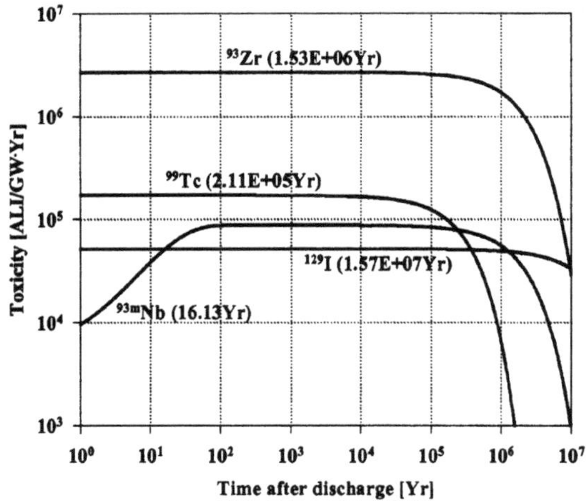

Fig. 5 Time behavior of toxicity of some FP in spent fuel.

Table 1 Basic neutronics characteristics of ^{93}Zr in different types of reactor environment

		Thermal reactor	Fast reactor	Fusion Neutron Source (FNS)	
				Carbon moderator	Beryllium moderator
Total flux (φ, $\times 10^{14}$ n/cm^{2}s)		1	10	4.8	3.8
^{93}Zr	Average cross-section (σ, b)	0.28	0.03	1.47	-
	Mean life-time (τ, year)	1.14×10^3	1.05×10^3	56.3	-

In the blanket configuration assumed for ^{93}Zr transmutation both flux and spectrum are influenced by zirconium content in transmutation zone. This effect is vividly demonstrated in Fig. 6. Despite of the increase in transmutation rate, variation of ^{93}Zr from 1 to 10 atomic percents by a factor of five makes the mean lifetime longer because of hardening the neutron spectrum. Since the FWL is limited to 1 MW/m^2, there is no way to preserve the transmutation characteristics but keeping the zirconium concentration at 1% that was reported in the previous section as a reference.

Another important fact is that the transmutation of ^{93}Zr results in accumulation of ^{94}Zr which is difficult to transmute. For the FWL of 1 MW/m^2 it converges to 80% of asymptotic level in no less than 5000 yr. To reduce its mean lifetime to one comparable with ^{93}Zr would require the FWL as high as 50 MW/m^2 that is unacceptable from the view-point of radiation damage. However, being neutronically inert, ^{94}Zr makes a little effect on ^{93}Zr lifetime as illustrated by Fig. 7.

This gives strong evidence of the possibility to compensate continuous increase in zirconium inventory (buildup of ^{94}Zr and ^{93}Zr feeding) by reducing the volume fraction of carbon. With careful selection of zirconium compounds in feeding stream (like zirconium hydride) it seems to be feasible to keep neutron spectrum and the dimensions of transmutations zone unchanged, thus to preserve the transmutation characteristics.

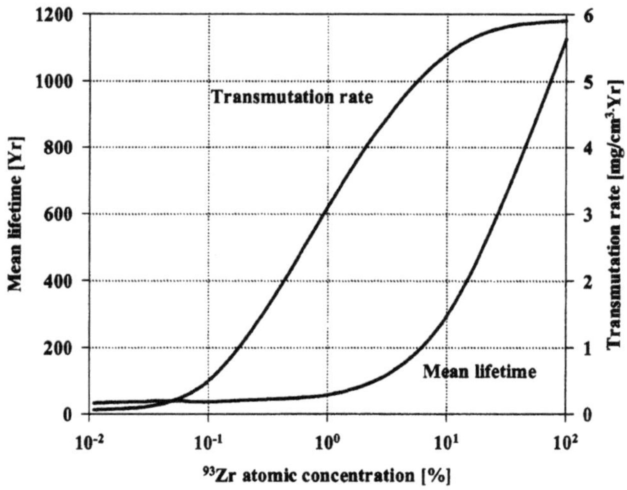

Fig. 6 Mean lifetime and transmutation rate of ^{93}Zr as a functions of its concentration in the transmutation zone.

Even total substitution of carbon by ^{94}Zr results in increment of mean lifetime of ^{93}Zr a little more than 10 yr. With ITER-like neutronics parameters this situation will occur at two thousands years of irradiation time. In other words, increase of ^{93}Zr accumulation could be cut and one transmuter could keep the ^{93}Zr inventory constant within the transmuter's life-time. It also gives the evidence that through appropriate optimization of the transmutation zone it would be possible to apply zirconium transmutation in its elemental form.

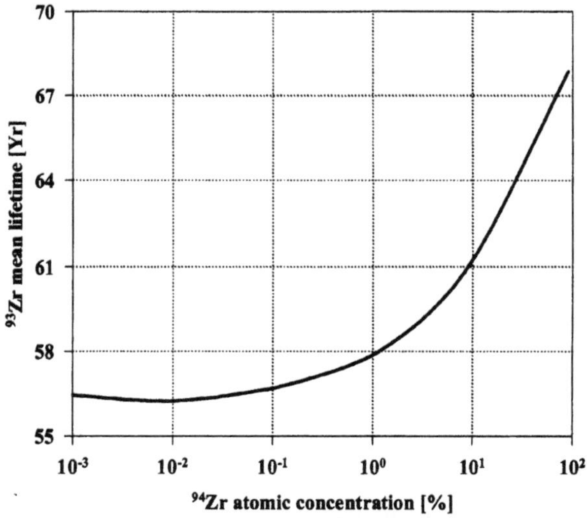

Fig. 7 Approach to equilibrium in FNS and accumulation of ^{94}Zr.

The transmutation efficiency for ^{93}Zr is shown in Table 2. The loading in transmuter was estimated by taking into account the specified geometry of transmutation and geometry of ITER torus with major and minor radii 8.1 and 3 m, respectively (ITER NSSR-2, 1998). The loading, along with neutronics characteristics, determines the mass transmuted annually. This in turn determines the amount of the FP discharged from the fission reactors that could be annually loaded in the transmuter. The reactor park supported by one fusion-based transmuter is obtained as large as 11 GWe for elemental ^{93}Zr. To avoid uncertainties in the thermodynamic efficiency of the FNS, fission utilization factor is prudent to estimate in terms of thermal power. The thermodynamics efficiency of 0.33 was assumed for fission reactors. Thermal power in a transmuter with ITER-like parameters is about 2 GW, so the fission utilization factor in this system will be as high as 95% in the case of ^{93}Zr transmutation. In other words, the associated burden of energy production is only within 5% of the total power produced in the system.

Table 2 Transmutation efficiency

	^{93}Zr
Mass loaded in the transmuter (t)	18.5
Time to approach equilibrium (90% asymptotic level) (year)	56
Transmutation rate at equilibrium (t/year)	022
PWR park supported (GWe)	11
Equilibrium mass (kg/GWe)	1000
Fission utilization factor (%)	95

5. CONCLUSIONS

We have proposed a so-called multi-component SCNES in which fission reactor systems can be combined with fusion reactor systems mainly for compensating the lack of enough capability for the transmutation of a regular SCNES in which fast reactor core is centered. Amongst long-lived fission products, little attention was given to radioactive ^{93}Zr so far, although its hazard appears to be rather substantial. The importance of ^{93}Zr transmutation is emphasized and the transmutation capability was examined with fusion neutron sources (FNS) by incorporating adequate moderation structures. We found that the soften FSN is very effective to transmute rather problematic long-live fission products like ^{93}Zr and resolve the problem of its accumulation within the time period of several decades. As a result, we have demonstrated that the fusion driven neutron source with high-flux blanket can be applied to transmute ^{93}Zr discharged from PWRs of which electricity generation scale is 11 GWe. Thus the multi-component SCNES with fusion reactor system can effectively resolve the problem with transmuting troublesome long-lived fission products, which is one of the prerequisite to the self-consistent nuclear energy system.

REFERENCES

Akatsuka, H., *et al.* (1995), *Prog. Nucl. Energy,* **29**, 477.

Artisyuk, V., *et al.* (1998), Kerntechnik, 63, 255.

Baetsle, L., *et al.* (1999), Status and Assessment Report on Actinide and Fission Product Partitioning and Transmutation, An OECD Nuclear Energy Agency Review, *Proc. Int. Conf on Future Nuclear Systems, GLOBAL '99*, Jackson Hole, Wyoming, August 29- September 3.

Breimester,J.B. (1993), LA-12625-M.

Fujii-e,Y., *et al.* (1992), Trans. Amer. Nucl Soc., **66**, 342.

ICPR Publication 30 (1979), "*Limits for Intakes of Radionuclides by Workers*", Pergamon Press, Oxford

ITER Non-Site Specific Safety Report 2 (1998)

Kosako, K.K., *et al.* (1994), JAERI-Data/Code 94-020.

Parish, T.A., Davidson, J.W. (1979), *Nucl. Technol.*, **47**, 324.

Pigford, T. (1990), UCB-NE-4176.

Saito, M., et al. (2000), *Abstracts of this seminar.*

Salvatores, M., *et al.* (1994), *Nucl. Sci. Eng.*, **116**, 1.

Sekimoto, H., Takagi, N. (1991), *J. Nucl Sci.Techn.*, **28**, 941.

Takibaev, A., *et al.* (2000), *J. Nucl. Sc. Techn.*, 37, 870.

Pergamon
www.elsevier.com/locate/pnucene

Progress in Nuclear Energy, Vol. 40, No. 3–4, pp. 629–636, 2002

Printed in Great Britain
0149-1970/02/$ - see front matter

PII: S0149-1970(02)00059-8

FUSION-DRIVEN TRANSMUTATION OF ELEMENTAL PALLADIUM-FISSION PRODUCT

A.SINTSOV, V.APSE, A.CHMELEV, M.SAITO* AND V.ARTISYUK*

Moscow Engineering Physics Institute, Moscow (Russia)
chmelev@nr.mephi.ru
*Tokyo Institute of Technology, Tokyo (Japan)
msaitoh@nr.titech.ac.jp

ABSTRACT

The paper deals with transmutation of elemental palladium separated from the spent fuel of fission reactors into the stable isotopes of silver and cadmium. The most efficient transmutation environment is achieved with resonance neutron spectrum in a blanket of fusion-driven transmuter. Three types of plasma conditioning are considered: DT-plasma with equal fractions of deuterium and tritium, DT-plasma with reduced tritium fraction and pure DD-plasma.

1. INTRODUCTION

Transmutation of long-lived fission products (FP) is a crucial problem in getting credit for the future successful nuclear power development. Fast reactor provides an appreciable neutron excess and, in the majority of transmutation studies, is considered as a leading candidate for a FP transmuter. Along with the neutron excess, transmutation environment is governed by neutron flux and spectrum. The transmutation environment in a typical fast reactor is sufficiently robust to attain for some long-lived FP on condition of their isotope separation (^{99}Tc, ^{129}I, ^{107}Pd) effective life-time in a neutron field as short as about 50 yr (Salvatores M., et al.,1998, Kobayashi K., et al.,1998). On the other hand, transmutation of long-lived FP without isotope separation in the most cases would require neutron excess that is unachievable within the fission technology. Fusion neutron source with its rich excess neutron-producing potential was put forward as a promising candidate for transmutation of FP in their elemental forms, thus to avoid the sophisticated, dangerous and expensive process of isotope separation (Saito M., et al.,1998). Several papers have been recently published on fusion-driven FP transmutation (Takibaev A., et al., 2000, Saito M., et al., 2000). They clearly exhibit that each target FP requires individual approach in view of unique nuclear properties of target, residual nuclides and chemical forms of bearing matrix. The present paper in particular addresses the approach to transmutation of elemental palladium. Fusion-driven palladium transmutation is considered for three plasma options: DT-plasma with equal fractions of deuterium and tritium, DT-plasma with reduced tritium fraction (henceforth denoted as DDT-plasma) and pure DD-plasma. This is done in order to estimate the effect of necessary tritium breeding on neutron balance and transmutation characteristics in a fusion-driven transmuter. The paper starts with discussion on properties of palladium discharged from fission reactors, turns to neutronics of the proposed fusion-driven transmuter, moves to analysis of three plasma options and demonstrate transmutation efficiency in terms of palladium transmutation rate and equilibrium palladium loading.

2. CHARACTERIZATION OF PALLADIUM-FISSION PRODUCT

Listed in Table 1 are isotope vectors of natural palladium and palladium accumulated in spent UOX fuel (33 GWd/tHM burnup, 150 days cooling), yields normalized per one fission of ^{235}U and ^{239}Pu at thermal point and per one average fission in the PWR core. These data are presented just to feature the palladium produced in the reactor fuel that, in contrast to the natural one, does not include stable ^{102}Pd. Another feature is that fission yields of isotopes with mass numbers less than 104 are negligibly small, and they, together with isotopes heavier than ^{110}Pd, having a very short lifetime, decay very quickly and are not present in the spent fuel after 150 days of cooling. As a result, palladium discharged from typical PWR consists of 5 stable isotopes and one long-lived radionuclide ^{107}Pd ($T_{1/2}$ =6.5 · 10^6 yr) with total elemental annual accumulation rate about 30 kg/GWe.

Table1 Characteristics of palladium accumulation

Isotope	Half-Life	Isotope vector, % Natural	Isotope vector, % Spent Fuel*	Yield normalized (at/fission) Cumulative yield** ^{235}U	^{239}Pu	Spent fuel*	Number of neutrons required for transmutation (n/fission)
102	Stable	1.02		1.63E-10	9.58E-13	-	-
103	16.9 d			2.82E-17	7.11E-15	-	-
104	Stable	11.14	17.53	7.36E-10	1.83E-08	0.014	0.07
105	Stable	22.33	30.2	0.009638	0.053514	0.024	0.097
106	Stable	27.33	18.7	0.004023	0.042856	0.015	0.045
107	$6.5 \cdot 10^6$ y		17.54	0.001403	0.03361	0.014	0.027
108	Stable	26.46	12.08	0.000669	0.021685	0.01	0.01
109	13.7 h	-	-	0.000345	0.018795	-	-
110	Stable	11.72	3.95	0.000304	0.005997	0.003	0.003
111	5.5 h	-	-	0.000199	0.002979	-	-
112	21 h	-	-	0.00016	0.001329	-	-
						Total:	0.25

* Spent Fuel: UOX in PWR, burnup - 33 GWd/tHM, 150 days cooling, obtained with ORIGEN code (Croff, A.G., 1980)
**Data at 0.0253 eV neutron energy taken from ENDF/B-VI (*ENDF/B-VI, 1991*)

Since contribution of 14-MeV neutrons into total flux in the blanket region is only several per cent (Takibaev A., et al., 2000, Saito M., et al., 2000), capture reactions are reasonably expected to be the dominant transmutation process. By data presented in Table 1 one can judge that complete chain of necessary palladium transmutation starts from stable ^{104}Pd and terminated by short lived ^{111}Pd ($T_{1/2}$ =5.5 h):

$$^{104}\mathrm{Pd}(n,\gamma)\,^{105}\mathrm{Pd}(n,\gamma)\,^{106}\mathrm{Pd}(n,\gamma)\,^{107}\mathrm{Pd}(n,\gamma)\,^{108}\mathrm{Pd}(n,\gamma)\,^{109}\mathrm{Pd}(n,\gamma)\,^{110}\mathrm{Pd}(n,\gamma)\,^{111}\mathrm{Pd}(n,\gamma).$$

This chain is disturbed by natural decay of short-lived ^{109}Pd ($T_{1/2}$ =13.7 h) so that the ratio of its natural decay rate to capture rate - $\lambda/(\sigma\varphi)$ becomes a chopper of the chain and effects the neutron consumption.

Neither ENDF/B nor JENDL files provide the capture cross-section for ^{109}Pd. To estimate the break-even flux at which the decay rate is equal to the rate of capture, the maximum cross-section among palladium isotopes (^{105}Pd, thermal point, 20 b) was assumed as a reference. This reveals the break-even flux of thermal neutrons as high as $\approx 10^{18}$ n/(cm^2·s) that is hardly achievable in any feasible technology with nuclear processes. So, neutron link between ^{109}Pd and ^{110}Pd is rather weak and, in the course of transmutation, ^{109}Pd is expected to decay immediately after formation. Consequently, ^{110}Pd will be incinerated alone, independently on strong neutronic link of ^{104}Pd-^{109}Pd.

In view of above discussion it is possible to estimate the low bound for neutron requirements to transmute palladium. Each ^{110}Pd atom requires at least one neutron to be transmuted through the following reactions:

$$^{110}\mathrm{Pd}(n,\gamma)\,^{111}\mathrm{Pd}(\beta,\ 23.4\ \mathrm{min})\,^{111}\mathrm{Ag}((\beta,\ 7.45\ \mathrm{d})^{111}\mathrm{Cd};$$

or, being normalized to one fission, neutron requirements are exactly governed by ^{110}Pd yield: $\eta^{tr}_{110}=1\times Y_{110}$ (henceforth, Y_i is the yield of *i*-isotope of palladium). That is rather different from the ^{104}Pd-^{109}Pd chain, in which the only neutron requirements of ^{108}Pd are singularly determined by its neutron yield, while the other isotopes requires neutrons according to how far they are remote in the chain from ^{109}Pd. In general, these could be estimated by the following formula:

$$\eta^{tr}_i = Y_i \times (6 - j),$$

where 6 is the maximum number of successive neutron capture reactions in the ^{104}Pd-^{109}Pd chain; j is the isotope number in the chain, starting from j=1 for ^{104}Pd. Estimated neutron requirements are included in the right column of Table 1. Totally they are as high as 0.25 n/f that is close to the low boundary estimation for neutron excess available in a fast reactor, being in the range 0.3-0.58 n/f (Salvatores, M., *et al.*,1994, Akatsuka, H., *et al.*,1995). However, the neutron requirements thus estimated represent just low limit. In addition to that, one should reasonably expect that in practice, introduction of additional material in a reactor would eventually degrade the transmutation environment (flux and spectra). For this reason, the present paper advocates fusion-driven transmutation, that is rather superior to fission reactors from the viewpoint of neutron excess and keeps the best transmutation environment.

3. THE CONCEPT OF TRANSMUTATION ZONE

It is a matter of fact that fusion power in tokamaks with DT plasma is mostly governed by 14 MeV fusion neutrons crossing the First Wall on their way from the plasma region into the blanket. This neutron current is often called the First Wall Load (FWL). It determines the neutron flux in the blanket as well as the radiation damage in structural materials – the main limitation on the FWL. In the plasma with reduced tritium inventory, the greater contribution of 2.45-MeV neutrons from DD fusion reaction is expected.

The concept of the blanket with characteristic dimensions is sketched in Fig.1. The main structural material should be one with low neutron capture cross section and good moderation properties (like beryllium). In general the following regions are to be distinguished:

- First Wall and shielding made of stainless steel;
- Neutron multiplier adjacent to the First Wall (lead);
- Layers of neutron moderator (beryllium);
- Some fraction of ^6Li to breed tritium introduced in the layer adjacent to the First Wall (to reduce parasitic capture of the softened neutrons in the First Wall) and in the layer before shielding.

It should be noted that the case of DD referred to in this paper is treated just a limiting option in rather flexible plasma content with respect to tritium fraction, thus, optimum design and material arrangement for transmuter oriented only for DD plasma might be different from that shown in Fig.1.

ΔR(cm): 250	50	2	15	2	4	2	5	20	20
Plasma	Vacuum	Stainless steel	99.2%Pb+0.8%Li	Transmutation zone 1 35%Pd+10%Be+55%Pb	Be	Transmutation zone 2 40%Pd+8%Be+52%Pb	Be	Transmutation zone 2 55%Pb+33%Li+12%Be	50%Be+50%Li

Fig.1 Blanket configuration

Dimensions and compositions of transmutation zones were defined separately for each type of fusion neutron source. Volume fraction of palladium was varied within the range 20-50%, thickness of beryllium layer – within the range 2-5 cm. In Fig. 2, radial distributions of neutron flux are presented for six energy groups in the blanket driven with DDT-neutron source. It can be seen that the region with increased flux of resonance neutrons is formed in transmutation Pd-Be-Pb zone.

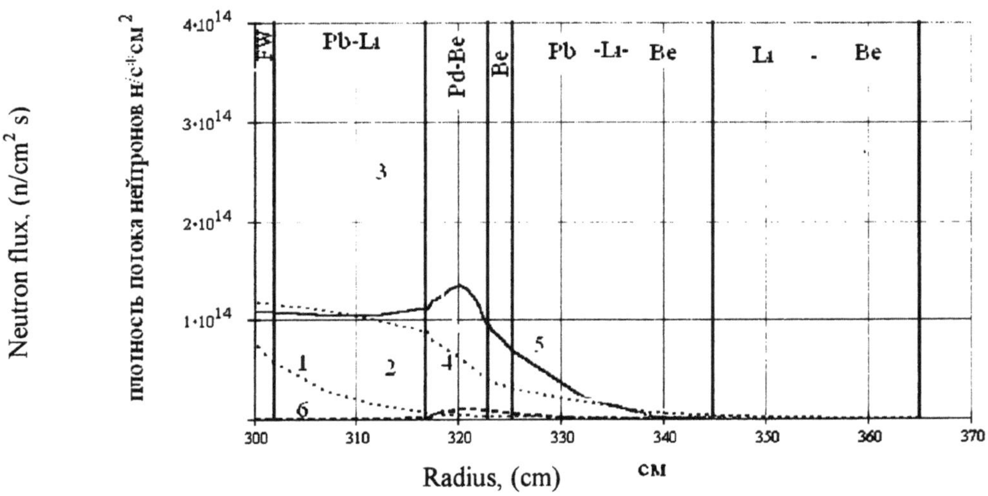

Fig. 2. Radial neutron flux distributions for six energy groups in DDT based transmuter
(1): 10-14 MeV; (2):1-10 MeV; (3): 0.1-1 MeV; (4): 10-100 keV;
(5): 5 eV-10 keV; (6): below 5 eV.

For considered fusion neutron sources the concentration ratio of tritium to deuterium n_t/n_d in plasma and reaction rate ratio R_{dt}/R_{dd} as functions of plasma temperature were evaluated using formulae from Ref.10. At the same time, reaction rate ratio R_{dt}/R_{dd} can be obtained from tritium balance equations:

$$R_{dt} - 0.5 * R_{dd} = \frac{1}{p_t} \times (k_t^{14} \times R_{dt} + 0.5 \times k_t^{2,45} \times R_{dd}); \qquad (1)$$

where k_t^{14}, $k_t^{2,45}$ - factors of tritium generation in blanket per one fusion neutron with energy 14 MeV and 2.45 MeV, respectively; p_t – necessary tritium breeding ratio (p_t = 1.05). Joint application of formulae from (Artsimovich L.A., 1961) and Eq.(1) allows one to calculate parameters of plasma and tritium breeding which are presented in Table 2.

Neutronic analyses were carried out with application of the computer code SCALE 4.3 (SCALE, 1997) and evaluated nuclear data file ENDF/B-6. Balance of neutron reactions, averaged one-group cross-sections and neutron flux were calculated for both transmutation zones. These parameters were further used for solving the burn-up equations of elemental palladium and thus for determination of equilibrium transmutation characteristics (equilibrium Pd-vectors and transmutation rates). It was assumed that palladium was transmuted under continuous feeding of fresh palladium and removal of transmutation products.

As a basic value for neutron loading on the first wall, the value adopted in the ITER-FDR design was used: 1 MW/m^2 or $4.5 \cdot 10^{13}$ $n/(cm^2 \cdot s)$ for current of 14-MeV fusion neutrons. For three plasma options, the factors were evaluated which characterize the damage dose rate: K_{dt}^{dpa} (E_n=14 MeV) and K_{dd}^{dpa} (E_n=2.45 MeV) for material of the First Wall (stainless steel SS316LN). These factors were determined as a rate of DPA accumulation per unit of fusion neutron current through the First Wall. Neutron cross-sections used for calculation of damage doses were taken from (Avramenko V.I. et al, 1984). Results are presented in Table 2. For DDT-plasma option, the factor of damage dose was evaluated taking into account real spectrum of neutrons emitted by plasma: linear combination of contributions from DT and DD-neutrons.

Table 2 Physical parameters of fusion neutron source

Type of plasma	DT	DDT	DD
k_t^{14} (E_n=14 MeV)	1,04	0,72	-
$k_t^{2,45}$ (E_n=2.45 MeV)	0	0,43	-
n_t / n_d	1	0,017	0
R_{dt} / R_{dd}	106	2,24	0
K^{dpa}, $10^{-21}\left[(DPA/s)/(n/(cm^2 \cdot s)\right]$	8,38	7,80	4,73
J^+, 10^{13} $n/(cm^2 \cdot s)$	4,50	4,86	7,97

It can be seen that DPA factors for DDT and DD-plasma options are less than those for DT-plasma. That is because of softened spectra of neutrons generated by the first two sources. It means that current of fusion neutrons through the First Wall may be increased in the case of DDT and DD-plasma as compared with that obtained with DT-plasma. Real values of fusion neutron current (and intensity of fusion neutron source, as a consequence) were determined from condition of equal DPA rate in considered options and in ITER project. It was assumed that plasma parameters of DT-option and the ITER-FDR design are close to each other. So, the data obtained for DT-option were used as reference data of ITER project.

4. CHARACTERISTICS OF PALLADIUM TRANSMUTATION

Some results of neutronic analyses with three fusion neutron source options are presented in Table 3.

Table 3 Balance of neutron reactions (per one fusion reaction)

Type of plasma	Yield of fusion neutrons plus (n,xn)-breeding	Absorption in lithium	Absorption in Pd	Absorption in Pd, per fusion neutron	Absorption in Pd per damage dose in FW
DT	1+0,69	1,04	0,50	0,5	1
DDT	0,764+0,362	0,48	0,54	0,7	1,5
DD*	0,5+0	0,18	0,28	0,56	1,95

*** DD-plasma with complete removal of generated tritium and helium.**

As follows from the Table 3, the DDT-plasma is the most preferable option from the viewpoint of neutron utilization for palladium transmutation (0.7 per one fusion neutron). As for the DT-plasma, it necessitates full tritium breeding thus reducing the number of neutrons available for palladium incineration to 0.5. Apparent drawback of pure DD-plasma is a poor neutron economy, since no neutron multiplication is possible for DD fusion neutrons, whose energy is below the threshold of (n,xn)-reactions for most of the nuclides. Parasitic neutron absorption in the First Wall is about 10% and leakage from the system is well within 2% for all three plasma options.

Table 4 lists one-group capture cross-sections of palladium isotopes, transmutation product ^{109}Ag, and neutron flux obtained in neutronics analysis of two transmutation zones in the case of DDT-plasma. Other options have similar neutronics properties. It can be seen that cross-sections of ^{105}Pd and ^{107}Pd are practically the same in both zones. It is interesting to note, that fraction of neutron consumption in ^{105}Pd changes in the course of transmutation, being significantly larger in a freshly charged palladium composition.

Table 4 Cross-sections of palladium isotopes and neutron flux

σ (n,γ), barn	FDT with DDT-plasma	
	Zone 1	Zone 2
^{104}Pd	0,573	0,87
^{105}Pd	1,69	2,86
^{106}Pd	0,36	0,5
^{107}Pd	1,72	2.94
^{108}Pd	0,58	1,09
^{110}Pd	0,35	0,54
^{109}Ag	8,49	20,91
φ, 10^{14} n/(cm²·s)	3,93	2,16

Absorption cross-section of ^{109}Ag is larger than that of ^{107}Pd. It would increase parasitic neutron loss during irradiation. On-line management with continuous Pd feeding and removal of transmutation products might reduce this effect. Important characteristic of transmutation process is the mean life-time of transmuted nuclide in neutron field. For each palladium isotopes and ^{109}Ag they are presented in Table 5.

Table 5 Effective life-time of palladium isotopes, 1/(σ φ) in years

Isotope	DT-plasma		DDT-plasma		DD-plasma	
	Zone 1	Zone 2	Zone 1	Zone 2	Zone 1	Zone 2
^{104}Pd	115	150	141	168	122,7	161,5
^{105}Pd	38,3	45,2	47,8	51,5	39,9	46,6
^{106}Pd	191	271	227	296	200,9	300
^{107}Pd	38,7	46,7	47	50	40,4	47,9
^{108}Pd	69,1	70	138	70,7	89,6	81,7
^{110}Pd	201	272	233	270	216,2	283,4
^{109}Ag	6,9	6,2	9,5	7	5,7	6

As one can see, for all options of fusion neutron sources the effective lifetime of ^{107}Pd is shorter than 50 years.

Figure 3 reflects the time behavior of palladium (^{106}Pd, ^{107}Pd and mixture of all the isotopes) concentrations while approaching equilibrium in the blanket Zone 1 with DDT-neutron source. Presented curves demonstrate that approach to equilibrium for ^{107}Pd is slowed down because of small rate of ^{106}Pd transmutation. It can be seen from Table 5 that predecessors of ^{107}Pd have different life-times in different transmutation zones.

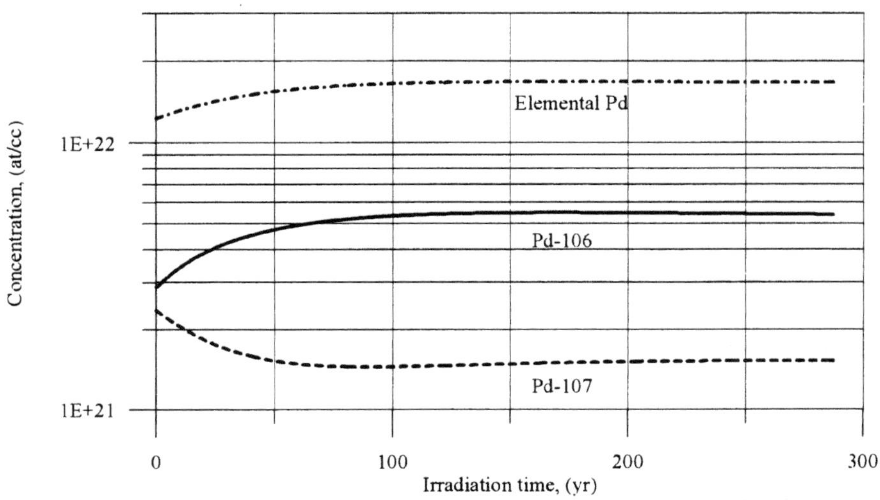

Fig. 3. Approach to equilibrium in the course of palladium transmutation (zone 1)

Table 6 shows the equilibrium Pd-vectors in two transmutation zones of the blanket with DDT-neutron source and one for spent fuel of power reactors. Equilibrium Pd-vectors are different from the feeding Pd-vector, and they are shifted towards stable palladium isotopes ^{106}Pd and ^{108}Pd. These isotopes are weak neutron absorbers, and this can lead to decrease in palladium transmutation rate.

Table 6 Equilibrium palladium vectors in FDT with DDT-plasma

Isotope	DDT-plasma		Composition of
	Zone 1	Zone 2	feeding Pd
^{104}Pd	0,0579	0,0622	0,1548
^{105}Pd	0,0541	0,0517	0,2744
^{106}Pd	0,407	0,4654	0,2335
^{107}Pd	0,1085	0,1008	0,1907
^{108}Pd	0,3563	0,3031	0,1209
^{110}Pd	0,0161	0,0166	0,0257

It can be concluded from data presented in Table 6 that, in equilibrium transmutation regime, fraction of ^{105}Pd in neutron absorption is less than that in feeding palladium. It allows one to increase total palladium loading into the blanket. So, equilibrium palladium loading was larger than initial one by a factor of 1,5 (see Fig. 3).

The data presented in Table 7 characterize the transmutation rate in a fusion-driven transmuter with 1000 m^2 First Wall surface area (close to ITER project). Park of LWRs supported by fusion transmuter is substantially determined by plasma parameters (intensity of neutron source) as well as by blanket structure. When transferring from equal-component DT-plasma to plasma with reduced tritium content, transmutation of elemental palladium is substantially improved.

Table 7 Parameters of equilibrium Pd-transmutation

	Per 1000 m^2 of the FW		
Type of plasma	DT	DDT	DD
Equilibrium Pd loading (t)	87,7	196,7	201,6
Park of serviced LWRs (GWe)	11,2	16	20,5
Transmutation rate of elemental Pd (kg/y)	337,3	480,4	615,5
Transmutation rate of ^{107}Pd (kg/y)	64,3	91,7	117,3
Production rate of ^{109}Ag (kg/y)	328,6	467,7	599,7

5. CONCLUSIONS

Several configurations of fusion-driven blanket intended for transmutation of elemental palladium fission product have been analysed for three cases of plasma composition: equal-component DT-plasma, DT-plasma with reduced tritium fraction, pure DD-plasma. These cases are characterized by different spectra of fusion neutrons.

Effective life-time of long-lived radionuclide ^{107}Pd is achieved as short as about 50 years. The neutrons escaping the transmutation zones towards blanket periphery and towards the first wall are utilized for tritium breeding (DT and DDT plasma options). The transmutation rate can be increased through reduction of tritium fraction in plasma composition. However, it will be accompanied by appropriate increase in the ignition temperature.

It was demonstrated that one fusion-driven transmuter with dimensions typical for ITER project, is able to support the park of uranium-fuelled LWRs with total power 11 GWe, 16 GWe and 20.5 GWe for cases of DT, DDT and DD-neutron source, respectively. In the case of DDT-plasma the power produced by FDT will not exceed 4% of total power of serviced LWRs.

REFERENCES

Akatsuka, H., et al. (1995), Scientific Feasibility of Incineration in SCNES, *Progr. Nucl. Energy, 29, 477.*

Artsimovich, L.,(1961), Controlled Fusion Reactions. Moscow.

Avramenko, V., et al. (1984), Neutron Cross-Sections for Determination of Radiation Damage Dose in Reactor Materials. – *Atomnaya Energiya, 56/ 3, 139.*

Croff, A. (1980), ORNL-5621.

ENDF/B-VI (1991), *Summary Documentation*. Report BNL-NCS-17541.

Kobayashi, K., et al. (1998), Applicability Evaluation to a MOX Fueled Fast Breeder Reactor for a Self-Consistent Nuclear Energy System, *Progr. Nucl. Energy, 32, 681.*

Saito, M., et al. (1998), Contribution of External Nuclear Sources in Excess Neutron Generation for SCNES, *Progr. Nucl. Energy, 32, 697.*

Saito, M., et al. (2000), Fusion-Driven Transmutation of Fission Product Cesium in its Elemental Form, *J. Nucl.Sci.Techn., 37, 1024.*

Salvatores, M., et al. (1994), A Global Physics Approach to Transmutation of Radioactive Nuclei, *Nucl. Sci. Eng., 116*, 1.

Salvatores, M., et al. (1998), The Transmutation of Long-Lived Fission Products by Neutron Irradiation. , *Nucl. Sci. Eng., 130, 309.*

SCALE (1997), A Modular Code System for Performing Standardized Computer Analyses for Licensing Evaluation, *ORNL/NUREG/CSD-2/R5.*

Takibaev, A., et al. (2000), Transmutation of Zirconium-93 in High-Flux Blanket of Fusion Neutron Source, *J. of Nucl. Sci. and Techn.* 37, *870.*

Pergamon

www.elsevier.com/locate/pnucene

Progress in Nuclear Energy, Vol. 40, No. 3–4, pp. 637–645, 2002

Printed in Great Britain

0149-1970/02/$ - see front matter

PII: S0149-1970(02)00061-6

RADIOLOGICAL HAZARD OF LONG-LIVED SPALLATION PRODUCTS IN ACCELERATOR-DRIVEN SYSTEM

V.ARTISYUK, M.SAITO, A.STANKOVSKII, Yu.KOROVIN*, A.SHMELEV**

Research Laboratory for Nuclear Reactors, Tokyo Institute of Technology
2-12-1 O-okayama, Meguro-ku, Tokyo 152, Japan
Tel.(03) 5734-3074, Fax: (03) 5734-2959
Email: artisyuk@nr.titech.ac.jp
*Obninsk Institute of Nuclear Power Engineering
Studgorodok 1, 249020, Obninsk, Russia
**Moscow Engineering Physics Institute
Kashirskoe Shosse 31, Moscow, Russia

ABSTRACT

The present paper focuses on analysis of radiological hazard of spallation products that appears to be an important factor that might affect the choice of beam/target performance in designing the accelerator-driven systems. The analysis is done in terms of toxicity expressed in units of Annual Limit on Intake (ALI). It reveals the significant contribution of alpha emitting rare earths (^{146}Sm, ^{148}Gd, ^{150}Gd, ^{154}Dy) into overall toxicity of spallation targets. © 2002 Published by Elsevier Science Ltd.

1. INTRODUCTION

Accelerator-Driven System (ADS) often assumes a clean image compared to conventional critical reactors. Emotionally it relates to the fact that ADS is considered as a promising candidate for waste transmutation. Scientifically it is based on the subcritical operation mode that gives a chance of more safe incineration of the transuranics whose fission releases more neutrons than conventional fuel, other terms being equal. Such a safe fissioning, in turn, heavily relies on neutron supply initiated through spallation reaction triggered by high-energy ions in so-called spallation target. Because of non-fission origin, these external neutrons being normalized to one hypothetical fission give a clear advantage of ADS in terms of available neutron excess relatively to the critical systems. That is very much attractive for transmutation of fission products (FP). Systematic analysis of neutron excess generation in ADS for FP transmutation has been already done in several studies (Salvatores, *et al*, 1994; Saito, *et al*, 1997), while radiological burden of spallation products (SP) associated with this excess received less attention so far. Up till now, the most concern was given to alpha emitting ^{210}Po ($T_{1/2}$ =138 d) whose accumulation through neutron capture reactions in lead and lead-bismuth spallation targets is broadly discussed. However, some rare earth (RE) elements presented among

SP (Samarium, Gadolinium and Dysprosium) have the isotopes unstable to alpha decay with half-lives ranging from several decades (^{148}Gd, $T_{1/2}$ =74.6 yr) to million years (^{146}Sm, $T_{1/2}$=1.03×10^8 yr).

Being scarcely presented in the waste stream of fission and fusion technologies alpha emitting RE are rather unique for nuclear engineering community. Because of their small mass number they cannot be transmuted through fission reaction in the same manner as alpha emitting transuranics. So, they are definitely the nuclear ash and their radiological burden is important to compare with activation products (like ^{210}Po). In view of lack of data on RE accumulation, their yields obtained by means of computer simulation remains to be an important source of information. The authors of the present paper did some careful validation of high-energy transport code CASCADE/INPE (Barashenkov, *et al*, 1999, Konobeev, *et al*, 1999) which is the main tool behind the present paper. The validation revealed underestimation of SP accumulation at the neutron number close to 82, where alpha emitting SP are centered (Stankovsky, *et al*, 2001). So, their evaluated toxicity is expected to be the lowest bound. Since the objective of this paper is to emphasize the role of RE in toxicity analysis, this underestimation does not disturb the course of the present discussion.

Radiological burden of various spallation targets was estimated in terms of toxicity (units of Annual Limit on Intake). It reveals significant contribution of RE into overall toxicity of spallation targets that, in the case of Pb and Pb-Bi targets is comparable with polonium toxicity. The ways to reduce their accumulation without compromising the neutron production in ADS are also discussed.

2. SPALLATION PRODUCT'S YIELD

Figure 1 gives generic representation of SP accumulation obtained with the CACADE/INPE-code in four different types of spallation target by giving the independent yields (nuclides are formed directly in a high-energy interaction between proton and target nuclide). Natural uranium, lead, tungsten and tin are selected to encompass the targets under consideration in various ADS projects (Cappiello, 1999; Rubbia, 1995; Takizuka, *et al.*, 1997, Khorosanov, *et al*, 1999). Three classes of SP might be identified from the shown distributions. One is represented by sharp peak of the light products dominated by tritium. The second comprises a wide list of intermediate products resulted from either fission and evaporation (in the case of heavy targets) or only evaporation (in the case of tungsten and tin) and the third is formed by extended range of nuclides sharply peaked at the mass numbers of initial target nuclei.

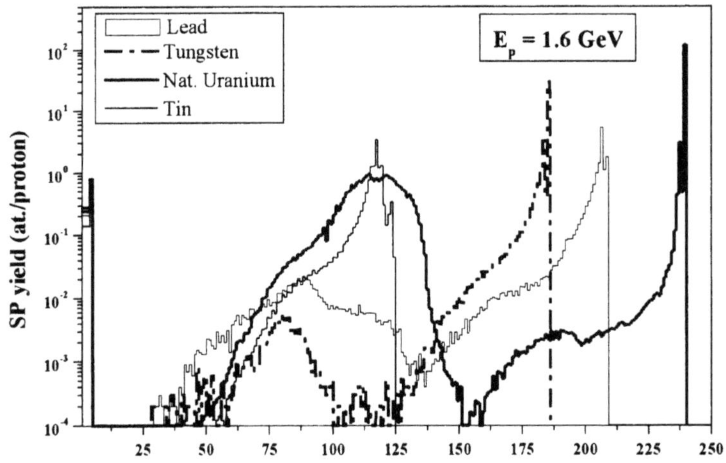

Fig. 1. Independent yield of spallation products in various targets irradiated by 1.6 GeV protons obtained with CASCADE/INPE code (Stankovsky, *et al*, 2001)

Figure 1 may shed a light on the mechanism of RE accumulation. Their yield is obviously tends to increase for target material with mass numbers declining to rare earths. Even in uranium targets, where fission process could contribute in RE accumulation, their yield is obviously inferior to Pb and W targets. That gives strong evidence that the basic mechanism of their generation is evaporation process following the intranuclear cascade reaction. Needless to say that in tin target there is no RE among the SP simply because the mass number of initial target material is less than RE.

Figure 2 shows the cumulative yield in comparison with independent one for some important RE spallation products. By definition, cumulative yield includes for each particular nuclide its own independent yield and independent yields of its short-lived predecessors within isobaric decay chain as follows:

$$Y_i^{cum} = Y_i^{ind} + \sum_{j=1}^{i-1} \prod_{k=j}^{i-1} (P_k \times Y_j^{ind}), \tag{1}$$

where P_k is the probability for j-nuclide to remain in the isobaric decay chain leading to i-isotope, superscripts "*ind*" and "*cum*" stand for independent and cumulative yields (Y, atoms per one incident proton). As one can judge from Fig.2, decay of short-lived isobaric predecessors adds significantly (factor two or three) into accumulation of alpha emitting RE isotopes compared to independent yield. It should be stressed that henceforth in toxicity analysis we use cumulative yield.

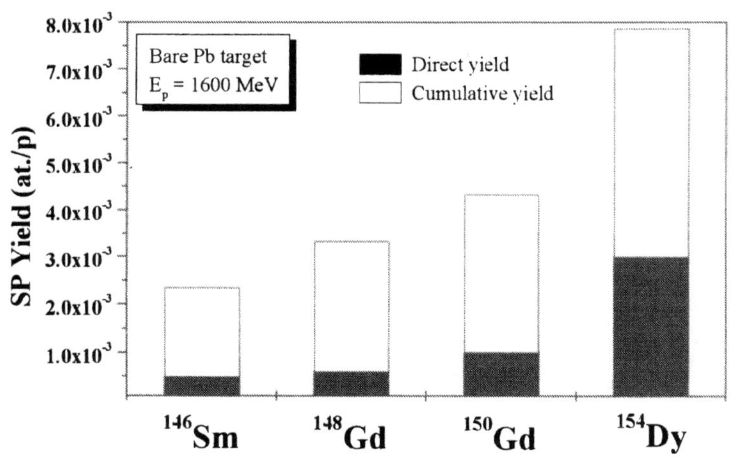

Fig.2. Cumulative and independent yields of alpha emitting RE spallation products

3. COMMENTS ON TOXICITY OF ALPHA EMITTING RARE EARTHS

Potential biological hazard from radionuclides is technically best explained in terms of toxicity. By general definition toxicity is a ratio of amount of radioactivity to specified radioactivity guide (Hebel, *et al*, 1987). Toxicity considered in this paper is based upon the concept of Annual Limit on Intake (ALI), which is the smaller value of intake (Bq) in a year, by the reference man that would result in a relevant committed dose limits. Taken the amount in activity units (Bq) and ALI as a radioactivity guide, the toxicity is simply treated in ALI units. Database for ALI is available from The Code of Federal Regulations (CFR, 1991). Characteristically enough, alpha emitting RE are poorly presented giving no data for ^{150}Gd and ^{154}Dy. Figure 1 explains the reason why RE received little attention so far. Mass numbers of RE are centered around 150 where yields seem to be insignificant compared to nuclides of general concern like, for example, tritium and polonium. Since for ^{150}Gd and ^{154}Dy cumulative yields are the highest among other RE (see Fig.2) it is quite important to make a reasonable assumption of their dose standards and that is the subject of this section.

The concept of ALI focuses on the committed dose that, by definition, represents the dose received during some period of time following the intake (usually 50 yr for adult). So, to estimate the radioactivity guide one should take into consideration decay energy and energy deposition (both determining the dose) as well as half-life and characteristics of retention in the body (both reflecting the dose accumulation after intake). It is worth to note that ALI values are specified for different types of intake (inhalation, ingestion), different ranges of clearance half-times (day, week, year), different chemical compounds and might be differently normalized (whole body, critical organs). To avoid an ambiguity, in the present paper the minimum ALI value was selected for each particular nuclide.

Table 1 gives a general picture of minimum ALI values derived from CFR, 1991, Part 20 for several classes of nuclides important in ADS analysis: Minor Actinides (accelerator-driven transmuters are designed for); Pb, Bi and Po isotopes (activation products of especial concern in spallation target designs); several FP (like Tc and I are the candidates for transmutation in ADS), RE (with known ALI) and tritium being presented merely for comparison. Also included in Table 1 are retention characteristics taken from ICRP, 1994, Publication 68 in terms of gut transfer factor – proportion of the intake transferred to body fluids in the gut. The information presented in Table 1 helps to make an assumption of ALI for ^{150}Gd and ^{154}Dy absent in the referred data sources.

Table 1. Toxicity characteristics of some selected radionuclides

Element	Isotope	Decay mode	Decay Energy (MeV)	Half-life	ALI (Bq)*	Gut Transfer Factor**
Am	240	β	1.38	2.12 d	7.4 x10^7	5.0×10^{-4}
	241	α	5.6	432.2	2.2 x10^2	
Np	237	α	4.96	2.14x10^6 yr	1.5 x10^2	5.0x10^{-4}
	238	β	1.3	2.12 d	2.2 x10^5	
Po	207	β	2.9	5.80 h	1.1 x10^8	0.1
	210	α	5.4	138 d	2.2 x10^4	
Bi	210m	α	5.3	3.0x10^6 yr	2.6 x10^4	0.05
	210	β	1.16	5.01 d	1.1 x10^5	
Pb	210	β	0.064	22.3 yr	1.5 x10^4	0.02
	205	β	0.051	1.43x10^7 yr	3.7 x10^7	
	202	β	0.05	5.25x10^4 yr	1.9 x10^5	
Gd	148	α	3.3	74.6 yr	1.1 x10^3	5.0x10^{-4}
Sm	146	α	2.53	1.0x10^8 yr	2.2 x10^3	5.0x10^{-4}
I	129	β	0.194	5.7x10^6 yr	1.9 x10^5	1***
Tc	99	β	0.3	2.1x10^5 yr	2.6 x10^7	0.8
H	3	β	0.019	12.33 yr	3.0 x10^9	1.0

* Derived from CFR, 1991, Part 20
** Taken from ICRP, 1994, Publication 68
*** Data for fast consumption through inhalation

It is worth to compare energy deposited (let's denote this entity as *ED*) during the 50-year period by alpha radiation from transuranics and rare earths. Assuming the same number intake and accounting for the same gut transfer factor this entity (and consequently ALI) will be governed by the following product:

$$ED = E_\alpha \times C, \tag{2}$$

where $C = [1 - \exp(-\lambda \times T)]$, $T = 50$ yr, E_α - decay energy.

ED for several nuclides is presented in the Table 2.

Table 2. Characteristics of alpha emitting nuclides essential in toxicity estimation

Nuclide	Half-life (yr)	λ (1/yr)	Decay Energy (MeV)	ED
Pu-239	2.41×10^4	2.88×10^{-5}	5.25	7.54×10^{-3}
Pu-240	87.7	7.9×10^{-3}	5.26	1.72
Am-241	4.32×10^6	1.6×10^{-7}	5.6	4.49×10^{-5}
Cm-244	18.1	3.83×10^{-2}	5.9	5.03
Cm-245	8.5×10^3	8.15×10^{-5}	5.6	2.28×10^{-2}
Sm-146	1.03×10^8	6.73×10^{-9}	2.53	8.51×10^{-7}
Gd-148	74.6	9.29×10^{-3}	3.3	1.23
Gd-150	1.79×10^6	3.87×10^{-7}	2.8	5.42×10^{-5}
Dy-154	3×10^6	2.31×10^{-7}	2.9	3.35×10^{-5}

As one can see, energy deposited during 50 yrs following the intake is almost the same for ^{241}Am, ^{150}Gd and ^{154}Dy. This is strong evidence that expected ALI values for ^{150}Gd and ^{154}Dy are in the range between 2.2×10^3 Bq (the same as for ^{146}Sm that is maximum among alpha emitting RE, see Table 1) and 2.2×10^2 Bq (data for ^{241}Am). In the present paper for ^{150}Gd and ^{154}Dy ALI = 2.2×10^3 Bq is assumed to specify the lower bound of toxicity.

4. TOXICITY OF RE SPALLATION AND PO ACTIVATION PRODUCTS

Toxicity analysis starts from differential equation for the time behavior of radioactive species:

$$\frac{dN_i}{dt} = I\,Y_i - (\lambda + \sigma\varphi)_i N_i = 0 \tag{3}$$

where the first term on the right side stands for SP accumulation rate associated with beam current *I* and the last term stands for depletion rate through both natural and artificial transmutation (herein the only neutron induced transmutation is envisioned). This equation might be rewritten in the form of time-dependent toxicity (*T*) as follows:

$$\frac{dT_i}{dt} = I\,T_i^{gen} - (\lambda + \sigma\varphi)_i T_i = 0, \tag{4}$$

where generated toxicity $T_i^{gen} = \left(\frac{\lambda Y}{ALI}\right)_i$ is just SP yield translated to toxicity units.

Fission neutrons from the blanket will effect the neutronics in the target. In the present analysis simulation of effect of these neutrons on SP and ^{210}Po behavior is simulated by neutronics condition corresponding to 50 mA current of 1 GeV protons, blanket performance is simulated by material composition of typical sodium fast reactor.

Figure 3 represents the dynamics of ^{210}Po and important SP accumulation expressed in units of quantity (atoms), activity (Bq) and toxicity (ALI). Polonium reaches its equilibrium state within 1 year – operation time which is common for nuclear facilities between two successive shutdowns for maintenance. Other nuclides presented require several decades. What important to conclude from Fig.3 is that equilibrium toxicity of alpha emitting rare earths is superior to that of ^{210}Po.

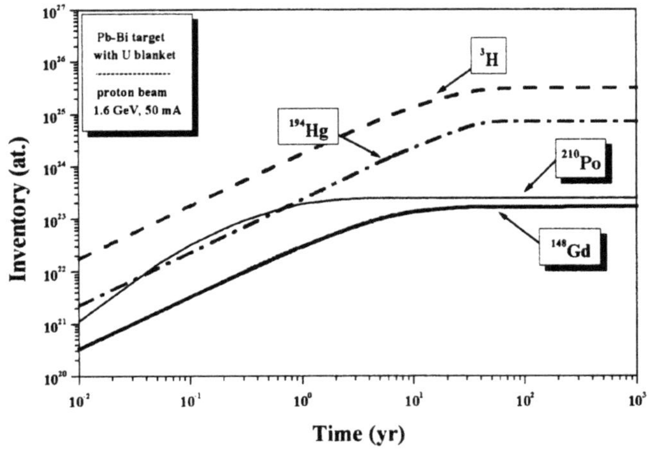

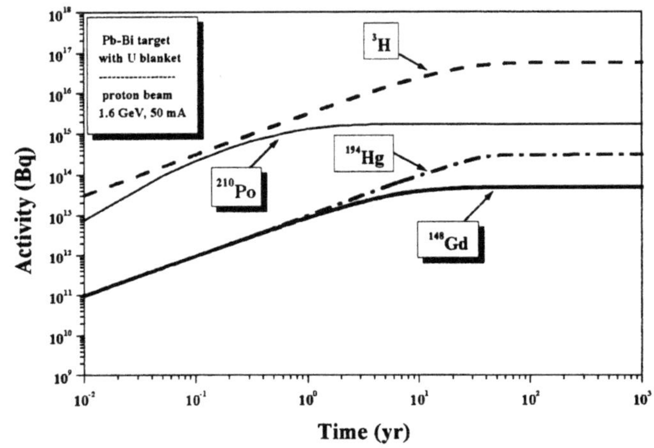

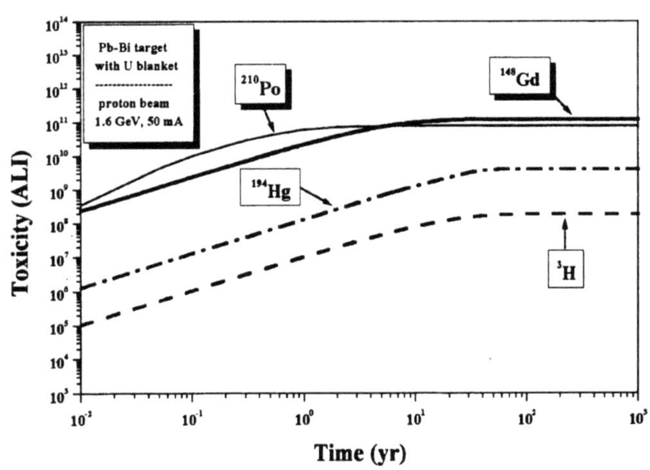

Fig.3. Inventory, activity and toxicity of key spallation and activation products in Pb-Bi target surrounded by fissionable blanket (incident proton energy is 1 GeV)

5. WAYS TO REDUCE ACCUMULATION OF RE IN LEAD TARGET

Changing the target material from lead (or lead-bismuth) to tin is a rather straightforward approach. The technology of liquid tin coolant is less known than that of lead-based alloys. For this reason it is worth to consider the ways of reducing the RE accumulation in lead based target. At first consider the beam energy. Figure 4 shows prominent decrease in accumulation of SP with atom numbers around 150 with decrease in proton energy. However this effect will confront with the overall decrease in neutron production. Another approach to reduce the RE accumulation is to apply the deuteron beam. It is well known fact that at the same initial energy, deuteron generates more neutrons than proton. Increase in neutron production is about 10% as follows from Fig.5a that comprises the energy dependent neutron production in spallation targets bombarded by protons and deuterons.

Pure outcome from energy adjustment might be seen in terms of equilibrium toxicity derived from Eq. (4):

$$T_i^{eq} = I\left(\frac{Y}{ALI \times (1+\sigma\varphi/\lambda)}\right)_i . \qquad (4)$$

Total equilibrium toxicity ($\sum_i T_i^{eq}$) related to one neutron escaped from spallation target is shown for tin and Pb-Bi targets in Fig. 5b. Below 0.8 GeV for lead target it is largely determined by ^{194}Hg ($T_{1/2}$ =520yr) and two long-lived isotopes ^{202}Pb ($T_{1/2}$ =5.25×10^{4}yr) and ^{205}Pb ($T_{1/2}$ =1.53×10^{7}yr) and, though contribution of alpha emitters vanishes, lead remains to be inferior to tin. This is the main result that might be obtained from energy adjustment for both proton and deuteron beams.

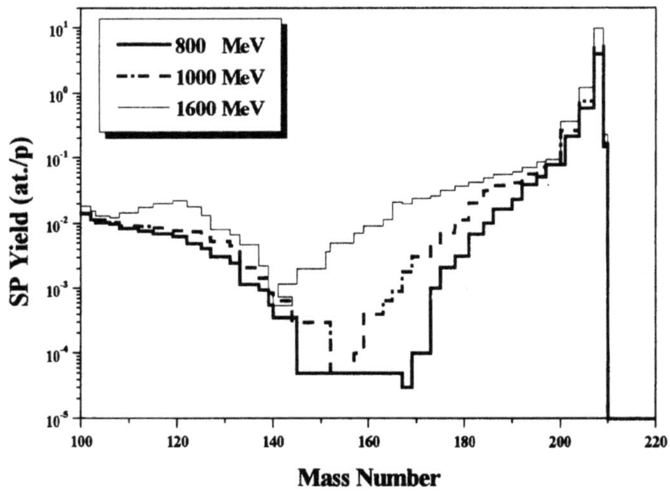

Fig. 4. Yield of spallation products in lead target irradiated by 0.8, 1.0, and 1.6 GeV protons

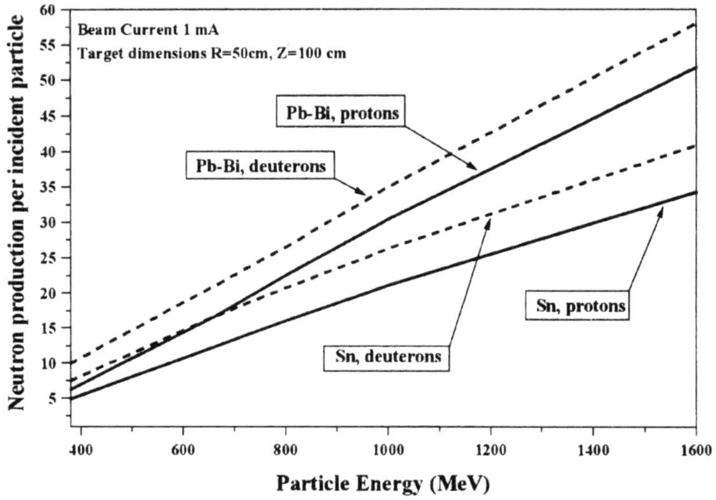

a) Neutron production per one incident particle

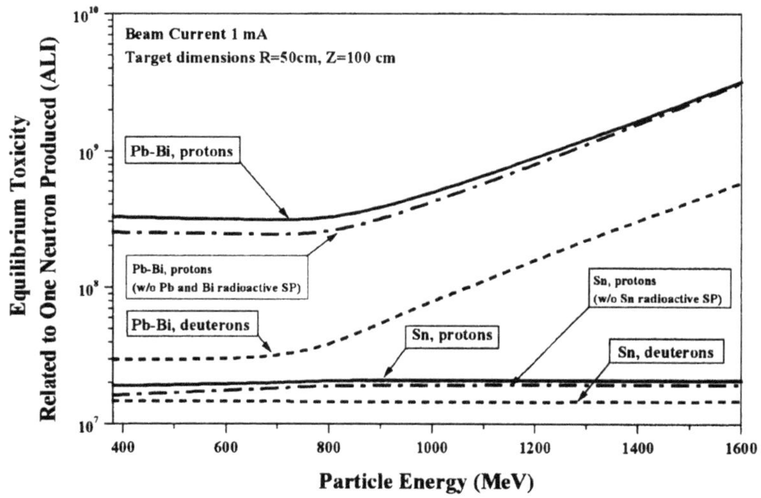

b) Equilibrium toxicity normalized to one neutron produced

Fig. 5. Effect of beam characteristics on neutron production and equilibrium toxicity in Pb-Bi and Sn spallation targets

6. CONCLUSIONS

The main outcome of the present paper is toxicity analysis of radionuclides accumulated in various spallation targets that reveals an important role of alpha emitting rare earths ^{146}Sm, ^{148}Gd, ^{150}Gd and ^{154}Dy. These nuclides give a significant contribution into overall toxicity of spallation targets with exception of tin, which is free from alpha emitters. At the same initial energy per incident hydrogen ion, deuteron beam appears to be advantageous since it gives reduction in toxicity of alpha emitting rare earths by almost one order of magnitude. This seems to be an important factor to make a design choice of accelerator-driven neutron source.

REFERENCES

Barashenkov V.S., Konobeyev A.Yu., Korovin Yu.A., Sosnin V.N. (1999), Computer Software Package "CASCADE/INPE", *Atomnaya Energiya, 87, 283.*

Cappiello M.W. (1999), *Proc. Third International Topical Meeting on Nuclear Applications of Accelerator Technology*, p.53. Long Beach, CA, November 14-18.

CFR (1991), The Code of Federal Regulations, Part 20: Standards for Protection Against Radiation, Appendix B.

Hebel L.C., *et al* (1987), Report to the American Physical Society by the Study Group on Nuclear Fuel Cycles, *Rev. Mod. Phys., 50, N° 1, Part II.*

ICRP (1994), International Commission on Radiological Protection, Publication 68, 72, Inhalation and Ingestion Data for Workers.

Khorosanov G.L., *et al* (1999), *Proc. 3-rd Int. Conf. On Accelerator-Driven Transmutation Technology and Applications (ADTTA '99).* Prague, Czech Republic, 7-11 June.

Konobeyev A.Yu., Korovin Yu.A., Sosnin V.N, Vecchi M. (1999), Study of Accelerator-Driven Reactor Systems, *Kerntechnik 64, 216.*

Rubbia C. (1995), CERN/AT/95-44(ET).

Saito M., *et al* (1997), *Progress in Nuclear Energy 32, 697.*

Salvatores M., *et al* (1994), *Nucl. Sci. Eng., 116, 1.*

Stankovsky A., *et al* (2001), *J. Nucl. Sci. Tech. 38, No.7, 503.*

Takizuka T., *et al* (1997), *Proc. Int. Conf. On Future Nuclear Systems GLOBAL '97*, p.422. Yokohama, Japan, October 5-10.

Pergamon
www.elsevier.com/locate/pnucene

Progress in Nuclear Energy, Vol. 40, No. 3–4, pp. 647–653, 2002

Printed in Great Britain
0149-1970/02/$ - see front matter

PII: S0149-1970(02)00062-8

A FEW COMMENTS BASED ON NUCLEAR PHYSICS

YASUKI NAGAI

Research Center for Nuclear Physics, Osaka University, 10-1 Mihogaoka,
Ibaraki, Osaka 567-0047, Japan

The workshop and symposium on "Accelerator-Driven Sub-Critical Reactor System and Nuclear Physics" were held in March 2000 at KEK (Tsukuba) and in October 2000 at Niigata University during the Autumn Meeting of the Japan Physical Society. This was the first joint meeting between the Japanese nuclear physics community members and the Japanese Atomic Energy Society members. About 150 participants working in various fields, such as nuclear physics, reactor physics, reactor engineering, and nuclear power plants in industries discussed the problems related to the nuclear transmutation. Young students also participated in the meetings. Based on the comments in these meetings and on my personal research experience on nuclear physics at university I would like to make a few comments in considering the energy problem in the 21st century.

I. INTRODUCTION

For Japanese nuclear physicists the project of the transmutation of the nuclear waste, first separating long-lived nuclides such as minor actinide and long-lived fission products from the waste stream and then converting them into either short-lived or stable ones, has been a matter of concern. However, most Japanese nuclear physicists have not well known the essential nuclear physics problems in the nuclear transmutation problem because of little communication between nuclear physicists and scientists working in the field of the atomic energy. Here it should be mentioned that there are a few nuclear physicists, who have been working on the subjects related to the nuclear transmutation. However, since they have been reporting their result at the meeting of Atomic Energy Society of Japan, but not at the meeting of the Physical Society of Japan, nuclear physicists have been missing an opportunity to learn the importance of their result. The atmosphere of this little communication is changing gradually since the High Energy Accelerator Research Organization (KEK) and the Japan Atomic Energy Research Institution (JAERI) have proposed a new joint project. Previously, KEK proposed the Japan Hadron Facility following the strong recommendation by the Japanese nuclear physics community and JAERI proposed the Neutron Science Project, respectively. The research project on the nuclear transmutation has been proposed by the JAERI group, and it is now one of the main research projects in the joint project. Since the nuclear transmutation problem may have the nuclear physics problems, the Japanese nuclear physics community members have decided to organize a workshop to learn the nuclear physics problems in the transmutation problem. Thus, the workshop on "Accelerator-Driven Sub-critical Reactor System and Nuclear Physics" was held at KEK with about 150 participants from various research fields, and the symposium on the same subject was held at Niigata University during the Autumn Meeting of the Japan Physical Society. The workshop was supported jointly by the Japanese nuclear physics committee, KEK and JAERI. They were the first joint workshop and symposium between the Japanese nuclear physics community members and the Japanese Atomic Energy Society members, and the participants exchanged the information they have. Based on the comments we had in these meetings and on my research experience I would like to make three comments for designing the plan of the energy problem in the 21st century. They are concerned with the personal training of young generations, a dream plan for solving the energy problem, and open and scientific discussions between people working in the different research fields.

First I speak about our research on nuclear astrophysics from the viewpoint of personal training. Since I have been working mostly at university, we are very much concerned with the personal training of young generations. When we make research plans, we consider the scientific importance of the plan and the scientific attractiveness for students. Recently many active students show much interest in nuclear astrophysics, and they have been making significant contributions to make the progress of our research as discussed below. Here it should be mentioned that the research subject is related to the nuclear transmutation in stars and has been performed by measuring a fast neutron induced reaction cross section of a nucleus using newly developed experimental methods and/or detectors. In the measurement it is extremely important to develop a new experimental method and/or a high sensitive detector system, and young students have contributed to these developments. The methods and detectors would play a crucial role for studying the nuclear physics problems in the nuclear transmutation problem.

II. keV NEUTRON INDUCED REACTION OF NUCLEI AND NUCLEAR ASTROPHYSICS

Accurate nuclear data, such as the cross section of neutrons and protons induced reaction of a nucleus in the energy of between thermal and a few GeV, are strongly required for designing the nuclear transmutation program based on the Accelerator-Driven Sub-critical Reactor System. In order to accurately measure the cross section of these reactions it is very important to develop an innovative experimental method and a high sensitive measuring system. So far nuclear physicists also have been making significant contributions to such developments, since these developments are quite important for them to obtain crucial information for deeply understanding nuclear physics problems and nuclear astrophysics problems. Here it should be mentioned that precise cross section data of the neutron-induced reaction of a nucleus are also required for studying nuclear astrophysics problems. From this astrophysical interest we have been measuring the cross sections of the keV neutron induced reaction of a nucleus, such as (n,γ) and (n,p) reactions, as discussed below.

Various elements observed, for example, in the solar system are synthesized either in the early universe or in star and the observed elemental abundance patterns give relevant information about the early history and chemical evolution of the universe. "published (Burbidge et al.,1957)". Our current interest is to trace the history of Big-Bang universe by studying stellar evolution and origin of the elements. Among various elements, light elements up to ^{7}Li are synthesized in the early universe and their observed abundance is in good agreement with the calculated value based the standard big-bang model. "published (Peebles, 1966)". The elements heavier than ^{12}C up to iron are synthesized by charged particle induced thermonuclear reactions in stars. "published (Burbidge et al.,1957)". While the elements heavier than iron are synthesized by the neutron capture reaction of a nucleus through both slow (s) and rapid (r) processes. "published (Burbidge et al.,1957)". Here, the s-process is assumed to occur during the He-burning phases of stellar evolution in Red Giant stars and/or in the Asymptotic Giant Branch Stars. Recently much interest has arisen in the abundance of the s-process isotope for various metallicity stars to construct models of the chemical evolution of Galaxy. "published (Wheeler, Sneden and Truran, 1989)". A lot of efforts involving both observations and yields estimation of these isotopes are being made for a wide range of metallicity and stellar masses to compare the chemical evolution models with the observational data. In order to construct the models of the chemical evolution of the s-process isotopes we have to know the cross section of neutrons induced reaction of a nucleus in stars, which can be simulated in a terrestrial laboratory using keV neutrons. The neutrons correspond to the temperature in stars ($\sim 10^{8}$ °K) in the He burning stage. Here it should be added that the experimental study of the keV neutron induced reaction of a nucleus is also crucial to construct models of the nucleosynthesis of the r-process isotopes.

The experiment to measure the fast neutron capture (n,γ) cross section of a nucleus has been measured by two different methods; activation and prompt γ-ray detection. We have developed the latter method, since it can be applied for all nuclei irrespective of the β-decay half-life of a neutron capturing nucleus, and the cross section can be measured as a function of the neutron energy using a time-of-flight (TOF) method, and background γ-rays induced by scattered neutrons can be discriminated from those due to keV neutrons by the TOF method, and a prompt discrete γ-ray characterizing the final nucleus can be detected, which plays a crucial role in determining the cross section unambiguously. "published (Nagai et al., 1991) and (Igashira et al., 1995)". Our experimental method is briefly discussed below. The pulsed keV neutrons were produced by the ^{7}Li(p,n)^{7}Be reaction using pulsed protons provided from the 3MV Pelletron Accelerator of the Research Laboratory for Nuclear Reactors at the Tokyo Institute of Technology. Discrete γ-rays emitted from a keV neutron capturing state by a nucleus was detected by a high sensitive anti-Compton NaI(Tl) γ-ray detector. "published (Ohsaki et al., 1999)". A schematic view of the experimental setup is shown in Fig.1. Here the incident neutron energy spectrum was measured by a TOF method with a ^{6}Li-glass scintillation counter. The compound ^{6}LiH plays a crucial role to effectively absorb neutrons scattered by the sample without emitting γ-rays. A typical γ-ray spectrum from the ^{16}O(n,γ)^{17}O reaction is shown in Fig.2, where we see clearly the discrete γ-rays characterizing a final nucleus of ^{17}O. "published (Igashira et al., 1995)". Our measured cross section 34 μbarn at the average temperature of 30 keV is unexpectedly large compared to an estimated value using a 1/v law, and it is about 170 times larger than the value reported previously. This large value has given strong impact for both theoretical astrophysicists and theoretical nuclear physicists. The astrophysical impact of the present result has been discussed by theoretical astrophysicists, who claimed that ^{16}O strongly affects the production yield of elements heavier than iron in low metallicity star. "published (Rayet and Hashimoto, 2000)". While theoretical nuclear physicists have shown great interest in the present unexpected large neutron capture cross section, which is explained nicely as the non-resonant direct p-wave neutron capture process. The process has not been observed in the neutron capture reaction of a light nucleus. "published (Otsuka et al., 1994) and (Kikuchi et al., 1999)".

We have also developed a new detector system for precisely measuring the cross section of the (n,p) reaction, which is shown in Fig.3. "published, (Shima et al., 1995) and (Kii et al., 1999)". The new detector has high sensitivity with a large acceptance of 4π to detect charged particles emitted from the neutron-induced reaction of a nucleus.

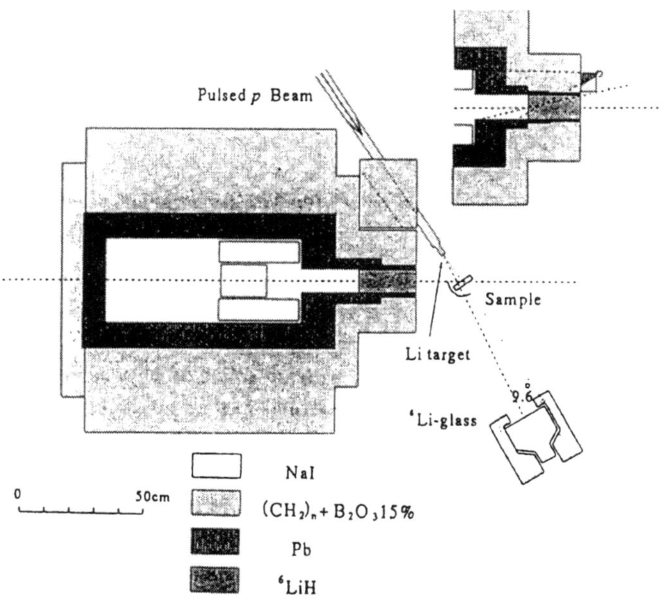

Fig. 1. A schematic view of the experimental setup. Prompt γ-rays from a neutron capturing state of a nucleus were detected by an anti-Compton NaI(Tl) spectrometer with heavy shields against neutrons scattered from the sample. The neutrons were absorbed either by the compound ^{6}LiH or the borated paraffin.

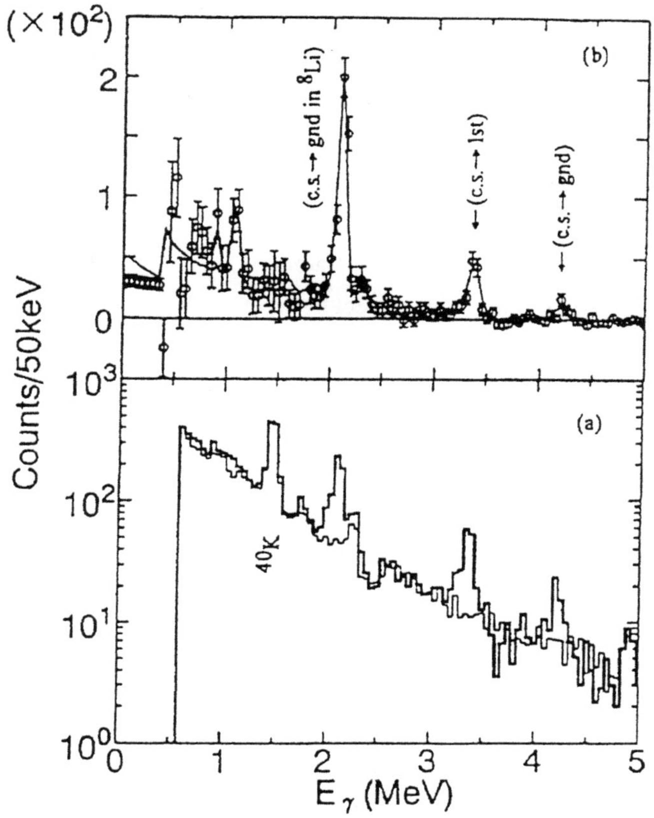

Fig. 2. Total (thick) and background (thin) spectra (a), and background-subtracted γ-ray spectrum (b) from the $^{16}O(n,\gamma)^{17}O$ reaction, respectively.

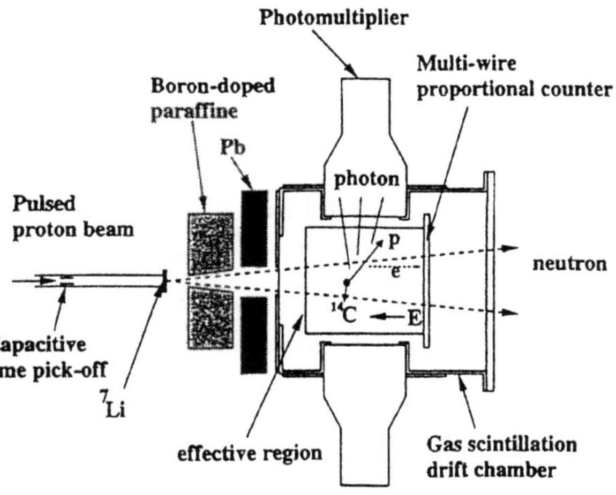

Fig. 3. A schematic view of the experimental arrangement with a newly developed gas scintillation detector.

III. JOINT WORKSHOP BETWEEN NUCLEAR PHYSICIST AND REACTOR PHYSICIST

As mentioned-above the joint workshop between the Japanese nuclear physics committee members and Japanese Atomic Energy Society members was held to discuss the subject related to the Accelerator-Driven Sub-Critical Reactor System (ADS).

We discussed following subjects at the workshop.

(1) Present status of researches on accelerator driven sub-critical reactor system.

(2) Research and developments for high efficiency accelerators.

(3) Research of transmutation by ADS in the joint project between KEK an JAERI.

(4) Nuclear transmutation research and nuclear physics information.

(5) Spallation reaction induced by GeV protons.

(6) General discussions.

We could summarize the dicussions of the workshop as follows.

First, most participants much appreciated open discussions about current and future problems on the nuclear energy program and scientific communication between nuclear physicists and reactor scientists. Second, nuclear physicists started to understand the importance of the accurate nuclear physics information for studying the nuclear transmutation problem, such as the cross sections of the neutron induced reactions of various nuclei and of the high-energy proton induced reactions of a neutron target of Bi-Pb or Hg etc. Third, we noticed that several outcomes, which have been obtained by nuclear physicists in studying nuclear physics problems, are often quite useful for the nuclear transmutation problems. Certainly the workshop and symposium are the first step for nuclear physicists to learn the nuclear physics problems in the transmutation problem. The next step would be to construct the network between nuclear physicists and reactor scientists for continuing such open and scientific discussions on the atomic energy problem and thereby for getting the public understanding on the problem.

An energy problem is certainly one of the important problems in the 21st century, and various development trials to solve the problem should be pursued together with young people. We see young students in university having strong motivation in serving a development project on the energy problem. In order to attract these young students and introduce them into this field developing new energy sources, one has to make a dream development program on the future energy source. When one makes such program, there may be two important factors. First, one has to make the grand concept of the future energy source by considering all possible energy sources. So far, fissile energy, hydroelectricity, nuclear energy, solar energy, wind-generated energy, hydrogen energy and geothermal energy have been proposed as a possible energy source. Of all these energy sources one should discuss merits and demerits, and possible ways to reduce the demerits and increase the merits. Through these discussions one can hope to obtain a national consensus of the role of the atomic energy in the near future, and then one can make a dream development program of the atomic energy, which certainly would require various kind of basic researches. Second, such basic researches should be performed in university, since there are many young students, who want to contribute to solve the energy problem in the 21st century. Hence in order to make great progress of the researches and also to make the personnel training of young generation, it is very important for our society to make clear picture how to support the research and education activities in universities.

IV. SUMMARY

The energy problem is one of the important issues in the 21st century, and therefore one has to make a grand concept of the future energy source by considering all possible energy sources. Through the discussions one can hope that the atomic energy may keep a position to be one of the practical energy sources for another many years with a condition of less production of nuclear waste. The construction of the concept is crucial to attract young generations and introduce them into our field, where nuclear physicists and reactor scientists are starting to make collaborative work for reducing the amount of the nuclear waste and for developing a new atomic energy source based the accelerator driven sub-critical reactor system.

REFERENCES

E.B.Burbidge et al. (1957), Rev. Mod. Phys. 29, 547

P.J.E.Peebles, (1966) Phys. Rev. Lett. 16, 410

J.C.Wheeler, C.Sneden and J.W.Truran. Jr., (1989) Ann.Rev.Astron.Astrophys. 27, 279

Y. Nagai et al., (1991) Astrophys.J 372, 683,

M. Igashira et al., (1995) Astrophys.J 441, L89

T. Ohsaki et al., (1999) Nucl. Instr. Meth. Phys. Res. A 425, 302

M. Rayet and M. Hashimoto, (2000) Astron. Astrophys. 354, 740

T. Otsuka et al., (1994) Phys. Rev. C49, R2289

T. Kikuchi et al., (1999) Phys. Rev. C 57, 2724

T.Shima et al., (1995) Nucl. Instr. Meth. Phys. Res. A356,347

T.Kii et al., (1999) Phys. Rev. C 59, 3397

Pergamon

www.elsevier.com/locate/pnucene

Progress in Nuclear Energy, Vol. 40, No. 3–4, pp. 655–660, 2002

Printed in Great Britain
0149-1970/02/$ - see front matter

PII: S0149-1970(02)00063-X

HIGH POWER ACCELERATORS

IN FUNDAMENTAL PHYSICS AND NUCLEAR TECHNOLOGY

LEONID I. PONOMAREV

Russian Research Center "Kurchatov Institute" and
Research Coordinative Center "MUCATEX",
ul. Rogova 9 – 2, Moscow, 123098, Russia; Tel. 947-27-90, FAX: 947-2787,
e-mail: mucatex.leonid@g23.relcom.ru

ABSTRACT

The trends of the high power accelerators development are outlined. The natural examples of their applications in nuclear physics and technology are discussed: muon physics, physics of rare decays, intense 14 MeV – neutron source based on muon catalyzed fusion (INS - MCF) and accelerator driven system (ADS) for nuclear waste incineration. The accelerator with power ~ 10 MW and particle energy ~1GeV/nucl is considered as the best candidate for these purposes.

1. INTRODUCTION

All the past century the accelerators were used mainly in subatomic physics research, especially in elementary particle physics. In fact all the last 50 years was the age of the elementary particle physics and the main instrument in this period was the accelerator.

The first technical application of accelerators was the isotope separation: the atomic bomb dropped on Hiroshima was manufactured with ^{235}U separated with the Lawrence's cyclotrons. Isotope production, another technological duty of accelerators has began in 1940 when Abelson and McMillan using Berkeley's cyclotron bombarded uranium with deuterium and have produced Np and Pu. Today the isotope production for medicine and technology is well-developed and profitable industry.

The suggestion to use accelerators in the nuclear industry for was formulated first in the late forties by Lawrence & McMillan. But in the past 50 years this idea has transformed several times beyond recognition:

1947 – plutonium production;
~1960 – neutron generation;
~1970 – subcritical reactor;
~1980 – plutonium burner;
~1990 – RAW transmutation.

In what follows we will outline the trends of the high power accelerators development in the nuclear physics and nuclear technology.

The competition of two accelerator types, linear and cyclic, took place from the early beginning and now it becomes especially essential in the ADS problem at choosing the type of the high power accelerator as driver. The parameters of some existing accelerators are presented in Table 1 (Clendenin J., et al., 1996). Most of them are cyclic machines and the power any of them do not exceed 1 MW. There are two ways to increase the accelerator power $W = E \cdot I$: to increase the particle energy E or to increase the beam current I.

Table 1. Existing proton beams available for neutron generation

	JINR Dubna, Russia	KENS KEK, Japan	INPS Argonne, USA	LANSCE Los Alamos, USA	ISIS RAL, UK	SINQ PSI, Switzerland
Accelerator	Cyclotron	Synchrotron	Synchrotron	Linac	Synchrotron	Cyclotron
Particle energy (MeV)	680	500	450	800	800	590
Beam current (μA)	3.2	4.6	15	1000	200	1500
Beam Power (kW)	2.2	2.3	6.8	800	160	885

The main stream of accelerator science and technology was directed till now by the requirement to achieve the highest particle energy. For example, the Fermi lab Tevatron at particle energy $E \sim 10^3$ GeV has the beam current $I \sim 0.1$ μA, i.e. its power $W \sim 0.1$ MW. The characteristic energy in nuclear physics are located at $E \sim 1$ GeV and to increase the accelerator power keeping this energy it is necessary to increase the beam current.

Table 2. Projects of the high power accelerators ($W > 1$ MW)

	SNS Oak Ridge, USA	ESS Europe	NSP JAERI, Japan	SCC KEK-JAERI Japan	MOS KEK, Japan	DC JINR, Dubna
Accelerator	Proton Linac	Proton Linac	Proton Linac	Proton Linac Synchrotron	Proton multy-orbit Synchrotron	Deuteron Cyclotron
Particle energy (GeV)	1.0	1.33	1.5	3.0	1.5	1.8
Beam current (mA)	1. (4.4)	3.7	1. (3.2)	0.33	6.0	≤10
Beam power (MW)	1. (4.4)	5.0	1.5 (5)	1.0	10.0	≤20

In the Table 2(Gardner I.S.K., 1998, Stammbach T.H., 1998) the parameters of the proposed high power accelerators are presented. The power almost all of them is equal $W \sim 5 \sim 10$ MW, particle energy is equal $E \sim 1$ GeV and the beam current is equal $I \sim 5 \sim 10$ mA.

2. MUON PHYSICS AND PHYSICS OF FUNDAMENTAL INTERACTIONS.

The most intensive beams of pions and muons are obtained now at PSI, Switzerland ($\sim 2\times10^9$ π^-/s, $\sim10^8$ μ^-/s) but even this intensity is not enough to check the fundamental statements of the grand unification theory, particularly to measure the branching ratio of the rare decays:

$$\mu \to e\gamma \text{ } (\mu e\gamma \text{ - conversion})$$
$$\mu \to eee \text{ } (\mu 3e \text{ - decay})$$
$$\mu^- Z \to e^- Z \text{ } (\mu e \text{ - conversion})$$

Using high power accelerators with the beam current I~10mA and super conductive solenoid (Lobashov V.M., et al., 1992) one can achieve the intensity $\sim 10^{16}$ π^-/s and $\sim 10^{15}$ μ^-/s.
This opportunity will change qualitatively our ability to formulate problems and possibilities to get answers.

One of the application of the intense muon flux was suggested by K.Nagamine (Workshop, 1999). Using the specific features of muon catalysis phenomenon the flux $\sim 10^3$ μ^-/s of slow muons can be obtained and used in studying the surface physics, including catalytic reactions and hydrogen dynamics reactions on surfaces.

Having the intense source of μ^+ we can plan also the real experiments to observe muonium – antimuonium conversion

$\mu^+ e^- \to \mu^- e^+$

as well as the different QED experiments, e.g. the precision measurements of 2s – 1s transition in muonium, etc.

The slow muons source can be used also as emitter for the high energy muon collider (Workshop, 1999).

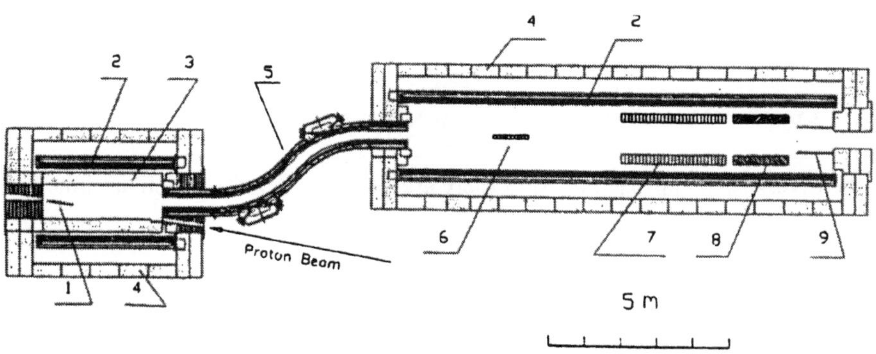

Fig 1. the principal layout of the intense muon source suggested by Lobashov et al, (1996):
1-pion production target, 2-superconductor solenoid, 3-radiactive shielding, 4-shilding, 5- superconductor solenoid, 6-target, 7-detector, 8- detector, 9- shielding.

3. INTENSE 14-MEV-NEUTRON SOURCE BASED ON MUON CATALYZED FUSION (INS-MCF).

The layout of INS-MCF considered in papers (Ponomarev L., 1990, Anissimov V.V., et al., 2001) is presented in Fig. 2.

The deutron beam (E_a = 1 GeV/nucl, I_a = 12 mA) comes to lithium target (150 cm long, 0.75 cm radius) and produces in it 0.163 π^-/d. 58% of pions are captured by magnetic field of superconductor solenoid (14 m long, 40cm diameter) with magnetic field $B_0 = 7T$ and mirror field $B_m = 17T$. 11% of muons produced in solenoid at $\pi^- \to \mu^- + \nu_\mu$ decay are stopped in synthesizer (D/T liquid mixture, ~5l volume) and every muon produces 117 neutrons with energy 14.1 MeV in the muon catalyzed cycles:

$$\mu^- \to (D/T) \to t\mu \to d\,t\mu \to {}^4He + n + \mu^-$$

Every deuteron produces $0.164\cdot0.58\cdot0.11\cdot117 \approx 1.2$ neutrons/deuteron and the deuteron beam I = 12 mA produces in the volume 5 l D/T target $0.6\cdot10^{19}\cdot12\cdot10^{-3}\cdot1.2 \approx 10^{17}\,n/s$ that corresponds to the neutron flux $\sim 10^{14}$ n/s • cm^2.

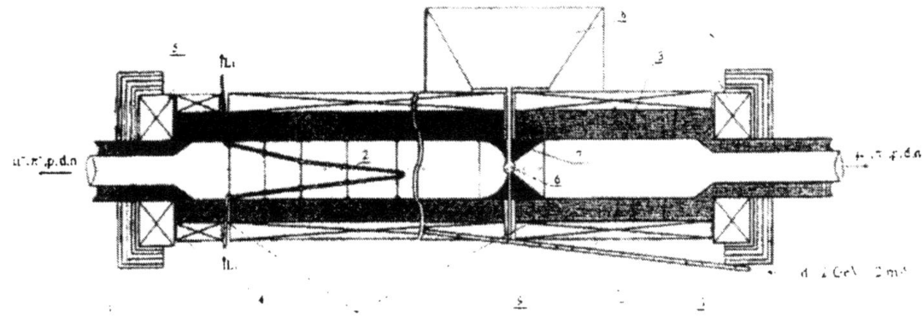

Fig.2 Conceptual design of INS-MCF.
1-deuteron beam, 2-pion producing target, 3-covertor $\pi^- \rightarrow \mu^-$, 4-superconductor solenoids, 5-magnetic mirrors, 6-synthesizer, 7-tested materials, 8-tritium handling system, 9-manipulator for the synthesizer substitution, 10-radiation shielding, 11-magnetic mirrors, 12-^{3}He blanket.

To justify the presented conceivable schema of INS-MCF the following calculations were performed: angular and energy distribution of π^- produced in Li-target, heat regime of the pion producing target and synthesizer; optimization of geometry of both targets and hydrodynamical regime of their cooling; parameters of the solenoid and magnetic field configuration; radiation shielding of the superconductive solenoid; working up of radioactive isotopes, etc.

The result of these calculations is the statement: INS-MCF has no physical and technological restriction preventing construction. There are only two problems:
Who is really interested now in the intense 14 MeV neutron source?
Who will take responsibility to construct it?

4. ACCELERATOR DRIVEN SYSTEM FOR NUCLEAR WASTE TRANSMUTATION.

Despite the diversity in approach to ADS there are several elements which are inherent in them. Any ADS always includes:

accelerator;
primary neutron producing target;
blanket (reactor).

Every incident proton with energy E_a produced in the primary target n_{sp} neutrons (Bauer G.S. and McDonalds A., 1983):

$$n_{sp} \approx 20 \cdot E_a \text{ (GeV)}.$$

The electric blanket power W_{bl} is equal:

$$W_{bl} = \frac{n_{sp} E_f k_{eff} \eta}{\nu_f \Delta k_{eff}} I_a,$$

where k_{eff} is the neutron multiplication factor, $E_f = 200$MeV is the fission energy release, $\Delta k_{eff} = 1 - k_{eff}$, ν_f is the neutron release per one blanket fusion, η is the efficiency of the heat-electricity conversion and I_a is the beam current of the accelerated protons. Assuming $\eta = 0.3 \sim 0.4$, and $\nu_f = 2.3 \sim 3.0$, we can obtain the relation between the beam power W_a and blanket power W_{bl} (Ponomarev L. and Artisyuk V., 1999):

$$W_a = (0.6 \sim 0.8)\Delta k_{eff} W_{bl}.$$

At W_{bl} = 1GeV and Δk_{eff} = 0.05 we have $W_a = 30 \sim 40$ MW. Just this beam power is choosen in almost all projects of ADS (See Table 3).

Table 3 ADS systems proposals-99 for RAW transmutation

Project	ATW LANL USA	TRASKO ENEA-INFN-Ansaldo Italy	GEDEON CEA-CNRS-FRAMATOME France	HYPER KAERI Korea	OMEGA JAERI Japan	CSMSR Kurchatov Int. Russia
Power, MW	2000	-	-	1000	820	1000
k_{eff}	0.92÷0.98	-	0.92÷0.98	0.97	0.95	0.95
Accelerator	40MW 1GeV-40mA proton linac	40MW 1GeV-40mA proton linac	1÷200MW R&D	16MW 1GeV-16mA proton linac	40MW 1.5GeV-25mA proton linac	8MW 1.5GeV-5mA deuteron cyclotron

This power is however beyond the today scope of accelerator science and technology (See Tables 1 and 2).

For the reduction of the accelerator beam power W_a it was suggested (Alexeev P.N., et al., 1995, GENES-2, 1996) to use in ADS the cascade scheme combined with the molten salt fuel. The schema of the cascade subcritical molten salt reactor (CSMSR) is presented at Fig. 3.

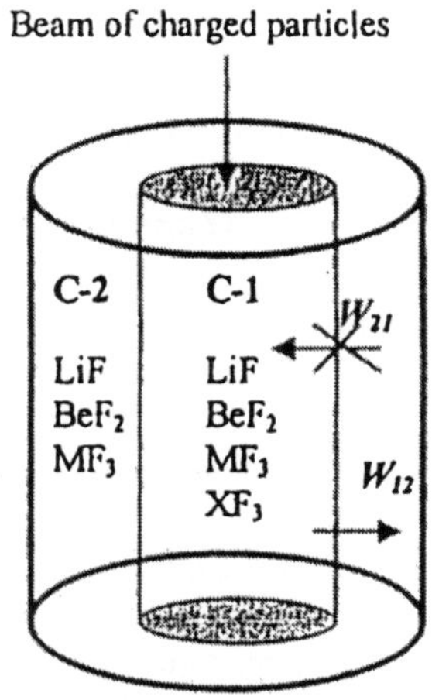

Fig. 3 Principal layot of CSMSR:
C-1 – neutron production core, C-2 – transmutation core, W_{12} – the probability of neutron diffusion from C-1 to C-2, W_{21} – the probability of the backward diffusion of the thermal neutrons.

The main feature of this scheme is the suppression the backward thermal neutron flow to the level W_{21}~10^{-3} from C-2 (blanket core) to C-1 (primary neutron produced core). This schema allows to reduce the accelerator power several times to the acceptable level $W_a \leq 10$ MW. The R&D of this ADS-conception is now in progress in the framework of the ISTC project # 1486 "Experimental and Theoretical Justification of the Cascade Scheme of the Subcritical Molten-Salt Reactor for Transmutation of Long-lived RW of the Nuclear Fuel Cycle".

5. CONCLUSION

The current possibilities of increasing the energy of accelerated particles are almost exhausted (The experience with the shut down of the SSC construction in USA essentially confirms this general statement).

In contrast the accelerator of power ($W \sim 10$ MW and particle energy $E \sim 1$GeV, i.e. with beam current $I \sim 10$mA) can be constructed basing the contemporary technology in the near future. It seems that high power cyclotron with the parameters $W \sim 10$ MW, $E_a \sim 1$ GeV, $I_a \leq 10$mA has the definite preferences (The construction cost is the last but not the least argument in favor of this choice (ADTTA, 1996).

The construction of such a machine will give the unique opportunity to get pion and muon sources for the fundamental physics research with intensity $\sim 10^{16}$~10^{15} s^{-1}, i.e. six order of magnitude higher than existing ones.

Using this machine it is possible also to develop 14-MeV INS-MCF with intensity $\sim 10^{17}$ n/s and flux $\sim 10^{14}$ n/s·cm^2.

Basing such an accelerator ADS and INS-MCF ideas can be combined in one installation, because 14-MeV INS-MCF uses only 30% of beam power and secondary particles scattered in the backward hemisphere. The energy of these particles is used for the production of pions and muons (see Fig.2). The other ~70% of incident beam can be used for feeding ADS, isotope production, etc.

Taking into account the instabilities of the proton accelerators (~1 trip/hour (NEA Workshop, 1998)) it is worth considering also the possibility of using the high power electron accelerators, especially in the cascade scheme of ADS (Abalin S.S., et al., 1994). Of course all these ideas need detailed R&D and I hope some of them at least will be carried out in the nearest future.

Most of the outlined problems can be naturally solved in the framework of the Joint KEK-JAERI Project (KEK Report, 1999).

REFERENCES

Abalin S.S., et al. (1994), *Proc. Int. Conf. on ADTT and applications,* Las Vegas, USA, July 25-29.

ADTTA (1996), *Proc. Int. Conf. on ADTT and applications*, Kalmar, Sweden, June 3-7.

Alexeev P.N., et al. (1995), Cascade subcritical reactor with the ultimate safety. *Atomnaya Energia 79,327.*

Anissimov V.V., et al. (2001), 14-MeV neutron source based on muon catalyzed fusion. An advanced design I-III. *Fusion Technology 39, 198.*

Bauer G.S. and McDonalds A., (1983), *Nuclear Technologies in a Sustainable Energy System*, Springer Verlag.

Clendenin J., Rinolfi L., Takata K., Warner D.J., (1996), *Compendium of Scientific Linacs*, CERN/PS 96-32 (DI).

Gardner I.S.K., (1998), A review of spallation neutron source accelerators, *Proc. of EPAC-98*, p. 98.

GENES-2 (1996), *Global Environment & Nuclear Energy System – 2, Proceedings, Eds. Ninokata H. and Sawada T.*, Tsuruga, Fukui, 1996, *Progr. Nucl. Energy 32, No.3/4 (1998).*

KEK Report (1999), *The Joint Project for High-Intensity Proton Accelerators*, KEK Rep. 99-4, JAERI-Tech 99-056, JHF-99-3, July, 1999.

Lobashov V.M., et al., (1992), *MELC-experiment to search for $\mu^- A \to e^- A$ process*, Preprint INR-786/92, Moscow.

NEA Workshop, (1998), *NEA Workshop on Utilization and Reliability of high power accelerators*, Mito, Japan, October 13-15, 1998.

Ponomarev L., (1990), *Muon Catalyzed Fusion, Contemporary Physics, 31, 219.*

Ponomarev L. and Artisyuk V., (1999), *Accelerator Driven Systems in Nuclear Power*, Ref.5, p. 83

Workshop, (1999), *Proc. of workshop on ultra-high intensity muon beam and muon science in the 21st Century, Ed.S.Sakamoto*, Tsukuba, Japan, March 23-25.

Stammbach T.H., (1998), High intensity problems, *Proc. of 15th Conf. On Cyclotrons and their applications*, 17-369, Caen (1998).

Pergamon

www.elsevier.com/locate/pnucene

Progress in Nuclear Energy, Vol. 40, No. 3–4, pp. 661–671, 2002

0149-1970/02/$ - see front matter

PII: S0149-1970(02)00064-1

EVOLUTIONARY MOBILE FUEL REACTORS

A. Gandini

University of Rome - Nuclear Engineering Faculty
DINCE - Corso Vittorio Emanuele II, 224 - 0186 Rome
Tel +39-06-3052-134, Fax +39-06-3048-4203, e-mail: gndna@tin.it

ABSTRACT

In the present paper issues relevant to an extended, long term use of nuclear energy are commented. In particular, the role of ADS and of the so called evolutionary mobile fuel reactors are considered. A note is presented on specific calculation methods to be adopted for intercomparison and survey analysis of these systems. © 2002 Published by Elsevier Science Ltd.

1. NEXT GENERATION REACTORS

The main lesson learnt so far in the four decades of pacific uses of atomic energy, after the two Geneva Conferences of 1955 and 1964, is that the accidents associated with nuclear energy posing major risk problems, i.e., greater than those perceived acceptable as an inevitable price to be paid for progress, are related to human factors. Such has been in fact the case with the events occurred at Three Mile Island, Chernobyl, and, more recently, Tokaimura. This latter accident, in particular, has broadened the perception of major possible risks outside the reactor system. Besides, although the quite positive experience with LWR reactors in OECD countries in the last decades cannot be forgotten, it is also perceived that a broader, world-wide nuclear energy expansion would pose a number of concerns related to the safety of reactor plants (power excursions, residual heat risk), as well as to those associated with the fuel flow (criticality accidents, fuel diversion, radiological risk, proliferation) and to the problems related to reprocessing (antiproliferation issues, fuel transportation risks). To find an answer to these issues in a long term scenario, the following objectives should be persued:

1. A nuclear energy reactor concept assuring optimal characteristics in relation to economy, safety, anti-proliferation, anti-diversion, radiological risk minimization, public acceptance;
2. A viable, smooth transition from today's nuclear energy production structure into that, at equilibrium, relevant to the new concept;
3. The possibility of closing the fission cycle, whenever a different important energy source is developed as a substitution, with minimal radio-toxicity from residual waste.

Three reactor concepts appear worth being considered as candidates for a scenario consistent with the above objectives :

- The molten salt reactor (MSR), a variety of which is the Th-fueled, feed and bleed concept TASSE (Slessarev et al., 1999);
- The evolutionary pebble bed reactor (EPBR), a PBR-like concept with an analogous feed and bleed logic;
- The molten metal fuel reactor (MMFR), proposed at ANL (Palmiotti, Feldman,1999).

All three concepts adopt mobile fuel so to maintain a quasi-constant multiplication factor. They may be considered evolutionary, in the sense that a strategy may be chosen, so to respond during their life to the objectives outlined above. A variety of options need being considered, affecting reactor type (critical or subcritical, thermal or fast), accelerator (linear, or cyclotron), processing (aqueous, or pyro-metallurgical techniques). Due to the uncertainty affecting these various options, in terms of performance and/or feasibility, it seems quite difficult at present a clear choice. Further, extensive research studies in these areas and international cooperation are required (a joint program on MFR's within the NERI initiative is being proposed).

The quasi constancy of the multiplication factor during the reactor life, associated (for the thermal option) with quasi-maximal reactivity configuration conditions, and then, consequently, the reassuring distance from prompt criticality, under operational, normal or accidental conditions, make these systems particularly attractive from th safety point of view. The low toxicity associated with the thorium cycle, and the enhanced possibility, at least during a first expanding phase, of avoiding reprocessing are other points worth being considered

1.1. The TASSE logic

As mentioned before, one of the main advantages with the (ADS, or critical) MFR systems is the possibility of drastically reducing the need of a reactivity (or current, in case of an ADS-like concept) reserve during operation and burnup. This removes, in particular, the risk of a reactivity accident consequent to an inadvertent control rod extraction. Critical MSR and PBR designs are generally well known, although there are different versions. As far as MSR systems are concerned, we mentioned above the TASSE concept (Slessarev et al., 1999), a modification of the MSR thermal design proposed by Bowman et al. (1992). With this concept it would be possible to avoid, in an energy expanding phase (doubling time of the order of 20÷30 years), the (economically penalizing) fuel reprocessing step, the same fuel being recycled in a feed and bleed mode. The bleed fuel at quasi-equilibrium composition (in the expanding phase) would be used for feeding a new system, when the doubling of the necessary mass is achieved.

As far as the PBR is concerned, we remind the old German (thermal) reactor project. As belonging to the MFR family, the PBR has the desirable property, mentioned above, of maintaining the multiplication factor quasi-constant during the reactor life. Small operation changes could be obtained by proper power adjustments of the coolant pump, exploiting the large (negative) temperature coefficients which characterize these systems. An important feature of the PBR design is its operation at very high temperatures (~ 850°C), which allows:

- on one side, to reach very high thermal efficiencies (of the order of 50%);
- on the other side, to consider these systems also for the production of process heat (for instance, for hydrogen production).

The same feed an bleed logic of the TASSE concept might be applied also to the PBR.

The potentiality of the PBR systems, also with respect to their possible role as ADS, may be also inferred from the results of a recent work by X. Raepsaet et al. (1999). In their paper it is demonstrated in fact that in a HTR of the GA type with fixed fuel elements, after 500 GWd/t, a K_∞ value is reached of the order of 1.01 with 66% PU+MA fissile (from PWR). In a corresponding PBR at equilibrium conditions with similar average burnup (and then similar K_{oo}) the oldest pebbles (ready for been definitely discharged) would have a burnup of the order of 80%. For a large, and then low leakage, reactor, possibly with an external neutron source, it would be possible, in principle, to adopt this concept for an effective actinide burner with a relatively reduced content of MA.

1.2. Transition, equilibrium and phasing out

In above long term scenario, a transition lasting several decades would be foreseen during which a gradual substitution of present reactors plants are replaced by these new concepts (IFR-like advanced fast reactors with fixed fuel elements, or MFR's). During this (presumably, expanding) phase, these reactors should be (at least partially) fed by existing stocks of spent fuel from old generation reactors, and/or by military plutonium. At the end, an equilibrium condition would be eventually reached in which the (average) feed fuel isotopic distribution would be quasi-constant in time.

At this equilibrium phase, the transition into, or the inception of a U-Th cycle would take place. The two cycles (U-Pu and U-Th) could coexist for an indefinite time. This gradual transition to the U-Th cycle would allow, on one side, to optimize the exploitation of natural resources and, on the other side, to minimize the long term radiological risk.

When the time arrives to start the phasing out of the fissile cycle (presumably, for the advent of a new, more advantageous source of bulk energy), a definite transition from the U-Pu cycle into the U-Th one should be established. During this phasing-out period, major quantities of TRU would be destroyed, until, at equilibrium, the fresh feed would consist only of thorium plus, possibly, some fissile (U^{235}), to upgrade the fuel. The amount of added fissile would depend on the system concept. In case of ADS, such fission content could be significantly reduced by an appropriate increase of the subcriticality, this in turn implying an increasing fraction of the produced energy for feeding the accelerator.

At the end of the fission cycle, a relatively small quantity of fissile material would remain, in terms of change of the overall radioactivity preexistent the fission energy exploitation.

2. SURVEY ANALYSIS METHODS

In order to arrive at identifying the concept, or concepts best candidates for the next generation of reactors, among the large variety of concepts proposed, it is important to have available efficient, fast methods capable of giving an adequate insight into the most crucial physical problems (relevant to safety, radiological risk, plant efficiency, etc.), and at the same time relatively simple and fast enough for allowing parametric studies for intercomparison analyses. One method responding to these requirements is that proposed by Salvatores et al. (1995) for the analysis of different fuel cycle strategies, in relation to the long term radiotoxicity associated with them. In the following, we shall make a discussion on methods aiming at a similar relative "simplicity" and efficiency. Attention will be given to three topics: the reactivity balance approach, extended to subcritical system analysis, the core evolution sensitivity analysis based on GPT methodology, in which the concepts of neutron importance relevant to the system power control and that of generalized reactivity are defined, and a simple, physically significant point kinetics approach based on such concepts.

2.1. Balance of power

Extensive studies on the ADS behavior under incidental conditions are presently made for verifying their claimed advantage, under the safety point of view, with respect to the critical reactors. An extensive, effective safety corresponding fast analysis of these systems was made in the late 80's at ANL (Wade, 1986), specifically with respect to the Integrated Fast Reactor (IFR). A synthetic, quite effective deterministic method was used based on a "balance of reactivity" approach. We can rewrite a similar methodology in relation to the ADS, making rather use of a "balance of power" approach. This would allow to estimate the behavior of the ADS systems at abnormal conditions. The balance of reactivity approach consists in writing a quasi-static balance of reactivity (ρ):

$$\rho = (P-1)A + (P/F-1)B + \delta T_{in}C + \delta\rho_{ext} = 0 \qquad (1)$$

where:

P and F are power and coolant flow (normalized at operating conditions, i.e., $P_o=1$, $F_o=1$)
δT_{in} is the change from normal coolant inlet temperature T_{in}
C is the inlet temp. reactivity coefficient
(A+B) is the reactivity coefficient experienced in going to full power and flow from zero power isothermal at coolant inlet temperature
B is the power/flow reactivity coefficient
$\delta\rho_{ext}$ is an external reactivity insertion.

In Eq. (1) it is assumed that convergence (criticality) has been reached asymptotically.

A similar approach may be followed in relation to ADS systems, starting, rather than from a balance of criticality approach (suitable for critical systems), from a balance of power one. To note that the formulation proposed shows to be quite general, being applicable to subcritical, as well as to critical systems.

Whereas in an IFR reactor the equilibrium condition after a (limited) incidental transient corresponds to a new criticality state, in an ADS system in a similar circumstance, due to the external source presence, and assuming that the incidental transient does not lead to criticality (an intrinsic prerequisite for this system safety), any subcritical steady state condition may be generally achieved. The following equations are then proposed, at equilibrium,

$$\rho = (P-1)A + (P/F-1)B + \delta T_{in}C + \delta\rho_{ext} \tag{2}$$

$$P = \alpha \frac{s_n + \delta s_n}{1 - K_{eff}(1 + \rho / K_{eff})} \tag{3}$$

K_{eff} being the multiplication coefficient of the subcritical system before the accidental event, ρ the reactivity associated with the deviation of the multiplication coefficient from K_{eff}, and δs_n a change of the external neutron source s_n, induced by a change δi of the accelerator current i. Coefficient α is determined considering operating conditions at which $P=P_o=1$. Indicating by $\bar{\rho}$ the subcriticality $(1-K_{eff})$, at operating conditions, the following equation may be obtained

$$P[\bar{\rho} - (P-1)A - (P/F-1)B - \delta T_{in}C - \delta\rho_{ext}] - \bar{\rho}(1 + \delta i / i) = 0 \tag{4}$$

where we have set $\delta i/i$, i.e., the fractional change of the accelerator current, in place of $\delta s_n/s_n$. This equation is quite general. In fact, if the external source term $\bar{\rho}(1+\delta i/i)$ vanishes (since for a critical system it is $K_{eff}=1$), we obtain again Eq.(1). Eq.(4) may be solved with respect to P. We easily obtain

$$P = \frac{\Delta - \sqrt{\Delta^2 - 4\bar{\rho}(A + B/F)(1 + \frac{\delta i}{i})}}{2(A + B/A)} \tag{5}$$

where $\Delta = \bar{\rho} + (A+B) - \delta T_{in}C - \delta\rho_{ext}$.

An important quantity to be analyzed is the coolant output temperature T_{out}. If by ΔT_c we denote the coolant temperature rise at nominal full power/flow ratio, the change δT_{out} is defined as

$$\delta T_{out} = \delta T_{in} + (P/F-1)\Delta T_c \quad . \tag{6}$$

A number of important incidental events may be analyzed and compared for different systems with this methodology, such as the TOP (transient of power), the LOF (loss of coolant flow), the HTWS (heat transfer without sink), the CIT (chilled inlet temperature) events. For exemplification, in the following we give a few results taken from a previous study (Gandini ed al., 1999). The ADS system considered was the Russian BREST fast reactor (Orlov and Slessarev, 1988).

<u>LOHS-WS event</u>. If the process of secondary heat exchange is arrested, in a critical reactor the inlet temperature starts increasing. The negative reactivity effect induced by the inlet temperature rise is compensated by the positive one relevant to the power decrease, up near zero-level. Assuming $F = F_o$, then $T_{in} \rightarrow T_{out}$. For an ADS system, there is no equilibrium in the outlet coolant temperature. The outlet temperature is increasing constantly because the ADS power cannot approach zero level, notwithstanding significant feed-backs. The power is sustained down to a lower limit proportional to $\bar{\rho}$. This means that the lower the K_{eff} value, and the higher the neutron source chosen, the higher will be the rate of the asymptotic outlet temperature increase. This in turn means that core intrinsic characteristics do not allow to achieve a deterministic safety level, in case of failure of the proton beam stop device. The results are in Table 1.

<u>LOF-WS event</u>. With this event the inlet temperature T_{in} is assumed not to change while the coolant flow will coast down to natural circulation. The consequent raising power to flow ratio induces an increase of the core

average temperature, this in turn inducing a negative reactivity feedback. This negative reactivity is compensated by a positive one induced by the power reduction. Asymptotically, a natural circulation flow F_{NC} will be established. For preliminary quantitative analysis one can set $F_{NC} \approx 0.15\ F_0$ at nominal core thermal parameters. Table 2 presents the evaluation of power change as well as the outlet lead temperature growth for ADS and for the corresponding critical reactor. The results show that the asymptotic temperature level for ADS is unacceptable.

The accidents considered, even if unlikely, may pose serious questions relevant to the safety of ADS. To overcame these difficulties, in a recent work (Gandini et al., 2000) an approach has been proposed by which the probability of such an event is drastically reduced. It is based on the coupling of the accelerator proton current, rather than with the external grid, to the electricity produced by the same reactor, except at startup conditions, for which an independent electricity source of a relatively low power would be used.

Table 1. LOHS-WS asymptotic parameters

ADS	Critical reactor
No equilibrium of outlet coolant temp.	$\delta T_{out} = \left(\frac{A+B}{C\Delta T_c} - 1\right)\Delta T_c$
For all K_o $T_{out} > 1000°C$ $W \rightarrow W_{asympt}$	$\delta T_{in} = 190°C$ $\delta T_{out} = 70°C$ $T_{out} = 610°C$ $W \rightarrow 0$

Table 2. LOF-WS asymptotic parameters

ADS	Critical reactor
Power decreases with decreasing $\bar{\rho}$ [See Eq.(5)]	$P = 1 + \frac{B(F_{NC}/F_o - 1)}{B + AF_{NC}/F_o}$ $\delta T_{out} = \frac{B(F_{NC}/F_o - 1)}{B + AF_{NC}/F_o}\Delta T_C$
$\bar{\rho} = 10\beta_{eff}$ $P = 0.93$ $\delta T_{out} = 625°C$ $T_{out} = 1165°C$	$P = 0.48$, $\delta T_{out} = 260°C$ $T_{out} = 800°C$

2.2. The HGPT Methodology for ADS

One first indication of the potential use of the HGPT methodology with respect to neutron kinetic analysis of critical and noncritical systems and to the possibility of analyzing integral experiments in reactor facilities at subcritical conditions was suggested in 1968 (Gandini). In particular, the neutron and precursor importances associated with a given response was considered. In subsequent articles (Gandini, 1976, 1981), the use of HGPT methods for time-dependent problems was again discussed. In particular, the composite neutron, precursor and multi-channel temperature field, generally in presence of external neutron and enthalpy sources, was suggested for application of the HGPT methodology in dynamic studies. The application of the HGPT methodology for the cycle life analysis of these systems was proposed in 1997 by Gandini basing on a previous procedure (Gandini 1987a, 1987b, 1988) developed for critical ones. Here we shall shortly review these topics in relation to their application to ADS systems. In particular, the role is discussed of the importance function associated with the power control, and the definition of the concept of "generalized reactivity". Basing on these results, a formulation is described of a point kinetic equation, with physically significant coefficients (Gandini 2001).

2.2.1. Core cycle analysis. To the neutron and fuel nuclide densities, represented by vectors $\mathbf{n}(\mathbf{r},t)$ and $\mathbf{c}(\mathbf{r},t)$, respectively, defined in the reactor cycle interval (t_o,t_F), a specified intensive control variable, $\rho(t)$, is associated so that the assigned, overall power history W(t) is maintained. We assume here that the power level is controlled by the source strength via the regulation of the accelerator current (this being one main advantage claimed for ADS). The following nonlinear governing equations may than be written

$$\mathbf{m}_{(n)}(\mathbf{n},\mathbf{c},\rho \mid \mathbf{p}) = -\frac{\partial \mathbf{n}}{\partial t} + B\mathbf{n} + \rho \mathbf{s}_n = 0 \qquad (7)$$

$$\mathbf{m}_{(c)}(\mathbf{n},\mathbf{c}\mid\mathbf{p}) = -\frac{\partial \mathbf{c}}{\partial t} + E\mathbf{c} = 0 \quad (8)$$

$$\mathbf{m}_{(\rho)}(\mathbf{n},\mathbf{c}\mid\mathbf{p}) = <\mathbf{c}, S\mathbf{n}> - W = 0 \quad (9)$$

where B is the neutron diffusion, or transport, operator, E the nuclide evolution matrix, $\mathbf{s}_n$ is a given source term, while, with obvious notation,

$$S = \gamma \begin{vmatrix} \sigma^1_{f,1} & \dots & \sigma^1_{f,G} \\ \dots & \dots & \dots \\ \sigma^J_{f,1} & \dots & \sigma^J_{f,G} \end{vmatrix} V, \quad (10)$$

γ being the amount of energy per fission and V the neutron velocity diagonal matrix. In quasi-static problems, as those of interest here, the derivative $\partial \mathbf{n}/\partial t$ is negligible. So, the time derivative at second member of Eq. (7) may be neglected in the course of the integration process.

Any response, functional of variables **n**, **c**, and ρ, could be considered for analysis. We limit here consideration to the response

$$Q \equiv \rho(t_F) = \int_{t_o}^{t_F} \delta(t - t_F)\rho(t)dt \quad (11)$$

which corresponds to the relative source strength required at t_F to assure the power level imposed. We may assume that, at unperturbed conditions, ρ(t)=1 in the interval (t_o,t_F). If a system parameter is altered, as in an optimization search analysis, it may be of interest to evaluate the corresponding change of ρ at the end of cycle, to make sure that given upper limit specifications of the source strength are non exceeded.

Along the HGPT methodology, the following sensitivity expression is obtained (Gandini, 1997), with respect to a generic system parameter p_i:

$$\frac{\partial \rho(t_F)}{\partial p_k} = \rho_F^*[<\bar{\mathbf{n}}_F^* \frac{\partial}{\partial p_k}(B\mathbf{n} + \mathbf{s}_n)> + \frac{\partial}{\partial p_k}(<\mathbf{c}, S\mathbf{n}> - W)]_{t_F}$$
$$+ \int_{t_o}^{t_F}[<\tilde{\mathbf{n}}^*, \frac{\partial}{\partial p_k}(B\mathbf{n} + \mathbf{s}_n)> + <\mathbf{c}^*, \frac{\partial E}{\partial p_k}\mathbf{c}>]dt + \int_{t_o}^{t_F} \tilde{\rho}^* \frac{\partial}{\partial p_k}(<\mathbf{c}, S\mathbf{n}> - W)dt) \quad (12)$$

where the importance functions $\mathbf{n}^*(\mathbf{r},t)$ $[\equiv \mathbf{n}_F^*\,\delta(t - t_F) + \tilde{\mathbf{n}}^*(\mathbf{r},t)]$, $\mathbf{c}^*$ and ρ^* obey equations

$$-\frac{\partial \mathbf{n}^*}{\partial t} = B^*\mathbf{n}^* + \Omega_c^*\mathbf{c}^* + S^T\mathbf{c}\rho^* \quad (13)$$

$$-\frac{\partial \mathbf{c}^*}{\partial t} = E^*\mathbf{n}^* + \Omega_n^*\mathbf{n}^* + S\mathbf{n}\rho^* \quad (14)$$

$$<\mathbf{n}^*, \mathbf{s}_n> + \delta(t-t_F) = 0 \,, \quad (15)$$

Ω_c^* and Ω_n^* being coupling operators Eq. (15) is an orthonormal condition for $\mathbf{n}^*$.

Function $\mathbf{n}_F^*$ obeys equation

$$B^*\mathbf{n}_F^* + S^T\mathbf{c}(t_F)\rho_F^* = 0 \quad (16)$$

with $\rho_F^* = -1/W$. It corresponds, then, to the importance function relevant to the relative power (equal to unity at nominal conditions). From these 'final' values, a recurrent calculation scheme may be defined starting from t_F and proceeding backward.

2.2.2. Stationary case. For a given subcritical system at stationary conditions (which may be interpreted at the beginning of its cycle life), we may consider the same system above in which the neutron source and the nuclide density are assumed time-independent during an arbitrary time interval (t_o,t_B). The governing equations can then be written, in case the power level is controlled by the source strength,

$$\mathbf{B}\mathbf{n}_o + \rho_o \mathbf{s}_{n,o} = 0 \tag{17}$$

$$<\mathbf{c}_o, \mathbf{S}\mathbf{n}_o> - W_o = 0 . \tag{18}$$

The following sensitivity expression is found

$$\frac{\partial \rho_o}{\partial p_k} = <\mathbf{n}_o^*, \frac{\partial}{\partial p_k}(\mathbf{B}\mathbf{n}_o + \mathbf{s}_{n,o})> + \frac{\partial}{\partial p_k}(<\mathbf{c}_o, \mathbf{S}\mathbf{n}_o> - W_o) \tag{19}$$

The neutron importance $\mathbf{n}_o^*$ obeys equation

$$\mathbf{B}^*\mathbf{n}_o^* + \mathbf{S}^T\mathbf{c}_o\rho_o^* = 0 \ . \tag{20}$$

with $\rho_o^* = 1/W_o$. If, rather than via the source strength, the power level reset control is assumed to be regulated via neutron absorption, so that the control ρ_o would enter into operator B, it can be shown that the sensitivity coefficient would be given always by Eq. (19), and $\mathbf{n}_o^*$ by (20), but with $\rho_o^* = -1/<\mathbf{n}_o^*, (\partial B/\partial \rho)\mathbf{n}>$. We might as well consider a (fictitious) control mechanism affecting the fission source, rather than the neutron absorption, i.e., we might choose as control a coefficient multiplying the fission matrix (F) and, therefore, entering into operator B ($\equiv A+\rho_o F$). The sensitivity coefficient would be given again by Eq. (19), but with $\rho_o^* = -1/<\mathbf{n}_o^*, F\mathbf{n}_o>$.

The sensitivity expression (19) may be generalized so that

$$\frac{d\rho_o}{dp_j} = -\frac{<\mathbf{n}_o^*, \frac{\partial \mathbf{m}_{(n,o)}}{\partial p_j}> + \frac{\partial}{\partial p_j}(<\mathbf{c}_o, \mathbf{S}\mathbf{n}_o> - W_o)}{<\mathbf{n}_o^*, \frac{\partial \mathbf{m}_{(n,o)}}{\partial \rho_o}>} \tag{21}$$

with $\mathbf{n}_o^*$ obeying Eq. (20) while $\mathbf{m}_{(n)}(\mathbf{n},\mathbf{c},\rho\,|\,\mathbf{p})$ and $\mathbf{m}_{(\rho)}(\mathbf{n},\mathbf{c}\,|\,\mathbf{p})$ have been defined previously. A corresponding perturbation expression, giving $\delta\rho_o$ in terms of parameter perturbations, defined as $\delta\mathbf{m}_{(n,o)}$, may then be obtained. $\delta\rho_o$ corresponds to the control change necessary to reestablish the power level existing before the perturbation $\delta\mathbf{m}_{(n,o)}$. We may as well say that the perturbation $\delta\mathbf{m}_{(n,o)}$ [and $\delta(S^T\mathbf{c}_o)$] would produce a power level change equivalent to that produced by a control change

$$\delta K_\rho = \frac{<\mathbf{n}_o^*, \delta\mathbf{m}_{(n,o)}> + <\mathbf{n}_o, \delta(S^T\mathbf{c}_o)>}{<\mathbf{n}_o^*, \frac{\partial \mathbf{m}_{(n,o)}}{\partial \rho_o}>} \tag{22}$$

which defines the "generalized ρ-mode reactivity".

In the case of the (fictitious) control on the neutron fission, setting λ in place of ρ to distinguish this peculiar case, we may explicitly write

$$\delta K_\lambda = \frac{<\mathbf{n}_o^*, \delta B \mathbf{n}_o>}{<\mathbf{n}_o^*, F\mathbf{n}_o>} + \frac{<\mathbf{n}_o^*, \delta \mathbf{s}_{n,o}>}{<\mathbf{n}_o^*, F\mathbf{n}_o>} + \frac{<\mathbf{n}_o, \delta(S^T \mathbf{c}_o)>}{<\mathbf{n}_o^*, F\mathbf{n}_o>} \tag{23}$$

The first term at the right side closely resembles the reactivity expression for critical systems. So, we shall call a quantity δK_λ as given by expression (23) a "λ-mode generalized reactivity". The second term may be defined the "source reactivity", whereas the last one a "direct term".

2.3. Point kinetics

Let us now consider equations governing the neutron flux ϕ ($\equiv V\mathbf{n}$) and precursor m_i ($i=1,2,...,I$) in a multigroup (G groups) neutron energy scheme:

$$V^{-1}\frac{d\phi}{dt} = A\phi + (1-\beta)\chi_P S_f \phi + \chi_D \mathbf{u} \sum_{i=1}^{I} \lambda_i m_i + \mathbf{s}_n \tag{24}$$

$$\frac{dm_i}{dt} = \beta_i \nu \Sigma_f^T \phi - \lambda_i m_i \tag{25}$$

where A is the transport, capture and scattering matrix operator, V the diagonal neutron velocity matrix, **u** is a unit (G component) vector and

$$S_f = \begin{vmatrix} \nu\Sigma_{f,1} & \cdots & \nu\Sigma_{f,G} \\ \cdots & \cdots & \cdots \\ \nu\Sigma_{f,1} & \cdots & \nu\Sigma_{f,G} \end{vmatrix}, \quad \begin{cases} \Sigma_f^T = \begin{vmatrix} \Sigma_{f,1} & \cdot\cdot & \Sigma_{f,G} \end{vmatrix} \\ \chi_Z = \mathrm{diag}\begin{vmatrix} \chi_{Z,1} & \cdot\cdot & \chi_{Z,G} \end{vmatrix} \end{cases}$$

The neutron and precursor importances, at unperturbed conditions, obey equations

$$A_o^* \mathbf{n}_{s,o}^* + \nu S_{f,o}^T \left[(1-\beta)\chi_P + \beta\chi_D\right] \mathbf{n}_{s,o}^* + \frac{\gamma}{W_o}\Sigma_{f,o} = 0 \tag{26}$$

$$m_{s,i,o}^* \equiv m_{s,o}^* = \mathbf{u}^T \chi_D \mathbf{n}_{s,o}^*, \tag{27}$$

γ being the number of energy units per fission and W_o the power at stationary, unperturbed conditions. Considering the "normalized power"* P and defining:

$$\ell_{eff} = \frac{<\mathbf{n}_{s,o}^*, V^{-1}\phi_o>}{<\mathbf{n}_{s,o}^*, \bar{\chi} S_f \phi_o>} \qquad \text{(effective prompt neutron lifetime)} \tag{28}$$

---

* For simplicity of notation, the normalized power P is intended here to be the ratio $\frac{W}{W_o(1+q)} \equiv \frac{\gamma <\Sigma_{f,o}, \phi>}{\gamma <\Sigma_{f,o}, \phi_o>}$, where W_o is the power at unperturbed, nominal conditions, γ is the number of energy units per fission, and $q = \frac{<\delta\Sigma_f, \phi>}{<\Sigma_{f,o}, \phi>} \cong \frac{<\delta\Sigma_f, \phi_o>}{<\Sigma_{f,o}, \phi_o>}$.
In order to have the correct power W, the resulting value P should be multiplied by $W_o(1+q)$.

$$\rho_{gen} = \frac{<\mathbf{n}^*_{s,o}, \left\{ \delta A + (1-\beta)\chi_P + \beta\chi_D]\delta S_f \right\} \phi_o >}{<\mathbf{n}^*_{s,o}, \bar{\chi} S_f \phi_o >} \quad \text{(generalized reactivity)} \tag{29}$$

$$\rho_{source} = \frac{<\mathbf{n}^*_{s,o}, \delta s_n >}{<\mathbf{n}^*_{s,o}, \bar{\chi} S_{f,o}\phi_o >}, \quad \alpha = \frac{<\mathbf{n}^*_{s,o}, \chi_D S_{f,o}\phi_o >}{<\mathbf{n}^*_{s,o}, \bar{\chi} S_{f,o}\phi_o >}, \quad \zeta = \frac{1}{<\mathbf{n}^*_{s,o}, \bar{\chi} S_{f,o}\phi_o >}, \quad \xi_i = \frac{<m^*_{s,o} m_i >}{<\mathbf{n}^*_{s,o}, \bar{\chi} S_{f,o}\phi_o >},$$

with $\xi_i = \beta_i/\lambda_i$ at steady state conditions, and $\bar{\chi} = (1-\beta)\chi_P + \beta\chi_D$, the following point kinetic equations result

$$\ell_{eff} \frac{dP}{dt} = (\rho_{gen} - \alpha\beta)P + \alpha \sum_{i=1}^{I} \lambda_i \xi_i + \zeta(1-P) + \rho_{source} \tag{30}$$

$$\frac{d\xi_i}{dt} = \beta_i P - \lambda_i \xi_i \, . \tag{31}$$

It is interesting also to note that, with the system approaching criticality, quantity ζ vanishes. Consequently, the third term at the right side of Eq. (30) also vanishes (whereas the space distribution of $\mathbf{n}^*_{s,o}$ approaches the standard adjoint flux ϕ^*_o (Gandini, 1997). In this case, Eqs. (30) and (31) reduce to the homogeneous, standard form of the point kinetics equations.

Asymptotically, if after the perturbation the system is still subcritical, a new (relative) power P_{as} will be reached, given by the expression $P_{as} = \frac{\zeta + \rho_{source}}{\zeta - \rho_{gen}}$ which, as expected, increases with ρ_{source} and ρ_{gen}.

Quantity ζ plays the role of a measure of the system subcriticality. In fact, if the multiplication coefficient

$$K_{sub} = \frac{<\mathbf{n}^*_{s,o}, \bar{\chi} S_{f,o}\phi_o >}{1 + <\mathbf{n}^*_{s,o}, \bar{\chi} S_{f,o}\phi_o >} \tag{32}$$

is defined, we may write

$$\zeta = (1 - K_{sub})/K_{sub} \, .$$

It was shown (Gandini, 1997) that for K_{sub} approaching unity, function $\mathbf{n}^*_{s,o}$ increases asymptotically, its space shape approaching that of the standard adjoint flux. Correspondingly, ρ_{gen} converges to the standard form of reactivity.

We have seen that the quantity ρ_{gen} plays a role analogous to that of the reactivity in the point kinetics equation for critical systems. We may also verify that this quantity, for the same parameter perturbation, gives a decreasing contribution to the power change with the system subcriticality increasing. This is due to the presence of the source-related term $\zeta(1\text{-}P)$ at the right side of Eq.(30), where ζ increases with the subcriticality.

As we have seen, the coefficients appearing in Eqs. (30) and (31) are all physically meaningful. The generalized reactivity, ρ_{gen}, in particular, may be determined by measurement. In fact, as shown in the previous section, it is given by the product of the generalized reactivity associated with the source control,

$$\rho_{gen,s} = <\mathbf{n}^*_{s,o}, (\delta A + \bar{\chi}\delta S_f)\phi_o > + \frac{\gamma}{W_o} < \delta\Sigma_f, \phi_o > \tag{33}$$

multiplied by the quantity ζ, given above. Since $\rho_{gen,s}$ corresponds to the source strength change necessary to reset the power level after the perturbation, it is clearly a measurable quantity. For what concerns ζ, this quantity doesn't seem easily amenable to experimental evaluation. It seems easier to determine its counterpart, with $\mathbf{n}^*_{s,o}$ replaced by standard adjoint function $\boldsymbol{\phi}^*_o$, for instance, via fundamental mode period measurements. ζ could be then evaluated by multiplying its calculated value by a bias factor, i.e.,

$$\zeta^{exp} = \zeta^{cal} \left(\frac{1-K_{eff}}{K_{eff}}\right)^{exp} \Bigg/ \left(\frac{1-K_{eff}}{K_{eff}}\right)^{cal} \tag{34}$$

In above expressions we have assumed, for simplicity of presentation, constant values for the delayed neutron fractions β and β_j. In reality these quantities are generally dependent on energy and space, in correspondence to the space distribution of the fuel elements composition and of the neutron energy spectrum. To take these dependencies into due account, in Eqs. (30) and (31) the following the "effective" values should be considered:

$$\beta_{i,eff} = \frac{\sum_{g=1}^{G}\sum_{j=1}^{J} \langle n^*_{s,o,g} c_j \chi^j_{D,g} \beta^j_{i,g} \nu\sigma^j_{f,g} \phi_g \rangle}{\sum_{g=1}^{G}\sum_{j=1}^{J}\sum_{i=1}^{I} \langle n^*_{s,o,g} c_j \chi^j_{D,g} \nu\sigma^j_{f,g} \phi_g \rangle}, \qquad \beta_{eff} = \sum_{i=1}^{I} \beta_{i,eff} \tag{35}$$

2.3.1. Example. A simple subcritical thermal system has been considered with: K_{eff}=0.95, $\ell_{eff} = 10^{-3}$, λ=0.3, β=0.007. A number of cases have been evaluated relevant to different reactivity insertions as shown in Figg. 1, 2, 3 and 4 (showing P vs. sec).

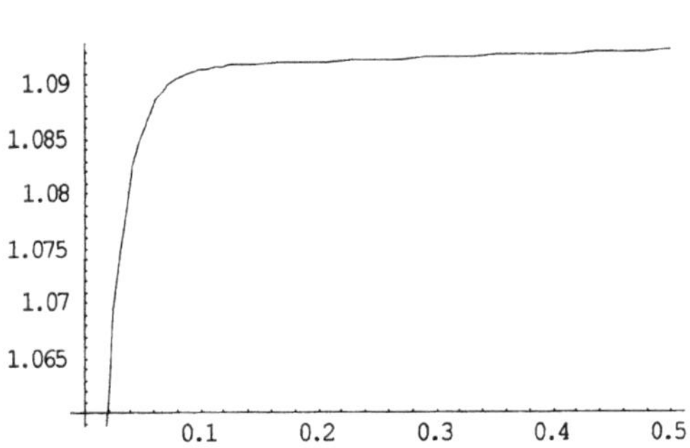

Fig. 1. ρ_{gen} = 0.005 (asymptotic value: P=1.11

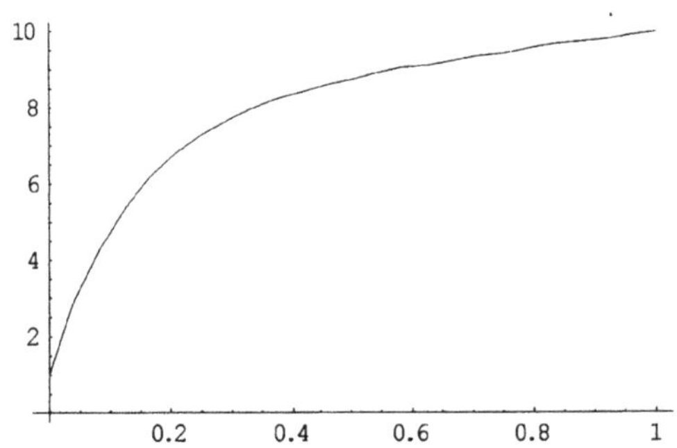

Fig. 2. ρ_{gen} = 0.0526 (critical conditions)

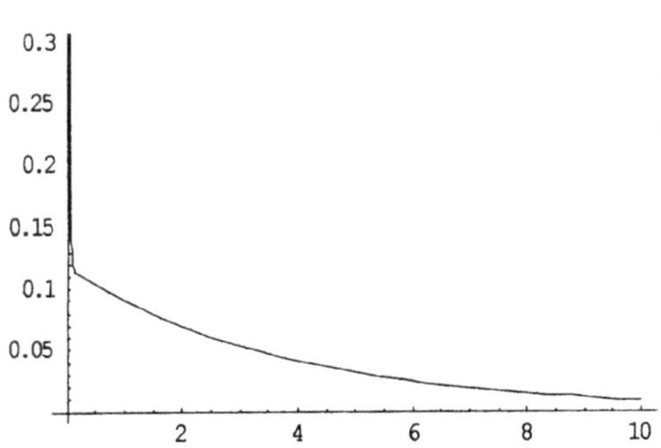

Fig. 3. ρ_{source} = -1 (source removal)

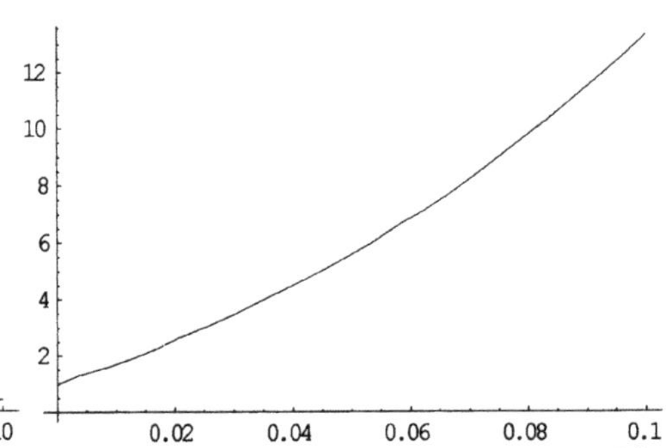

Fig. 4. ρ_{gen} = 0.07 (over prompt critical)

REFERENCES

Bowman C. (1992) *Nucl.Instr.Meth,.* **A320**, 336

Gandini, A. (1967) *J. Nucl.Energy*, **21**, 755.

Gandini A. (1969) *Nucl.Sci.Eng.*, **35**, 141.

Gandini A. (1976) *Nucl.Sci.Eng.*, **59**, 60.

Gandini A. (1981) *Nucl.Sci.Eng.*, **77**, 316.

Gandini A. (1983) *Trans.Am.Nucl.Soc.*, **45**, 325.

Gandini A. (1987a) *Adv.Nucl.Sci.Techn.*, **19**, 205.

Gandini A. (1987b) *Ann. Nucl. Energy*, **14**, 273.

Gandini A. (1988) *Ann. Nucl. Energy*, **15**, 327.

Gandini A. (1997) *Ann. Nucl. Energy*, **24**, 1241,and IAEA -TC-903.3 p377 [Proc. Madrid Conf. on ADS].

Gandini A., M. Salvatores M., I. Slessarev I. (1999) *Annals of N.E.,* **27** (1), 71.

Gandini A., M. Salvatores M., I. Slessarev I. (2000) *Annals of N.E.,* **27**(13) 1147.

Gandini A. (2001) *Ann. Nucl. Energy,* **28**, 1193 [see also: *Ann. Nucl. Energy,* "On the Evaluation of ADS Subcriticality" (in print)].

Orlov V.,Slessarev I. (1988) *Proc.Int.Conf. "Safety New Generat.Power React",* Seattle, v.1, p.742.

Palmiotti P. Feldman E., *Trans. ANS* , **81**, 279 (1999)

Raepsaet X.,Damian F.,Lecompte M.,Lenain R. (1999) "Fuel Cycle in HTR" Global'99, Tokyo

Salvatores M., Slessarev.I, Uematsu M. (1995) Nucl.Sci.Eng., **116**, 1.

Slessarev I., Berthou V., Salvatores M. (1999) "Concept of the Thorium Fuelled ADS", Prague Conf'99.

Wade D., "LMR Core Design for Inherent Safety", *NEACRP Meeting Proceed.,* Paris, Sept. 1986.

Pergamon

www.elsevier.com/locate/pnucene

Progress in Nuclear Energy, Vol. 40, No. 3–4, pp. 673–680, 2002

Printed in Great Britain
0149-1970/02/$ - see front matter

PII: S0149-1970(02)00065-3

DATABASE DEVELOPMENT FOR ANALYSIS OF ACCELERATOR-DRIVEN SYSTEMS

Yu.A.KOROVIN, A.Yu.KONOBEYEV, P.E.PERESLAVTSEV

Obninsk Institute of Nuclear Power Engineering (INPE)
Studgorodok 1, 249020, Obninsk, Russia
Tel: +7-08439-38320 Fax: 7-08439-36931
Email: korovin@iate.obninsk.ru

ABSTRACT

Nuclear data libraries for study of transmutation, activation and neutron transport in materials of accelerator-driven systems are advocated. The libraries include nuclear data describing processes in materials irradiated by intermediate and high-energy particles and cover wide range of target nuclei and incident particle energies. © 2002 Published by Elsevier Science Ltd.

1. INTRODUCTION

In accelerator-driven subcritical systems for fission energy production and/or nuclear waste transmutation, as well as in intermediate-energy accelerator driven neutron sources, ions and neutrons with energies beyond 20 MeV will interact with materials. For analyses of these interactions it is necessary to elaborate new databases containing information about nuclear reaction characteristics in the energy range above 20 MeV. Such data libraries must include the data for structural materials as well as the most important actinides considered for transmutation in sub-critical systems. One of the first files covering the energy range 0 – 50 MeV for incident neutrons has been elaborated in 1995 at the Institute of Nuclear Power Engineering (INPE) for ^{56}Fe. This and similar libraries (Koning *et al*, 1998) for other nuclides are currently applied in engineering calculations.

The present paper describes the recent progress in elaboration of intermediate energy databases in INPE, as well as the principles and methods to obtain the data.

To study particle transport and radiation characteristics for materials irradiated by fast neutrons a set of nuclear data files covering the energy region 0-50 MeV has been prepared. Also, the data library BISERM-2 (Konobeyev *et al*, 1992) has been elaborated to study radiation damage and gas production in structural materials irradiated by high-energy neutrons. It contains the displacement cross-sections, p-, ^3He- and α-production cross-sections for the energy range up to 1 GeV.

To perform activation calculations the data library IEAF-2000 has been created. The library includes data for neutron induced reaction cross-sections for elements with Z = 1 – 84 at energies up to 150 MeV. As the basis the data from the EAF-97 data library (Sublet *et al*, 1997) was taken in the energy range below 20 MeV.

For transmutation studies the data library WIND containing cross-sections for neutron and proton induced reactions on actinides at energies up to 100 MeV has been prepared.

2. DATA FILES TO STUDY NEUTRON TRANSPORT IN MATERIALS AT THE INTERMEDIATE ENERGIES

To study neutron transport, nuclear heating, gas production and radiation damage for materials irradiated by fast neutrons the nuclear data files have been prepared recently. The files include information about total cross-sections, elastic and inelastic scattering cross-sections, cross-sections for threshold reactions, energy and angular distributions for secondary neutrons, protons and α-particles and, for some elements, spectra of photons produced in the reactions. These characteristics have been obtained for ^{12}C, ^{16}O, ^{23}Na, ^{28}Si, ^{39}K, ^{51}V, ^{52}Cr, ^{56}Fe, ^{208}Pb, ^{209}Bi and for fissile nuclides: ^{232}Th, ^{233}Pa, ^{233}U, ^{238}U, ^{239}Pu, ^{240}Pu at the energies 0 – 50 MeV and for ^{235}U in the energy range 0 – 300 MeV.

Total cross-section, differential cross-sections for elastic and discrete inelastic scattering were obtained on the basis of coupled-channels model realized in the ECIS code (Raynal, 1988). The optical model parameters were taken from different origins (Koning *et al*, 1998; Korovin *et al*, 1997) and slightly corrected to include the discrete levels in calculations and to achieve an agreement with available neutron data at energies above 20 MeV.

Neutron induced threshold reaction cross-sections and double differential cross-sections for nonelastic reactions were obtained using the geometry dependent hybrid exciton and evaporation models. The numerical calculations were performed by the modified ALICE code (Blann, 1988). The principal changes were done to incorporate the algorithms for description of pre-equilibrium cluster emission (d ,t, ^{3}He, α) (Konobeyev and Korovin, 1994; Konobeyev and Korovin, 1996,a), pre-compound γ-ray emission (Oblozinsky, 1988; Konobeyev *et al*, 1997,a) and for calculation of the nuclear level density according to the different approaches (Konobeyev et al, 1997,b).

The calculations of pre-equilibrium nucleon spectra were performed on the basis of GDH model taking into account the multiple nucleon emission using the approach from Blann and Vonach (1983). Exciton state density was calculated according to the Ericson-Strutinsky formula considering the Pauli principle and pairing of nuclear levels. To define intranuclear transition rate (λ_+) nucleon-nucleon interaction cross-sections corrected for the Pauli principle were used. Normalization factor for calculating (λ_+) was taken to be unity. Inverse reaction cross-sections were calculated through the optical model. For the state with three excitons the level density calculations were performed using the formula (Betak and Dobes, 1976) taking into account the final depth of potential well.

The non-equilibrium complex particle emission spectra were calculated in the framework of coalescence pick-up model (Iwamoto and Harada, 1982; Sato *et al*, 1982) combined with the hybrid exciton model as shown in (Konobeyev and Korovin, 1994). Contribution of direct processes to deuteron emission spectra has been obtained by phenomenological model developed in (Konobeyev and Korovin, 1996,a).

Equilibrium spectra were calculated using evaporation Weisskopf-Ewing model. The generalized superfluid nuclear model (Ignatyuk *et al*, 1979) has been applied to describe the level density.

The method for threshold reaction cross-section calculations was tested using experimental data for neutron induced reactions contained in EXFOR and available measured cross-sections for proton induced reactions (p,xnypzα).

The empirical (Yao Lishan, 1992; Yao Lishan and Jin Juling, 1992; Qaim and Wolfle, 1978) and the semi-empirical systematics (Konobeyev *et al*, 1994; Konobeyev and Korovin, 1995; Konobeyev *et al*, 1996; Dityuk *et al*, 1996; Konobeyev and Korovin, 1996,b) at the energy around 14.5 MeV have been used to evaluate the cross-sections for some reactions. Additional correction has been made so that the obtained data agree with the data from known data libraries at low energies.

Single radiative transitions (Oblozinsky, 1988) were assumed to be the main origin for non-equilibrium γ-rays. To obtain the possible contribution of the "quasi-deuteron" mechanism for γ-emission the energy distributions ($d\sigma^{qd}/d\varepsilon_\gamma$) for photons emitted in the elementary $n+p \to d+\gamma$ interactions of primary particle and nucleons of the nucleus have been calculated. Then the corresponding emission rate and γ-spectra have been obtained using the formalism of the hybrid exciton model. The calculation of $d\sigma^{qd}/d\varepsilon_\gamma$ has been performed taking into account Pauli exclusion principle and other effects connected with the influence of the nuclear matter on the n+p interactions. The detailed description of the approach is given in (Konobeyev *et al*, 1997).

For particular reactions (n,n'), (n,p), (n,α) at the energies up to (n,2n) reaction threshold the ALICE code has been used to obtain recoil spectra for residual nuclei ($d\sigma/d\varepsilon_R$). At the energies above 11 MeV the DISCA code (Konobeyev and Korovin, 1992; Artisyuk *et al*, 1993; Konobeyev and Korovin, 1993; Konobeyev and Korovin, 1998; Korovin *et al*, 1992; Robinson, 1968) based on modified intranuclear cascade evaporation model has been applied to calculate $d\sigma/d\varepsilon_R$.

The data are recorded in the ENDF-6 format. File MF=1, 2, 3, 4, 5, 6, 12, 14 and 15 are used to record the data.

Summary for intermediate energy neutron transport data available for ADS project is presented in Table 1.

Table 1. Neutron transport files for ADS project

Nuclide	Energies up to 50 MeV	Energies up to 150 MeV
^{16}O	INPE, LA	LA
^{28}Si	INPE,LA	LA
^{29}Si	LA	LA
^{30}Si	LA	LA
^{50}Cr	LA	LA
^{52}Cr	INPE, LA	LA
^{53}Cr	LA	LA
^{54}Cr	LA	LA
^{54}Fe	LA	LA
^{56}Fe	INPE, LA	LA
^{57}Fe	LA	LA
^{58}Ni	LA	LA
^{60}Ni	LA	LA
^{61}Ni	LA	LA
^{62}Ni	LA	LA
^{64}Ni	LA	LA
^{206}Pb	LA	LA
^{207}Pb	LA	LA
^{208}Pb	INPE, LA	LA
^{209}Bi	INPE	INPE
^{235}U	INPE	INPE (up to 300 MeV)
^{238}U	INPE, LA, IPPE	IPPE
^{239}Pu	INPE	-
^{240}Pu	INPE	-

3. DATA LIBRARY BISERM-2 TO STUDY RADIATION CHARACTERISTICS OF IRRADIATED MATERIALS

Cross-section data library BISERM-2 to study radiation effects induced by intermediate energy neutrons has been elaborated. The library contains neutron displacement cross-sections, hydrogen and helium production cross- sections for 259 stable nuclei from ^{27}Al to ^{209}Bi. Hydrogen and helium production cross-sections are given at the energies from the threshold of (n,p), (n,α) and (n,^{3}He) reactions up to 1 GeV. Displacement cross-sections are recorded from 10 MeV to 1 GeV.

The first version of BISERM is described in (Konobeyev and Korovin, 1992; Korovin *et al*, 1992).

The Robinson function (Robinson, 1968) has been used for displacement cross-section calculations. Neutron displacement cross-sections for elastic scattering (σ_{del}) were calculated up to the energy of 150 MeV. At the energies above 150 MeV, the approximate approach developed in (Konobeyev *et al*, 1991) was used to obtain elastic displacement cross-sections for neutrons.

Displacement cross-sections for nonelastic interactions (σ_{dnon}) were obtained with the help of the modified intranuclear cascade evaporation model using the DISCA code.

The calculation of hydrogen (σ_H) and helium (σ_{He}) production cross-sections at the energies up to 100 MeV has been performed on the basis of the geometry dependent hybrid exciton and evaporation models (Blann and Vonach, 1983). The precompound α-particle spectra were obtained using the coalescence pick-up model.

To obtain σ_H and σ_{He} values at the energies up to 1 GeV, the intranuclear cascade evaporation model has been used. The nonequilibrium α-particle production was taken into account using the approach advocated by Denisov and Mekhedov (1972).

To obtain σ_H and σ_{He} values in agreement with experimental data below 20 MeV the correction of the calculated (n,p) and (n,α) cross-sections has been performed using the EXFOR data and the systematics predictions at 14.5 MeV. For some nuclei the (n,p) and (n,α) cross-sections were taken from ADL-3 (Grudzevich *et al*, 1993) and other libraries (Konobeyev *et al*, 1992, Shubin *et al*, 1995) at the energies below the maximum of excitation functions. Calculated cross-sections for the reaction (n,^{3}He) were corrected on the basis of the available experimental data.

The Fig.1 shows the example of data presented in the BISERM-2 library.

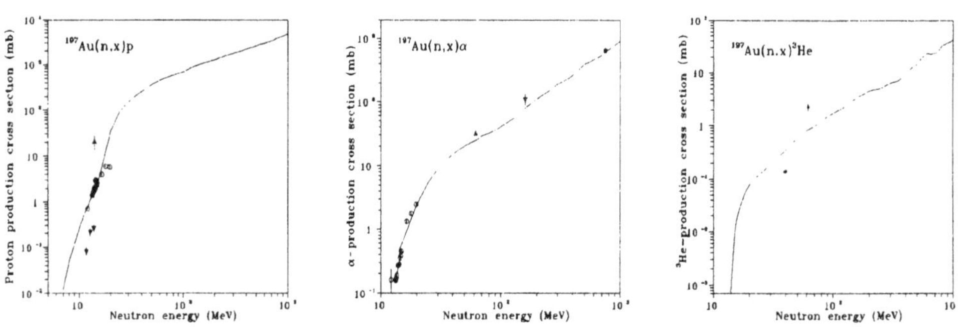

Fig.1. Examples of proton, alpha and ^{3}He production cross-sections presented in BISERM-2 (Konobeyev et al, 1992) library.

4. DATA LIBRARY IEAF-2000 FOR ACTIVATION STUDIES

The library is intended for activation and transmutation study for materials irradiated by intermediate energy neutrons. IEAF-2000 contains evaluated neutron induced reaction cross-sections at the energies 0-150 MeV for 675 stable and unstable nuclei from H to Po.

Approximately 50.000 excitation functions are included in the library.

The earlier library versions have been described in (Konobeyev *et al*, 1992; Shubin *et al*, 1995).

The basic differences of IEAF-2000 from MENDL-1 and MENDL-2 activation data libraries (Konobeyev *et al*, 1992; Shubin *et al*, 1995) are following

- the unification of the results of evaluation above 20 MeV with EAF-97 data files (Sublet *et al*, 1997) at the lower energies
- including in the library the radiative capture cross-sections from EAF-97
- performing of the new evaluation for complex particle emission based on pick-up and knock-out models (Konobeyev *et al*, 1995)
- using different models for the level density calculation appropriate for different mass regions
- including in the files of IEAF-2000 the proton, deuteron, triton, ^{3}He and α-particle production cross-sections
- performing of the evaluation for nuclei with atomic number Z = 1-12 not included in MENDL-1 and MENDL-2 versions

The IEAF-2000 data are written in the ENDF-6 format combining MF=3,6 MT=5 data recording.

5. DATA LIBRARY WIND TO STUDY TRANSMUTATION OF ACTINIDES

To study activation and transmutation of uranium and transuranics irradiated by fast neutrons the library WIND (Konobeyev *et al*, 1995; Korovin *et al*, 1995; Korovin *et al*, 1994) has been created. The library contains evaluated neutron induced fission cross-sections, cross-sections for threshold reactions (n,xn), (n,pxn) and (n,αxn) for uranium, neptunium and plutonium isotopes at energies from 0 to 100 MeV. The WIND includes the cross-sections for 576 reactions taking place in neutron irradiation of ^{232}U, ^{233}U, ^{234}U, ^{235}U, ^{236}U, ^{237}U, ^{238}U, ^{237}Np, ^{239}Np, ^{236}Pu, ^{237}Pu, ^{238}Pu, ^{239}Pu, ^{240}Pu, ^{241}Pu, ^{242}Pu, ^{243}Pu, ^{244}Pu. Also the library includes the special file for ^{239}Pu fission products at the energies up to 1 GeV (Korovin *et al*, 1996).

The detailed description of method to obtain cross-sections contained in the library is given in (Konobeyev et al, 1994).

6. NUCLIDE COMPOSITION OF Pb-Bi COOLANT IRRADIATED IN 80 MW SUBCRITICAL REACTOR

Computer code package CASCADE/INPE (Barashenkov et al, 1999; Konobeyev et al, 1999) including new version of CASCADE code, MCNP/4B code (Briesmeister, 1997) and other routines has been used for the particle spectra calculation for the sub-critical unit.

The data described in Chapter 3 were used for nuclide composition and activity calculations performed with the help of the SNT code (Korovin et al, 1992).

The amounts of bismuth and lead radioactive isotopes are presented in Table 2.

Table 2. Mass of Bi and Pb radioactive isotopes (grams) produced after 1 year of irradiation in Pb-Bi coolant in all reactor. Data correspond to the end of irradiation.

Bismuth isotopes	Mass, g	Lead isotopes	Mass, g
83 210	2.009E+02	82 210	3.095E-06
83 210m	4.018E+03	82 205	1.434E+03
83 208	1.672E+02	82 203	3.128E-01
83 207	5.967E+01	82 202	2.815E+01
83 206	8.078E-01	82 201	2.902E-02
83 205	1.754E+00	82 200	5.642E-02
83 204	3.210E-02		
83 203	3.020E-02		

7. NEUTRON MULTIPLICATION FACTOR AND NEUTRON YIELD CALCULATIONS

Neutron multiplication factor has been calculated with the help of MCNP/4B code for EAP-80 facility using different neutron data libraries. Neutron data for MCNP were prepared in the present work with the help of NJOY/94 code. The results of k_{eff} estimation are given in Table 3.

Comparison of the data presented in Table 3 shows the big discrepancies in k_{eff} values calculated with the help of different data libraries. For this reason the choice of appropriate data library for neutron transport calculations is very much essential in criticality analysis.

The k_{eff} *value* calculated with the help of the different data libraries

Library	Full data set is used for calculation	Data for Pb and Bi isotopes are taken from below mentioned libraries, other ones are from ENDF/B-VI
ENDF/B-VI	0,918 ± 0,001	
ENDF/B-V	0,896 ± 0,001	
BROND-2.2	0,906 ± 0,001	0,921 ± 0,001
ENDL-85	0,865 ± 0,001	0,889 ± 0,001
JENDL-3.2	0,913 ± 0,001	0,911 ± 0,001

8. CONCLUSIONS

Different intermediate energy nuclear data libraries are elaborated and briefly reported.

To study neutron transport, heating, gas production and radiation damage for materials irradiated by intermediate energy neutrons the nuclear data files for ^{12}C, ^{16}O, ^{23}Na, ^{28}Si, ^{39}K, ^{51}V, ^{52}Cr, ^{56}Fe, ^{208}Pb, ^{209}Bi, ^{232}Th, ^{233}Pa, ^{233}U, ^{238}U and ^{239}Pu have been prepared at the energies 0-50 MeV.

The cross-section data library BISERM containing neutron displacement cross-sections and hydrogen and helium production cross-sections for 259 stable nuclei from ^{27}Al to ^{209}B at the energy up to 1 GeV has been elaborated.

The library IEAF-2000 intended for activation and transmutation study for materials irradiated by intermediate energy neutrons has been realized. The library contains evaluated neutron induced reaction cross-sections at the energies 0-150 MeV for 675 stable and unstable nuclei from H to Po.

To study activation and transmutation of transactinides irradiated by neutrons of energies up to 100 MeV the library WIND has been created.

The data obtained were used for the nuclide composition calculation for an 80 MW subcritical reactor.

Special study has been performed to obtain neutron multiplication factors l using different nuclear data libraries. The results show the big discrepancies between k_{eff} *values* calculated with ENDF/B, BROND and ENDL data.

REFERENCES

Artisyuk V.V., Konobeyev A.Yu. and Korovin Yu.A. (1993), *Kerntechnik 58, 174.*

Barashenkov V.S., Konobeyev A.Yu., Korovin Yu.A., Sosnin V.N. (1999), Computer Software Package "CASCADE/INPE", *Atomnaya Energiya, 87, 283.*

Betak E., and Dobes J. (1976), *Z.Phys. A279, 319.*

Blann M. and Vonach H.K. (1983), *Phys. Rev. C28, 1475.*

Briesmeister Ju.F. (1997), Report LA-12625-M, March 1997.

Denisov F.P. and Mekhedov V.N. (1972), *Nuclear Reactions at High Energies*, Atomizdat, Moscow.

Dityuk A.I., Konobeyev A.Yu., Lunev V.P., Shubin Yu.N. (1996), *Voprosy Atomnoi Nauki i Techniki. Ser.: Yadernije Konstanti (Problems of Nuclear Science and Technology. Series: Nuclear Data), 1, 129.*

Grudzevich O.T., Zelenetsky A.V., Ignatyuk A.V., Pashchenko A.B. (1993), *Voprosy Atomnoi Nauki i Techniki. Ser.: Yadernije Konstanti (Problems of Nuclear Science and Technology. Series: Nuclear Data) 3-4, 3.*

Ignatyuk A.V., Istekov K.K. and Smirenkin G.N. (1979), *Yadernaja Fizika 29, 875.*

Iwamoto A. and Harada K.(1982), *Phys. Rev. C26, 1821.*

Koning A.J., Delaroche J.-P., and Bersilon O. (1998), Nuclear data for accelerator driven systems: Nuclear models, experiments and data libraries. *Nucl. Instr. Meth. A , 49.*

Konobeyev A.Yu. and Korovin Yu.A. (1992), *J. Nucl. Mater, 195, 286.*

Konobeyev A.Yu. and Korovin Yu.A. (1993), *Nucl. Instr. Meth. Phys. Res. B82, 103.*

Konobeyev A.Yu. and Korovin Yu.A. (1994), *Kerntechnik 59, 72.*

Konobeyev A.Yu. and Korovin Yu.A. (1995), *Nucl. Instr. Meth, B103, 15.*

Konobeyev A.Yu. and Korovin Yu.A. (1996,a), *Kerntechnik 61, 45.*

Konobeyev A.Yu. and Korovin Yu.A. (1996,b), *Izvesitja Vuzov. Ser.: Yadernaja Energetika (Transactions of High School. Ser.:Nuclear Power Engineering), 2, 14.*

Konobeyev A.Yu. and Korovin Yu.A. (1998), *Kerntechnik 63, 124.*

Konobeyev A.Yu., Korovin Yu.A and Pereslavtsev P.E. (1994), *Nucl. Instr. Meth, B93, 409.*

Konobeyev A.Yu., Korovin Yu.A. and Sosnin V.N (1991), *Vestsi Akademii Navuk BSSR. Ser. Fiz.-Energ. Navuk (Transactions of Byelorussian Academy of Sciences, Phys. and Energ. Series) 1, 29.*

Konobeyev A.Yu., Korovin Yu.A., and Sosnin V.N. (1992), *J. Nucl. Mater. 186, 117.*

Konobeyev A.Yu., Korovin Yu.A., Lunev V.P, Masterov V.S, Shubin Yu.N. (1992), *Voprosy Atomnoi Nauki i Techniki. Ser.: Yadernije Konstanti (Problems of Nuclear Science and Technology. Series: Nuclear Data) 3-4, 55.*

Konobeyev A.Yu., Korovin Yu.A., Pereslavtsev P.E. (1997,a), *Izvestija Vuzov. Ser.: Yadernaja Energetika (Transactions of High School. Ser.:Nuclear Power Engineering) 1, 2.*

Konobeyev A.Yu., Korovin Yu.A., Pereslavtsev P.E. (1997,b), *Code ALICE/ASH for Calculation of Excitation Functions, Energy and Angular Distributions of Emitted Particles in Nuclear Reactions*, Report INPE, Obninsk, February 1997.

Konobeyev A.Yu., Korovin Yu.A., Pereslavtsev P.E., Plyaskin V.I., Stankovsky A.Yu. (1995), Report IAEA, INDC(CCP)-384.

Konobeyev A.Yu., Korovin Yu.A., Plyaskin V.I. (1994), *Kerntechnik 59, 87.*

Konobeyev A.Yu., Korovin Yu.A., Sosnin V.N, Vecchi M. (1999), Study of Accelerator-Driven Reactor Systems, *Kerntechnik 64, 216.*

Konobeyev A.Yu., Lunev V.P. and Shubin Yu.N. (1995), *Acta Physica Slovaca 45, 705.*

Konobeyev A.Yu., Lunev V.P. and Shubin Yu.N. (1996), *Nucl. Instr. Meth, B108, 233.*

Korovin Yu.A, Konobeyev A.Yu., Pereslavtsev P.E. (1992), *Voprosy Atomnoi Nauki i Techniki. Ser. Yadernije Konstanti, (Problems of Nuclear Science and Technology, Series: Nuclear Data) 3-4, 117.*

Korovin Yu.A., Konobeyev A.Yu. and Sosnin V.N. (1992), *Kerntechnik 57, 188.*

Korovin Yu.A., Konobeyev A.Yu., Lunev V.P., Pereslavtsev P.E., Stankovsky A.Yu. (1996), *Neutron Activation Data for ^{239}Pu at the Energies up to 1 GeV*, Report INPE, Obninsk.

Korovin Yu.A., Konobeyev A.Yu., Pereslavtsev P.E., Plyaskin V.I. (1994), In: *Proc. Int. Conf. Nuclear Data for Science and Technology*, p.655. Gatlinburg, USA, May.

Korovin Yu.A., Konobeyev A.Yu., Pereslavtsev P.E., Plyaskin V.I., Stankovsky A.Yu. (1995), *Prog. Nucl. Energy 29 (Supplement), 297.*

Korovin Yu.A., Konobeyev A.Yu., Pereslavtsev P.E., Stankovsky A.Yu., Broeders C., Broeders I., Fischer U., v.Möllendorff U., Wilson P., Woll D. (1997), *Proc. Int. Conf. Nuclear Data for Science and Technology*, p.851. Trieste, Italy, May 9-24.

M.Blann (1988), Report IAEA-NDS-93 REV.O.

P.Oblozinsky (1988), *Phys. Lett. B215, 597.*

Qaim S.M. and Wolfle R. (1978), *Nucl. Phys. A295, 150.*

Raynal J. (1988), *Proc. Workshop on Applied Nuclear Theory and Nuclear Model Calculations for Nuclear Technology Applications*, p.506. Trieste, Italy.

Robinson M.T. (1968), *Philos. Mag. 17, 639*; and in: *Radiation Induced Voids in Metals*, eds.J.W.Corbett, L.C.Ianiello (National Technical Information Services, CONF-710601, 1972) p.397.

Sato N, Iwamoto A. and Harada K. (1982), *Phys. Rev. C28, 1527.*

Shubin Yu.N., Lunev V.P., Konobeyev A.Yu., Dityuk A.I. (1995), Report IAEA, INDC (CCP)-385.

Sublet J.-Ch., Kopecky J., and Forrest R.A. (1997), The European Activation File: EAF-97. *Proc. Int. Conf. Nuclear Data for Science and Technology*, p.1145. Trieste, Italy, May 9-24.

Yao Lishan (1992), *Communication of Nuclear Data Progress 7, 85.*

Yao Lishan and Jin Yuling (1992), *Communication of Nuclear Data Progress 7, 95.*

Pergamon

www.elsevier.com/locate/pnucene

Progress in Nuclear Energy, Vol. 40, No. 3–4, pp. 681–688, 2002

0149-1970/02/$ - see front matter

PII: S0149-1970(02)00066-5

DEEP UNDERGROUND TRANSMUTOR (PASSIVE HEAT REMOVAL OF LWR WITH HARD NEUTRON ENERGY SPECTRUM)

HIROSHI TAKAHASHI

Brookhaven National Laboratory, Upton, New York, 11973

Tel 631-344-4099,Fax 631-344-7650, e-mail takahash@bnl.gov

ABSTRACT

To run a high conversion reactor with Pu –Th fueled tight fueled assembly which has a long burn-up of a fuel, the reactor should be run in a deep under-ground . Putting the reactor in a deep under ground a reactor heat can be removed passively not only in steady state run and also in an emergency case of loss of coolant and loss of on-site power, and safety of reactor can be improved. Also we can minimize an evacuation area, and can run the reactor near consumer area. This reduces the cost of generating electricity by eliminating the container building and shortening the transmission line. © 2002 Published by Elsevier Science Ltd.

INTRODUCTION

A high conversion light water reactor using high concentration of Pu fuel tight lattice has been proposed[1]. This reactor has a hard neutron energy spectrum closed to Na cooled fast reactor , and high burn up of fuel can be obtained. The reactor with uranium fertile material has positive water coolant void coefficient, so that to get a negative void coefficient it requires the pan-cake type flat core configuration or fuel assembly with neutron streaming void section which reduces the neutron economy. Use of thorium fertile material [2] provides the negative void coefficient without having neutron leaky core configuration, the neutron economy is improved and the higher burn-up of fuel can be obtained more than the reactor with uranium fertile materials. However, the pumping power of water coolant has to be substantially increased to remove the high density heat from tight latticed fueled core. In the steady operation, coolant flow can be maintained by increase of pumping power several time of the regular LWR. When the emergency cases such as outage of on-site power or loss of coolant , the heat removal becomes serious problem, and a detailed analysis of this accident scenario have been studied and the experimental study for heat removal from tight lattice has been planed in Japanese program.

Due to the hard neutron energy spectrum of high concentration of Pu fuel tight lattice and good neutron economy, this reactor can be used for transmuting MA and LLFP. MA has a similar neutronics property as the fertile material of thorium. by capturing the neutrons, MA will be converted into fissile material, and it contributes to long burn up of fuel. To transmute LLFP of Tc which has low neutron capture cross section , the moderator such as zirconium hydride is used in the transmutor cooled by Na or Pb-Bi, however, the neutron can be effectively moderated by light water, the configuration of the light water cooled transmutor becomes simple.

PASSIVE HEAT REMOVAL

To withstand the emergency case of loss of pumping power, the passive cooling system such as heat removal using the natural circulation of coolant has been studied. We proposed here the tight latticed water reactor which is embedded in deep under-ground location so that it is cooled by natural circulation of water. The high pressure difference between inlet and outlet in the narrow water channel of the tight lattice is provided by the difference of gravity force between the low density of boiled water and the high density water condensed after the steam go trough out the steam turbine. To get the high pressure difference the vacuum condenser is located far above the boiling point of the water. The pumping pressure difference to circulate water in the regular BWR and PWR are respectively 2 atm and 1.5 atm which equivalent to 20 –15 meter water height difference, but for our high conversion (HC) LWR with tight lattice, the pumping power difference increases several times, the water height of more than 80-60 meter is needed to naturally circulate coolant water. By putting deep under ground, we can provide enough height space to get high pressure difference between inlet and outlet using the density difference between steam section and the water which is condensed after go through steam turbine and steam condenser which are located far above the reactor vessel.

By deepening the reactor location, the pressure imposed to the pressure vessel is increased by gravitational force of surrounding earth. The water pressure of 100 atm and 150atm for BWR and PWR can be provided respectively by the earth pressure in depth of 400 and 600 meters.

In the passive cooling system using natural circulation which is conventionally proposed, it is operated in the environment which is not enough pressure, the steam - water state is not well defined state, some instability is created, and it is not necessary to be safe operation even in the passive operation. By operating the high enough pressure these nonlinear effect can be eliminated we can obtain the safe operation of the reactor in the deeper under ground. From this point of view, there is many advantage on the concept of the deep underground reactor. Due to earth gravity pressure, the pressure vessel can be made by thin pressure vessel, thus the reactor become much lighter than the regular LWR operated in on the earth surface. It does not requires the hugh heavy crane to move this reactor and can be constructed in the modular type design.

It has been proposed to use super-critical steam for gaining high efficiency of electric generation[3] , this reactor requires 250 atm water pressure, this can be achieved by the earth pressure in the 1000 meter deep under ground: 260 atm. The more high water pressure can be achieved by providing thick pressure vessel .

DEEP UNDER-GROUND FACILITY

Study of deep underground geological storage of a high level waste has been conducted. Although the depth of the Yucca mountaine deep storage is about 300 meter depth, making the tunnel of more than tens of miles are planned.

To measure the neutrino oscillation, the super-kamiokande detector with 50 meter height and 40 meter diameter water tank is installed 1000 meters depth in the mountain in Kamioka mine in Japan, many other high energy facilities such as the Grand Sasso (Italy) have been used for high energy experiments.

The cost of the digging large hall in under ground is not so expensive, we are informed that to cost of digging the about 10x20 meter size tunnel is about 10,000 dollars per 1 meter depth in Japan, although it depend on the geological condition, but the cost of the digging in the hard rock condition in the deep under-ground is the cheaper than the shallow but fractured rock[4].

Figure 1 and 2 show the conceptual lay out of the BWR and PWR installed in deep under ground . the emergency cooling can be installed in the high above reactor, which provide high pressure to cooled the decay heat. In the PWR version the heat exchanger with primary boiling is used to get the high pressure difference between inlet and out let of narrow water channel.

To get high gravitational force, the deep tunneling such as the case of Yucca Mt program can be utilized instead of making deep vertical hole, the reactor can be embedded in the tunneling wall, we can get high pressure of water coolant by earth gravity and putting the steam turbine and vacuum condenser far above the reactor we can get high pressure difference between inlet and out in the narrow cooling water channel so that heat can be removed by natural circulation of water. Due to light reactor without thick pressure vessel, when the fuel exchange (which is infrequent due to long burn up of fuel in the Pu fueled with thorium fertile with tight latticed reactor) is needed , the reactor can be taken out the tunneling wall , exchange of fuel can be done enough space adjacent the tunneling wall, although the disconnection and connection of the tubes of steam and condensed water are required, but it can be done without difficulty.

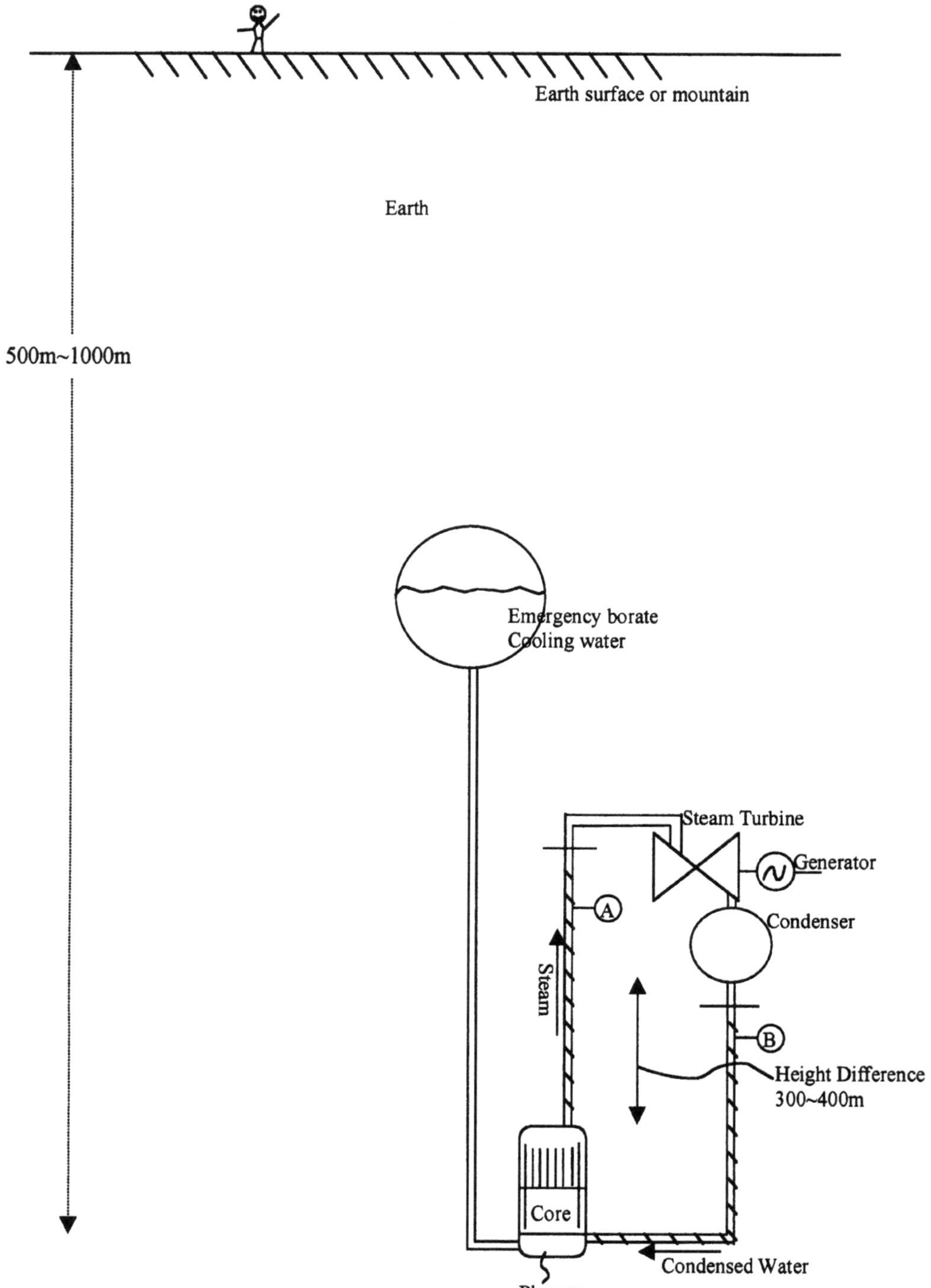

Figure 1 The conceptual lay out of the BWR installed in deep under ground.

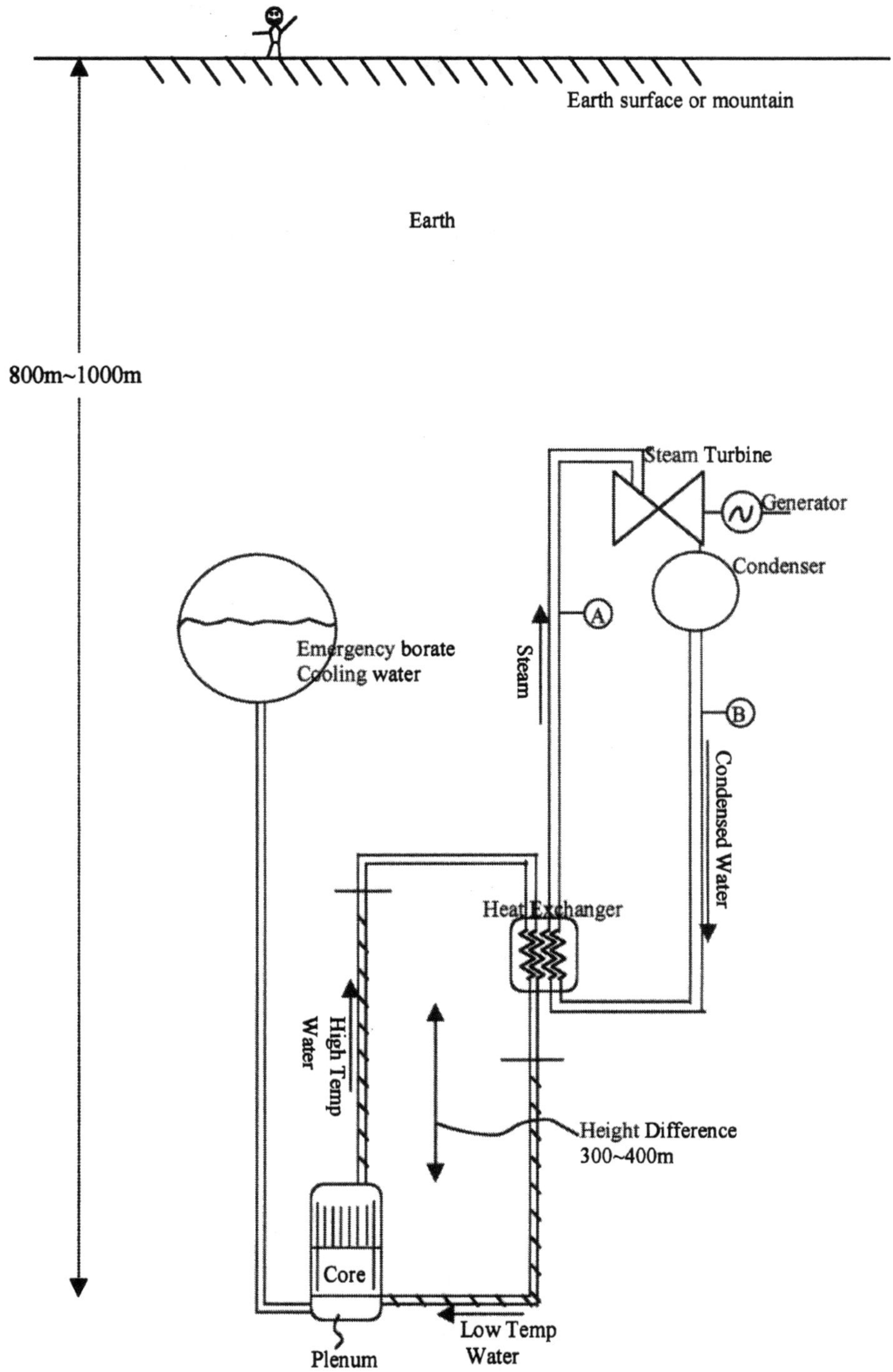

Figure 2 The conceptual lay out of the PWR installed in deep under ground.

ECONOMY OF DEEP UNDER-GROUND REACTOR , TRANSMISSION LINE,CONTAINER BUILDING EMERGENCY COOLING SYSTEM AND EVACUATION

To protect the public from radiation hazard from fall out the radioactive release from regular nuclear power plant, the container building of reactor is provided. By putting reactor in the deep under ground, the radiation field becomes very small we can minimize the evacuation of people.

Thus we can build NP near consumer area and the transmission line can be shorten and make substantial deduction of electricity generation cost , the cost of transmission line is very substantial, especially to make line in the densely populated area and the cost of land is high, it was estimated the construction of transmission line more than 400 Km is more than the cost of power plant .

Although by installing many facility such as the steam turbine, vacuum condenser is more costly than the building on the earth surface, however, as discussed later not constructing the long transmission line and the double walled container building and the other facility associated with seismically strong building structure, cost of the constructing the reactor in deep under ground might be not so expensive although very detailed cost evaluation is required.

To remove the decay heat after accident, the emergency cooling borated water is store in the container building, in the case of the tight lattice reactor the height of the cooling water should be very high to get high pressure it is very vernable for seismic protection.

When the reactor is in deep under ground the emergency borated water can be stored also in the deep under ground where the seismic movement of earth quake is smaller than the shallow under ground , and there enough room to store a huge amount of water in the high above the reactor and high pressure of water can be provided to cool the reactor.

Also for the protection of the re-criticality of the core due to melting of core, the large pool of water can be provided under reactor without difficulty in this deep under-ground reactor.

To protect the public from radiation hazard, the reactor are installed in the expensive containment building made of thick pre-stressed concrete together with emergency cooling borated water.

The installation of the nuclear facility in under deep under-ground gives good protection of public, we can eliminate these facility of containment building and the seismic hazard can be reduced due to avoiding strong earth surface motions. The area of emergency of evacuation , which killed the Shoreham NP near BNL in the decade years ago, also be minimized, thus there is possibility to construct the NP near consumer area, and that the expensive construction of high voltage electricity transmission line can be shorten. Therefore , building the deep underground reactor might becomes more economical than the building the nuclear power plant on the above ground. Although it might be very difficults to get public acceptance due to the present political climate, it is wise to built the future reactor in the deep under ground.

ACCELERATOR DRIVEN REACTOR

We discussed the power generating reactor here , but this reactor with hard neutron energy spectrum can also be used for the transmuting the minor actinide and LLFP. When the neutron energy spectrum becomes harder we get high neutron economy, but the negative void coefficient might be small and the delayed neutron portion also becomes small. To run this type of reactor in more safe mode it is desirable to run it in the sub-critical conditions.

For transmuting minor actinide (MA) and long lived fission product (LLFP), several kinds of nuclear reactor concepts such as the Na- , Pb-, Pb-Bi cooled fast reactor in critical operation, and in sub-critical operation driven by spallation neutrons have been proposed. For transmutor of Tc which has a low neutron capture cross section in intermediate energy, to get high transmutation rate, the high energy neutron is thermalized by moderator such as Zr hydride. In this water cooled reactor excellent moderator of water is in the core it is not necessary to install another moderator.

Accelerator driven reactor [5] which the author propose to run the reactor in sub-critical condition by providing the spallation neutrons created by medium energy proton. When the spallation target has equipped with beam window, it should be protected from radiation damage by expanding the proton beam. By locating the reactor in the deep underground, we have enough space to install the beam expansion section .

CONCLUSION

Construction of the nuclear power plant in underground was proposed by Sacharov and E. Teller [6] for protection for radiation hazard and Russian Pu and Electric Generation nuclear power plant is operated in Enisei river[7]. However, our proposal for deep underground reactor is used an earth gravity force to provides passive heat removal using natural circulation of reactor coolant and the nuclear power plant can be operated in more safely. High pressure required for heat removal coolant can be supplemented use of earth gravity also it provided pressure difference requires to run for removal of fission heat from the nuclear fuel with natural circulation of gravity force. This natural water coolant circulation can remove the concern of on-site electricity black out, also the storage facility of emergency cool system can be built far above the reactor due to enough space available in deep under ground installation.

For also defense purpose for protecting the people from nuclear hazard by smart bombing, which we would dislike to discuss, the future reactor should be built in the deep under ground.

Here, we discussed mainly on the light water reactor, this concept can be applied to gas cooled reactor which requires high pressure, it will apply many other type of the rectors. As we said future reactor should be constructed in deep under ground or deep under- sea area.

Especially for the HC reactor with tight latticed fuel assembly, the capability of removing heat not only steady state operation but also in the accidental condition of a loss of coolant or an outage in on-site power is essential. By putting the reactor in deep underground and remove the heat passively is essential for the reactor and public safety.

ACKNOWLEDGEMENT

The author would like to express his thanks to Drs.J.Herczeg, U. Rohatgi, S.Matsuura, Profs.T.Yamazaki, K. Takayama, T.Downar,J.Lee for their valuable discussions .

REFERENCE

[1] T. Iwamura et. al.(1999) " Research on Reduced-Moderation Water Reactor (RMWR) JAERI_research Report, 99-058 (In Japanese).

[2] Hiroshi Takahashi , Upendra Rohatgi , and Thomas Dawnar (2000) "A Proliferation Resistant Hexagonal Tight Lattice LWR Fueled Core for Increased Burn-up and Reduced Fuel Storage Requirements" Progress report of NERI: Aug.1999 to May, 2000

[3] Y. Oka (1998) " Physics of supercritical- pressured light water cooled reactors " 1998 Frederic Joliot Summer School Aug. 17-Aug.26 1998 Cadarache , France.

[4] U. S. Department of Energy (1999) Office of Civilian Radioactive Management DOE/EISW 0250D " Draft Environment Impact Statement for a Geological Repository for the Disposal of Spent Nuclear Fuel and High- Level Radioactive Waste at Yucca Mountain Nye County, Nevada." ACC; MOL. 19990816.0240.

[5] Hiroshi Takahashi (1989) " A fast breeder and incinerator assisted by a proton accelerator" Fusion Technology 20,657, 1991

[6] Edward Teller, Muriel Ishikawa,Lowell Wood, Roderick Hyde and John Nukolls (1996) " Completely automated nuclear reactors for long term operation II. ICENES 96.

[7] W. Kane (2000) , Private communication

www.ingramcontent.com/pod-product-compliance
Ingram Content Group UK Ltd.
Pitfield, Milton Keynes, MK11 3LW, UK
UKHW050948290926
15470UKWH00005B/1715

9 780080 441733